Lecture Notes in Computer Science 8644

Commenced Publication in 1973
Founding and Former Series Editors:
Gerhard Goos, Juris Hartmanis, and Jan van Leeuwen

Hendrik Decker   Lenka Lhotská
Sebastian Link   Marcus Spies
Roland R. Wagner (Eds.)

Database and Expert Systems Applications

25th International Conference, DEXA 2014
Munich, Germany, September 1-4, 2014
Proceedings, Part I

 Springer

Volume Editors

Hendrik Decker
Instituto Tecnológico de Informática
Valencia, Spain
E-mail: hendrik@iti.upv.es

Lenka Lhotská
Czech Technical University in Prague, Faculty of Electrical Engineering
Prague, Czech Republic
E-mail: lhotska@fel.cvut.cz

Sebastian Link
The University of Auckland, Department of Computer Science
Auckland, New Zealand
E-mail: s.link@auckland.ac.nz

Marcus Spies
LMU University of Munich, Knowledge Management
Munich, Germany
E-mail: marcus.spies@lrz.uni-muenchen.de

Roland R. Wagner
University of Linz, FAW
Linz, Austria
E-mail: rrwagner@faw.at

ISSN 0302-9743   e-ISSN 1611-3349
ISBN 978-3-319-10072-2   e-ISBN 978-3-319-10073-9
DOI 10.1007/978-3-319-10073-9
Springer Cham Heidelberg New York Dordrecht London

Library of Congress Control Number: 2014945726

LNCS Sublibrary: SL 3 – Information Systems and Application,
incl. Internet/Web and HCI

Typesetting: Camera-ready by author, data conversion by Scientific Publishing Services, Chennai, India

Printed on acid-free paper

Springer is part of Springer Science+Business Media (www.springer.com)

Preface

This book comprises the research articles and abstracts of invited talks presented at DEXA 2014, the 25th International Conference on Database and Expert Systems Applications. The conference was held in Munich, where DEXA had already taken place in 2001. The papers and abstracts show that DEXA faithfully continues to cover an established core of themes in the areas of databases, intelligent systems and applications, but also boosts changing paradigms and emerging trends.

For the 2014 edition, we had called for contributions in a wide range of topics, including

- Acquisition, Modeling and Processing of Data and Knowledge
- Authenticity, Consistency, Integrity, Privacy, Quality, Security
- Big Data
- Constraint Modeling and Processing
- Crowd Sourcing
- Data Clustering and Similarity Search
- Data Exchange, Schema Mapping
- Data Mining and Warehousing
- Data Provenance, Lineage
- Data Structures and Algorithms
- Database and Information System Architecture
- Database Federation, Cooperation, Integration, Networking
- Datalog 2.0, Deductive, Object-Oriented, Service-Oriented Databases
- Decision Support
- Dependability, High Availability, Fault Tolerance, Performance, Reliability
- Digital Libraries
- Distributed, Replicated, Parallel, P2P, Grid, Cloud Databases
- Embedded Databases, Sensor Data, Streaming Data
- Graph Data Management, Graph Sampling
- Incomplete, Inconsistent, Uncertain, Vague Data, Inconsistency Tolerance
- Information Retrieval, Information Systems
- Kids and Data, Education, Learning
- Linked Data, Ontologies, Semantic Web, Web Databases, Web Services
- Metadata Management
- Mobile, Pervasive and Ubiquitous Data
- NoSQL and NewSQL Databases
- Spatial, Temporal, High-Dimensional, Multimedia Databases
- Social Data Analysis
- Statistical and Scientific Databases
- Top-k Queries, Answer Ranking
- User Interfaces

– Workflow Management
– XML, Semi-structured, Unstructured Data

In response to this call, we received 159 submissions from all over the world, of which 37 are included in these proceedings as accepted full papers, and 46 more as short papers. We are grateful to all authors who submitted their work to DEXA. Decisions on acceptance or rejection were based on at least 3 (on average more than 4) reviews for each submission. Most of the reviews were meticulous and provided constructive feedback to the authors. We owe many thanks to all members of the Program Committee and also to the external reviewers, who invested their expertise, interest, and time to return their evaluations and comments.

The program of DEXA 2014 was enriched by two invited keynote speeches, presented by distinguished colleagues:

– Sourav S. Bhowmick: "DB ⋈ HCI"
– Dirk Draheim: "Sustainable Constraint Writing and a Symbolic Viewpoint of Modeling Languages"

In addition to the main conference track, DEXA 2014 also featured 6 workshops that explored a wide spectrum of specialized topics of growing general importance. The organization of the workshops was chaired by Marcus Spies, A. Min Tjoa, and Roland R. Wagner, to whom we say "many thanks" for their smooth and effective work.

Special thanks go to the host of DEXA 2014, the Ludwig-Maximilians University in Munich, Germany, where, under the guidance of the DEXA 2014 general chairpersons Marcus Spies and Roland R. Wagner, an excellent working atmosphere was provided.

Last, but not at all least, we express our gratitude to Gabriela Wagner. Her patience, professional attention to detail, skillful management of the DEXA event as well as her preparation of the proceedings volumes, together with the DEXA 2014 publication chairperson Vladimir Marik, and Springer's editorial assistant, Elke Werner, are greatly appreciated.

September 2014   Hendrik Decker
                                                        Lenka Lhotská
                                                        Sebastian Link

Organization

General Chair

Marcus Spies   Ludwig-Maximilians Universität München, Germany

Roland R. Wagner   Johannes Kepler University Linz, Austria

Program Committee Co-chairs

Hendrik Decker   Instituto Tecnológico de Informática, Spain
Lenka Lhotska   Czech Technical University, Czech Republic
Sebastian Link   The University of Auckland, New Zealand

Program Committee

Slim Abdennadher	German University at Cairo, Egypt
Witold Abramowicz	The Poznan University of Economics, Poland
Hamideh Afsarmanesh	University of Amsterdam, The Netherlands
Riccardo Albertoni	Institute of Applied Mathematics and Information Technologies - Italian National Council of Research, Italy
Eva Alfaro	UPV, Spain
Rachid Anane	Coventry University, UK
Annalisa Appice	Università degli Studi di Bari, Italy
Mustafa Atay	Winston-Salem State University, USA
Spiridon Bakiras	City University of New York, USA
Jie Bao	University of Minnesota at Twin Cities, USA
Zhifeng Bao	National University of Singapore, Singapore
Ladjel Bellatreche	ENSMA, France
Nadia Bennani	INSA Lyon, France
Morad Benyoucef	University of Ottawa, Canada
Catherine Berrut	Grenoble University, France
Debmalya Biswas	Iprova, Switzerland
Athman Bouguettaya	RMIT, Australia
Danielle Boulanger	MODEME, University of Lyon, France
Omar Boussaid	University of Lyon, France

Bernhard Thalheim	Christian Albrechts Universität Kiel, Germany
Jean-Marc Thevenin	University of Toulouse 1 Capitole, France
Helmut Thoma	Thoma SW-Engineering, Basel, Switzerland
A. Min Tjoa	Vienna University of Technology, Austria
Vicenc Torra	IIIA-CSIC, Spain
Traian Marius Truta	Northern Kentucky University, USA
Theodoros Tzouramanis	University of the Aegean, Greece
Maria Vargas-Vera	Universidad Adolfo Ibanez, Chile
Krishnamurthy Vidyasankar	Memorial University of Newfoundland, Canada
Marco Vieira	University of Coimbra, Portugal
Jianyong Wang	Tsinghua University, China
Junhu Wang	Griffith University at Brisbane, Australia
Qing Wang	The Australian National University, Australia
Wei Wang	University of New South Wales at Sydney, Australia
Wendy Hui Wang	Stevens Institute of Technology, USA
Gerald Weber	The University of Auckland, New Zealand
Jef Wijsen	Université de Mons, Belgium
Andreas Wombacher	Nspyre, The Netherlands
Lai Xu	Bournemouth University, UK
Ming Hour Yang	Chung Yuan Christian University, Taiwan
Xiaochun Yang	Northeastern University, China
Haruo Yokota	Tokyo Institute of Technology, Japan
Zhiwen Yu	Northwestern Polytechnical University, China
Xiao-Jun Zeng	University of Manchester, UK
Zhigang Zeng	Huazhong University of Science and Technology, China
Xiuzhen (Jenny) Zhang	RMIT University, Australia
Yanchang Zhao	RDataMining.com, Australia
Xiaofang Zhou	University of Queensland, Australia
Qiang Zhu	The University of Michigan, USA
Yan Zhu	Southwest Jiaotong University, China

External Reviewers

Samhaa R. El-Beltagy	Nile University, Egypt
Christine Natschläger	Software Competence Center Hagenberg GmbH, Austria
Badrinath Jayakumar	Georgia State University, USA
Janani Krishnamani	Georgia State University, USA
Kiki Maulana Adhinugraha	Monash University, Australia
Deepti Mishra	Atilim University, Turkey
Loredana Caruccio	University of Salerno, Italy
Valentina Indelli Pisano	University of Salerno, Italy
Claudio Gennaro	ISTI-CNR, Italy

Fabrizio Falchi   ISTI-CNR, Italy
Franca Debole   ISTI-CNR, Italy
Lucia Vadicamo   ISTI-CNR, Italy
Domenico Grieco   University of Bari, Italy
Ozgul Unal Karakas   University of Amsterdam, The Netherlands
Naser Ayat   University of Amsterdam, The Netherlands
Qiuyu Li   Huazhong University of Science and
                            Technology, China
Yuanyuan Sun   Huazhong University of Science and
                            Technology, China
Konstantinos Nikolopoulos City University of New York, USA
Bin Mu   City University of New York, USA
Erald Troja   City University of New York, USA
Dino Ienco   TETIS-IRSTEA, France
Vasil Slavov   University of Missouri-Kansas City, USA
Anas Katib   University of Missouri-Kansas City, USA
Shaoyi Yin   Paul Sabatier University, France
Hala Naja   Lebanon University, Lebanon
Souad Boukhadouma   University of Algiers, Algeria
Nouredine Gasmallah   University of Souk Ahras, Algeria
José Carlos Fonseca   Polytechnic Institute of Guarda, Portugal
Bruno Cabral   University of Coimbra, Portugal
Xianying Liu   The University of Michigan at Dearborn, USA
Cheng Luo   Coppin State University, USA
Gang Qian   University of Central Oklahoma, USA
Kotaro Nakayama   University of Tokyo, Japan
Tomoki Yoshihisa   Osaka University, Japan
Emir Muñoz   Fujitsu Ireland Ltd. & National University of
                            Ireland, Ireland
Jianhua Yin   Tsinghua University, China
Yuda Zang   Tsinghua University, China
Rocio Abascal Mena   UAM, Mexico
Erick Lopez Ornelas   UAM, Mexico
Manel Mezghani   MIRACL lab, Sfax, Tunisia
André Péninou   IRIT, France
Bouchra Soukarieh   CNRS, France
Qiong Fang   The Hong Kong University of Science &
                            Technology, Hong Kong
Jan Vosecky   The Hong Kong University of Science &
                            Technology, Hong Kong
Zouhaier Brahmia   University of Sfax, Tunisia
Hemank Lamba   IBM Research, India
Prithu Banerjee   IBM Research, India

Tymoteusz Hossa	The Poznan University of Economics, Poland
Jakub Dzikowski	The Poznan University of Economics, Poland
Shangpu Jiang	University of Oregon, USA
Nunziato Cassavia	ICAR-CNR, Italy
Pietro Dicosta	Unical, Italy
Bin Wang	Northeastern University, China
Frédéric Flouvat	University of New Caledonia, New Caledonia
Saeed Samet	Memorial University, Newfoundland, Canada
Jiyi Li	Kyoto University, Japan
Alexandru Stan	Romania
Anamaria Ghiran	Romania
Cristian Bologa	Romania
Raquel Trillo	University of Zaragoza, Spain
María del Carmen Rodríguez-Hernández	University of Zaragoza, Spain
Julius Köpke	University of Klagenfurt, Austria
Christian Koncilia	University of Klagenfurt, Austria
Henry Ye	RMIT, Australia
Xumin Liu	RMIT, Australia
Azadeh Ghari-Neiat	RMIT, Australia
Hariton Efstathiades	University of Cyprus, Cyprus
Andreas Papadopoulos	University of Cyprus, Cyprus
Meriem Laifa	Bordj Bouarreridj University, Algeria
Zhifeng Bao	University of Tasmania, Australia
Jianxin Li	Swinburne University of Technology, Australia
Md. Saiful Islam	Swinburne University of Technology, Australia
Tarique Anwar	Swinburne University of Technology, Australia
Mikhail Zymbler	South Ural State University, Russian Federation
Constantin Pan	South Ural State University, Russian Federation
Timofey Rechkalov	South Ural State University, Russian Federation
Ruslan Miniakhmetov	South Ural State University, Russian Federation
Pasqua Fabiana Lanotte	University of Bari, Italy
Francesco Serafino	University of Bari, Italy
Xu Zhuang	Southwest Jiaotong University, China
Paul de Vrieze	Bournemouth University, UK
Lefteris Kalogeros	Ionian University, Greece
Sofia Stamou	Ionian University, Greece
Dingcheng Li	Mayo Clinic, USA
Zhe He	Columbia University, USA
Jun Miyazaki	Tokyo Institute of Technology, Japan
Yosuke Watanabe	Nagoya University, Japan

Jixue Liu	University of South Australia, Australia
Selasi Kwashie	University of South Australia, Australia
Christos Kalyvas	University of the Aegean, Greece
Stéphane Jean	LIAS, France
Brice Chardin	LIAS, France
Selma Khouri	ESI, Algeria
Selma Bouarar	LIAS, France
Rima Bouchakri	ESI, Algeria
Idir Ait Sadoune	Supelec, France

Table of Contents – Part I

XML Keyword Search

Skyline Queries

Graph Algorithms

Information Retrieval

Classification and Clustering

Queries

Table of Contents – Part II

Social Computing

Similarity Search

Ranking

Data Mining

Big Data

Approximations

Privacy

Data Exchange

Data Integration

Web Semantics

Repositories

Partitioning

Business Applications

DB ⋈ HCI: Towards Bridging the Chasm between Graph Data Management and HCI

Sourav S. Bhowmick

School of Computer Engineering, Nanyang Technological University, Singapore
assourav@ntu.edu.sg

Abstract. Visual query interfaces enable users to construct queries without special training in the syntax and semantics of a query language. Traditionally, efforts toward such interface design and devising efficient query processing techniques are independent to each other. This is primarily due to the chasm between HCI and data management fields as since their inception, rarely any systematic effort is made to leverage techniques and principles from each other for superior user experience. In this paper, we lay down the vision of bridging this chasm in the context of visual graph query formulation and processing. Specifically, we present the architecture and novel research challenges of a framework for querying graph data visually where the visual interface management is *data-driven* and query processing and performance benchmarking are HCI *driven*.

1 Introduction

"Thirty years of research on query languages can be summarized by: we have moved from SQL to XQuery. At best we have moved from one declarative language to a second declarative language with roughly the same level of expressiveness. It has been well documented that end users will not learn SQL; rather SQL is notation for professional programmers."

Abiteboul et al. [1]

"A picture is worth a thousand words. An interface is worth a thousand pictures."

B. Shneiderman, 2003

It is widely acknowledged that formulating a textual query using a database query language (*e.g.*, XQuery, SPARQL) often demands considerable cognitive effort from end users [1]. The traditional approach to alleviate this challenge is to improve the user-friendliness of the task by providing a visual querying scheme to replace data retrieval aspects of a query language. This has paved the way to a long stream of research in visual query languages in the context of relational databases [4, 30], object-oriented databases [10, 15], the Web [5, 13], semi-structured and XML databases [9, 14, 23], and graph databases [7, 8, 12]. These approaches leverage principles from the human-computer interaction (HCI) field to build visual query interfaces that involve a sequence of tasks ranging from *primitive* operations such as pointing and clicking a mouse button, keyboard entry to higher-level tasks such as selection of menu items, selection and

H. Decker et al. (Eds.): DEXA 2014, Part I, LNCS 8644, pp. 1–11, 2014.
© Springer International Publishing Switzerland 2014

dragging of target objects, etc. In this paper, we leverage on this grand desire to provide user-friendly visual querying interface to lay down the vision of *integrating* DB and HCI (hereafter, referred to as *HCI-aware data management*) techniques and tools towards superior consumption and management of data.

Specifically, we discuss HCI-aware data management in the context of graph-structured data. This is because many modern applications (*e.g.,* drug discovery, social networks, semantic Web) are centered on such data as graphs provide a natural way of modeling data in a wide variety of domains. For example, searching for *similar* chemical compounds to aid drug design is essentially searching for graphs as it can be used to represent atoms and bonds in chemical compounds. Consequently, there is a pressing need to build user-friendly visual framework (*e.g.,* [12, 21]) on top of a state-of-the-art graph query system that can support user-friendly formulation of various types of graph queries such as subgraph search, reachability queries, homeomorphic queries, etc. Our vision can be easily extended to other types of data such as relational, XML, etc.

A visual interface for graph query formulation is typically composed of several panels such as a panel to display the set of labels of nodes or edges of the underlying data graph(s), a panel to construct a graph query graphically, a panel containing *canned patterns* to aid query formulation, and a panel to display query results in an intuitive manner. For example, Figure 1 depicts the screenshot of a real-world visual interface provided by PubChem[1] for substructure (subgraph) search for chemical compounds. Specifically, Panel 3 provides a list of chemical symbols that a user can choose from to assign labels to nodes of a query graph. Panel 2 lists a set of *canned patterns* (*e.g.,* benzene ring) which a user may drag and drop in Panel 4 during visual query construction. Note that the availability of such patterns greatly improves usability of the interface by enabling users to quickly construct a large query graph with fewer clicks compared to constructing it in an "edge-at-a-time" mode. For instance, the query graph in Panel 4 can be constructed by dragging and dropping two such canned patterns from Panel 2 instead of taking the tedious route of constructing 9 edges iteratively.

One may observe that the above efforts toward visual query interface design and devising efficient query processing techniques are traditionally independent to each other for decades. This is primarily due to the fact that the two key enablers of these efforts, namely HCI and database management, have evolved into two disparate and vibrant scientific fields, rarely making any systematic effort to leverage techniques and principles from each other towards superior realization of these efforts. Specifically, data management researchers have striven to improve the capability of a database in terms of both performance and functionality, often devoting very little attention to the HCI aspects of the solutions. On the other hand, the HCI community has focused on human factors, building sophisticated models of various types of visual tasks and menu design that are orthogonal to the underlying data management system.

We believe that the chasm between these two vibrant fields sometimes create obstacles in providing superlative visual query formulation and data management services to end users. On the one hand, as visual query interface construction process is traditionally *data-unaware*, it may fail to generate flexible, portable, and user-friendly query interface. For instance, reconsider the canned patterns in Panel 2 in Figure 1. Given the

---

[1] http://pubchem.ncbi.nlm.nih.gov/edit2/index.html?cnt=0

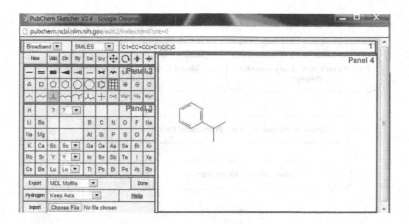

Fig. 1. GUI for substructure search in PubChem

space constraint in the GUI, the selection of a limited set of patterns that are to be displayed on it are not "data-driven" but typically carried out manually by domain experts. An immediate aftermath of such manual selection is that the set of canned patterns may not be sufficiently diverse enough to support a wide range of graph queries as it is unrealistic to expect a domain expert to have comprehensive knowledge of the topology of the entire graph dataset. Consequently, an end user may not find the canned patterns in Panel 2 useful in formulating certain query graphs. Similar problem may also arise in Panel 3 where the labels of nodes may be manually added instead of automatically generated from the underlying data. Additionally, the visual interface is "static" in nature. That is, the content of Panels 2 and 3 remain static even when the underlying data evolves. As a result, some patterns (*resp.* labels) in Panel 2 (*resp.* Panel 3) may become obsolete as graphs containing such patterns (*resp.* labels) may not exist in the database anymore. Similarly, some new patterns (*resp.* labels), which are not in Panel 2 (*resp.* Panel 3), may emerge due to the addition of new data graphs. Furthermore, such visual query interface lacks of portability as the same interface cannot be seamlessly integrated on a graph database in a different domain (*e.g.*, computer vision, protein structure). As the contents of Panels 2 and 3 are domain-dependent and remain static, the GUI needs to be reconstructed from scratch when the domain changes in order to accommodate new domain-specific canned patterns and labels.

On the other hand, traditionally query processing techniques are only invoked once a user has completed her visual query formulation as the former is completely decoupled from the latter. For instance, during the formulation of a visual query in Panel 4, the underlying query processing engine remains idle and is only initiated after the Run icon is clicked. That is, although the final query that a user intends to pose is revealed gradually during visual query construction, it is not exploited by the query processor prior to clicking of the Run icon. Consequently, valuable opportunities to significantly improve the *system response time* (SRT)[2] by initiating query processing during visual

---

[2] The SRT is the duration between the time a user presses the Run icon to the time when the user gets the query results.

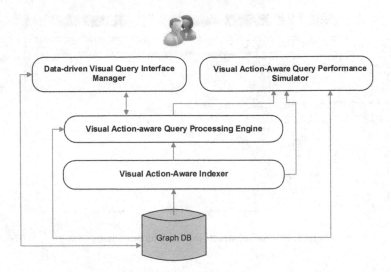

Fig. 2. The architecture

query construction are wasted. Note that in contrast to the traditional paradigm where the SRT is the time taken to process the *entire* query, by bridging the chasm between visual query formulation and query processing activities, the SRT is reduced to processing a part of the query that is yet to be evaluated (if any). Additionally, due to this chasm, opportunities to enhance usability of graph databases by providing relevant *guidance* and *feedback* during query formulation are lost as efficient and timely realization of such functionalities may require prefetching of candidate data graphs during visual query construction. For instance, whenever a newly constructed edge makes a graph query fragment yield empty answer, it can be immediately detected by processing the prefetched data graphs. It is not efficient if it is only detected at the *end* of query formulation as a user may have wasted her time and effort in formulating additional constraints.

In this paper, we lay down the vision of bridging the long-standing chasm between traditional data management and HCI in the context of querying graph-structured data. Specifically, we propose an HCI-*aware visual graph querying* framework that aims to encapsulate several novel and intriguing research challenges toward the grand goal of bridging this chasm. Realization of these challenges entail significant rethinking of several long-standing strategies for visual interface construction and data management.

2   Novel Research Challenges

In this section, we first present the generic architecture of the HCI-aware visual graph querying framework to realize our vision. Next, we identify the key novel research challenges that need to be addressed to realize this framework. To facilitate exposition of these challenges, we assume that the graph database consists of a large number of small or medium-sized graphs. However, these challenges are not limited to such collection of graphs as they are also pertinent to querying large networks such as social networks, biological networks, etc.

2.1 Architecture

Figure 2 depicts the generic architecture of the framework for realizing our vision of bridging the chasm between HCI and graph data management. The *Data-driven Visual Query Interface Manager* component provides a framework to construct various panels of the visual query interface in a *data-driven* manner. It also provides an interactive visual environment for query formulation without the knowledge of complex graph query languages as well as a framework for intelligent guidance and *interruption-sensitive* feedback to users to further ease the cognitive overhead associated with query formulation. The *Visual Action-aware Query Processing Engine* embodies our vision of blending query processing with visual query formulation by utilizing the latency offered by the GUI actions. The *Visual Action-aware Query Performance Simulator* module aims to provide a comprehensive framework for large-scale empirical study of the query processor by simulating the paradigm of blending query formulation and query processing. Finally, the *Visual Action-aware Indexer* component provides an array of *action-aware indexes* to support efficient query processing as well as query performance simulation in our proposed paradigm.

2.2 Data-Driven Visual Interface Management

Data-driven visual query interface construction. Recall from Section 1 the limitations associated with the manual construction of panels of the visual query interface. There is one common theme that runs through these limitations: *the visual query interface construction is not data-driven.* Specifically, the GUI does not exploit the underlying graph data to automatically generate and maintain the content of various panels. Hence, it is necessary to rethink the traditional visual query interface construction strategy by taking a novel data-driven approach for visual interface construction. While the unique set of labels of nodes of the data graphs (Panel 3) can be easily generated by traversing them, automatically generating the set of canned patterns is computationally challenging. These patterns should not only be able to *maximally cover* the underlying graph data but should also minimize *topological similarity* (redundancy) among themselves so that a diverse set of canned patterns is available to the user. Note that there can be prohibitively large number of such patterns. Hence, the size of the pattern set should not be too large due to limited display space on the GUI as well as users' inability to absorb too many patterns for query formulation.

As some of the canned patterns may be frequent in the graph database, at first glance it may seem that they can be generated using any frequent subgraph mining algorithm (*e.g., gSpan* [27]). However, this is not the case as it is not necessary for *all* canned patterns to be frequent. It is indeed possible that some patterns are frequently used by end users to formulate visual queries but are infrequent in the database. Also, graph summarization techniques [25], which focus on grouping nodes at different resolutions in a large network based on user-selected node attributes and relationships, cannot be deployed here as we generate concise canned pattern set by maximizing coverage while minimizing redundancy under the GUI constraint.

Data-driven visual query suggestion and feedback. Any visual querying interface should intelligently guide users in formulating graph queries by (a) helping them to formulate their desired query by making appropriate suggestions and (b) providing appropriate feedback whenever necessary in a timely manner. In order to realize the former, it is important to devise efficient techniques which can predict relevant node labels and canned patterns that may be of interest to a user during query formulation. Also, given a partial query already drawn by a user, it is important to provide suggestion of a concise set of fragments (canned patterns) that she is likely to draw in the next step. Successful realization of these tasks require analyzing topological features of the underlying data as well as query log (if any) w.r.t to the partially constructed query. On the other hand, query feedbacks are essential during query construction as a user may not know if the query she is trying to formulate will return any results. Hence it is important to observe user's actions during query formulation and notify her if a query fragment constructed at a particular step fails to return any results and possibly advise the user on which subgraph in the formulated query fragment is the "best" to remove in order to get non-empty results.

Interruption-sensitive notifications. Addressing aforementioned challenges improve the usability of visual querying systems by enabling users to construct graph queries without special training in the syntax and semantics of a query language. They also guide users into correct query construction by notifying them of a variety of problems with a query such as incorrect syntax, alerts for an empty result, etc. The state-of-the-art algorithms that notify users of such alerts and suggestions are, however, overly aggressive in their notifications as they focus on *immediate* notification regardless of its impact on the user. More specifically, they are insensitive to the cognitive impact of *interruptions* caused by notifications sent at *inopportune times* that disrupt the query construction process.

Many studies in the cognitive psychology and HCI communities have demonstrated that interrupting users engaged in tasks by delivering notifications inopportunely can negatively impact task completion time, lead to more errors, and increase user frustration [6, 16, 22]. For instance, suppose a user is notified (with a pop-up dialog box) of an empty result (due to previously formulated condition) when she is undertaking a task such as dragging a canned pattern from Panel 2 and preparing to drop it in Panel 4. This interruption may frustrate her as mental resources allocated for the current task are disrupted because she is forced to leave it to acknowledge and close the dialog box. Although, such inopportune interruption adversely affects the usability of visual querying systems, traditional data management techniques have devoted very little attention to the cognitive aspect of a solution. Here we need to ensure that a solution to the aforementioned issues is "cognitive-aware" by devising models and techniques to deliver a notification *quickly but at an appropriate moment* when the mental workload of the user is minimum. Detecting such an opportune moment should be transparent to the user and must not seek explicit input from her.

A promising direction in achieving this goal is to seamlessly integrate *defer-to-breakpoint* strategy [18, 19] with visual query formulation tasks when reasoning about when to notify the user. This will also entail to leverage work in HCI to build quantitative models for time available to complete various data management tasks (*e.g.*, detection of empty results, query suggestion) in order to ensure notification delivery at optimal breakpoints that lower the interruption cost. A keen reader may observe that making aforementioned solutions interruption-sensitive essentially questions the holy grail of database research over the past forty years: *whether faster evaluation is always better?* As remarked earlier, the HCI and cognitive psychology communities have observed that slowing or deferring some activities (*e.g.*, notification delivery) can increase usability. Hence, by making visual query formulation process cognitive-aware, we essentially aim to slow down a small part of the system in order to increase its usability.

2.3 Visual Action-Aware Query Processing

Visual action-aware graph indexing. A host of state-of-the-art graph indexing strategies have been proposed since the last decade to facilitate efficient processing of a variety of graph queries (*e.g.*, [28]). While several of these techniques are certainly innovative and powerful, they cannot be directly adopted to support our vision. These techniques are based on the conventional paradigm that the *entire* query graph must be available *before* it can leverage the indexes for processing the query. However, in our proposed framework we aim to initiate query evaluation as soon as a fragment of the query graph is visually formulated. Hence, the indexes need to be aware of visual actions taken by a user and accordingly filter negative results after every action taken by her. Recall that a user may construct a single edge or a canned pattern at a particular step during query formulation. She may also modify it by deleting an existing canned pattern or an edge. Hence, the size of a partially-constructed query graph may grow or shrink by $k \geq 1$ at each step. Furthermore, a query fragment may evolve from a frequent pattern to infrequent one and vice versa. In this case, it is important to devise efficient indexing schemes to support identification of the data graphs that match (exact or approximate) the partially constructed query graph at each step. Note that since subgraph isomorphism testing is known to be NP-complete, the indexing scheme should minimize expensive candidate verification in order to retrieving these partial results.

Visual action-aware query matching. Our goal is to utilize the latency offered by the GUI actions to retrieve partial candidate data graph. Specifically, when a user draws a new edge or canned pattern on the query canvas, candidate data graphs containing the current query fragment need to be efficiently retrieved and monitored by leveraging the indexes. If the candidate set is non-empty at a specific step then high-quality suggestions to complete the construction of the subsequent steps may be provided at an opportune time. On the other hand, if the candidate set is empty then there does not exist any data graphs that match the query fragment at a specific step. In this scenario, the user needs to be notified appropriately in an interruption-sensitive manner and guided to modify the query appropriately. The above process is repeated until the user clicks on the Run icon to signify the end of the query formulation step. Consequently, the final query results are generated from the prefetched candidate data graphs by performing verification test whenever necessary. Note that as this step invokes the subroutine for subgraph isomorphism test, the verification process needs to be minimized by judiciously filtering as many false candidates as possible without any verification test.

Non-traditional design issues. Observe that in order to realize the aforementioned visual query processing paradigm, we need a query processor that incorporates the following three non-traditional design issues.

- First, is the need for materialization of intermediate information related to all partial candidate graphs matching the query at each step during query construction. While this has always been considered as an unreasonable assumption in traditional databases, materialization of all intermediate results is recently supported to enhance database usability [11]. Note that this issue is particularly challenging here due to computational hardness of subgraph isomorphism test. Hence judicious strategy to minimize candidate verification while retrieving partial candidates is required.

- Second, materialization of partial candidates needs to be performed efficiently within the available GUI latency. This is pivotal as inefficient materialization can slow down generation of candidate graphs at each query construction step eventually adversely affecting the SRT. Ideally, we should be able to materialize candidate graphs of a query fragment *before* the construction of the succeeding edge (or canned pattern). Consequently, it is paramount to accurately and systematically estimate the time taken by a user for constructing a query fragment (edge or pattern) as this latency is exploited by the query processing paradigm to prefetch candidate matches. Here it is important to drew upon the literature in HCI (*e.g.*, [2,3]) to quantitatively model the time available to a user to perform different visual tasks such as selection of canned patterns. This will enable us to quantify the "upper bound" of materialization time and seek efficient solution accordingly.

- Third, the query processor needs to support *selectivity-free* query processing. Selectivity-based query processing, that exploits estimation of predicate selectivities to optimize query processing, has been a longstanding approach in classical databases. Unfortunately, this strategy is ineffective in our proposed framework as users can formulate low and high selective fragments in any arbitrary sequence of actions. As query processing is interleaved with the construction (modification) of each fragment, it is also not possible to "push-down" highly selective fragments. Similarly, query feedbacks such as detection and notification of empty result are intertwined with the order of constructed query conditions as it must be delivered at an opportune time. Hence, it needs to operate in a selectivity-free environment as well. The only possible way to bypass this stumbling block in this environment is to ensure that the sequence of visual actions formulated by a user is ordered by their selectivities. However, users cannot be expected to be aware of such knowledge and it is unrealistic to expect them to formulate a query in a "selectivity-aware" order.

2.4   HCI-Driven Performance Simulation

The aforementioned framework must have adequate support to evaluate visual query performance at a large scale. In contrast to traditional paradigm where the runtime performance of a large number of graph queries can be easily measured by automatically extracting a random collection of query graphs from the underlying data and executing them, each query in the proposed framework must be formulated by a set of real users. Furthermore, each query can follow many different query formulation sequences. Consequently, it is prohibitively expensive to find and engage a large number of users who are willing to formulate a large number of visual queries. Hence, there is a need for a performance measurement framework that can simulate formulation of a large number of visual graph queries without requiring a large number of users.

The performance benchmarking framework needs to address two key challenges. First, is the automated generation of a large number of queries of different types (frequent and infrequent), topological characteristics, and result size. The action-aware graph indexing framework can be leveraged here to extract frequent and infrequent subgraphs from the underlying data graphs that satisfy different constraints. Second, is to simulate the formulation of each generated query following different query formulation sequences. Note that different users may take different time to complete each visual action during query formulation. For instance, time taken to move the mouse to Panel 2 (Figure 1) and select a canned pattern may be different for different users. It is important to accurately and systematically estimate the time taken by a user for each of these actions as this latency is exploited by the query processing paradigm to prefetch candidate matches. Similar to the materialization of candidate graphs, here it is important to drew upon the literature in HCI (*e.g.*, [2, 3]) to quantitatively model the time available to a user to perform different visual tasks. Then for each step in the query construction process, the query simulation algorithm *waits* for appropriate amount of time to simulate the execution of the task by a user before moving to the next step.

2.5 Extension to Massive Graphs

Recall that the aforementioned research challenges to realize our vision assume that the database contains a large collection of small or medium-sized graphs. However, in recent times graphs with millions or even billions of nodes and edges are commonplace. As a result, there is increasing efforts in querying massive graphs by exploiting distributed computing [24]. We believe that the aforementioned challenges need to be addressed as well to realize our vision on such massive graph framework. However, the solution to these challenges in this framework differ. For example, the data-driven visual interface construction needs to generate the canned patterns and labels in a distributed manner as graph data is stored in multiple machines. Similarly, each visual action during query processing needs to be judiciously processed in a distributed environment by selecting relevant slaves that may contain the candidate matches as well as minimizing communication costs among the machines.

3 Early Efforts

In [7, 7, 17, 20, 21], we took the first step to implement the visual action-aware query processing module for subgraph containment and subgraph similarity search queries for a set of small or medium-sized graphs as well as for large networks. Our study demonstrates that the paradigm of blending visual query formulation and query processing significantly reduces the SRT as well as number of candidate data graphs compared to state-of-the-art techniques based on traditional paradigm. However, in these efforts we assume that the visual query is formulated using an "edge-at-a-time" approach (canned patterns are not used), making it tedious to formulate large queries. Consequently, the underlying indexing schemes and query processing strategies need to be adapted to support such query construction. Furthermore, it is interesting to investigate whether the proposed paradigm can efficiently support a wider variety of graph queries such as reachability queries, supergraph containment queries, homeomorphic graph queries, etc. Lastly, the data-driven visual interface management and the HCI-aware performance benchmarking framework are open research problems that are yet to be addressed.

Visual action-aware query processing is also studied in the context of XML query processing [26, 29]. Similar to visual graph querying, here we proposed a visual XML query system called XBLEND which blends visual XML query formulation and query processing in a novel way. It is built on top of a relational framework and exploits the latency offered by the GUI-based query formulation to prefetch portions of the query results by issuing a set of SQL queries.

4   Conclusions

This paper contributes a novel vision of HCI-aware graph data management to bridge the chasm between HCI and data management fields. We present a framework for querying graph data visually where the visual interface management is data-driven and query processing and performance benchmarking are HCI-driven. Addressing the research challenges associated with this vision entail a multi-disciplinary effort drawing upon the literature in HCI, cognitive psychology, and data management. Although, in this paper we focused on graph querying, it is easy to see that our vision can be extended to other visual querying environments.

Acknowledgement. Sourav S Bhowmick was supported by the Singapore-MOE AcRF Tier-1 Grant RG24/12. His travel expenses were supported by the Marsden fund council from Government funding, administered by the Royal Society of New Zealand. The author would also like to thank Changjiu Jin, H. Hung, and B. Q. Trung for implementing several features of the vision.

References

1. Abiteboul, S., Agrawal, R., Bernstein, P., et al.: The Lowell Database Research Self-Assessment. Communication of the ACM (2005)
2. Accot, J., Zhai, S.: Refining Fitts' Law Models for Bivariate Pointing. In: ACM SIGCHI (2003)
3. Ahlstrom, D.: Modeling and Improving Selection in Cascading Pull-Down Menus Using Fitt's Law, the Steering Law, and Force Fields. In: CHI (2005)
4. Angelaccio, M., Catarci, T., Santucci, G.: QBD*: A Graphical Query Language with Recursion. IEEE Trans. Soft. Engg. 16(10), 1150–1163 (1990)
5. Atzeni, P., Mecca, G., Merialdo, P.: To Weave the Web. In: VLDB (1997)

6. Bailey, B.P., Konstan, J.A.: On the Need for Attention Aware Systems: Measuring Effects of Interruption on Task Performance, Error Rate, and Affective State. Journal of Computers in Human Behavior 22(4) (2006)
7. Bhowmick, S.S., Choi, B., Zhou, S.: VOGUE: Towards a Visual Interaction-aware Graph Query Processing Framework. In: CIDR (2013)
8. Blau, H., Immerman, N., Jensen, D.: A Visual Language for Querying and Updating Graphs. Tech. Report 2002-037, Univ. of Mass., Amherst (2002)
9. Braga, D., Campi, A., Ceri, S.: XQBE (XQuery By Example): A Visual Interface to the Standard XML Query Language. TODS 30(2) (2005)
10. Carey, M., Haas, L., Maganty, V., Williams, J.: PESTO: An Integrated Query/Browser for Object Databases. In: VLDB (1996)
11. Chapman, A., Jagadish, H.V.: Why Not? In: SIGMOD (2009)
12. Chau, D.H., Faloutsos, C., Tong, H., et al.: GRAPHITE: A Visual Query System for Large Graphs. In: ICDM Workshop (2008)
13. Comai, S., Damiani, E., Posenato, R., Tanca, L.: A Schema-Based Approach to Modeling and Querying WWW Data. In: Andreasen, T., Christiansen, H., Larsen, H.L. (eds.) FQAS 1998. LNCS (LNAI), vol. 1495, pp. 110–125. Springer, Heidelberg (1998)
14. Comai, S., Damiani, E., Fraternali, P.: Computing Graphical Queries Over xml Data. ACM TOIS 19(4), 371–430 (2001)
15. Cruz, I.F., Mendelzon, A.O., Wood, P.T.: A Graphical Query Language Supporting Recursion. In: ACM SIGMOD (1987)
16. Cutrell, E., et al.: Notification, Disruption, and Memory: Effects of Messaging Interruptions on Memory and Performance. In: Proc. of the IFIP TC 13 Int. Conf. on HCI, Tokyo, Japan (2001)
17. Hung, H., Bhowmick, S.S., et al.: QUBLE: Towards Blending Interactive Visual Subgraph Search Queries on Large Networks. The VLDB Journal 23(3) (May 2014)
18. Iqbal, S.T., Bailey, B.P.: Understanding and Developing Models for Detecting and Differentiating Breakpoints during Interactive Tasks. In: CHI (2007)
19. Iqbal, S.T., Bailey, B.P.: Effects of Intelligent Notification Management on Users and Their Tasks. In: CHI (2008)
20. Jin, C., et al.: GBLENDER: Towards Blending Visual Query Formulation and Query Processing in Graph Databases. In: ACM SIGMOD (2010)
21. Jin, C., et al.: PRAGUE: A Practical Framework for Blending Visual Subgraph Query Formulation and Query Processing. In: ICDE (2012)
22. Monk, C.A., Boehm-Davis, D.A., Trafton, J.G.: The Attentional Costs of Interrupting Task Performance at Various Stages. In: Proc of the Human Factors and Ergonomics Society (2002)
23. Papakonstantinou, Y., et al.: QURSED: Querying and Reporting Semistructured Data. In: ACM SIGMOD (2002)
24. Sun, Z., et al.: Efficient Subgraph Matching on Billion Nodes Graphs. In: VLDB (2013)
25. Tian, Y., et al.: Efficient Aggregation for Graph Summarization. In: SIGMOD (2008)
26. Truong, B.Q., Bhowmick, S.S.: MUSTBLEND: Blending Visual Multi-Source Twig Query Formulation and Query Processing in RDBMS. In: Meng, W., Feng, L., Bressan, S., Winiwarter, W., Song, W. (eds.) DASFAA 2013, Part II. LNCS, vol. 7826, pp. 228–243. Springer, Heidelberg (2013)
27. Yan, X., et al.: gSpan: Graph-based Substructure Pattern Mining. In: ICDM (2002)
28. Zhao, X., et al.: A Partition-Based Approach to Structure Similarity Search. In: VLDB (2013)
29. Zhou, Y., Bhowmick, S.S., et al.: Xblend: Visual xml Query Formulation Meets Query Processing. In: ICDE (2009)
30. Zloof, M.M.: Query-By-Example: A Data Base Language. IBM Syst. J. 16(4), 324–343 (1977)

Sustainable Constraint Writing and a
Symbolic Viewpoint of Modeling Languages
(Extended Abstract)

Dirk Draheim

University of Innsbruck, Technikerstr. 23, A-6020 Innsbruck, Austria
draheim@acm.org

Constraints form an important part of models to capture experts' domain knowledge. Sustainable constraint writing is about making constraints robust in model evolution scenarios, i.e., it is about saving the knowledge expressed by data constraints against model updates. As an important example for sustainable constraint writing, we walk through the semantics of power type constructs. Power types are used to model sets of sets. Modeling sets of sets of objects is important, because it arises naturally in many expert domains. We will see that the intuitively intended meaning of power type constructs is not merely about structuring information but is about establishing type-generic constraints. We give a precise semantics for power types based on sustainable constraints. In general, sustainable constraints need to contain meta data parts, i.e., they are reflective constraints. In accordance to the notion of sustainable constraint writing, we introduce a symbolic viewpoint on model manipulation that complements current mainstream viewpoints. The symbolic viewpoint is about denying a model/data level divide. We discuss sustainable constraint writing with respect to further issues in concrete technologies and tools for transparent database access layers (IMIS), meta modeling (AMMI) and generative programming (GENOUPE).

Let us have a look at a standard power type example [21] in Listing 1 modelling trees and tree species. There are three concrete tree species, i.e., sugar maple, apricot tree and American elm, which are modeled each as a sub class of the tree class. The concept of tree species is made explicit in the model as a class. The relation between the tree class and the tree species class follows the type object pattern [22]. An attribute of a tree species object contains a property that is common to all objects of a concrete tree species, e.g., the leaf pattern in our

---

Listing 1. Standard power type example exploiting subtypes and the types-object pattern

---

```
class Tree{ treeSpecies:TreeSpecies ... }
class SugarMaple extends Tree{ ... }
class Apricot extends Tree{ ... }
class AmericanElm extends Tree{ ... }
...
class TreeSpecies{ leafPattern:LeafPattern ... }
```

H. Decker et al. (Eds.): DEXA 2014, Part I, LNCS 8644, pp. 12–19, 2014.
© Springer International Publishing Switzerland 2014

example. Yet, with respect to its intended meaning, the model is not complete. Some crucial constraints are missing. In order to meet the intended meaning we need to express that the same tree species object must be assigned to all objects of a concrete tree species class. We give appropriate constraints to express just this in a tiny pseudo constraint language in Eqn. 1.

$$\exists s \lhd TreeSpecies.\forall t \lhd SugarMaple.(t.treeSpecies = s)$$
$$\exists s \lhd TreeSpecies.\forall t \lhd Apricot.(t.treeSpecies = s)$$
$$\exists s \lhd TreeSpecies.\forall t \lhd AmericanElm.(t.treeSpecies = s) \tag{1}$$
$$\ldots$$

With the relation $o \lhd C$ in (1) we express that an object o is an instance of class C. Next, we also need to express that objects of different tree species classes are assigned to different tree species objects, i.e., we want to establish a 1–1-correspondence:

$$\forall t_1 \lhd SugarMaple.\forall t_2 \lhd Apricot.(t_1.treeSpecies \neq t_2.treeSpecies)$$
$$\forall t_1 \lhd SugarMaple.\forall t_2 \lhd AmericanElm.(t_1.treeSpecies \neq t_2.treeSpecies)$$
$$\forall t_1 \lhd Apricot.\forall t_2 \lhd AmericanElm.(t_1.treeSpecies \neq t_2.treeSpecies) \tag{2}$$
$$\ldots$$

Note, that the class model in Listing 1 is a correct UML (Unified Modeling Language) model, because it adheres to the abstract syntax of the UML specification in [21]. Therefore it is possible to give an OCL (Object Constraint Language) [19] version of the constraints (1) and (2) – see Appendix A.

Sustainable Constraint Writing

Do we grasp all the knowledge that we have about the tree species domain by the model in Listing 1 and the constraints (1) and (2)? Do we already grasp the essential meaning of the power type construction in the example? Not really. What we actually want to express is that all potential subclasses of the tree class must adhere to similar constraints as those given in (1) and (2). The constraints should hold not only for the tree species of the current version of the model, i.e., for sugar maples, apricot trees and American elms, but for all tree species that we had in the past and might add in the future. With respect to this, the constraints (1) and (2) are not appropriate, they are merely instances of more general constraint schemes. The constraints are vulnerable to model updates and we call them ephemeral constraints therefore. The constraints given in (3) and (4) solve this issue. They generalize the constraints in (1) and (2) by reflecting on the user-defined types:

$$\forall C < Tree.\exists s \lhd TreeSpecies.\forall t \lhd C.(t.treeSpecies = s) \tag{3}$$

$$\forall C_1 < Tree.\forall C_2 < Tree.\forall t_1 \lhd C_1.\forall t_2 \lhd C_2.(t_1.treeSpecies \neq t_2.treeSpecies) \tag{4}$$

The constraints (1) and (2) are pure data-level constraints. The constraints (3) and (4) are still data-level constraints, however they are *reflective*, i.e., they mix in access to meta data, i.e., access to the model. Note, that in this discussion we clearly distinguish between the model-level and the object-level. In accordance to database terminology, we call the object-level also the data-level and, furthermore, we call the model level the meta-data level – see Table 1. Now, an expression of the form $C < T$ in (3) and (4) stands for bounded type quantification, with C being a type variable and T being a user-defined type. As a quick win, the constraints (3) and (4) are much less verbose than the constraints (1) and (2). But the crucial point is that we have achieved sustainable constraints, i.e., constraints that are robust against updates at the model level.

The sustainable constraints we have seen so far are data constraints; they do not impose constraints on the models. Model and data constraints can be teamed together to give semantics to modeling constructs. For example, with the so-called generalization sets, the UML provides extra syntax to model power type scenarios. In Listing 1 we can group the generalizations from the three sub classes to the tree super class into a UML generalization set. Then we can assign the tree species class as a power type property to this generalization set. In [21] the intended meaning of the UML power type construction is given informally, by examples and detailed explanations in terms of data and constraints. With sustainable constraints we are able to give formal semantics to UML generalization sets and power types, because we can simply further generalize the constraints in (3) and (4) from the *Tree* type to all user-defined types. What we need for this purpose is a language that supports sufficiently rich meta specification capabilities, i.e., sufficiently rich support for *reification, introspection* and *reflection*. We define a reflective extension of OCL for this purpose – see Appendix C. We analyze sustainable constraint writing from the perspective of meta programming and exploit existing work on generative programming languages, i.e., GENOUPE [16,18] and FACTORY [15,12], for this purpose.

There are many other alternative means to adequately model sets of sets. The alternatives discussed so far are oriented towards the usual power type constructions, which combine subtypes with a kind of type-object pattern. However, neither of these constructs is needed to adequately model sets of sets. The essence always lays in the established constraints.

The Symbolic Viewpoint of Modeling Languages

In *form-oriented analysis* [14,13] we have taken an approach that is free from metaphors to characterize conceptual modeling and information system modeling. For us, information system modeling is the school of structuring and maintaining information. A deep investigation of the intended meaning of information is not needed for improving modeling languages, tools and technologies. For us, this is the pragmatic reference point for all efforts, including semantic considerations. Table 1 summarizes different existing viewpoints on information system modeling. Each of the viewpoints is important. Each of the viewpoints grasps important issues in pragmatics of information system design and operations.

In the database viewpoint the schema is clearly separated from the data. It is assumed that the schema is almost fixed, whereas the data is not. The data is continuously manipulated. The schema is the structure in which we can capture and maintain data. A crucial feature of databases is to support the enforcement of constraints. Schema updates can occur. However, they are regarded as cost-intensive. Whenever a schema update occurs, this triggers a data migration step. This data migration step can be very complex, because the existing data must be re-shaped [6,3,9].

The OO modeling viewpoint, which encompasses the UML viewpoint, distinguishes between models and objects. Again, models are considered relatively fix as compared to objects. In the OO modeling viewpoint, models are expressions of the modeling language. Objects are not. Usually, objects are considered to have semantics in the so-called real world. Sometimes, objects are immediately considered to be part of the real world. Objects are somehow considered semantics or effect of a model. Models are also considered to have semantics in the real world. Instance specifications are considered expressions of the modeling language. They reside at the model level. They represent objects. Do we need to rephrase our object constraints for instance specifications to avoid inconsistencies– see Appendix B? The clear distinction that we have between meta-data and data in the database viewpoint is sometimes obfuscated in the OO modeling perspective. For example, a pitfall in the semantics of power types is in confusing the level-crossing UML instantiation relation with the set membership relation $\in$ in the intended domain. Care must be taken. With respect to the design of tools and technologies, it is important that we always clearly distinguish between concepts of our modeling framework and concepts in the domain we model.

In the symbolic viewpoint, class models and objects are expressions of the same modeling language. This means, that the model/object divide is explicitly overcome. Class models and objects are pure information. Class models and objects together intend meaning [4,5]. Class models and objects together can be made subject to constraint writing. Class models are different from objects, as per definition. What is a model? You may say that only class models are models. You may say that a class model together with some objects is a also model. Then, you may say that all parts of a model, also the objects, model something. Less likely, you would say that an object is a model. In practice, a slack usage of the model terminology might not harm. However, in precise modeling projects, in formal endeavors, in the design of new modeling tools and information system technologies, it can harm a lot. Therefore, we prefer to talk about expressions of a modeling language in formal endeavors.

In the symbolic viewpoint the model-object instantiation mechanism is characterized in terms of constraints. In a purely symbolic modeling environment each instantiation relation must be accompanied by appropriate constraints, which makes a purely symbolic viewpoint bulky for practical purposes. What is needed is an appropriate reflective constraint specification language or constraint programming approach, on the basis of a clear distinction between data

Table 1. Viewpoints on modeling

OO Modeling Languages	UML	Semantic Modeling	Database Systems	Symbolic Viewpoint
meta-model	M2, UML		syntax specification	meta-model, grammar
model	M1, user model	ER model	schema, meta data	class models, objects (data, words)
objects	M0, runtime instances	entities, relationships	data	
real world				intended meaning

and meta-data. The symbolic viewpoint is helpful in the clarification of semantic questions and the design of emerging tools [7,8,17,2] and technologies [10,11,1].

A   Example OCL Constraints

Specification of the contraints (1) and (2) in OCL:

$$SugarMaple.allInstances \rightarrow treeSpecies \rightarrow asSet \rightarrow size = 1$$
$$Apricot.allInstances \rightarrow treeSpecies \rightarrow asSet \rightarrow size = 1 \tag{5}$$
$$AmericanElm.allInstances \rightarrow treeSpecies \rightarrow asSet \rightarrow size = 1$$

B   Lifting Constraints to Instance Specifications

OCL constraint that establishes a minimum wage:

$$Employee.allInstances \rightarrow forAll(salary > 15.00) \tag{6}$$

The following constraint lifts (6) from UML objects to UML instance specifications:

$$InstanceSpecification.allInstances$$
$$\rightarrow select(classifier \rightarrow includes(type.name = \text{“}Employee\text{”})).slot$$
$$\rightarrow select(definingFeature.name = \text{“}salary\text{”})$$
$$\rightarrow forAll(\tag{7}$$
$$\quad value \rightarrow size = 1$$
$$\quad and\ value.realValue() > 15.00$$
$$)$$

Whereas (6) is a pure object constraint, (7) is a pure model constraint. Constraint (7) is written in terms of types of the UML meta model. There is no need
for the reification and animation constructs introduced in Appendix C.

C   Reflective Extension of OCL

We re-specify the constraints in (3) and (4) in an appropriate reflective extension
of OCL:

$\langle Class \rangle \uparrow .allInstances$
$\rightarrow select(self.superClass.name = \text{``Tree''})$ (8)
$\rightarrow forAll(\langle self \rangle \downarrow .allInstances.treeSpecies \rightarrow asSet \rightarrow size = 1)$

We introduce two operators $\langle \_ \rangle \uparrow$ and $\langle \_ \rangle \downarrow$ that mitigate between the meta-level and the model-level. Constraint (8) is an object constraint. We say that object constraints are model-level constraints. We say that model constraints are meta-level constraints. OCL offers access to the meta-level in object constraints. Unfortunately, the meta-level acess is very limited to a couple of introspective features. With $\langle \_ \rangle \uparrow$ we introduce full introspective access. With $\langle \_ \rangle \uparrow$ we can mix in access to arbitrary types of the UML meta model level. The full OCL language is than available for introspection. With $\langle \_ \rangle \downarrow$ we introduce an operator that *animates* a meta-level expression at the model level, i.e., enacts its *reflection* as an object constraint. With the introduction of only two operators we can achieve full support for introspection and reflection on the basis of the full UML meta-model.

Based on the reflective extension of OCL we can give a general semantics for UML power types. For example, we can generalize constraint (3) to all user-defined types:

$\langle Class \rangle \uparrow .allInstances.forAll(super, sub, power |$
$\quad \langle GeneralizationSet \rangle \uparrow .allInstances \rightarrow exists(gs |$
$\quad gs.powertpye = power$
$\quad and \quad gs.generalization.general \rightarrow includes(super)$
$\quad and \quad gs.generalization.specific \rightarrow includes(sub)$
$)$
$implies($
$\quad super.ownedAttribute \rightarrow select(type = power) \rightarrow size = 1$ (9)
$\quad and$
$\quad super.ownedAttribute \rightarrow select(type = power) \rightarrow forall(p |$
$\qquad \langle sub \rangle \downarrow .allInstances.\langle p \rangle \downarrow \rightarrow asSet \rightarrow size = 1$
$\quad)$
$)$
$)$

Constraint (9) needs to decide upon semantic issues. For example, with (9) the modeler is allowed to give an arbitrary name to a supertype-to-powertype relation. Consequently, (9) requires supertype-to-powertype relations to be unique.

References

1. Atkinson, C., Draheim, D.: Cloud Aided-Software Engineering – Evolving Viable Software Systems through a Web of Views. In: Mahmood, Z., Saeed, S. (eds.) Software Engineering Frameworks for Cloud Computing Paradigm. Springer (2013)
2. Atkinson, C., Bostan, P., Draheim, D.: A Unified Conceptual Framework for Service-Oriented Computing - Aligning Models of Architecture and Utilization. In: Hameurlain, A., Küng, J., Wagner, R. (eds.) Transactions on Large-Scale Data- and Knowledge-Centered Systems, vol. 7. Springer (December 2012)

3. Bordbar, B., Draheim, D., Horn, M., Schulz, I., Weber, G.: Integrated Model-Based Software Development, Data Access, and Data Migration. In: Briand, L.C., Williams, C. (eds.) MoDELS 2005. LNCS, vol. 3713, pp. 382–396. Springer, Heidelberg (2005)
4. Brentano, F.: Psychologie vom empirischen Standpunkt. Duncker & Humblot (1874)
5. Brentano, F.: Psychology from an Empirical Standpoint. Routledge (1995)
6. Draheim, D., Horn, M., Schulz, I.: The Schema Evolution and Data Migration Framework of the Environmental Mass Database IMIS. In: Proceedings of SSDBM 2004 – 16th International Conference on Scientific and Statistical Database Management. IEEE Press (2004)
7. Himsl, M., Jabornig, D., Leithner, W., Regner, P., Wiesinger, T., Küng, J., Draheim, D.: An Iterative Process for Adaptive Meta- and Instance Modeling. In: Wagner, R., Revell, N., Pernul, G. (eds.) DEXA 2007. LNCS, vol. 4653, pp. 519–528. Springer, Heidelberg (2007)
8. Draheim, D., Himsl, M., Jabornig, D., Leithner, W., Regner, P., Wiesinger, T.: Intuitive Visualization-Oriented Metamodeling. In: Bhowmick, S.S., Küng, J., Wagner, R. (eds.) DEXA 2009. LNCS, vol. 5690, pp. 727–734. Springer, Heidelberg (2009)
9. Draheim, D., Natschläger, C.: A Context-Oriented Synchronization Approach. In: Electronic Proceedings of the 2nd International Workshop in Personalized Access, Profile Management, and Context Awarness: Databases (PersDB 2008) in Conjunction with the 34th VLDB Confercence, pp. 20–27 (2008)
10. Draheim, D.: Business Process Technology – A Unified View on Business Processes, Workflows and Enterprise Applications. Springer (September 2010)
11. Draheim, D.: Smart Business Process Management. In: Fischer, L. (ed.) Social Software - 2011 BPM and Workflow Handbook, Digital Edition. Future Strategies, Workflow Management Coalition (February 2012)
12. Draheim, D., Lutteroth, C., Weber, G.: Factory: Statically Type-Safe Integration of Genericity and Reflection. In: Proceedings of the ACIS 2003 – the 4th Intl. Conference on Software Engineering, Artificial Intelligence, Networking, and Parallel/Distributed Computing (2003)
13. Draheim, D., Weber, G.: Form-Oriented Analysis – A New Methodology to Model Form-Based Applications. Springer (October 2004)
14. Draheim, D., Weber, G.: Modelling Form-Based Interfaces with Bipartite State Machines. Journal Interacting with Computers 17(2), 207–228 (2005)
15. Draheim, D., Lutteroth, C., Weber, G.: Generative Programming for C#. ACM SIGPLAN Notices 40(8) (August 2005)
16. Draheim, D., Lutteroth, C., Weber, G.: A Type System for Reflective Program Generators. In: Glück, R., Lowry, M. (eds.) GPCE 2005. LNCS, vol. 3676, pp. 327–341. Springer, Heidelberg (2005)
17. Draheim, D., Himsl, M., Jabornig, D., Küng, J., Leithner, W., Regner, P., Wiesinger, T.: Concept and Pragmatics of an Intuitive Visualization-Oriented Metamodeling Tool. Journal of Visual Languages and Computing 21(4) (August 2010)
18. Lutteroth, C., Draheim, D., Weber, G.: A Type System for Reflective Program Generators. Science of Computer Programming 76(5) (May 2011)

19. Object Management Group. Object Constraint Language, version 2.3.1 OMG (2012)
20. Object Management Group. OMG Unified Modeling Language – Infrastructure, version 2.4.1. OMG (August 2011)
21. Object Management Group. OMG Unified Modeling Language – Superstructure, version 2.4.1. OMG (August 2011)
22. Johnson, R., Woolf, B.: Type Object. In: Pattern Languages of Program Design, vol. 3. Addison-Wesley (1997)

Get a Sample for a Discount
Sampling-Based XML Data Pricing

Ruiming Tang[1], Antoine Amarilli[2], Pierre Senellart[2], and Stéphane Bressan[1]

[1] National University of Singapore, Singapore
{tangruiming,steph}@nus.edu.sg
[2] Institut Mines–Télécom; Télécom ParisTech; CNRS LTCI. Paris, France
{antoine.amarilli,pierre.senellart}@telecom-paristech.fr

Abstract. While price and data quality should define the major trade-off for consumers in data markets, prices are usually prescribed by vendors and data quality is not negotiable. In this paper we study a model where data quality can be traded for a discount. We focus on the case of XML documents and consider completeness as the quality dimension. In our setting, the data provider offers an XML document, and sets both the price of the document and a weight to each node of the document, depending on its potential worth. The data consumer proposes a price. If the proposed price is lower than that of the entire document, then the data consumer receives a sample, i.e., a random rooted subtree of the document whose selection depends on the discounted price and the weight of nodes. By requesting several samples, the data consumer can iteratively explore the data in the document. We show that the uniform random sampling of a rooted subtree with prescribed weight is unfortunately intractable. However, we are able to identify several practical cases that are tractable. The first case is uniform random sampling of a rooted subtree with prescribed size; the second case restricts to binary weights. For both these practical cases we present polynomial-time algorithms and explain how they can be integrated into an iterative exploratory sampling approach.

1 Introduction

There are three kinds of actors in a data market: data consumers, data providers, and data market owners [14]. A data provider brings data to the market and sets prices on the data. A data consumer buys data from the market and pays for it. The owner is the broker between providers and consumers, who negotiates pricing schemes with data providers and manages transactions to trade data.

In most of the data pricing literature [4–6, 9], data prices are prescribed and not negotiable, and give access to the best data quality that the provider can achieve. Yet, data quality is an important axis which should be used to price documents in data markets. Wang et al. [15, 18] define dimensions to assess data quality following four categories: intrinsic quality (believability, objectivity, accuracy, reputation), contextual quality (value-added, relevancy, timeliness, ease

H. Decker et al. (Eds.): DEXA 2014, Part I, LNCS 8644, pp. 20–34, 2014.

of operation, appropriate amount of data, completeness), representational quality (interpretability, ease of understanding, concise representation, consistent representation), and accessibility quality (accessibility, security).

In this paper, we focus on contextual quality and propose a data pricing scheme for *XML trees* such that *completeness* can be traded for discounted prices. This is in contrast to our previous work [17] where the *accuracy* of *relational data* is traded for discounted prices. Wang et al. [15, 18] define completeness as "the extent to which data includes all the values, or has sufficient breadth and depth for the current task". We retain the first part of this definition as there is no current task defined in our setting. Formally, the data provider assigns, in addition to a price for the entire document, a *weight* to each node of the document, which is a function of the potential worth of this node: a higher weight is given to nodes that contain information that is more valuable to the data consumer. We define the completeness of a rooted subtree of the document as the total weight of its nodes, divided by the total weight of the document. A data consumer can then offer to buy an XML document for less than the provider's set price, but then can only obtain a rooted subtree of the original document, whose completeness depends on the discount granted.

A data consumer may want to pay less than the price of the entire document for various reasons: first, she may not be able to afford it due to limited budget but may be satisfied by a fragment of it; second, she may want to explore the document and investigate its content and structure before purchasing it fully.

The data market owner negotiates with the data provider a pricing function, allowing them to decide the price of a rooted subtree, given its completeness (i.e., the weight). The pricing function should satisfy a number of axioms: the price should be non-decreasing with the weight, be bounded by the price of the overall document, and be *arbitrage-free* when repeated requests are issued by the same data consumer (arbitrage here refers to the possibility to strategize the purchase of data). Hence, given a proposed price by a data consumer, the inverse of the pricing function decides the completeness of the sample that should be returned. To be fair to the data consumer, there should be an equal chance to explore every possible part of the XML document that is worth the proposed price. Based on this intuition, we sample a rooted subtree of the XML document of a certain weight, according to the proposed price, uniformly at random.

The data consumer may also issue repeated requests as she is interested in this XML document and wants to explore more information inside in an iterative manner. For each repeated request, a new rooted subtree is returned. A principle here is that the information (document nodes) already paid for should not be charged again. Thus, in this scenario, we sample a rooted subtree of the XML document of a certain weight uniformly at random, without counting the weight of the nodes already bought in previously issued requests.

The present article brings the following contributions:

- We propose to realize the trade-off between quality and discount in data markets. We propose a framework for pricing the completeness of XML data,

based on uniform sampling of rooted subtrees in weighted XML documents. (Section 3)
- We show that the general uniform sampling problem in weighted XML trees is intractable. In this light, we propose two restrictions: sampling based on the number of nodes, and sampling when weights are binary (i.e., weights are 0 or 1). (Section 4)
- We show that both restrictions are tractable by presenting a polynomial-time algorithm for uniform sampling based on the size of a rooted subtree, or on 0/1-weights. (Section 5)
- We extend our framework to the case of repeated sampling requests where the data consumer is not charged twice the same nodes. Again, we obtain tractability when the weight of a subtree is its size. (Section 6)

2   Related Work

Data Pricing. The basic structure of data markets and different pricing schemes were introduced in [14]. The notion of "query-based" pricing was introduced in [4, 6] to define the price of a query as the price of the cheapest set of pre-defined views that can determine the query. It makes data pricing more flexible, and serves as the foundation of a practical data pricing system [5]. The price of aggregate queries has been studied in [9]. Different pricing schemes are investigated and multiple pricing functions are proposed to avoid several pre-defined arbitrage situations in [10]. However, none of the works above takes data quality into account, and those works do not allow the data consumer to propose a price less than that of the data provider, which is the approach that we study here.

The idea of trading off price for data quality has been explored in the context of privacy in [8], which proposes a theoretic framework to assign prices to noisy query answers. If a data consumer cannot afford the price of a query, she can choose to tolerate a higher standard deviation to lower the price. However, this work studies pricing on accuracy for linear relational queries, rather than pricing XML data on completeness. In [17], we propose a relational data pricing framework in which data accuracy can be traded for discounted prices. By contrast, this paper studies pricing for XML data, and proposes a tradeoff based on data completeness rather than accuracy.

Subtree/Subgraph Sampling. The main technical result of this paper is the tractability of uniform subtree sampling under a certain requested size. This question is related to the general topic of subtree and subgraph sampling, but, to our knowledge, it has not yet been adequately addressed.

Subgraph sampling works such as [3, 7, 13, 16] have proposed algorithms to sample small subgraphs from an original graph while attempting to preserve selected metrics and properties such as degree distribution, component distribution, average clustering coefficient and community structure. However, the distribution from which these random graphs are sampled is not known and cannot be guaranteed to be uniform.

Other works have studied the problem of uniform sampling [2, 11]. However, [2] does not propose a way to fix the size of the samples. The authors of [11] propose a sampling algorithm to sample a connected sub-graph of size k under an approximately uniform distribution; note that this work provides no bound on the error relative to the uniform distribution.

Sampling approaches are used in [19, 12] to estimate the selectivity of XML queries (containment join and twig queries, respectively). Nevertheless, the samples in [19] are specific to containment join queries, while those in [12] are representatives of the XML document for any twig queries. Neither of those works controls the distribution from which the subtrees are sampled.

In [1], Cohen and Kimelfeld show how to evaluate a deterministic tree automaton on a probabilistic XML document, with applications to sampling possible worlds that satisfy a given constraint, e.g., expressed in monadic second-order logic and then translated into a tree automaton. Note that the translation of constraints to tree automata itself is not tractable; in this respect, our approach can be seen as a specialization of [1] to the simpler case of fixed-size or fixed-weight tree sampling, and as an application of it to data pricing.

3   Pricing Function and Sampling Problem

This paper studies data pricing for tree-shaped documents. Let us first formally define the terminology that we use for such documents.

We consider trees that are unordered, directed, rooted, and weighted. Formally, a tree t consists of a set of nodes $V(t)$ (which are assumed to carry unique identifiers), a set of edges $E(t)$, and a function w mapping every node $n \in V(t)$ to a non-negative rational number $w(n)$ which is the *weight* of node n. We write root(t) for the root node of t. Any two nodes n_1, $n_2 \in V(t)$ such that $(n_1, n_2) \in E(t)$ are in a *parent-child relationship*, that is, n_1 is the parent of n_2 and n_2 is a child of n_1.

By children(n), we represent the set of nodes that have parent n. A tree is said to be *binary* if each node of the tree has at most two children, otherwise it is *unranked*. Throughout this paper, for ease of presentation, we may call such trees "XML documents".

We now introduce the notion of *rooted subtree* of an XML document:

Definition 1. *(Subtree, rooted subtree) A tree t' is a **subtree** of a tree t if $V(t') \subseteq V(t)$ and $E(t') \subseteq E(t)$. A **rooted subtree** t' of a tree t is a subtree of t such that root(t) = root(t'). We name it **r-subtree** for short. The weight function for a subtree t' of a tree t is always assumed to be the restriction of the weight function for t on the nodes in t'.*

For technical reasons, we also sometimes talk of the *empty* subtree that contains no node.

Example 1. Figure 1 presents two example trees. The nodes $\{n_0, n_2, n_5\}$, along with the edges connecting them, form an r-subtree of the tree in Figure 1(a).

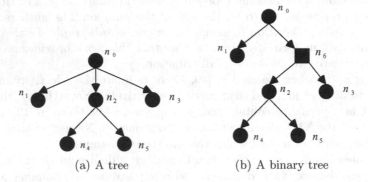

(a) A tree   (b) A binary tree

Fig. 1. Two example trees (usage of the square node will be introduced in Section 5.2)

Likewise, the nodes $\{n_2, n_4, n_5\}$ and the appropriate edges form a subtree of that tree (but not an r-subtree). The tree of Figure 1(b) is a binary tree (ignore the different shapes of the nodes for now).   □

We now present our notion of data quality, by defining the completeness of an r-subtree, based on the weight function of the original tree:

Definition 2. *(Weight of a tree) For a node $n \in V(t)$, we define inductively* weight$(n) = w(n) + \sum_{(n,n') \in E(t)}$ weight(n'). *With slight abuse of notation, we note* weight$(t) =$ weight$(\text{root}(t))$ *as the **weight** of t.*

Definition 3. *(Completeness of an r-subtree) Let t be a tree. Let t' be an r-subtree of t. The **completeness** of t' with respect to t is $c_t(t') = \frac{\text{weight}(t')}{\text{weight}(t)}$. It is obvious that $c_t(t') \in [0, 1]$.*

We study a framework for data markets where the data consumer can buy an incomplete document from the data provider while paying a discounted price. The formal presentation of this framework consists of three parts:

1. An XML document t.
2. A pricing function φ_t for t whose input is the desired completeness for an r-subtree of the XML document, and whose value is the price of this r-subtree. Hence, given a proposed price pr_0 by a data consumer, the completeness of the returned r-subtree is decided by $\varphi_t^{-1}(pr_0)$.
3. An algorithm to sample an r-subtree of the XML document uniformly at random among those of a given completeness.

We study the question of the sampling algorithm more in detail in subsequent sections. For now, we focus on the pricing function, starting with a formal definition:

Definition 4. *(Pricing function) The **pricing function** for a tree t is a function $\varphi_t : [0, 1] \to \mathbb{Q}^+$. Its input is the completeness of an r-subtree t' and it returns the price of t', as a non-negative rational.*

A healthy data market should impose some restrictions on φ_t, such as:

Non-decreasing. The more complete an r-subtree is, the more expensive it should be, i.e., $c_1 \geqslant c_2 \Rightarrow \varphi_t(c_1) \geqslant \varphi_t(c_2)$.

Arbitrage-free. Buying an r-subtree of completeness $c_1 + c_2$ should not be more expensive than buying two subtrees with respective completeness c_1 and c_2, i.e., $\varphi_t(c_1) + \varphi_t(c_2) \geqslant \varphi_t(c_1 + c_2)$. In other words, φ_t should be sub-additive. This property is useful when considering repeated requests, studied in Section 6.

Minimum and maximum bound. We should have $\varphi_t(0) = pr_{\min}$ and $\varphi_t(1) = pr_t$, where $pr_{\min}$ is the minimum cost that a data consumer has to pay using the data market and pr_t is the price of the whole tree t. Note that by the non-decreasing character of φ_t, $pr_t \geqslant pr_{\min} \geqslant 0$.

All these properties can be satisfied, for instance, by functions of the form $\varphi_t(c) = (pr_t - pr_{\min})c^p + pr_{\min}$ where $p \leqslant 1$; however, if $p > 1$, the arbitrage-free property is violated.

Given a proposed price pr_0 by a data consumer, $\varphi_t^{-1}(pr_0)$ is the set of possible corresponding completeness values. Note that φ_t^{-1} is a relation and may not be a function; φ_t^{-1} is a function if different completeness values correspond to different prices. Once a completeness value $c \in \varphi_t^{-1}(pr_0)$ is chosen, the weight of the returned r-subtree is fixed as $c \times \mathsf{weight}(t)$.

Therefore, in the rest of the paper, we consider the problem of uniform sampling an r-subtree with prescribed weight (instead with prescribed completeness). Let us now define the problem that should be solved by our sampling algorithm:

Definition 5. *(Sampling problem) The problem of **sampling an r-subtree**, given a tree t and a weight k, is to sample an r-subtree t' of t, such that $\mathsf{weight}(t') = k$, uniformly at random, if one exists, or fail if no such r-subtree exists.*

4 Tractability

4.1 Intractability of the Sampling Problem

We now turn to the question of designing an algorithm to solve the sampling problem defined in the previous section. Sadly, it can be shown that this problem is already NP-hard in the general formulation that we gave.

Proposition 1. *Given a tree t and a weight x, it is NP-hard to sample an r-subtree of t of weight x uniformly at random.*

Proof. Consider the simpler problem of deciding if such an r-subtree of t exists. Further restrict t such that $\mathsf{w}(n) = 0$ unless n is a leaf. In this case, deciding whether there exists an r-subtree of weight x is equivalent to the subset-sum problem, which is known to be NP-complete. Now there is a PTIME-reduction from this existence problem to the sampling problem, as an algorithm for sampling can be used to decide whether there exists an r-subtree of the desired weight (the algorithm returns one such) or if none exists (the algorithm fails). □

4.2   Tractable Cases

We now define restricted variants of the sampling problem where the weight function is assumed to be of a certain form. In the next section, we show that sampling for these variants can be performed in PTIME.

Unweighted Sampling. In this setting, we take the weight function $w(n) = 1$ for all $n \in V(t)$. Hence, the weight of a tree t is actually the number of nodes in t, i.e., its size, which we write $\text{size}(t)$.

In this case, the hardness result of Proposition 1 does not apply anymore. However, sampling an r-subtree with prescribed size uniformly at random is still not obvious to do, as the following example shows:

Example 2. Consider the problem of sampling an r-subtree t' of size 3 from the tree t in Figure 1(a). We can enumerate all such r-subtrees: $\{n_0, n_1, n_2\}$, $\{n_0, n_1, n_3\}$, $\{n_0, n_2, n_3\}$, $\{n_0, n_2, n_4\}$ and $\{n_0, n_2, n_5\}$, and choose one of them at random with probability $\frac{1}{5}$. However, as the number of r-subtrees may be exponential in the size of the document in general, we cannot hope to perform this approach in PTIME. Observe that it is not easy to build a random r-subtree node by node: it is clear that node n_0 must be included, but then observe that we cannot decide to include n_1, n_2, or n_3, uniformly at random. Indeed, if we do this, our distribution on the r-subtrees will be skewed, as n_1 (or n_3) occurs in $\frac{2}{5}$ of the outcomes whereas n_2 occurs in $\frac{4}{5}$ of them. Intuitively, this is because there are more ways to choose the next nodes when n_2 is added, than when n_1 or n_3 are added.   □

We show in the next section that this problem can be solved in PTIME.

0/1-weights Sampling. In this more general problem variant, we require that $w(n) \in \{0, 1\}$ for all $n \in V(t)$, i.e., the weight is binary.

We show in Section 5.3 that this problem can also be solved in PTIME using an adaptation of the unweighted sampling algorithm.

5   Algorithms for Tractable Uniform Sampling

In this section, we present a PTIME algorithm for the unweighted sampling problem, namely the problem of sampling an r-subtree of size k from an XML document, uniformly at random.

We first describe the algorithm for the case of *binary* trees, in Section 5.1. Next, we adapt the algorithm in Section 5.2 to show how to apply it to arbitrary trees. Last, we study the more general case of 0/1-weights in Section 5.3, showing that the algorithm for unweighted sampling can be adapted to solve this problem.

Algorithm 1. Algorithm for unweighted sampling problem on binary trees

---

Input: a binary tree t and an integer $k \geqslant 0$
Result: an r-subtree t' of t of size$(t') = k$ uniformly at random
// Phase 1: count the number of subtrees
1 $D \leftarrow$ SubtreeCounting(t);
   // Phase 2: sample a random subtree
2 **if** $k \leqslant$ size(t) **then**
3 $\quad$ **return** UniformSampling$($root$(t), D, k)$;
4 **else**
5 $\quad$ fail;

Algorithm 2. SubtreeCounting(t)

Input: a binary tree t
Result: a matrix D such that $D_i[k]$ is the number of r-subtrees of size k rooted
$\qquad$ at n_i for all i and k

---

1 $D_{\mathsf{NULL}}[0] \leftarrow 1$;
   // We browse all nodes in topological order
2 **foreach** non-NULL node n_i accessed bottom-up **do**
3 $\quad$ $n_j \leftarrow$ first child of n_i (or NULL if none exists);
4 $\quad$ $n_g \leftarrow$ second child of n_i (or NULL if none exists);
5 $\quad$ $D_i[0] \leftarrow 1$;
6 $\quad$ $T \leftarrow D_j \oplus D_g$;
7 $\quad$ **for** $j \in [0, |T| - 1]$ **do**
8 $\quad\quad$ $D_i[j+1] \leftarrow T[j]$;
9 **return** D;

5.1 Unweighted Sampling for Binary Trees

In this section, we provide an algorithm which proves the following theorem:

Theorem 1. *The unweighted sampling problem for binary trees can be solved in* PTIME.

Our general algorithm to solve this problem is given as Algorithm 1. The algorithm has two phases, which we study separately in what follows. For simplicity, whenever we discuss binary trees in this section, we will add special NULL children to every node of the tree (except NULL nodes themselves), so that all nodes, including leaf nodes, have exactly two children (which may be NULL). This will simplify the presentation of the algorithms.

First phase: Subtree Counting (Algorithm 2). We start by computing a matrix D such that, for every node n_i of the input tree t and any value $0 \leqslant k \leqslant$ size(t), $D_i[k]$ is the number of subtrees of size k rooted at node n_i. We do so with Algorithm 2 which we now explain in detail.

There is only one subtree rooted at the special NULL node, namely the empty subtree, with size 0, which provides the base case of the algorithm (line 1). Otherwise, we compute D_i for a node n_i from the values D_j and D_g of D for its children n_j and n_g (which may be NULL); those values have been computed before because nodes are considered following a topological ordering, in a bottom-up fashion.

Intuitively, the only r-subtree of size 0 is the empty subtree, hence we must have $D_i[0] = 1$ (line 5). Otherwise, any r-subtree of size $k > 0$ rooted at n_i is obtained by retaining n_i, and choosing two r-subtrees t_j and t_g, respectively rooted at n_j and n_g (the children of n_i), such that $\text{size}(t_j) + \text{size}(t_g) = k - 1$ (which accounts for the size of the additional node n_i). The number of such choices is computed by the *convolution* of D_j and D_g in line 6, defined as:

$$\text{For } 0 \leqslant p \leqslant |D_j| + |D_g| - 2, \quad (D_j \oplus D_g)[p] := \sum_{m=\max(0,p-|D_g|-1)}^{\min(p,|D_j|-1)} D_j[m] \times D_g[p-m].$$

Lines 7 and 8 account for the size of the retained node n_i.

Example 3. Let t be the tree presented in Figure 1(b) (again, ignore the different shapes of nodes for now). Starting from the leaf nodes, we compute $D_1 = D_4 = D_5 = D_3 = (1,1)$, applying lines 5 to 8 on $D_{\text{NULL}} \oplus D_{\text{NULL}} = (1)$. This means that there are two r-subtrees rooted at leaf nodes: the empty subtree, and the subtree with just that leaf.

Now, when computing D_2, we first convolve D_4 and D_5 to get the numbers of pairs of r-subtrees of different sizes at $\{n_4, n_5\}$, i.e., $D_4 \oplus D_5 = (1,2,1)$, so that $D_2 = (1,1,2,1)$. When computing D_6, we first compute $D_2 \oplus D_3 = (1,2,3,3,1)$, so that $D_6 = (1,1,2,3,3,1)$. Finally, $D_0 = (1,1,2,3,5,6,4,1)$. □

We now state the correctness and running time of this algorithm.

Lemma 1. *Algorithm 2 terminates in time polynomial in* $\text{size}(t)$ *and returns D such that, for every i and k, $D_i[k]$ is the number of r-subtrees of size k rooted at node n_i.*

Proof. Let us first prove the running time. All arrays under consideration have size at most n ($n = \text{size}(t)$), so computing the convolution sum of two such arrays is in time $O(n^2)$ (computing each value of the convolution sum is in time $O(n)$). The number of convolution sums to compute overall is $O(n)$, because each array D_i occurs in exactly one convolution sum. The overall running time is $O(n^3)$.

Let us now show correctness. We proceed by induction on the node n_i to prove the claim for every k. The base case is the NULL node, whose correctness is straightforward. Let us prove the induction step. Let n_i be a node, and assume by induction that $D_j[k']$ is correct for every k' for every child n_j of n_i. Let us fix k and show that $D_i[k]$ is correct. To select an r-subtree at n_i, either $k = 0$ and there is exactly one possibility (the empty subtree), or $k > 0$ and the number of possibilities is the number of ways to select a set of r-subtrees at the children of n_i so that their size sum to $k - 1$. This is the role of lines 5 to 8. Now, to

---

Algorithm 3. UniformSampling(n_i, D, x)

Input: a node n_i (or NULL), the precomputed D, and a size value x
Result: an r-subtree of size x at node n_i

1   **if** $x = 0$ **then**
2     $\lfloor$ **return** $\emptyset$;
3   $n_j \leftarrow$ first child of n_i (or NULL if none exists);
4   $n_g \leftarrow$ second child of n_i (or NULL if none exists);
5   **for** $0 \leqslant s_j \leqslant \text{size}(n_j), 0 \leqslant s_g \leqslant \text{size}(n_g)$ s.t. $s_j + s_g = x - 1$ **do**
6     $\lfloor$ $p(s_j, s_g) \leftarrow D_j[s_j] \times D_g[s_g]$;
7   Sample an (s_j, s_g) with probability $p(s_j, s_g)$ normalized by $\sum_{s_j, s_g} p(s_j, s_g)$;
8   $L \leftarrow$ UniformSampling(n_j, D, s_j);
9   $R \leftarrow$ UniformSampling(n_g, D, s_g);
10 **return** the tree rooted at n_i with child subtrees L and R;

---

enumerate the ways of choosing r-subtrees at children of n_i whose size sum to $k-1$, we can first decide the size of the selected r-subtree for each child: the ways to assign such sizes form a partition of the possible outcomes, so the number of outcomes is the sum, over all such assignments of r-subtree sizes to children, of the number of outcomes for this assignment. Now, for a fixed assignment, the subtrees rooted at each children are chosen independently, so the number of outcomes for a fixed assignment is the product of the number of outcomes for the given size for each child, which by induction hypothesis is correctly reflected by the corresponding $D_j[k']$. Hence, for a given k, either $k = 0$, or $k > 0$, in which case $D_i[k]$ is $(D_j \oplus D_g)[k-1]$ by lines 5 to 8, which sums, over all possible subtree size assignments, the number of choices for this subtree size assignment. Hence, by induction, we have shown the desired claim. $\square$

Second phase: Uniform Sampling (Algorithm 3). In the second phase of Algorithm 1, we sample an r-subtree from t in a recursive top-down manner, based on the matrix D computed by Algorithm 2. Our algorithm to perform this uniform sampling is Algorithm 3. The basic idea is that to sample an r-subtree rooted at a node n_i, we decide on the size of the subtrees rooted at each child node, biased by the number of outcomes as counted in D, and then sample r-subtrees of the desired size recursively.

Let us now explain Algorithm 3 in detail.

If $x = 0$, we must return the empty tree (lines 2 and 2). Otherwise we return n_i and subtrees t_j and t_g rooted at the children n_j and n_g of n_i. We first decide on the size s_j and s_g of t_j and t_g (lines 6 to 7) before sampling recursively a subtree of the prescribed size, uniformly at random, and returning it.

The possible size pairs (s_j, s_g) must satisfy the following conditions to be possible choices for the sizes of the subtrees t_j and t_g:

1. $0 \leqslant s_j \leqslant \text{size}(n_j)$ and $0 \leqslant s_g \leqslant \text{size}(n_g)$ (of course $\text{size}(\text{NULL}) = 0$)
2. $s_j + s_g = x - 1$ (which accounts for node n_i)

Intuitively, to perform a uniform sampling, we now observe that the choice of the size pair (s_j, s_g) partitions the set of outcomes. Hence, the probability that we select one size pair should be proportional to the number of possible outcomes for this pair, namely, the number of r-subtrees t_j and t_g such that $\mathsf{size}(t_j) = s_j$ and $\mathsf{size}(t_g) = s_g$. We compute this from D_j and D_g (line 6) by observing that the number of pairs (t_j, t_g) is the product of the number of choices for t_j and for t_g, as every combination of choices is possible.

Example 4. Follow Example 3. Assume we want to sample an r-subtree t' of $\mathsf{size}(t') = 3$ uniformly. We first call UniformSampling$(n_0, 3)$. We have to return n_0. Now n_0 has two children, n_1 and n_6. The possible size pairs are $(0, 2)$ and $(1, 1)$, with respective (unnormalized) probabilities $p(0, 2) = D_1[0] \times D_6[2] = 1 \times 2 = 2$ and $p(1, 1) = D_1[1] \times D_6[1] = 1 \times 1 = 1$. The normalized probabilities are therefore $\frac{2}{3}$ and $\frac{1}{3}$. Assume that we choose $(0, 2)$. We now call recursively UniformSampling$(n_1, 0)$ and UniformSampling$(n_6, 2)$.

UniformSampling$(n_1, 0)$ returns $\emptyset$. We proceed to UniformSampling$(n_6, 2)$. We have to return n_6. Now n_6 has two children, n_2 and n_3. The possible size pairs for this call are $(1, 0)$ and $(0, 1)$ with probabilities $\frac{1}{2}$ and $\frac{1}{2}$. Assume that we choose $(1, 0)$. We now call recursively UniformSampling$(n_2, 1)$ and UniformSampling$(n_3, 0)$.

UniformSampling$(n_3, 0)$ returns $\emptyset$. We proceed to UniformSampling$(n_2, 1)$. n_2 is selected. n_2 has two children, n_4 and n_5. There is only one possible size pair for this call $(0, 0)$ with probability 1. We can only choose $(0, 0)$ and call UniformSampling$(n_4, 0)$ (which results in $\emptyset$) and UniformSampling$(n_5, 0)$ (which results in $\emptyset$). Hence, the end result is the r-subtree whose nodes are $\{n_0, n_6, n_2\}$ (and whose edges can clearly be reconstituted in PTIME from t). □

We now show the tractability and correctness of Algorithm 3.

Lemma 2. *For any tree t, node $n_i \in \mathsf{V}(t)$ and integer $0 \leqslant x \leqslant \mathsf{size}(n_i)$, given D computed by Algorithm 2, UniformSampling(n_i, D, x) terminates in polynomial time and returns an r-subtree of size x rooted at n_i, uniformly at random (i.e., solves the unweighted sampling problem for binary trees).*

Proof. Let us first prove the complexity claim. On every node n_i of the binary tree t, the number of possibilities to consider is linear in the tree size because every node has exactly two children, and for each possibility the number of operations performed is constant (assuming that drawing a number uniformly at random is constant-time). The overall running time is quadratic, i.e., $O(\mathsf{size}(t)^2)$.

We now show correctness by induction on n_i. The base case is $n_i = $ NULL, in which case we must have $x = 0$ and we correctly return $\emptyset$. Otherwise, if n_i is not NULL, either $x = 0$ and we correctly return $\emptyset$, or $x > 0$ and, as in the proof of Lemma 1, the set of possible outcomes of the sampling process is partitioned by the possible assignments, and only the valid ones correspond to a non-empty set of outcomes. Hence, we can first choose a size pair, *weighted by the proportion of outcomes which are outcomes for that assignment*, and then choose an outcome for this pair. Now, observe that, by Lemma 1, D correctly represents the number of outcomes for each child of n_i, so that our computation of p (which mimics that of Algorithm 2) correctly represents the proportion of outcomes that are outcomes for every size pair. We then choose an assignment according to p, and then observe that choosing an outcome for this assignment amounts to choosing an outcome for each child of n_i whose size is given by the assignment. By induction hypothesis, this is precisely what the recursive calls to UniformSampling(n_i, D, x) perform. This concludes the proof. □

5.2 Algorithm for Sampling an Unranked Tree

In this section, we show that the algorithm of the previous section can be adapted so that it works on arbitrary unranked trees, not just binary trees.

We first observe that the straightforward generalization of Algorithm 1 to trees of arbitrary arity, where assignments and convolutions are performed for all children, is still correct. However, it would not run in polynomial time anymore as there would be a potentially exponential number of size pairs to consider.

Fortunately, there is still hope to avoid enumerating the size pairs over all the children, thanks to the *associativity of convolution sum*. Informally, assume we have three children $\{n_1, n_2, n_3\}$, we do the following: we treat $\{n_1\}$ as a group and $\{n_2, n_3\}$ as the second group, then enumerate size pairs over $\{n_1\}$ and $\{n_2, n_3\}$; once a size pair, in which a positive integer is assigned to $\{n_2, n_3\}$, is selected, we can treat $\{n_2\}$ and $\{n_3\}$ as new groups and enumerate size pairs over $\{n_2\}$ and $\{n_3\}$. This strategy can be implemented by transforming the original tree to a binary tree. From this intuition, we now state our result:

Theorem 2. *The unweighted sampling problem can be solved in* PTIME, *for arbitrary unranked trees.*

Proof (Sketch). The proof proceeds by encoding arbitrary trees to *encoded trees*, that are binary trees whose nodes are either regular nodes or *dummy* nodes. Intuitively, the encoding operation replaces sequences of more than two children by a hierarchy of dummy nodes representing those children; replacing dummy nodes by the sequence of their children yields back the original tree. The encoding is illustrated in Figure 1, where the tree in Figure 1(b) is the encoded tree of the one in Figure 1(a) (dummy nodes are represented as squares).

It can then be shown that, up to the question of keeping or deleting the dummy nodes with no regular descendants (we call them *bottommost*), there is a bijection between r-subtrees in the original tree and r-subtrees in the encoded tree. Hence, we can solve the unweighted sampling problem by choosing an r-subtree in the encoded tree with the right number of regular nodes, uniformly at random, and imposing the choice of keeping bottommost dummy nodes.

There only remains to adapt Algorithms 2 and 3 to run correctly on encoded trees, that is, mananging dummy nodes correctly.

We change Algorithm 2 by replacing lines 5 to 8 with $D_i \leftarrow D_j \oplus D_g$ when n_i is a dummy node (as it must always be kept, and does not increase the size of the r-subtree).

We change Algorithm 3 by imposing at line 2 the condition that n_i is either NULL or a regular node (otherwise we cannot return $\emptyset$ as we must keep dummy nodes). Also, we change line 6 so that, when node n_i is a dummy node, we require $s_j + s_g = x$ (rather than $x - 1$), as we do not count the dummy node in the size of the resulting subtree.

The correctness and running time of the modified algorithms can be proved by straightforward adaptations of Lemma 1 and Lemma 2.   □

5.3   Uniform Sampling under 0/1-Weights

Our tractability result extends to trees with binary weights:

Theorem 3. *The 0/1-weights sampling problem can be solved in* PTIME, *for arbitrary unranked trees.*

Proof (Sketch). The proof proceeds by modifying the unweighted sampling algorithm to handle nodes of weight 0 (in addition to the modifications described in the previous section).

In Algorithm 2, when n_i is a weight-0 node, lines 7 to 8 must be replaced by $j \in [1, |T| - 1], D_i[j] \leftarrow T[j]$ (the node has no weight); line 5 is removed and set $D_i[0] \leftarrow 1 + T[0]$ (we can keep or discard the weight-0 node, unlike the dummy nodes of the previous section).

In Algorithm 3, when $x = 0$ and n_i is a weight-0 node, the empty subtree should be returned with probability $\frac{1}{D_i[0]}$; otherwise, continue the execution of the algorithm to return a random non-empty r-subtree of size 0. This is because the introduction of weight-0 nodes, unlike weight-1 and dummy nodes, leads to multiple ways to sample a r-subtree of size 0. If n_i is a weight-0 node, line 6 should be changed to require $s_j + s_g = x$ rather than $x - 1$, as for dummy nodes.

The correctness and running time of the modified algorithms can be proved by straightforward adaptations of Lemmas 1 and 2.   □

6   Repeated Requests

In this section, we consider the more general problem where the data consumer requests a *completion* of a certain price to data that they have already bought. The motivation is that, after having bought incomplete data, the user may realize that they need additional data, in which case they would like to obtain more incomplete data that is not redundant with what they already have.

A first way to formalize the problem is as follows, where data is priced according to a known subtree (provided by the data consumer) by considering that known nodes are free (but that they may or may not be returned again).

Definition 6. *The problem of sampling an r-subtree of weight k in a tree t conditionally to an r-subtree t' is to sample an r-subtree t'' of t uniformly at random, such that* $\mathsf{weight}(t'') - \sum_{n \in (\mathsf{V}(t') \cap \mathsf{V}(t''))} \mathsf{w}(n) = k$.

An alternative is to consider that we want to sample an *extension* of a fixed size to the whole subtree, so that all known nodes are part of the output:

Definition 7. *The problem of sampling an r-subtree of weight k in a tree t that extends an r-subtree t′ is to sample an r-subtree t″ of t uniformly at random, such that (1) t′ is an r-subtree of t″; (2)* weight(t″) − weight(t′) = k.

Note that those two formulations are not the same: the first one does not require the known part of the document to be returned, while the second one does. While it may be argued that the resulting outcomes are essentially equivalent (as they only differ on parts of the data that are already known to the data consumer), it is important to observe that they define different distributions: though both problems require the sampling to be uniform among their set of outcomes, the additional possible outcomes of the first definition means that the underlying distribution is not the same.

As the uniform sampling problem for r-subtrees can be reduced to either problem by setting t′ to be the empty subtree, the NP-hardness of those two problems follows from Proposition 1. However, we can show that, in the unweighted case, those problems are tractable, because they reduce to the 0/1-weights sampling problem which is tractable by Theorem 3:

Proposition 2. *The problem of sampling an r-subtree of weight k in a tree t conditionally to an r-subtree t′ can be solved in* PTIME *if t is unweighted. The same holds for the problem of sampling that extends another r-subtree.*

Proof (Sketch). For the problem of Definition 6, set the weight of the nodes of t′ in t to be zero (the intuition is that all the known nodes are free). The problem can then be solved by applying Theorem 3.

For the problem of Definition 7, set the weight of the nodes of t′ in t to be zero but we have to ensure that weight-0 nodes are always returned. To do so, we adapt Theorem 3 by handling weight-0 nodes in the same way as handling dummy nodes in the previous section.   □

7 Conclusion

We proposed a framework for a data market in which data quality can be traded for a discount. We studied the case of XML documents with completeness as the quality dimension. Namely, a data provider offers an XML document, and sets both the price and weights of nodes of the document. The data consumer proposes a price but may get only a sample if the proposed price is lower than that of the entire document. A sample is a rooted subtree of prescribed weight, as determined by the proposed price, sampled uniformly at random.

We proved that if nodes in the XML document have arbitrary non-negative weights, the sampling problem is intractable. We identified tractable cases, namely the unweighted sampling problem and 0/1-weights sampling problem, for which we devised PTIME algorithms. We proved the time complexity and correctness of the algorithms. We also considered repeated requests and provided PTIME solutions to the unweighted cases.

The more general issue that we are currently investigating is that of sampling rooted subtrees uniformly at random under more expressive conditions than size restrictions or 0/1-weights. In particular, we intend to identify the tractability boundary to describe the class of tree statistics for which it is possible to sample r-subtrees in PTIME under a uniform distribution.

Acknowledgment. This work is supported by the French Ministry of European and Foreign Affairs under the STIC-Asia program, CCIPX project.

References

1. Cohen, S., Kimelfeld, B., Sagiv, Y.: Running tree automata on probabilistic XML. In: PODS, pp. 227–236 (2009)
2. Henzinger, M.R., Heydon, A., Mitzenmacher, M., Najork, M.: On near-uniform URL sampling. Computer Networks 33(1-6) (2000)
3. Hübler, C., Kriegel, H.-P., Borgwardt, K., Ghahramani, Z.: Metropolis algorithms for representative subgraph sampling. In: ICDM (2008)
4. Koutris, P., Upadhyaya, P., Balazinska, M., Howe, B., Suciu, D.: Query-based data pricing. In: PODS (2012)
5. Koutris, P., Upadhyaya, P., Balazinska, M., Howe, B., Suciu, D.: QueryMarket demonstration: Pricing for online data markets. PVLDB 5(12) (2012)
6. Koutris, P., Upadhyaya, P., Balazinska, M., Howe, B., Suciu, D.: Toward practical query pricing with QueryMarket. In: SIGMOD (2013)
7. Leskovec, J., Faloutsos, C.: Sampling from large graphs. In: SIGKDD (2006)
8. Li, C., Li, D.Y., Miklau, G., Suciu, D.: A theory of pricing private data. In: ICDT (2013)
9. Li, C., Miklau, G.: Pricing aggregate queries in a data marketplace. In: WebDB (2012)
10. Lin, B.-R., Kifer, D.: On arbitrage-free pricing for general data queries. PVLDB 7(9), 757–768 (2014)
11. Lu, X., Bressan, S.: Sampling connected induced subgraphs uniformly at random. In: Ailamaki, A., Bowers, S. (eds.) SSDBM 2012. LNCS, vol. 7338, pp. 195–212. Springer, Heidelberg (2012)
12. Luo, C., Jiang, Z., Hou, W.-C., Yu, F., Zhu, Q.: A sampling approach for XML query selectivity estimation. In: EDBT (2009)
13. Maiya, A.S., Berger-Wolf, T.Y.: Sampling community structure. In: WWW (2010)
14. Muschalle, A., Stahl, F., Löser, A., Vossen, G.: Pricing approaches for data markets. In: Castellanos, M., Dayal, U., Rundensteiner, E.A. (eds.) BIRTE 2012. LNBIP, vol. 154, pp. 129–144. Springer, Heidelberg (2013)
15. Pipino, L., Lee, Y.W., Wang, R.Y.: Data quality assessment. Commun. ACM 45(4) (2002)
16. Ribeiro, B.F., Towsley, D.F.: Estimating and sampling graphs with multidimensional random walks. In: Internet Measurement Conference (2010)
17. Tang, R., Wu, H., Bao, Z., Bressan, S., Valduriez, P.: The price is right. In: Decker, H., Lhotská, L., Link, S., Basl, J., Tjoa, A.M. (eds.) DEXA 2013, Part II. LNCS, vol. 8056, pp. 380–394. Springer, Heidelberg (2013)
18. Wang, R.Y., Strong, D.M.: Beyond accuracy: What data quality means to data consumers. J. of Management Information Systems 12(4) (1996)
19. Wang, W., Jiang, H., Lu, H., Yu, J.X.: Containment join size estimation: Models and methods. In: SIGMOD (2003)

Discovering *non-constant* Conditional Functional Dependencies with Built-in Predicates

Antonella Zanzi and Alberto Trombetta

Dipartimento di Scienze Teoriche e Applicate
Università degli Studi dell'Insubria, via Mazzini 5, 21100 Varese, Italy
{antonella.zanzi,alberto.trombetta}@uninsubria.it

Abstract. In the context of the data quality research area, Conditional Functional Dependencies with built-in predicates (CFD^{P}s) have been recently defined as extensions of Conditional Functional Dependencies with the addition, in the patterns of their data values, of the comparison operators. CFD^{P}s can be used to impose constraints on data; they can also represent relationships among data, and therefore they can be mined from datasets. In the present work, after having introduced the distinction between *constant* and *non-constant* CFD^{P}s, we describe an algorithm to discover *non-constant* CFD^{P}s from datasets.

Keywords: Functional Dependencies, Data Constraints, Data Quality, Data Mining.

1 Introduction

Conditional Functional Dependencies with built-in predicates (CFD^{P}s) have been defined in [3] as extensions of Conditional Functional Dependencies (CFDs) [8] (which have been proposed in the data quality field as extensions of Functional Dependencies – FDs).

FDs and their extensions, capturing data inconsistency, can be used to evaluate the quality of a dataset and – to a certain extent – also for data cleaning purposes. For example, the use of FDs for data cleaning purposes in relational databases is described in [16], where data dirtiness is equaled to the violation of FDs, and in [5] CFDs have been proposed as a method for inconsistency detection and repairing.

This approach is used, for example, in *Semandaq* [7], a tool using CFDs for data cleaning purposes. Another tool, called *Data Auditor*, is presented in [10] and supports more types of constraints (i.e., CFDs, *conditional inclusion dependencies*, and *conditional sequential dependencies*) used to test data inconsistency and completeness.

In a previous work [19] – along with other types of constraints and dependencies, such as FDs, CFDs, order dependencies and existence constraints – we used CFD^{P}s in the context of data quality evaluation. In particular, we developed a tool to check a dataset against a set of data quality rules expressed with the XML markup language.

H. Decker et al. (Eds.): DEXA 2014, Part I, LNCS 8644, pp. 35–49, 2014.

CFD^{P}s can potentially express additional constraints and quality rules that cannot be expressed by FDs and CFDs and thus be useful in the data quality field. However, their identification is not often straightforward just looking at the data. For this reason a tool supporting the discovery of CFD^{P}s can be useful to identify rules to be used in the evaluation of the quality of a dataset.

In the present work, after having distinguished between *constant* and *non-constant* CFD^{P}s, we describe an algorithm for discovering *non-constant* CFD^{P}s.

2   CFD^P Definition

CFDs specify constant patterns in terms of equality, while CFD^{P}s are CFDs with built-in predicates ($\neq, <, >, \leq, \geq$) in the patterns of their data values. It is assumed that the domain of an attribute is totally ordered if $<, >, \leq$ or $\geq$ is defined on it.

Syntax. Given a relation schema R and a relation instance r over R, a CFD^P φ on R is a pair $R(X \to Y, T_p)$, where: (1) $X, Y \subseteq R$; (2) $X \to Y$ is a standard FD, referred to as the FD embedded in φ; (3) T_p is a tableau with attributes in X and Y, referred to as the pattern tableau of φ, where, for each A in $X \cup Y$ and each tuple $t_{p_i} \in T_p$, $t_{p_i}[A]$ is either an unnamed variable '_' that draws values from $dom(A)$ or '$op\ a$', where op is one of $=, \neq, <, >, \leq, \geq$, and '$a$' is a constant in $dom(A)$.   □

Semantics. Considering the CFD^P $\varphi{:}R(X \to Y, T_p)$, where $T_p = t_{p_1}, \ldots, t_{p_k}$, a data tuple t of R is said to match $LHS(\varphi)$, denoted by $t[X] \asymp T_p[X]$, if for each tuple t_{p_i} in T_p and each attribute A in X, either (a) $t_{p_i}[A]$ is the wildcard '_' (which matches any value in dom(A)), or (b) $t[A]\ op\ a$ if $t_{p_i}[A]$ is '$op\ a$', where the operator op ($=, \neq, <, >, \leq$ or $\geq$) is interpreted by its standard semantics.

Each pattern tuple t_{p_i} specifies a condition via $t_{p_i}[X]$, and $t[X] \asymp T_p[X]$ if $t[X]$ satisfies the conjunction of all these conditions. Similarly, the notion that t matches $RHS(\varphi)$ is defined, denoted by $t[Y] \asymp T_p[Y]$. An instance I of R satisfies the CFD^P φ, if for each pair of tuples t_1, t_2 in the instance I, if $t_1[X]$ and $t_2[X]$ are equal and in addition they both match the pattern tableau $T_p[X]$, then $t_1[Y]$ and $t_2[Y]$ must also be equal to each other and must match the pattern tableau $T_p[Y]$.   □

2.1   *Constant* and *non-constant* CFD^{P}s

Extending the definition introduced for CFDs in [9], we distinguish between *constant* and *variable* – or *non-constant* – CFD^{P}s, calling:

- *constant*, the CFD^{P}s having in their pattern tableaux only operators and constant values (that is, without any unnamed variable '_');
- *non-constant*, the CFD^{P}s having, for the attributes in its right-hand side, an unnamed variable '_' in each pattern tuple of its pattern tableau.

Examples of *constant* (φ_1 and φ_2) and *non-constant* (φ_3, φ_4 and φ_5) CFD^{P}s for the *Iris* dataset[1] are shown in table 1: φ_1 indicates that when the length

---

[1] From the UCI Machine Learning Repository (http://archive.ics.uci.edu/ml).

Table 1. Examples of *constant* - and *non-constant* CFD^{P}s for the *Iris* dataset

φ_1: iris(petalLength $\rightarrow$ class, T_1)

T_1:

petalLength	class
< 2	*Iris setosa*

φ_2: iris(petalWidth, petalLength $\rightarrow$ class, T_2)

T_2:

petalWidth	petalLength	class
> 1.7	> 4.8	*Iris virginica*

φ_3: iris(sepalLength, petalWidth $\rightarrow$ class, T_3)

T_3:

sepalLength	petalWidth	class
< 5.9	–	–

φ_4: iris(sepalLength, petalLength $\rightarrow$ class, T_4)

T_4:

sepalLength	petalLength	class
$\neq$ 6.3	$\neq$ 4.9	–

φ_5: iris(petalLength, sepalWidth $\rightarrow$ class, T_5)

T_5:

petalLength	sepalWidth	class
$\neq$ 4.8	–	–
$\neq$ 5.1	–	–

of the petal is less than 2 cm then the class of the flower corresponds to *Iris setosa*; φ_2 expresses that when the width of the petal is greater than 1.7 cm and – at the same time – the length of the petal is greater than 4.8 cm then the class of the flower corresponds to *Iris virginica*; φ_3 expresses that the FD *sepalLength, petalWidth* $\rightarrow$ *class* holds on the subset of the relation tuples having the length of the sepal less than 5.9 cm; φ_4 expresses that the FD *sepalLength, petalLength* $\rightarrow$ *class* holds if the length of sepal is different from 6.3 cm and the length of the petal is different from 4.9 cm; finally, φ_5 expresses that the FD *petalLength, sepalWidth* $\rightarrow$ *class* holds if the length of the petal is different from 4.8 cm and from 5.1 cm.

3   Discovering CFD^{P}s

CFD^{P}s can be used to add information on data as exemplified in [3], in which case, the dependencies cannot be detected from the analysis of the dataset. However, the CFD^{P}s characterizing a dataset can be discovered analyzing the tuples contained in it.

We propose an algorithm for discovering from a dataset a subset of the existing CFD^{P}s satisfiyng the requirements to be *non-constant*, to have in their right-hand side only one attribute[2] and to have, in their pattern tableaux, conditions with operators only for numerical attributes.

---

[2] Without loss of generality because of the Armstrong decomposition rule: if $X \rightarrow YZ$, then $X \rightarrow Y$ and $X \rightarrow Z$.

More formally, the algorithm looks for CFD^{P}s that can be written as R($LHS \to RHS$, T_p), where:

- $LHS \to RHS$ is the FD embedded in the CFD^P;
- RHS contains a single attribute $A \in R$;
- $LHS \cap A = \emptyset$;
- $X, T \subset R$, $T \neq \emptyset$, $LHS = X \cup T$ and $X \cap T = \emptyset$;
- $\forall B \in T\ dom(B)$ is numeric;
- T_p is a pattern tableau with attributes in LHS and RHS;
- $t_p[A] = \text{`\_'}$;
- $\forall Z \in X$ and $\forall$ tuple $t_{p_i} \in T_p$, $t_{p_i}[Z]$ is an unnamed variable '_' that draws values from $dom(Z)$;
- $\forall B \in T$ and $\forall$ tuple $t_{p_i} \in T_p$, $t_{p_i}[B]$ is '$op\ b$', where 'b' is a constant in $dom(B)$ and op is one of the following operators: $<, >, \leq, \geq, \neq, =$.

In the following, we will refer to the attributes in X as *variable attributes*, to the attributes in T (for which conditions are searched) as *target attributes*, and to the conditions in the pattern tableau T_p as *target conditions*.

The algorithm is based on the selection of the tuples that do not satisfy a target dependency and on the use of the values of these tuples to build the conditions to obtain valid dependencies.

The algorithm accepts the following input parameters:

- *maxSizeLHS* – setting the maximum number of attributes that the dependencies have to contain in their LHS;
- *sizeT* – setting the size of the set T containing the *target attributes*;
- *maxNumConditions* – an optional parameter setting the maximum number of conditions that can be present in a dependency (i.e., the number of rows in the dependency pattern tableau);
- *depSupport* – an optional parameter indicating, in percentage respect to the dataset tuples, the support required for the resulting dependency (i.e., the minimum number of tuples satisfying the dependency).

The first step performed by the algorithm is the generation of candidates for the target dependencies, in the form $LHS \to A$ with the attributes in LHS divided in the *variable attributes* set X and in the *target attributes* set T.

To generate the candidates, we have adopted the small-to-large search approach, which has been successfully used in algorithms to discover traditional FDs and in many data mining applications, starting to compute dependencies with a number of attributes equal to the size of the set T in their left-hand side and then proceeding adding *variable attributes* in the set X.

In order to reduce the time spent by the algorithm producing the candidates, some pruning approaches have been introduced. A relevant reduction in the number of the generated candidates applies when a FD $Y \to A$, with $Y \subset R$ and $A \in R$, holds on the dataset. In this case, it is not necessary to build any candidates of the form $Z \to A$, with $Z \subset R$ and $Y \subseteq Z$.

The number of generated candidates is reduced also: (1) in the presence of attributes having the same value for all the tuples in the dataset – such attributes

Data: An instance relation r over the schema R
Input parameters: $maxSizeLHS$, $sizeT$, $maxNumConditions$, $depSupport$
Result: CFD^{P}s
$resultSet = \emptyset$;
$RHS = \{\{A\}|\forall A \in R\}$;
$numSupportTuples = \text{computeSupportTuples}(depSupport)$;
for $Y \in RHS$ **do**
   | $LHSattr_{init} = \{\{B\}|\forall B \in (R-Y)\}$;
   | $LHSattr_1 = \text{pruneSet}(LHSattr_{init}, Y)$;
   | $l = 1$;
   | **while** $l \leq (maxSizeLHS)$ **do**
          $candidateSet = \text{generateCandidates}(LHSattr_l, Y)$;
          for $candidate \in candidateSet$ **do**
              $patternTableauSet = \text{findTargetConditions}(candidate,$
              $numSupportTuples)$;
              if $patternTableauSet \neq \emptyset$ **then**
                 for $patternTableau \in patternTableauSet$ **do**
                    if $acceptResults(candidate, patternTableau,$
                    $maxNumConditions, numSupportTuples)$ **then**
                      | $resultSet \mathrel{+}= \text{buildCFDp}(candidate, patternTableau)$;
                    end
                 end
          end
          end
   | $LHSattr_{l+1_{init}} = \text{computeSetNextLevel}(LHSattr_l, l)$;
   | $LHSattr_{l+1} = \text{pruneSet}(X_{l+1_{init}}, Y)$;
   | $l = l + 1$;
    end
end

Pseudocode 1. Algorithm main steps

are not included in the candidate generation process; (2) in the presence of attributes having distinct values for each tuple in the dataset – such attributes are excluded from the RHS of the candidate when the support required for the dependency in greater than 1. Furthermore, the input parameters *maxSizeLHS* and *sizeT* contribute in reducing the number of generated candidates and thus the execution time of the algorithm.

After having determined a candidate, it is necessary to verify if it can be a CFD^{P} and determine which are the values for the attributes in the set T that have to be excluded to obtain a valid CFD^{P}. To perform this step, the algorithm proceeds in computing the tuple equivalence sets[3] for the set of attributes present in the candidate.

The algorithm selects the sets to be excluded and the sets to be accepted in order to obtain a valid dependency: the sets with the same values for the

---

[3] Two tuples t_1 and t_2 are equivalent respect to a set Y of attributes if $\forall\ B \in Y$ $t_1[B]=t_2[B]$.

```
procedure generateCandidates(LHSattr, Y)
    candidateSet = ∅;
    for S ∈ LHSattr do
        setT = buildSetT(S, sizeT);
        for T ∈ setT do
            setXattr = S − T;
            setX = buildSetX(setXattr);
            for X ∈ setX do
                candidateSet += buildCandidate(X, T, Y);
            end
        end
    end
    return candidateSet;

procedure pruneSet(S, Y)
    newSet = ∅;
    for Z ∈ S do
        if Z → Y holds on R then
            newSet = S − Z;
        end
    end
    return newSet;
```

Pseudocode 2. Algorithm procedures

attributes in LHS but different values of the attribute A are excluded. The values of the *target attributes* (the attributes in the set T) of the tuples contained in the excluded sets will be used to build the conditions for the dependency pattern tableau.

At this step, to reduce useless computation, the input parameter *depSupport* – when present – is used to filter out the candidates having in their selected sets a number of tuples less greater than the required support.

Then, as a preliminary step in the determination of the intervals, for every *target attribute*, the minimum distance among the values on the attribute domain is computed. It will be used to determine if the values are contiguous or not and thus to decide for each value if it has to be part of an interval condition or if it will generate an inequality condition.

Afterwards, the algorithm builds a set with the values of the *target attributes* for all the tuples contained in the excluded sets; this last set is used by the algorithm to compute the interval (or intervals) for which the candidate is a valid dependency. Instead of an interval, an equality condition is generated when an open interval contains only one value between the extreme values; e.g., when the interval is $(x − 1, x + 1)$ then the conditions "$> x − 1$" and "$< x + 1$" are replaced by the condition "$= x$".

Because of the semantics of the CFD^{p}s stating that a tuple has to satisfy the conjunction of all the conditions in a pattern tableau, if more than one interval

procedure *findTargetConditions(candidate, numSupportTuples)*
    patternTableauSet = ∅;
    equivalentSetsList = computeEquivalentSets(*candidate*);
    acceptedSetsList = selectSetsToBeKept(*equivalentSetsList*);
    if *countTuples(acceptedSetsList)* ≥ *numSupportTuples* **then**
       excludedSetsList = selectSetsToBeExcluded(*equivalentSetsList*,
       acceptedSetsList);
       patternTableauSet = computeConditions(*acceptedSetsList*,
       excludedSetsList);
    end
    return *patternTableauSet*;

procedure *acceptResults(candidate, patternTableau, maxNumConditions,*
numSupportTuples)
    if *size(patternTableau)* <= *maxNumConditions* **then**
       numTuples = countTuples(*candidate, patternTableau*);
       if *numTuples* >= *numSupportTuples* **then**
          return *true*;
       end
    end
    return *false*;

Pseudocode 3. Algorithm procedures

is identified for a candidate, it is necessary to build different pattern tableaux for that candidate.

If the input parameters *maxNumConditions* and *depSupport* have been set, the last step consists in the acceptance or rejection of the dependency according to the values of these parameters: a dependency is accepted if the number of conditions in its pattern tableau is less than or equal to the *maxNumConditions* parameter and if it is satisfied by a number of tuples greater than or equal to the support required by the *depSupport* parameter.

4   Testing the Algorithm

The algorithm has been implemented using the Java programming language and the PostgreSQL DBMS. The first test of the algorithm has been performed using some of the datasets provided by the UCI Machine Learning Repository [2], such as the Iris, Seeds, Escherichia Coli, BUPA Liver disorder[4], Yeast[5] and Wisconsin breast cancer[6] datasets.

To show some examples of the results produced by the algorithm, we use the following datasets:

---

[4] In the BUPA Liver disorder dataset duplicate rows have been excluded.

[5] In the Yeast dataset duplicate rows have been excluded.

[6] In the Wisconsin breast cancer dataset the attribute called Sample Code Number and the rows containing empty attributes have been excluded.

Table 2. Results of the execution of the algorithm on the *BUPA Liver* dataset

φ_1: BUPA-liver(alkphos, sgpt, drinks $\rightarrow$ selector, T_1)

	alkphos	sgpt	drinks	selector
T_1:	≥ 23	–	–	–
	$\neq 85$	–	–	–
	≤ 138	–	–	–

φ_2: BUPA-liver(gammagt, mcv, alkphos $\rightarrow$ sgot, T_2)

	gammagt	mcv	alkphos	sgot
T_2:	> 5	–	–	–
	< 297	–	–	–

φ_3: BUPA-liver(gammagt, mcv, alkphos $\rightarrow$ sgpt, T_3)

	gammagt	mcv	alkphos	sgpt
T_3:	> 5	–	–	–
	< 297	–	–	–

φ_4: BUPA-liver(sgpt, mcv, gammagt $\rightarrow$ drinks, T_4)

	sgpt	mcv	gammagt	drinks
T_4:	≥ 4	–	–	–
	$\neq 9$	–	–	–
	≤ 155	–	–	–

φ_5: BUPA-liver(sgpt, mcv, gammagt $\rightarrow$ alkphos, T_5)

	sgpt	mcv	gammagt	alkphos
T_5:	≥ 4	–	–	–
	$\neq 9$	–	–	–
	≤ 155	–	–	–

– The *Iris* dataset, which has 5 attributes respectively called Petal Length, Petal Width, Sepal Length, Sepal Width, and Class.
– The *BUPA Liver* dataset, which contains the following 7 attributes (all of them with values in the domain of the integer numbers): Mean Corpuscular Volume (*mcv*), Alkaline Phosphotase (*alkphos*), Alamine Aminotransferase (*sgpt*), Aspartate Aminotransferase (*sgot*), Gamma-Glutamyl Transpeptidase (*gammagt*), number of half-pint equivalents of alcoholic beverages drunk per day (*ndrinks*), and a field used to split data into two sets (*selector*).
– The *Wisconsin breast cancer* dataset, which contains the following 10 attributes (all of them with values in the domain of the integer numbers): Clump Thickness, Uniformity of Cell Size, Uniformity of Cell Shape, Marginal Adhesion, Single Epithelial Cell Size, Bare Nuclei, Bland Chromatin, Normal Nucleoli, Mitoses, and Class; the first 9 attributes have values in the range 1-10, while the Class attribute can have two values: 2 for "benign", 4 for "malignant".

Table 2 shows the CFD^{P}s resulting from the execution of the algorithm on the *BUPA Liver* dataset with the following values for the input parameters:

Table 3. Results of the execution of the algorithm on the *Wisconsin breast cancer* dataset

φ_1: wbc(uniformityCellShape, singleEpithelialCellSize, bareNuclei, normalNucleoli $\rightarrow$ class, T_1)

	uniformityCellShape	singleEpithelialCellSize	bareNuclei	normalNucleoli	class
T_1:	≥ 1.0	–	–	–	–
	< 7.0	–	–	–	–

φ_2: wbc(bareNuclei, clumpThickness, uniformityCellSize, uniformityCellShape $\rightarrow$ class, T_2)

	bareNuclei	clumpThickness	uniformityCellSize	uniformityCellShape	class
T_2:	≥ 1.0	–	–	–	–
	< 10.0	–	–	–	–

φ_3: wbc(bareNuclei, uniformityCellShape, marginalAdhesion, singleEpithelialCellSize $\rightarrow$ class, T_3)

	bareNuclei	uniformityCellShape	marginalAdhesion	singleEpithelialCellSize	class
T_3:	≥ 1.0	–	–	–	–
	< 10.0	–	–	–	–

maxSizeLHS equal to 3 , *sizeT* equal to 1, *maxNumConditions* equal to 3 and *depSupport* equal to 0.98. Table 3 shows the CFD$^\text{P}$s resulting from the execution of the algorithm on the *Wisconsin breast cancer* dataset with the following input parameters: *maxSizeLHS* equal to 4, *sizeT* equal to 1, *maxNumConditions* equal to 2 and *depSupport* equal to 0.8. Table 4 shows the CFD$^\text{P}$s resulting from the execution of the algorithm on the *Iris* dataset with the following input parameters: *maxSizeLHS* equal to 3, *sizeT* equal to 1, *maxNumConditions* equal to 3 and *depSupport* equal to 0.6. Finally, table 4 shows the CFD$^\text{P}$s resulting from the execution of the algorithm on the *Iris* dataset with the following input parameters: *maxSizeLHS* equal to 2, *sizeT* equal to 2, *maxNumConditions* equal to 5 and *depSupport* equal to 0.98.

Depending on the values assigned to the input parameters (in particular to the dependency support parameter), on the number of attributes and tuples in the relation, and, of course, on the type of data, the number of generated CFD$^\text{P}$s can vary greatly.

Table 6 reports the number of CFD$^\text{P}$s identified by the algorithm on different datasets provided by the UCI Machine Learning Repository with different values for the support input parameter *depSupport*. The results shown in the table have been computed with the following input parameters: *maxSizeLHS* equal to 4, *sizeT* equal to 1 and *maxNumConditions* equal to 4; while the values used for the dependency support parameter – called k – are specified in the table.

Furthermore, table 7 reports the number of CFD$^\text{P}$s identified by the algorithm on the same datasets using different values for the input parameter *maxSizeLHS* – the maximum number of attributes in the *LHS* of the dependency. In this case, the results have been computed with the dependency support equal to 0.5 and

Table 4. Results of the execution of the algorithm on the *Iris* dataset

φ_1: iris(petalLength, sepalLength $\rightarrow$ class, T_1)

	petalLength	sepalLength	class
T_1:	≥ 1.0	–	–
	$\neq 4.9$	–	–
	≤ 6.9	–	–

φ_2: iris(sepalLength, petalLength $\rightarrow$ class, T_2)

	sepalLength	petalLength	class
T_2:	≥ 4.3	–	–
	$\neq 6.3$	–	–
	≤ 7.9	–	–

φ_1: iris(sepalWidth, petalLength $\rightarrow$ class, T_1)

	sepalWidth	petalLength	class
T_3:	> 2.8	–	–
	≤ 4.4	–	–

φ_1: iris(petalWidth, sepalLength $\rightarrow$ class, T_1)

	petalWidth	petalLength	class
T_4:	≥ 0.1	–	–
	$\neq 1.8$	–	–
	≤ 2.5	–	–

φ_1: iris(petalLength, petalWidth $\rightarrow$ class, T_1)

	petalLength	petalWidth	class
T_5:	≥ 1.0	–	–
	$\neq 4.8$	–	–
	≤ 6.9	–	–

sizeT equal to 1 but without any limit on the maximum number of conditions allowed in the resulting pattern tableaux.

The results show that the number of the CFD^{P}s identified by the algorithm increases when the maximum size of *LHS* increases and – as expected – decreases at the increasing of the dependency support required through the input parameter. The high numbers of dependencies found when the input parameter for the dependency support is not specified is mainly determined by the presence of CFD^{P}s satisfied by a single tuple.

The approach to generate, during the same step, different tableaux for a candidate – producing disjoint intervals – determines that the dependencies generated for the same candidate are not redundant. However, redundant CFD^{P}s can be generated when there exist:

- two CFD^{P}s φ_a:$R(Z_1 \rightarrow A, T_{p_1})$ and φ_b:$R(Z_2 \rightarrow A, T_{p_2})$, with $Z_1 \subset Z_2$, $Z_1 = X_1 \cup T_1$, $Z_2 = X_2 \cup T_2$, $T_1 = T_2$ and $X_1 \subset X_2$: if the conditions in T_2 are subsumed by the conditions in T_1 then φ_b is redundant.

Table 5. Results of the execution of the algorithm on the *Iris* dataset

φ_1: iris(sepalLength, petalLength $\rightarrow$ class, T_1)

	sepalLength	petalLength	class
T_1:	≥ 4.3	≥ 1.0	–
	$\neq 6.3$	$\neq 4.9$	–
	≤ 7.9	≤ 6.9	–

φ_2: iris(petalLength, petalWidth $\rightarrow$ class, T_2)

	petalLength	petalWidth	class
T_2:	≥ 1.0	≥ 0.1	–
	$\neq 4.8$	$\neq 1.8$	–
	≤ 6.9	≤ 2.5	–

φ_3: iris(sepalWidth, petalLength $\rightarrow$ class, T_3)

	sepalWidth	petalLength	class
	≥ 2.0	≥ 1.0	–
T_3:	$\neq 2.7$	$\neq 5.1$	–
	$\neq 2.8$	$\neq 4.8$	–
	≤ 4.4	≤ 6.9	–

Table 6. Results from the execution of the algorithm with different values of the input parameter *depSupport* (k)

Dataset name	$|R|$	$|r|$	number of CFD^{P}s			
			k not defined	$k \geq 0.1$	$k \geq 0.5$	$k \geq 0.8$
Iris	5	150	274	72	19	8
BUPA Liver	7	341	1413	596	228	126
Seeds	8	210	78	48	38	27
E. Coli	9	336	3307	699	174	139
Wisconsin breast cancer	10	683	6578	1160	72	40
Yeast	10	1462	11236	1540	253	194

Table 7. Results from the execution of the algorithm with different values of the input parameter *maxSizeLHS* (max$|LHS|$)

Dataset name	$|R|$	$|r|$	number of CFD^{P}s			
			max$|LHS|$=2	max$|LHS|$=3	max$|LHS|$=4	max$|LHS|$=5
Iris	5	150	12	22	32	–
BUPA Liver	7	341	7	135	286	328
Seeds	8	210	76	91	91	91
E. Coli	9	336	66	210	413	558
Wisconsin breast cancer	10	683	0	6	74	198
Yeast	10	1462	17	143	382	659

Table 8. Results of the execution of the algorithm on the *Iris* dataset

φ_1: iris(petalWidth $\rightarrow$ class, T_1)

	petalWidth	class
T_1:	≥ 0.1	–
	< 1.4	–

φ_2: iris(petalWidth, sepalLength $\rightarrow$ class, T_2)

	petalWidth	sepalLength	class
T_2:	≥ 0.1	–	–
	< 1.4	–	–

φ_3: iris(sepalLength, petalWidth $\rightarrow$ class, T_3)

	sepalLength	petalWidth	class
T_3:	≥ 4.3	–	–
	< 5.9	–	–

φ_4: iris(sepalWidth, petalLength $\rightarrow$ class, T_4)

	sepalWidth	petalLength	class
T_4:	> 2.8	–	–
	≤ 4.4	–	–

- two CFD^{P}s φ_a:$R(Z_1 \rightarrow A, T_{p_1})$ and φ_b:$R(Z_2 \rightarrow A, T_{p_2})$, with $Z_1 \subseteq Z_2$, $Z_1 = X_1 \cup T_1$, $Z_2 = X_2 \cup T_2$, $T_1 \subset T_2$: if the conditions in T_2 are subsumed by the conditions in T_1 then φ_b is redundant.

However, the support of the CFD^{P}s can be different, and it can be higher for the dependency φ_b.

An example of the first case is shown in table 8 with the results from the execution of the algorithm on the *Iris* dataset (the following input parameters have been used: *maxSizeLHS* equal to 2, *sizeT* equal to 1, *maxNumConditions* equal to 2 and *depSupport* equal to 0.5), in particular the CFD^{P}s φ_1 and φ_2; whereas an example of the second case can be observed comparing table 4 and table 5.

5   Related Work

For the discovery of *non-constant* CFD^{P}s, to date and to our knowledge, there are no published algorithms.

Similarities between CFD^{P}s and approximate functional dependencies[7] [12] can be highlighted: in both cases a dependency holds excluding a subset of the set of tuples. However, the process to find a CFD^{P} requires the identification of the *target conditions* contained in the pattern tableau, while in the case of

---

[7] An approximate FD is a FD that does not hold over a small fraction of the tuples; specifically, $X \rightarrow Y$ is an approximate FD if and only if the $error(X \rightarrow Y)$ is at most equal to an error threshold ϵ ($0 < \epsilon < 1$), where the error is measured as the fraction of tuples that violate the dependency.

approximate dependencies it is sufficient to determine the number of tuples non-satisfying the dependency.

Several algorithms for the discovery of FDs have been proposed since 1990s and more recently for CFDs.

Examples of algorithms developed to discover traditional FDs are: TANE [11], Dep-miner [14], Fast-FD [17], FD_Mine [18].

For the discovery of general CFDs the following algorithms have been proposed: an algorithm based on the attribute lattice search strategy is presented in [4]; Fast-CFD [9] is inspired by the Fast-FD algorithm; CTANE [9] extends the TANE algorithm; CFD-Mine [1] is also based on an extension of the TANE algorithm. Moreover, some algorithms for the discovery of only constant CFDs have been proposed: CFDMiner [9] is based on techniques for mining closed item sets and finds a canonical cover of k-frequent minimal constant CFDs; an algorithm that extends the notion of non-redundant sets, closure and quasi-closure is described in [6]; in [13] new criteria to further prune the search space used by CFDMiner to discover the minimal set of CFDs are proposed.

6   Conclusions and Future Work

In this work we have introduced an algorithm to discover *non-constant* CFD^{P}s from datasets. Aim of the developed algorithm is the identification of a subset of the existing *non-constant* CFD^{P}s characterized by the requirements mentioned in section 3, without looking specifically for CFDs, for which dedicated algorithms already exist. The algorithm implements the approach of selecting the tuples that do not satisfy a dependency and using the values of the attributes of the identified tuples to build the *target conditions* to obtain valid dependencies.

The results of the first algorithm test, which has been executed on datasets from the UCI Machine Learning Repository, show that the number of CFD^{P}s generated by the algorithm can vary greatly depending on the values assigned to the input parameters, on the number of attributes and tuples in the relation, and – of course – on the type of data. When too many CFD^{P}s are retrieved from a dataset, the input parameters – in particular the *depSupport* and *maxSizeLHS* parameters – help in decreasing the number of identified dependencies. A high value of the *depSupport* parameter determines also the identification of the most interesting dependencies to be practically used in the data quality context.

As in the case of the algorithms for discovering FDs [15], the worst case time complexity of the developed algorithm, with respect to the number of attributes and tuples in the relation, is exponential. The criteria used by the algorithm to prune the number of candidates and the input parameters help in improving the algorithm efficiency as in reducing the number of identified dependencies.

As future work we plan to test the algorithm on other datasets and to experiment with other candidate pruning approaches to improve the algorithm efficiency. We are also studying the feasibility of an extension to the algorithm in order to include non-numeric attributes in the *target attribute* set T, considering the alphanumeric ordering or a semantic ordering defined on the domains of the relation attributes.

References

1. Aqel, M., Shilbayeh, N., Hakawati, M.: CFD-Mine: An efficient algorithm for discovering functional and conditional functional dependencies. Trends in Applied Sciences Research 7(4), 285–302 (2012)
2. Bache, K., Lichman, M.: UCI Machine Learning Repository (2013), http://archive.ics.uci.edu/ml
3. Chen, W., Fan, W., Ma, S.: Analyses and validation of conditional dependencies with built-in predicates. In: Bhowmick, S.S., Küng, J., Wagner, R. (eds.) DEXA 2009. LNCS, vol. 5690, pp. 576–591. Springer, Heidelberg (2009)
4. Chiang, F., Miller, R.: Discovering data quality rules. Proceedings of the VLDB Endowment 1(1), 1166–1177 (2008)
5. Cong, G., Fan, W., Geerts, F., Jia, X., Ma, S.: Improving data quality: Consistency and accuracy. In: Koch, C., et al. (eds.) International Conference on Very Large Data Bases (VLDB 2007), pp. 315–326. ACM (2007)
6. Diallo, T., Novelli, N., Petit, J.M.: Discovering (frequent) constant conditional functional dependencies. Int. Journal of Data Mining, Modelling and Management 4(5), 205–223 (2012)
7. Fan, W., Geerts, F., Jia, X.: Semandaq: A data quality system based on conditional functional dependencies. Proceedings of the VLDB Endowment 1(2), 1460–1463 (2008)
8. Fan, W., Geerts, F., Jia, X., Kementsietsidis, A.: Conditional functional dependencies for capturing data inconsistencies. ACM Transactions on Database Systems (TODS) 33(2), 94–115 (2008)
9. Fan, W., Geerts, F., Li, J., Xiong, M.: Discovering conditional functional dependencies. IEEE Transactions on Knowledge and Data Engineering (TKDE) 23(5), 683–697 (2011)
10. Golab, L., Karloff, H., Korn, F., Srivastava, D.: Data Auditor: Exploring data quality and semantics using pattern tableaux. Proceedings of the VLDB Endowment 3(2), 1641–1644 (2010)
11. Huhtala, Y., Karkkainen, J., Porkka, P., Toivonen, H.: TANE: An efficient algorithm for discovering functional and approximate dependencies. Computer Journal 42(2), 100–111 (1999)
12. Kivinen, J., Mannila, H.: Approximate inference of functional dependencies from relations. Theoretical Computer Science 149(1), 129–149 (1995)
13. Li, J., Liu, J., Toivonen, H., Yong, J.: Effective pruning for the discovery of conditional functional dependencies. The Computer Journal 56(3), 378–392 (2013)
14. Lopes, S., Petit, J.-M., Lakhal, L.: Efficient discovery of functional dependencies and Armstrong relations. In: Zaniolo, C., Lockemann, P.C., Scholl, M.H., Grust, T. (eds.) EDBT 2000. LNCS, vol. 1777, pp. 350–364. Springer, Heidelberg (2000)
15. Mannila, H., Raiha, K.J.: On the complexity of inferring functional dependencies. Discrete Applied Mathematics 40, 237–243 (1992)
16. Pivert, O., Prade, H.: Handling dirty databases: From user warning to data cleaning — Towards an interactive approach. In: Deshpande, A., Hunter, A. (eds.) SUM 2010. LNCS, vol. 6379, pp. 292–305. Springer, Heidelberg (2010)

17. Wyss, C., Giannella, C., Robertson, E.: FastFDs: A heuristic-driven, depth-first algorithm for mining functional dependencies from relation instances - extended abstract. In: Kambayashi, Y., Winiwarter, W., Arikawa, M. (eds.) DaWaK 2001. LNCS, vol. 2114, pp. 101–110. Springer, Heidelberg (2001)
18. Yao, H., Hamilton, H.: Mining functional dependencies from data. Journal Data Mining and Knowledge Discovery 16(2), 197–219 (2008)
19. Zanzi, A., Trombetta, A.: Data quality evaluation of scientific datasets: A case study in a policy support context. In: International Conference on Data Management Technologies and Applications (DATA 2013), pp. 167–174. SciTePress (2013)

Summary-Based Pattern Tableaux Generation for Conditional Functional Dependencies in Distributed Data

Soror Sahri, Mourad Ouziri, and Salima Benbernou

Université Paris Descartes, Sorbonnes Paris Cité, France
{firstname.lastname}@parisdescartes.fr

Abstract. Conditional Functional Dependencies (CFD) are an extension of Functional Dependencies (FDs) that capture rules about the data consistency. Existing work on discovering CFDs focused on centralized data. Here, we extend this work to horizontally distributed relations. Given an embedded functional dependency, we generate a pattern tableau that represents a CFD. The original feature of our work is generating CFD pattern tableaux from a distributed relation, without merging all the distributed tuples in a centralized relation. We propose a distributed algorithm based on the concept of pattern summary that minimizes data shipping between the sites of distributed relation.

1 Introduction

Conditional Functional Dependencies (CFDs) have recently been proposed for cleaning relational data [2]. Whereas a Functional Dependency (FD) establishes a relationship that holds for all possible values of the involved attributes, a CFD is true only when some conditional constraints hold on a subset of the relation. We illustrate the concept with an example borrowed from [10] and presented in Table1. *Sales* is a relation that stores purchase records of an international retailer and is specified by the schema:
$Sales(tid, name, type, country, price, tax)$
R is a relation instance of *Sales* and the semantics of data is expressed by the following FD:
$fd_1 : [name, type, country] \rightarrow [price, tax]$, where fd_1 requires that the product record and its type with the same country give the price and tax.

It can be augmented as a CFD on the instance of *Sales* with additional constraints, denoted by $cfd_1 : ([name, type, country ='UK'] \rightarrow [price, tax])$ and $([name, type ='book', country ='France'] \rightarrow [price, tax = 0])$. cdf_1 asserts that in the *UK*, product name and type determine price and tax, and asserts this only for books purchased in *France*, but giving the additional information that the tax is then zero. We cannot generalize these CFDs to FDs because in France, product name and type do not determine their price and tax, unless it is a book. Similarly, a book bought in a country different than *France* might be assessed a sales tax.

A CFD is a FD that holds on a fragment of the original instance relation, where the fragment relation is given by constraints applied to attributes. These constraints represent a selection operation on the relation. Thus, a CFD is given by a FD $X \rightarrow Y$ and a

H. Decker et al. (Eds.): DEXA 2014, Part I, LNCS 8644, pp. 50–65, 2014.

Table 1. R: an instance of *Sales* relation

tid	Name	Type	Country	Price	Tax
t1	Harry Potter	book	France	10	0
t2	Harry Potter	book	France	10	0
t3	Harry Potter	book	France	12	0.02
t4	Harry Potter	book	UK	12	0
t5	Harry Potter	book	UK	12	0
t6	The Lord of the Rings	book	France	25	0
t7	The Lord of the Rings	book	France	25	0
t8	Armani slacks	clothing	UK	250	0
t9	Armani slacks	clothing	UK	250	0
t10	Prada shoes	clothing	France	500	0.05
t11	Harry Potter	DVD	France	8	0
t12	Harry Potter	DVD	UK	14	0
t13	Spiderman	DVD	UK	19	0
t14	Star Wars	DVD	UK	29	0.03
t15	Star Wars	DVD	UK	29	0.03
t16	Terminator	DVD	France	25	0.08
t17	Terminator	DVD	France	25	0
t18	Terminator	DVD	France	20	0
t19	Harry Potter	VHS	UK	5	0

pattern tableau T_p that determines which tuples must obey the FD. Conversely, a FD is a CFD with a pattern tableau consisting of a single row with a single wildcard (represented by '_') for each attribute. The pattern tableau of cfd_1 are $(-, -, 'UK' \parallel -, -)$ and $(-, 'book'], 'France' \parallel -, 0)$ and the FD is fd_1. A CFD attempts to discover additional semantic meaning and is used to flag probable inconsistent data. Thus, we do not want to derive CFDs that hold without exceptions, but those that have few exceptions. Hopefully, those exceptions correspond to faulty data.

Recent work on CFDs focused mainly on reasoning about CFDs and repairing tuples in a given relation instance that violate a given set of CFDs [2,3,5]. Most of these studies assume that CFDs already exist. Discovery of CFDs in a relation instance received considerable interest [4,7,10]. Proposed approaches on CFDs discovery subsume discovery of embedded FDs and pattern tableaux. In [10], it is exposed how to discover CFDs for a given FD by generating a useful pattern tableau. However, these all approaches are working on centralized relations, and what has not been addressed is how to create useful pattern tableaux from *distributed relations*.

In fact, nowadays due to the recent interests to cloud computing and big data, it is increasingly common to find data partitioned horizontally or vertically and distributed across different sites. This is why it is a challenging issue to study CFDs on distributed data.

For example, Table 2 gives horizontal partitions R_1 and R_2 of *Sales* relation on two different sites. While we can use techniques for single site relations by shipping all of data, we are interested in minimizing data shipments between sites.

Table 2. Horizontal partitions of R on $Country$ attribute

tid	Name	Type	Country	Price	Tax
t4	Harry Potter	book	UK	12	0
t5	Harry Potter	book	UK	12	0
t8	Armani slacks	clothing	UK	250	0
t9	Armani slacks	clothing	UK	250	0
t12	Harry Potter	DVD	UK	14	0
t13	Spiderman	DVD	UK	19	0
t14	Star Wars	DVD	UK	29	0.03
t15	Star Wars	DVD	UK	29	0.03
t19	Harry Potter	VHS	UK	5	0

(R_1)

tid	Name	Type	Country	Price	Tax
t1	Harry Potter	book	France	10	0
t2	Harry Potter	book	France	10	0
t3	Harry Potter	book	France	12	0.02
t6	The Lord of the Rings	book	France	25	0
t7	The Lord of the Rings	book	France	25	0
t10	Prada shoes	clothing	France	500	0.05
t11	Harry Potter	DVD	France	8	0
t16	Terminator	DVD	France	25	0.08
t17	Terminator	DVD	France	25	0
t18	Terminator	DVD	France	20	0

(R_2)

The contributions of the paper are summarized as follows:

- We propose a distributed algorithm to generate pattern tableau for horizontally distributed data. Given an embedded FD. we generate a pattern tableau that represents a CFD based on support and confidence thresholds.
- In order to minimize data shipment between the sites of the distributed relation, we introduce the concept of *pattern summary* to generate the pattern tableaux.
- A set of experiments is discussed enforcing the proposed algorithm for generating pattern tableaux in a distributed manner.

The rest of this paper is organized as follows. Section 2 reviews CFDs and measures for CFDs. In section 3, we formalize the problem of pattern tableau generation for distributed data. Section 4 presents the tableau generation algorithm for horizontally distributed data in its basic case. Section 5 presents the Summary algorithm based on the concept of pattern summary for minimizing data shipment. In Section 6, we discuss the soundness of the proposed algorithms. Section 7 summarizes experimental results. Section 8 reviews related work, and Section 9 concludes the paper.

2   Definitions

In this section, we recall useful definitions for the rest of the paper related to conditional functional dependencies (CFDs) and CFDs measures according to the context of distributed data.

2.1   Conditional Functional Dependencies

A conditional functional dependency (CFD) Φ on R is a pair $(X \rightarrow Y, T_p)$, where $X \rightarrow Y$ is a standard FD, referred to as the embedded FD; and T_p is a pattern tableau that defines over which rows of the table the embedded FD applies. Each entry $t_p \in T_p$ specifies a pattern over $X \cup Y$, so for each attribute in A $\in X \cup Y$, either $t_p[A] = a$, where a is a constant in the domain of A, or else $t_p[A] = | - |$, for the special wildcard symbol $| - |$. We separate the X and Y attributes in a pattern tuple with $| \, \| \, |$.

A tuple t matches a pattern t_p of tableau T_p for attributes A, denoted by $t[A] \asymp t_p[A]$, if either $t[A] = t_p[A]$, or else $t_p[A] = |-|$. The pattern consisting of only '-'matches the entire relation. The CFD Φ holds on R if:

$$\forall i, j, p, t_i[X] = t_j[X] \asymp t_p[X] \Rightarrow t_i[Y] = t_j[Y] \asymp t_p[Y]$$

Example 1. Let's consider the CFD cfd_1 given in the introduction. It can be formally expressed as follows:
$cfd_1 : (fd_1 = [name, type, country] \rightarrow [price, tax],$
$t_{p_1} = (-, -, |UK| \| -, -), t_{p_2} = (-, |book|, |France| \| -, 0))$
    The FD $fd_1 = [name, type, country] \rightarrow [price, tax]$ is not valid in R but the CFD $cfd_1 = (fd_1, t_{p_1}, t_{p_2})$ holds on R. That is, the FD fd_1 is valid on tuples matching the pattern t_{p_1} (namely, $t_4, t_5, t_8, t_9, t_{12}, t_{13}, t_{14}, t_{15}, t_{19}$) and those matching the pattern t_{p_2} (namely, t_1, t_2, t_6, t_7).

2.2   Measures for CFDs

To avoid returning unnecessarily large number of CFDs and find out interesting CFDs, several quality measures for CFDs are used. We consider particularly support and confidence measures. We use the same definitions in [10] to define support and confidence and adapt these definitions in the context of distributed data.

Support: Given a CFD $\phi = (R : X \rightarrow Y, T_p)$ and a distributed relation $R = \bigcup_{i \in [1,n]} R_i$, let $cover(p)$ be the set of tuples in R matching pattern p from T_p:

$$cover(p) = \{t \in Dom(R) \text{ and } t[X] \asymp p[X]\}$$

In the context of distributed data, we define pattern support and tableau support in a single relation partition and a distributed relation:
• **In a single partition:**
A *single-partition pattern support* of a pattern p is the fraction of tuples, in one partition R_i of the distributed relation R, that match the antecedent of p (covered by). It is defined as below:

$$Support_i(p) = \frac{|cover_i(p)|}{N_i}$$

where $cover_i(p)$ is the set of tuples in R_i matching the pattern p and N_i is the number of tuples in R_i.
    A *single-partition tableau support* of a pattern tableau T_p is the fraction of tuples, in one partition R_i of the distributed relation R, that match the antecedent of at least one pattern of T_p. It is defined as below:

$$Support_i(T_p) = \frac{1}{N_i} | \bigcup_{p \in T_p} cover_i(p) |$$

• **In a distributed relation:**
A *multi-partition* pattern support of a pattern p is the fraction of tuples, of the whole distributed relation R, that match the antecedent of p. It is defined as below:

$$disSupport(p) \;=\; \frac{\sum_i |cover_i(p)|}{\sum_i N_i}$$

A *multi-partition tableau support* of T_p is the fraction of tuples, of the whole distributed relation R, that match at least one pattern T_p. It is defined as below:

$$disSupport(T_p) \;=\; \frac{1}{\sum_i N_i} \sum_i \Big|\bigcup_{p \in T_p} cover_i(p)\Big|$$

Example 2. Consider the relation R horizontally partitioned into R_1 and R_2 from Table 2 and the CFD cfd_1 from Example1. Let us compute the support according to the definitions above:

The *single-partition pattern supports* of t_{p_1} in R_1 and R_2 are respectively 1 and 0; and the ones of t_{p_2} in R_1 and R_2 are respectively in 0 and $\frac{1}{2}$.

The *single-partition tableau support* of T_p in R_1 is 1 and $\frac{1}{2}$ in R_2 .

The *multi-partition pattern supports* of t_{p1} and t_{p2} are respectively:

$disSupport(t_{p1}) \;=\; \frac{|cover_1(t_{p1})|+|cover_2(t_{p1})|}{9+10} = \frac{9+0}{19} = \frac{9}{19}$

$disSupport(t_{p2}) \;=\; \frac{|cover_1(t_{p2})|+|cover_2(t_{p2})|}{9+10} = \frac{0+5}{19} = \frac{5}{19}$

Confidence: Given a CFD $\phi = (R : X \rightarrow Y, T_p)$ and a distributed relation $R = \bigcup_{i \in [1,n]} R_i$. Let $keepers(p)$ be the maximum set of tuples satisfying the functional dependency $X \rightarrow Y$ amongst those of $cover(p)$ [10]:

$$keepers(p) = \bigcup_{x \in dom(X)} \{t : t[XY] \asymp p[XY] = xy_x\}$$

where $\forall x \in dom(X), y_x = argmax_y |\{t : xy = t[XY] \asymp p[XY]\}$

The confidence of a pattern p is computed from its keepers. It corresponds to the fraction of tuples satisfying the functional dependency $X \rightarrow Y$ amongst those of $cover(p)$:

$$Confidence(p) \;=\; \frac{|keepers(p)|}{|cover(p)|}$$

In the context of our work, we distinguish between *single-partition confidence* (or local confidence) and *multi-partition confidence* (or global confidence). The focus of this paper is indeed to propose an efficient algorithm for computing this multi-partition confidence from all single-partition confidences. The process will be developed in the next section.

3   Problem Statement

The problem of tableau generation in distributed setting introduces several challenges that were not encountered when generating pattern tableaux in a centralized database.

Unlike centralized databases where all the tuples of a relation are made available for processing, tuples in a distributed relation could not be transferred into a single relation for various considerations such as storage, privacy, etc. Consequently, the tableau pattern of a distributed relation has to be generated from its distributed fragments.

We formalize the tableau generation problem for distributed data as follow:

Definition 1. *Given a distributed relation R horizontally partitioned into fragments $(R_1, ..., R_n)$ that are distributed across distinct sites $(S_1, ..., S_n)$, and a FD $R : X \rightarrow Y$, we want to find a pattern tableau T_p such that the CFD $(R : X \rightarrow Y, T_p)$ satisfies a confidence $\hat{c}$ and a support $\hat{s}$ thresholds.*

From a functional dependency $X \rightarrow Y$ and a distributed relation $R = (R_1, ..., R_n)$, the proposed algorithm computes a pattern tableau T_p according to a confidence $\hat{c}$ and a support $\hat{s}$ thresholds. The algorithm should satisfy the following properties:

Property 1. Patterns holding on a distributed relation hold necessarily on at least one partition of this relation.
For instance, for the given FD $fd_3 : [name, country] \rightarrow [price]$ and thresholds $\hat{s}=0.5$ and $\hat{c}=0.8$, the pattern $(-, |France| \parallel -)$ holds on R (cf. Table 1) as well on R_2 (cf. Table 2).

Property 2. Patterns of a relation have necessarily their children as patterns of the same relation (This property is trivial).

In the next section, we detail the steps of our proposed algorithm and define the parent and children patterns.

4   Distributed Pattern Tableaux Generation Algorithm

The algorithm generates a pattern tableau from a distributed relation. The steps of the algorithm are detailed as follows:

Step 1 - Single-partition pattern candidates generation: The algorithm generates, in a distributed way at each site S_i, the pattern tableau T_{p_i} of each partition R_i using the on-demand algorithm [10]. According to *Property 1*, the patterns of each tableau T_{p_i} will constitute pattern candidates of the whole distributed relation R.

Step 2 - parent-child based graph structuring: The algorithm organizes the generated pattern candidates, from Step 1, into a parent-child directed graph $G_p = \langle N, A \rangle$, where $N = p_i \in \bigcup_k T_{p_k}$: is a set of nodes that represents pattern candidates of the distributed relation R and $A = \{(p_i, p_j) | p_i, p_i \in \bigcup_k T_{p_k}\}$: is a directed edge such that p_i is a parent pattern of p_j. p_i is defined as a parent of p_j if it has exactly one of p_j value replaced with '- '.

Step 3 - Multi-partition pattern candidate generation: The algorithm goes over the graph G_p based on breadth-first approach [15] to check whether each pattern candidate of G_p holds on the distributed relation R. According to *Property 2*, if a pattern candidate holds, then its children will not be checked and will be eliminated from G_p.

A pattern candidate holds on R if its *multi-partition confidence* meets the given confidence threshold ĉ. The patterns of the graph are checked until the multi-partition tableau support does not meet the support threshold ŝ.

The *multi-partition confidence* of a pattern p corresponds to the confidence of p in the whole distributed relation R. It is also called global confidence. It is computed following two cases: the existence and non-existence of the partitioning attribute A in the antecedent X of the given FD. The key idea of the identification of these two cases is minimizing data shipment:

(1) $A \in X$, we can avoid data shipment between the distributed partitions of R, to compute the multi-partition keepers and then the multi-partition confidence. (2) $A \notin X$, one needs to gather the single-partition keepers in order to compute the global one. In this case, the algorithm works with minimizing data shipment, as it is described in Algorithm 2 and presented in section 5.

To illustrate these two cases of our distributed algorithm, let us consider the distributed relation R and its horizontal fragments R_1 and R_2 given in Table 2. The partitioning attribute of R is the *Country* attribute. We consider the following functional dependencies according to the belonging of the *Country* attribute in their antecedent:

fd_1 : $[name, type, country] \rightarrow [price, tax]$

fd_2 : $[name, type] \rightarrow [price]$

4.1 The Ground Algorithm

In the case of the partitioning attribute is in the antecedent of the given FD, there is no need of data shipment to generate the pattern tableau of distributed relation. The multi-partition (or global) confidence of a pattern p in R is computed as follow:

• Compute the *single-partition keepers* of p, $keepers_i(p)$, and the *single-partition cover* of p, $cover_i(p)$, on each partition R_i (as shown in section 2.2) and get their size $|Keepers_i(p)|$ and $|cover_i(p)|$ respectively.

• Compute the *multi-partition keepers* of p $disKeepers(p)$, and the *multi-partition cover* $disCover(p)$ (as shown in section 2.2) and get their size $|disKeepers(p)|$ and $|disCover(p)|$ respectively. As the partitioning attribute appears in the antecedent of p, then $\bigcap_i keepers_i(p) = \emptyset$. Consequently:

$$disKeepers(p) = \bigcup_i keepers_i(p) \text{ and } |disKeepers(p)| = \sum_i |keepers_i(p)|$$

• Compute the *multi-partition confidence* of p in R as follows:

$$disConfidence(p) = \frac{|disKeepers(p)|}{|disCover(p)|} = \frac{\sum_i |Keepers_i(p)|}{\sum_i |cover_i(p)|}$$

For instance, let us consider the distributed relation R with the given functional dependency fd_1. The *single-partition keepers* of the pattern $p = ('Harry\ potter', 'book', -\ -\|-, -)$ in R_1 and R_2 are respectively $Keepers_1(p)$ and $Keepers_2(p)$ as presented in Table 3.

The *single-partition covers* of the same pattern p are $cover_1(p) = 2$ and $cover_2(p) = 3$ in R_1 and R_2, respectively.

Table 3. Single-partition keepers of p in R_1 and R_2 for fd_1

tid	Name	Type	Country		Price	Tax
t4	Harry Potter	book	UK		12	0
t5	Harry Potter	book	UK		12	0

$Keepers_1(p)$

tid	Name	Type	Country		Price	Tax
t1	Harry Potter	book	France		10	0
t2	Harry Potter	book	France		10	0

$Keepers_2(p)$

Consequently, the *multi-partition confidence* of p is:

$$disConfidence(p) = \frac{|keepers_1(p)| + |keepers_2(p)|}{|cover_1(p)| + |cover_2(p)|} = \frac{2+2}{2+3} = \frac{4}{5}$$

5 Summary-Based Distributed Generation of CFD

In the case where the partitioning attribute is not in the antecedent of the given FD, one needs to gather single-partitions keepers in a single site in order to compute the multi-partition keepers. Following, we explain through an example why it is necessary to gather tuples of the distributed partitions in a single site. We also show how to minimize data shipment between distributed sites, based on partition summaries.

5.1 Problem Statement

Let us consider the example of the functional dependency
fd_2 $(name, type \rightarrow price)$ and the pattern candidate
$p = ('Harry\ potter',' book' \| -)$. The single-partition covers of p in R_1 and R_2 are respectively $cover_1(p)$ and $cover_2(p)$ as depicted in Table 4.

Table 4. Single-partition covers of p in R_1 and R_2 for fd_2

tid	Name	Type	Price
4	Harry Potter	book	12
5	Harry Potter	book	12

$Cover1(p)$

tid	Name	Type		Price
1	Harry Potter	book		10
2	Harry Potter	book		10
3	Harry Potter	book		12

$Cover2(p)$

The *single-partition keepers* of each partition are tuples t_4 and t_5 for R_1 and tuples t_1 and t_2 for R_2, from Table 4. The *multi-partition keepers* could not be computed from these single-partition keepers. That is the single-partition keeper of R_1 eliminates a keeper (tuple t_3) in the distributed relation. Below, we compute the *multi-partition confidence* from the retrieved single-partition keepers:

$$disConfidence(p) = \frac{|Keepers_1(p)| + |Keepers_2(p)|}{|cover_1(p)| + |cover_2(p)|} = \frac{2+2}{3+2} = \frac{4}{5}$$

However, the exact global confidence on the whole distributed relation is:
$Confidence(p) = \frac{3}{5}$

Consequently, to determine the multi-partition keepers of a pattern p for the whole distributed relation R, we should gather all matching tuples to a single site referred to as a coordinator.

5.2   The Summary Algorithm

We have shown that the multi-partition confidence can be calculated only after grouping tuples of all partitions of a distributed relation. We introduce the concept of *partition summary* to avoid this centralization of the distributed relation.

Definition 2. *A partition summary $S_i(p)$, of a pattern p in a partition R_i, is the frequency of each tuple t_k in the covers of R_i:*
$S_i(p) = \langle t_k, f_k \rangle$ *where,* $t_k \in dom(R_i)$, *and* $f_k = \{t_j \in cover(p)|t_j[XY] = t_k[XY]\}$

Based on this concept, the *multi-partition confidence* of a pattern p in a distributed relation R is computed according to the following steps:

1. **Compute the partition summaries.** The *partition summary* of p from each partition R_i is computed following the definition above. The partition summaries of the pattern $p = ('Harry\ potter', -\|-)$ in the partitions R_1 and R_2 are respectively $S_1(p)$ and $S_2(p)$ as presented in Table 5.
   The algorithmic computing of summaries is given in the following Algorithm 1:

Algorithm 1. PARTITION SUMMARY COMPUTING ALGORITHM

Require: R_i – a partition relation
            $X \rightarrow Y$ – A functional dependency
            pc – A pattern
Ensure: Summary of pc on R_i
      $SummaryTuples \leftarrow \{t, t \in R_i\ and\ t[XY] = R_i[XY]\}$
   2. **for** $eacht_i \in SummaryTuples$ **do**
      $occ_i \leftarrow |\{t/t \in SummaryTuples\ and\ t = t_i\}|$ // the occurrence number of $t_i \in R_i$
   4. $summary \leftarrow summary \cup (t_i, occ_i)$
      end for
   6. **print** $summary$

---

2. **Merge the partition summaries.** Once the partition summaries are computed, they are transferred to a single site referred to as *coordinator*. The coordinator is chosen to be the site that contains the largest summary (in the example, either the sites of S_1 or S_2 may be coordinator as they have the same summary size). The partition summaries $S_i(p)$ are then aggregated in the coordinator by calculating the global frequencies of summaries:
   $S(p) = \bigcup_i S_i = \{< t_j, F_j >\}$ where $t_j \in \bigcup_i S_i$ and
   $$F_j = \sum_{t_k \in \bigcup_i S_i | t_i[XY] = t_k[XY]} F_k$$

In other words, for $t_i \in S_i(p)$ and $t_j \in S_j(p)$, if $t_i[XY] = t_j[XY]$, then $S(p)$ is the set of $< t_i = t_j, F_i + F_j >$. Here, F is the frequency of identical tuples t_i in the global relation R.

The aggregation of the summaries $S_1(p)$ and $S_2(p)$ of Table 5 results to the multi-partition summaries (or global summaries) in Table 6.

3. **Compute the multi-partition confidence.** The multi-partition confidence of a pattern p is calculated from the *multi-partition keepers* and the *multi-partition cover* of p as follows:

$$disConfidence(p) = \frac{|disKeepers(p)|}{|disCover(p)|}$$

If $disConfidence(p)$ meets the confidence threshold $\hat{c}$, then p can be considered as a valid pattern and added to the pattern tableau T. The cover and keepers of a pattern are calculated from the multi-partition summaries $S(p)$ as described in the following:

– *Compute the multi-partition cover.* From the multi-partition summaries $S(p) = \{< t_j, F >\}$, the size of covers of p is $|cover(p)| = \sum_{<t_k, f_k> \in S(p)} f_k$.
   For instance, from the multi-partition summaries of Table 6, the cover' size of the pattern $p = ('Harry\ potter',$
   $- ||-)$ is: $|cover(p)| = 8$

– *Compute the multi-partition keepers.* The multi-partition keepers of a pattern p correspond to the aggregated summaries $S(p)$ computed on step 2 without summaries violating the functional dependency $X \to Y$. That is, we only keep tuples with the maximum of frequency.

$$disKeepers(p) = \{< t_i, \max_{<t_k, f_k> \in S(p)} f_k \mid t_i[X] = t_k[X,] \, t_i[Y] \neq t_k[Y] >\}.$$

For instance, the multi-partition keepers of summary $S(p)$ computed above is: $disKeepers(p) = \{< ('Harry\ potter', 'book' \| 12), 3 >, < ('Harry\ potter', 'dvd' \| 8), 1 >, < ('Harry\ potter', 'vhs' \| 3), 1 >\}$

The size of the global keepers is the sum of frequencies of the retained summaries:

$$|disKeepers(p)| = \sum_{<t_k, f_k> \in disKeepers(p)} f_i$$

Consequently, the multi-partition confidence of the pattern $p = ('Harry\ potter', - ||-)$ is:

$$disConfidence(p) = \frac{|disKeepers(p)|}{|disCovers(p)|} = \frac{5}{8}$$

Algorithm 2. THE SUMMARY DISTRIBUTED ALGORITHM FOR CFD GENERATION

Require: $\hat{s}, \hat{c}$ – Support and confidence thresholds
$\{R_1, \ldots, R_n\}$ – n partitions of the distributed relation
$FD = X \rightarrow Y$ – Functional dependency
Ensure: Pattern tableau of the distributed relation R

2. Step 1: Generating the pattern candidates PC_i of each partition R_i using onDemand algorithm
   Step 2: Generating the parent-child graph G_p
4. Step 3: Validating the patterns
   while G_p is not empty AND the marginal support of the tableau is less than $\hat{s}$ **do**
6.   Extracting the next pattern pc from G_p following breadth-first visiting
       Distributed computing of the local summaries Sm_i of pc on each partition R_i
8.   Aggregating the local summaries Sm_i
       Computing the global keeper $Keeper_{pc}$ and $Cover_{pc}$ cover of pc from the aggregated summaries
10.   Compute the global confidence $Conf_{pc} = \frac{Keeper_{pc}}{Cover_{pc}}$ of pc
       if $Conf_{pc} \geq \hat{c}$ **then**
12.   Add the pattern pc to the tableau pattern
           Delete children of pattern pc from the graph G_p
14.   Add the marginal support of pc to the marginal of the tableau
       else
16.   Compute $child(pc, R_k)$, the children of pc from its originating partition R_k
           Add $child(pc, R_k)$ to the parent-child graph G_p
18.   **end if**
       end while

Table 5. Partition Summaries of p in R_1 and R_2

sid	Tuple			Frequency
	Name	Type	Price	
s_{11}	Harry Potter	book	12	2
s_{12}	Harry Potter	DVD	14	1
s_{13}	Harry Potter	VHS	5	1

$S1(p)$

sid	Tuple			Frequency
	Name	Type	Price	
s_{21}	Harry Potter	book	10	2
s_{22}	Harry Potter	book	12	1
s_{23}	Harry Potter	DVD	8	1

$S2(p)$

6   Algorithm Soundness

6.1   Discussion on Support and Confidence Thresholds

The proposed algorithms compute a CFD pattern tableau according to a support $\hat{s}$ and a confidence $\hat{c}$ thresholds. Those thresholds cannot be used to generate the local pattern candidates.

Let us consider the example of the following two partitions R_{H1} and R_{H2} (depicted in Table 7.a and Table 7.b resp). Given the functional dependency fd ($name, type \rightarrow$

Table 6. Multi-partition Summaries $S(p)$ in R

sid	Tuple			Frequency
	Name	Type	Price	
s_1	Harry Potter	book	10	2
s_2	Harry Potter	book	12	3
s_3	Harry Potter	DVD	8	1
s_4	Harry Potter	DVD	14	1
s_5	Harry Potter	VHS	5	1

$price$) and the pattern candidate $p = ('Harry\ potter', 'book' \| -)$. Given the thresholds: $\hat{s} = 0.7$ and $\hat{c} = 0.8$, the pattern p does not hold neither on R_{H1} nor on R_{H2}, because its confidence ($= 4/6$) in R_{H1} does not meet $\hat{c}$, and its support ($= 4/6$) in R_{H2} does not meet $\hat{s}$. However, p holds on $R_{H1} \cup R_{H2}$ as its support (=10/12) and confidence (=8/10) meet the given thresholds.

To cope with this issue, the local pattern candidates are generated according to either the support or the confidence thresholds. Consequently, the local thresholds are set to ($\hat{s}$,0) or (0, $\hat{c}$). In fact, the pattern p holds on R_{H1} and on R_{H2} with (0.7,0) and (0,0.8) respectively.

6.2   Optimality of Generated Pattern Tableaux

The proposed distributed algorithms generate optimal pattern tableau. We say that a pattern tableau is optimal in a relation R if all its patterns are optimal. A pattern is optimal if its parents do not hold on R.

Let us consider the generated pattern tableau T_p, we prove for each pattern p in T_p that its parents do not hold on the distributed relation $R = \bigcup_i R_i$.

Let us consider pp a parent of p generated from R_i. We prove that pp does not hold neither on R_i nor on R.

- p is optimal on the partition R_p: p is generated using on-demand algorithm that generates optimal pattern tableaux [10]. Therefore, pp does not hold on R_p.
- pp does not hold on the other partitions $R_{i|i\neq p}$. We prove it by contradiction. We guess that pp holds in a partition R_i, then pp is a pattern candidate in the parent-child graph. If pp holds on R, then p its child p should not added to pattern tableau T_p. This is a contradiction with the hypothesis: p is in T_p.

7   Experiments

For the experiments, we used Microsoft SQL Server on five machines Intel Core i3 CPU @ 2.2 Ghz and 8GB of main memory. The algorithm is implemented with the language C. Our experiments use an extension of the *Sales* relation in Table 1:

Table 7. R_{H1} and R_{H2} partitions of R

tid	Name	Type	Country	Price	Tax
t1	Harry Potter	book	UK	12	0
t2	Harry Potter	book	UK	12	0
t3	Harry Potter	book	UK	12	0
t4	Harry Potter	book	UK	12	0
t5	Harry Potter	book	UK	14	0
t6	Harry Potter	book	UK	14	0

R_{H1}

tid	Name	Type	Country	Price	Tax
t7	Harry Potter	book	France	12	0
t8	Harry Potter	book	France	12	0
t9	Harry Potter	book	France	12	0
t10	Harry Potter	book	France	12	0
t11	Harry Potter	DVD	France	14	0
t12	Harry Potter	DVD	France	15	0

R_{H2}

Sales(tid,name,type,country,price,tax). To populate the *Sales* relation, we wrote and executed a program that generates synthetic records. Next, we horizontally partition these relations according to the *Country* attribute.

The FDs considered in our experiments are those used in the illustration of our algorithm:
$fd_1 : [name, type, country] \rightarrow [price, tax]$ and $fd_2 : [name, type] \rightarrow [price]$

We conducted the following experiments to evaluate our algorithm. Our experiments represent averages over five trials.

7.1 Scalability Evaluation

We studied the scalability of our algorithm with respect to (i) the number of partitions of the distributed relation, and (ii) the size of the partitions.

Scalability on the number of partitions. We study the scalability of our algorithm by varying the number of the distributed partitions. We fix the size of the distributed *Sales* relation and increase the number of its horizontal partitions. The results in Figure 2 (left) illustrate the linear time response with respect to the number of partitions for the ground and the summary algorithms. We think that the response time will tend to stabilize from a certain number of partitions. Indeed, the most important time consuming of the two distributed algorithms resides on generating single-partition patterns using the on-demand algorithm, which is executed in parallel on partitions.

Scalability on data size. We evaluate the impact of the data size on the response time of our algorithm. We increase the size of partitions from 200k to 1000k. Figure 1 (a) shows the response times for the ground algorithm, and Figure 1 (b) for the summary algorithm. Both algorithms scale linearly. Compared to the on-demand algorithm running on a centralized relation, the response time of our algorithm is slightly higher. However, this overhead tends to decrease while the relation size increases. Notice that the ground algorithm outperforms the summary algorithm. This is due to the data shipment in the summary algorithm.

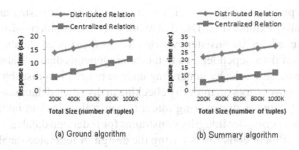

(a) Ground algorithm   (b) Summary algorithm

Fig. 1. Scalability experiments on data size

7.2 Data Shipment Rate Evaluation

This experiment estimates the rate of data transferred from partitions to the coordinator. This rate corresponds to the number of transferred partition summaries on the size of the distributed relation. In this experiment, we need to consider the homogeneity of data. Figure 2 (right) shows this experiment on two different data sets with regards to their homogeneity. Tuples are homogeneous according to a given FD if they are similar by considering only FD attributes. Data set 1 corresponds to more homogeneous tuples which leads to lower rate of data shipment. Data set 2 corresponds to less homogeneous tuples which leads to higher rate data shipment. Notice that the rate depends proportionally and only on the homogeneity of tuples. The results are sound as more tuples are homogeneous more they are aggregated in summaries. Therefore, the size of summaries decreases as well as data shipment.

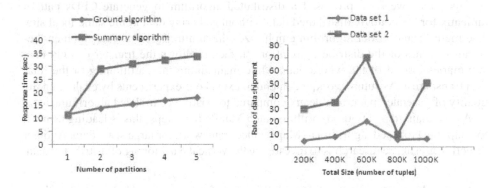

Fig. 2. (left) Scalability experiments, (right) Rate of the partition summaries experiment

8 Related Work

Most of existing works on CFDs were applied to centralized data [2,4,5,7,10]. To the best of our knowledge only work in [8,9,13] study CFDs in distributed setting. In [8], techniques for detecting violations of CFDs in distributed relations while minimizing data shipment were proposed. This work is extended in [9] by proposing an incremental approach. Recently, extended CFDs, that work over the problem of multi valued attributes, are also used for inconsistency detection in distributed databases [13]. In [14], the authors adapt data dependencies to the heterogeneous data in dataspaces. Our work is different for those works by generating tableau pattern in a distributed manner.

There have been many works on integrity checking in distributed databases [1,11,12]. Related work close to ours is discovering integrity constraints in data integration systems. In [6], the authors consider integrity constraints for federated databases. They propose the integration of local constraints during the design of federated databases based on the integration of the extensions (or validating range) they belong to. The constraints are divided into object constraints applied only to one object and class constraints that can be checked on at least two objects. Then, some rules are defined to deal with constraints mainly the ones that generate an overlapping conflict. In [16],the authors propose a distributed mining framework for discovery FDs from distribute databases. They develop a theorem that can prune candidate FDs effectively and extend the partition based approach for distributed databases. All above works are related to FDs on distributed setting while in this paper we are generating pattern tableaux for CFDs in distributed data. In [16],the authors propose a distributed mining framework for discovery FDs from distribute databases. They develop a theorem that can prune candidate FDs effectively and extend the partition based approach for distributed databases. All above works are related to FDs on distributed setting while in this paper we are generating pattern tableaux for CFDs in distributed data.

9   Conclusion

In this paper, we have proposed a distributed algorithm to generate CFDs pattern tableaux for horizontally distributed data without gathering data in the same local site. The main feature of our algorithm emphasizes the minimization of data shipment between the sites of the distributed relation. In fact, it allows the transfer of only data summaries instead of the whole data, while maintaining the partitioning of the relation transparent. As future work, we intend to extend our experiments by evaluating the quality of generated pattern tableaux compared to existing centralized algorithms.

As the main steps of our algorithm match MapReduce steps, this is leading us next, to adapt the proposed algorithm to MapReduce framework for large scale data. Another direction as a future research is to investigate the vertical partitioning of distributed data.

Acknowledgements. We thank Divesh Srivastava for relevant and helpful suggestions.

References

1. Agrawal, S., Deb, S., Naidu, K.V.M., Rastogi, R.: Efficient detection of distributed constraint violations. In: ICDE, pp. 1320–1324 (2007)
2. Bohannon, P., Fan, W., Geerts, F., Jia, X., Kementsietsidis, A.: Conditional functional dependencies for data cleaning. In: ICDE, pp. 746–755 (2007)

3. Bravo, L., Fan, W., Ma, S.: Extending dependencies with conditions. In: VLDB, pp. 243–254 (2007)
4. Chiang, F., Miller, R.J.: Discovering data quality rules. PVLDB 1(1), 1166–1177 (2008)
5. Cong, G., Fan, W., Geerts, F., Jia, X., Ma, S.: Improving data quality: Consistency and accuracy. In: VLDB, pp. 315–326 (2007)
6. Conrad, S., Schmitt, I., Türker, C.: Considering integrity constraints during federated database design. In: Embury, S.M., Fiddian, N.J., Gray, W.A., Jones, A.C. (eds.) BNCOD 1998. LNCS, vol. 1405, pp. 119–133. Springer, Heidelberg (1998)
7. Fan, W., Geerts, F., Lakshmanan, L.V.S., Xiong, M.: Discovering conditional functional dependencies. In: ICDE, pp. 1231–1234 (2009)
8. Fan, W., Geerts, F., Ma, S., Müller, H.: Detecting inconsistencies in distributed data. In: ICDE, pp. 64–75 (2010)
9. Fan, W., Li, J., Tang, N., Yu, W.: Incremental detection of inconsistencies in distributed data. In: ICDE, pp. 318–329 (2012)
10. Golab, L., Karloff, H.J., Korn, F., Srivastava, D., Yu, B.: On generating near-optimal tableaux for conditional functional dependencies. PVLDB 1(1), 376–390 (2008)
11. Gupta, A., Sagiv, Y., Ullman, J.D., Widom, J.: Constraint checking with partial information. In: PODS, pp. 45–55 (1994)
12. Huyn, N.: Maintaining global integrity constraints in distributed databases. Constraints 2(3/4), 377–399 (1997)
13. Reyya, S., Prameela, M., Yadav, G.V., Rani, K.S., Bhargavi, A.V.: An efficient extension of conditional functional dependencies in distributed databases. Database 5, 6
14. Song, S., Chen, L., Yu, P.S.: Comparable dependencies over heterogeneous data. VLDB J. 22(2), 253–274 (2013)
15. Wyss, C.M., Giannella, C.M., Robertson, E.L.: FastFDs: A heuristic-driven, depth-first algorithm for mining functional dependencies from relation instances - extended abstract. In: Kambayashi, Y., Winiwarter, W., Arikawa, M. (eds.) DaWaK 2001. LNCS, vol. 2114, pp. 101–110. Springer, Heidelberg (2001)
16. Ye, F., Liu, J., Qian, J., Xue, X.: A framework for mining functional dependencies from large distributed databases. In: Proceedings of the 2010 International Conference on Artificial Intelligence and Computational Intelligence, AICI 2010, vol. 03, pp. 109–113. IEEE Computer Society, Washington, DC (2010)

Motivation System Using Purpose-for-Action

Noriko Yokoyama, Kaname Funakoshi, Hiroyuki Toda, and Yoshimasa Koike

NTT Service Evolution Laboratories, NTT Corporation, Japan

Abstract. There are many web services from which we can acquire the information needed for action planning. One of the goals of those services is to suggest a target action and encourage the user to perform it. If this is to succeed, three factors must be emphasized: motivation, ability, and trigger. However, traditional web services focus on giving information related to the ability to perform the behavior (cost, place etc.) and motivation is not considered; without effective motivation the target action is unlikely to be performed. Our goal is to identify an effective motivation approach. To achieve this goal, we collect purpose-for-action. In this paper, we propose a method to extract purpose-for-action from social media texts by using clue expressions and modification structure. Moreover, we conduct a user experiment whose results confirm that showing purpose-for-action yields effective motivation.

Keywords: Text Mining, Purpose-for-Action Extraction, Motivation System.

1 Introduction

There are many web services from which we can acquire the information needed to form an action plan. For example, before going out for lunch, we can use restaurant sites like "Yelp"[1] and when we want to travel, we can use travel sites like "Tripadvisor"[2]. Moreover, location-based services like "Foursquare"[3] are increasing and we can get, at one time, information related to many types of action near the current area or the target area.

One of the goals of those services is to offer information suitable for the user and that encourages the user to perform a target action. In traditional services, categorizing facilities by area and offering precise information is thought to make it easier to initiate the action. Although these services suit users who have decided the target action ("I want to eat something" or "I want to go to museum" etc.), they aren't suitable for users who have no particular action in mind ("I want to do something but I'm not sure what").

We address this problem by using the Fogg behavior model [1], a behavior model common in the field of psychology. This model asserts that for a behavior to

---

[1] http://www.yelp.com/
[2] http://www.tripadvisor.com/
[3] https://foursquare.com/

H. Decker et al. (Eds.): DEXA 2014, Part I, LNCS 8644, pp. 66–80, 2014.

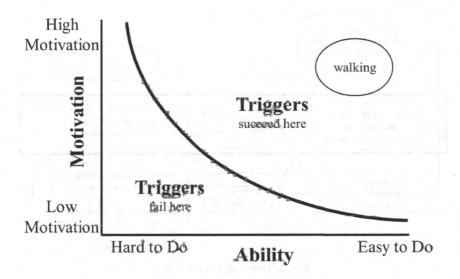

Fig. 1. Fogg behavior model

occur, the person must have sufficient motivation, sufficient ability, and an effective trigger. In other words, (1) motivation and (2) ability are necessary when (3), a trigger, is offered.

For example, when the target action of a web service is *walking*, users who want to do something refreshing (high motivation) are good candidates because *walking* will not be seen as a difficult task (high ability) (Fig. 1). Conversely, lack of motivation or barriers to action are typical causes of action failure. Two key issues are (1) increasing motivation and (2) eliminating perceived barriers.

Examples of perceived barriers to an action are time and cost, and it is difficult to lower these barriers quantitatively. However, a barrier might be over-rated, and information that helps the user rate the barrier correctly might lead to action initiation. The cost of gathering information is another barrier. "Yelp" offers facilities information like distance and price which lowers the barrier to action initiation.

Current solutions to the other main issue, motivation, are limited. Examples of motivation factors are pleasure and hope. Coupons target economic motivation; they are seen as effective and are used frequently. However, the application range of coupons is limited and a more comprehensive method is needed.

This research focuses on providing motivation through information to encourage action initiation. Motivation can be achieved by showing purpose-for-action (PfA hereafter) and triggering information cascade [2] by showing others' experiences. PfA and experience information can be collected from the various actions possible and can solve the motivation problem. For example, the user can be introduced to appropriate facilities and their PfA, which should encourage user action.

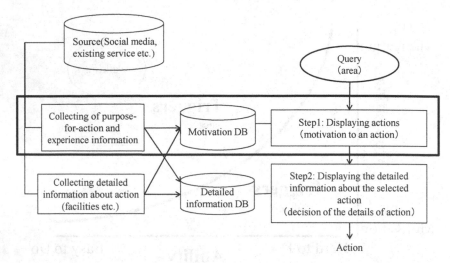

Fig. 2. Target area of this paper

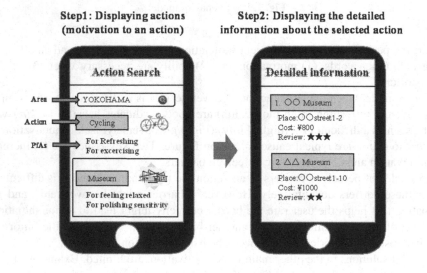

Fig. 3. Example of application image

Though information services come in various forms, we assume here a user-location-centered information service that has two steps as shown in Fig. 2. A typical application image is shown in Fig. 3. In the first step, the service displays near-by actions with PfAs and experience information aiming at increasing the user's motivation. In the second step, it displays details of the selected action. The second step is the same as ordinary information services. The focus of this paper is the first step which is indicated by the frame in Fig. 2, and the goal is to motivate people into demanding details of the target action of interest or actually performing the action. As an initial step in achieving this goal, this paper collects PfAs.

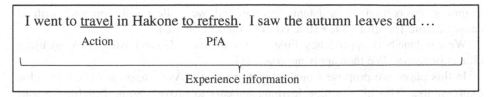

Fig. 4. Extracting PfA from experience information

We collect PfAs from experience information which contains a large variety of PfAs as shown in Fig. 4. Social media texts are used as the source material as they contain descriptions of large numbers of a very wide variety of experiences and PfAs. We also conduct a user experiment and evaluate whether showing PfA yields effective motivation.

2   Related Work

2.1   PfA Extraction Method

Two approaches can be adopted to extract PfAs from texts.

(1) Identify keywords that directly express purpose, e.g. "loose weight" etc. are registered in a dictionary.
(2) Identify expressions that indicate purpose but do not use keywords, e.g. Sentences that contain expressions such as "in order to" are processed and the word/words associated with the expressions are used as PfA.

Approach (1) is effective when the purpose is expressed by keywords, even if the relation between purpose and action isn't explicitly written. But it requires the building a dictionary. Approach (2) is far more flexible but it requires that the relation between purpose and action be explicitly written.

Research based on approach (1) includes a study that extracts PfAs from travel blogs [3]. In this research, only travel actions are considered, so the dictionary must, for our purpose, be extended to extract PfAs related to many other kinds of actions. One study constructs a dictionary related to a large variety of actions by targeting long term goals [4]. In this study, PfA topics are extracted by using LDA from twitter including "new year resolution". While they extract purpose itself, we try to extract purpose-action pairs. Purpose-action pairing is related to the extraction of cause information.

The techniques of (2) are widely used to extract cause information from news articles [5][6][7]. For example, cause information is extracted by using statistical information and initial clue expressions [5]. All these studies assume that the source material consists of formal sentence like news articles. Moreover, they don't distinguish cause/reason and purpose. We target social media which will certainly contain grammar errors and inconsistent spelling and we collect only purpose without cause/reason. Therefore, the extracted PfAs must be filtered.

We combine both approaches. First, we use approach (2) and extract PfAs by using clue expressions. We then apply approach (1).

In this paper, we propose a method of extracting PfAs by approach (2), using clue expressions. Though machine learning appears to provide some benefit, we start with a rule based method and leave machine learning as future work.

2.2   Support of Action Planning

There are a lot of web services from which we can acquire the information needed to form an action plan. For example, by using location-based services like "foursquare", we can get information near the current area. These services offer facilities information like distance, which may lower the barrier to action initiation. Moreover, showing others' experiences can trigger information cascade and may also provide motivation.

However, PfAs are not explicitly written and the number of experiences is limited. Our method can strengthen the rise in motivation and encourage user action.

From another viewpoint, there are studies that aim to offer the information suitable for users' preference and contexts. As examples of considering the users' preference, "Google"[4] uses search history and "Amazon"[5] uses similar users' information and recommends contents and items. With regard to studies on context-aware recommendation, there are many studies considering time, companion etc. [8]. Though recommendations based on the users' preference and contexts may be one form of motivation, it requires the users' information. Moreover, these techniques focus on the selection of items and don't consider additional information such as PfAs. In this paper, we propose a method that dispenses with users' information as the first step. Combining these techniques and our method is future work.

3   Problem Setting

This study extracts the PfAs of a specified action from texts containing grammar errors and inconsistent spelling. Here, we define PfA as the action or statement the writer wants to achieve through action or the effect of an action that is clear to the reader (Fig. 5). For example, if a sentence is "I ran to refresh", "refresh" is PfA because it is the goal of "run". If the sentence is "I ran and felt refreshed", "refreshed" is also PfA because "refreshed" clearly seems to be the result of "ran". The target actions considered here are those associated with whether we act or not from the viewpoint of behavior selection support. As one of all possible action areas, this paper focuses on leisure activity due to its sheer size (especially, activities outside the house). We define the words used in this paper as follows.

---

[4] https://www.google.co.jp/
[5] http://www.amazon.com/

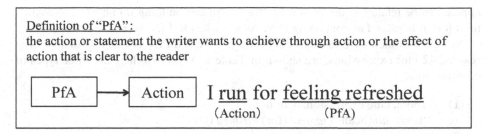

Fig. 5. Definition of PfA

Definition 1: Clue expressions
Expressions necessary to understand the relation between purpose and action in a sentence. We call expressions connecting purpose and action such as "in order to" clue expressions and use them in extracting PfAs.

Definition 2: PfA class
If several extracted PfAs have similar purposes, we bring them together to create a PfA class. For example, the given two sentences of "for health maintenance" and "for maintenance of health", we group them together as "for health maintenance" class.

4   Preliminary Analysis

In this chapter, we explain the dataset used and describe the results of an analysis of the clue expressions used in the selected dataset. We describe work done on a dataset in Japanese, but the analysis and method can be applied to other languages.

4.1   Dataset

The key requirement of the dataset is that it contains a large number of experiences and PfAs, both of which are described clearly. From the candidates of microblogs, Q&A sites, and blogs, we chose blogs because they have a large amount of information in one place. We are planning to enhance the dataset in the future.

4.2   Analysis of Dataset (Usage of Clue Expressions)

First, we confirm the clue expressions. Previous work pointed out that the use of clue expressions changes with the dataset. Therefore, we calculated the accuracy rate of common clue expressions found in this dataset.

First, ten very common actions were chosen from leisure activities [9], and 100 usages of the actions were found in 100 blogs.   If PfAs were used with these actions in the same sentence, the sentences were manually extracted.

It is sometimes difficult to judge whether an expression is a PfA. Our criterion was that a PfA was present if the sentence appeared to be related to one of the 12 reasons used in an opinion poll [10]. For example, the sentence "I have grown in strength" is appears to be related to the reason of "for health and building up physical strength", therefore it is judged as containing PfA. As a result, 430 PfAs were extracted.

Next, we extracted expressions connecting action and PfA as clue expressions. The results, 42 clue expressions, are shown in Table 1. As shown there are three types of constructions.

(1) purpose, clue expression, action
  (e.g. "*kenko no* (health) *tameni* (for) *hashiru* (run)")
(2) action, clue expression, purpose
  (e.g. "*hashi* (run) *tara* (and) *kibun tenkan dekita* (feel refreshed)")
(3) order of action and purpose can be switched.

Constructions (1) and (2) exhibited almost equal usage frequency. As the initial investigation, we consider only expressions of type (1) construction.

Table 1. Clue expressions extracted from blogs

① purpose ⇒ action	*wo kanete, tame, tameni, no tameni, wo motomete, to omoukara, to omotte, niha, nanode, takute, toiu kataniha, to omoi, to iukotonina-ri, karatte kotode, yoto, ga, wo, ni,* These are the meaning of "for ", "due to", "in order to" etc.
② action ⇒ purpose	*ha xx no tame, ha xx ga mokuteki, tara, demo, ga, toki no, mo, te misemasu, kotode, shitari, dakedemo, mo, noga, kedo* These are the meaning of "and ", "the purpose of xx is" etc.
③both	*de, kara, ha, node, shite, te, to, shi, no, kotode* These are the meaning of "so" etc.

Table 2. Accuracy rate of clue expressions

Clue expression	The number of examples	The number of correct examples	Accuracy rate
wo kanete	2	2	1.00
tame	8	4	0.50
tameni	**15**	**12**	**0.80**
no tameni	9	7	0.78
wo motomete	1	1	1.00
to omotte	1	0	0.00
niha	**69**	2	*0.03*
nanode	**24**	*1*	*0.04*
takute	3	3	1.00
to omoi	1	1	1.00
yoto	2	2	1.00
Total	137	35	0.26

To raise the accuracy, we deleted clue expressions that consisted of a single character leaving us with 15 clue expressions. We extracted "clue expression + action" sentences and judged whether the words immediately before the clue expression described a PfA. We processed the roughly one million blogs entered on one day. From among the above-mentioned ten actions, we focused on five (travel, dining out, museum, amusement park, and cycling). Following the definition in 3, the sentence *"ame ga hutta tame saikuringu ha chushi* (cycling was canceled due to rain)" describes reason thus not a PfA. The result is shown in Table 2. Clue expressions that yielded zero examples are not listed in the table.

The two requirements for the clue expressions we want to find are that it is used in many sentences and its accuracy rate is high. From this viewpoint, *"niha* (for)" *"nanode* (for)" are used in many sentences, but their accuracy rate is low. *"tameni* (for)" has the second largest number of examples and its accuracy rate is high. From this analysis, we selected *"tameni* (for)" as the clue expression.

5   PfA Extraction Method

In this chapter, we detail the automated PfA extraction method using clue expression *"tameni* (for)". The proposed method is composed of two parts, extracting PfA and categorization. We explain each part.

5.1   Extracting PfA

We extract PfA by using the clue expression and modification structure. Here, "bunsetu" is a basic block in Japanese. As shown in Fig. 6, the PfA extraction method is composed of the following 4 steps.

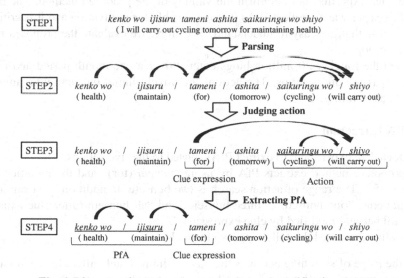

Fig. 6. PfA extraction using clue expression and modification structure

Step 1. The sentences including clue expression and target action are extracted.

Step 2. The sentences are parsed by the Japanese dependency analyzer, JDEP [11].

Step 3. A bunsetu (we call this first bunsetu) modified by a clue expression is extracted and the nearest bunsetu modifying the first bunsetu is also extracted as the second bunsetu. We extract first and second bunsetu and judge whether the two bunsetu include a target action.

Step 4. If the answer of step3 is *yes*, the first bunsetu that modifies the clue expression is extracted and the nearest bunsetu that modifies the first bunsetu is extracted as the second bunsetu. We extract the first and second bunsetu as PfA.

The number of bunsetu (two) was decided so as to extract the expression of "verb + object". As mentioned in Chapter 2, the extracted PfA must be filtered. Section 5.2 introduces one filtering method. We examine the results and propose other rule-based methods in 6.1.

5.2   Categorization (Improvement of Precision)

We categorize PfAs and adopt the PfA classes with high frequency to reduce the noise in the extracted PfAs. For categorization, each is decomposed into morphemes, and nouns and verbs are extracted from the morphemes. The PfAs with same nouns and verbs are categorized into a PfA class. The class name is the PfA with the highest frequency or the PfA consisting of the largest number of characters. The PfA classes are arranged in order of decreasing frequency and the classes of high rank or classes that include more than two PfAs are adopted. This method can display PfAs that are intuitively understandable to users.

6   PfA Extraction Experiment

We conducted experiments to confirm the validity of the proposed method. The purpose of this paper was to collect PfAs that are suitable for motivation and it is critical to extract PfAs that are easy to understand. Therefore, we evaluate the proposed method by precision.

We used the roughly 20 million blogs entered over a one month period and used five actions as the query terms. The actions are travel, dining out, museum, amusement park, and cycling.

6.1   PfA Extraction

We conducted an experiment to extract PfAs related to the five actions.

The proposed method extracts PfA by using "*tameni* (for)" and the modification structure of 5.1. The range of action search is two bunsetu. In addition, we examined 3 other patterns, "one bunsetu", "three bunsetu" and "all bunsetu (after clue expression and till bunsetu modified by clue expression)".

As baselines, we used the following two methods.

BL1: the range of searching action is the bunsetu immediately after clue expression
BL2: the range of searching action is all bunsetu after clue expression

BL2 is the softest judgment because the extent of action search is the largest. The range of PfA extraction is two bunsetu with the aim of extracting "verb+object" in both proposed methods and baselines.

The PfAs extracted by each method were manually judged as being correct or not and precision was calculated. When both action and PfA were correct, the PfA was judged as correct. Five actions (travel, dining out, museum, amusement park, and cycling) were judged to be correct when the queries were meant to extract information related to traveling, dining out, going to museum, going to amusement park, or cycling. For example, the sentence, "*kenko no tameni gaisyoku wo hikaeru* (refrain from dining out for health)" is judged as incorrect. PfA is judged as correct when the purpose is an action or statement that the writer wants to achieve or the effect of an action is something that the reader can readily understand. We show the results in Table 3 and Table 4.

The proposed methods ("1 bunsetu" and "2 bunsetu") demonstrated higher precision than the baselines in Table 3. As the result of BL1 shows, the PfAs aren't always correct when the action query immediately follows the clue expression. The sentence "*konohi no tameni ryoko no toki ni omiyage wo kata* (I bought souvenir at the traveling for this day)" could be excluded by using this knowledge. "1 bunsetu" has the highest precision but the number of correct examples is very small (Table 4). Therefore, we think that "2 bunsetu" is the best method.

According to the action query, the precision of each action varies as shown in Table 3. The precision of dining out is especially low because the sentences of dining out included many negative sentences.

In PfA extraction, the proposed method wrongly identified sentences as containing PfAs. We call this mistake False positive (FP). We analyzed FP and found the following causes.

Table 3. Precision

Action	BL1	BL2	1 bunsetu	2 bunsetu	3 bunsetu	All bunsetu
Travel	0.54	0.35	0.68	0.59	0.56	0.46
Dining out	0.60	0.28	*0.63*	*0.49*	0.45	0.38
Museum	0.69	0.47	1.00	0.83	0.78	0.64
Amusement park	0.73	0.65	1.00	0.85	0.86	0.88
Cycling	1.00	0.67	0.75	0.83	0.83	0.80
Average	0.58	0.37	**0.70**	**0.62**	0.58	0.49

Table 4. The number of correct examples

Action	BL1	BL2	1 bunsetu	2 bunsetu	3 bunsetu	All bunsetu
Travel	58	245	55	172	180	194
Dining out	18	31	19	27	28	29
Museum	9	44	6	29	31	35
Amusement park	8	15	3	11	12	15
Cycling	3	20	6	15	15	16
Average	96	355	*89*	**254**	266	289

(1) Problem of clue expression
  • a cause expression is extracted (e.g. *ame no tameni* (due to   rain))
  • a fixed form expression is extracted (e.g. *nanno tamni* (for what purpose))

(2) Problem of action extraction
  • the PfA of a negative action is extracted (e.g. refrain from dining out)
  • extracted action is wrong (e.g. I apply for travel insurance)
  • mistake in modification structure (e.g. cases in which actions are written in parallel)

(3) Problem of PfA extraction
  • PfA is incomplete (e.g. *miru tameni* (for seeing))

(4) Problem of preprocessing
  • spam

Then, as mentioned in 5.1, we applied a further method to remove these FP. Table 5 shows the result of this initial improvement policy. For example, to remove cause/reason errors, we conducted the following tense judgment and the paraphrase judgment because linguistic studies show that the verbs associated with cause/reason tend to be past tense and the cause/reason tends to be negative.

Table 5. Inital improvement policies

Error factor	Rough improvement policy
Cause/reason	Tense judgment and the paraphrase judgment
Stable sentence	Stop word ("or what purpose")
Negative action	Dictionary (the PfA of negative action are stored in a dictionary)
Wrong action	Connecting noun (remove the case that noun is immediately after query)
Incomplete PfA	Stop word (for xx and yy) The object before the PfA (remove the case that the word before PfA is object)

Table 6. The change in the number of errors

Error factor	The number of errors	The number of errors (after removal)
Cause/reason	15	4
Stable sentence	3	0
Negative action	30	15
Wrong action	62	21
Modification structure	10	8
Incomplete PfA	20	12
Spam	11	2
(Total)	151	__62__

Table 7. Categorization

Action	The number of correct examples	Precision(P)	P@10	P@Nc≧2
Travel	154	0.80	**0.9**	**0.94**(15/16)
Dining out	23	0.62	0.6	1.00(2/2)
Museum	25	0.81	0.9	1(3/3)
Amusement park	9	0.82	0.8	0
Cycling	12	0.92	0.9	0

Step 1 (tense judgment).

    if the tense of PfA is past, it is removed as cause/reason

Step 2 (paraphrase judgment).

    if the clue expression of "*seide* (due to)", which shows cause/reason, is connected to the PfA, the number of examples of the sentence in the dataset is calculated.

        if the number of examples is higher than that of examples wherein connection is "*tameni* (for)", the PfA is removed as cause/reason.

As a result, the number of mistakes is greatly decreased as shown in Table 6 and the precision increases to 78% from 62%. Although we discuss after the next section whether this precision is enough or not, one approach to increasing precision further is reviewing how actions are judged.

6.2   Categorization

The proposed method can output FP as shown in 6.1. To remove these mistakes, we bring together similar PfAs. We categorized the results of 6.1 and created PfA classes; for these we calculated $P@10$ (the average precision of the top 10 PfAs) and $P@Nc \geq 2$ (the average precision of the classes with more than two PfAs) using the correct/incorrect label used in 6.1. The result is shown in Table 7.

    The number of cases of amusement park, cycling, dining out and museum is small and it is difficult to evaluate the results. We intend to increase the data size and conduct new evaluations. As for travel, an improvement in precision was confirmed. The precision is over 90% which is thought to be sufficient for practical use. Examples taken from the PfA classes include "food" "meeting people" "congratulation" and so on; many kinds of PfAs were extracted.

7   User Evaluation Experiment

We conducted a questionnaire survey to evaluate the effect of motivation.

7.1   Experiment setup

Content of questions
Question 1.
We told the respondents to assume the situation in which an information search service (service in which some actions are recommended when a place is input) recommends an action ("How about running?" etc.). The respondent was asked to judge whether he or she would take the recommended action.

Question 2.
We told the respondents to assume the same situation but a PfA was shown together with the action ("How about running for health?" etc.). The respondent was asked if his/her motivation for doing the action was increased (five level scale was used) (Table 8).

If the person answered "yes" in question (1), we let him or her assume the situation where he or she didn't want to do the action and then showed them question (2).

This is because our aim is to motivate people who don't want to do the "action" when the only action is displayed.

Actions and PfAs
We employed 11 actions (running, cycling, cinema, museum, dining out, karaoke, travel, amusement park, shopping, barbeque, esthetic) including the five actions used in Chapter 6. These actions were selected from five subcategories written in a white paper on leisure [9].

With regard to PfAs, we manually categorized the PfAs extracted by the method of 6.1 and created PfA classes from the roughly 240 million blogs entered over a one year period. We used the affinity diagram technique to gradually raise the level of categorization [12].We got about 20~50 PfA classes for each action and used them in the questionnaire.

Respondents
2400 respondents were shown 2 actions. Therefore, 400 people responded per action.

Table 8. Examples of question 2

Running	Motivated or not				
↓ PfA	Increase a lot	Increase	Unchanged	Decrease	Decrease a lot
Exercise	☐	☐	☐	☐	☐
Maintain health	☐	☐	☐	☐	☐
Lost in thought	☐	☐	☐	☐	☐
...	☐	☐	☐	☐	☐

Table 9. Evaluation of the effect of motivation

Action	Motivated or not		
	Increase a lot + Increase	Unchangeable	Decrease a lot + Decrease
Running	.36	.56	.08
Cycling	.37	.53	.10
Cinema	.36	.55	.09
Museum	.41	.49	.10
Dining out	.35	.56	.10
Karaoke	.28	.58	.14
Travel	.43	.48	.09
Amusement park	.42	.49	.10
Shopping	.33	.57	.11
Barbeque	.43	.49	.09
Esthetic	.37	.52	.12
Average	.37	.53	.10

Table 10. Examples of PfA scores (Running)

Ranking	PfA	Motivated or not		
		Increase a lot + Increase	Unchangeable	Decrease a lot + Decrease
1	Health	0.54	0.43	0.03
2	Improvement in physical strength	0.52	0.45	0.03
3	Improvement in basal metabolism	0.51	0.45	0.04
4	Calorie consumption	0.47	0.48	0.05
5	Refreshing	0.46	0.47	0.07
⋮				
41	Physical strength consumption	0.22	0.60	0.18
42	Killing time	0.21	0.63	0.17
43	Event (date, travel etc.)	0.19	0.67	0.14
44	Get away from loneliness	0.16	0.68	0.17
45	Preparing for a role	0.12	0.71	0.17

7.2   Results and Discussion

The average of each action raised in question 2 is shown in Table 9. The average positive score (increase a lot + increase) is about 37%, which confirms the effectiveness of displaying PfAs in raising motivation. From this result, we confirm that showing PfAs extracted from blogs can encourage people to perform the target action.

The positive scores (increase a lot + increase) range from 7% to 75% for each PfA so clearly different PfAs have different levels of effectiveness. Examples of PfA scores are shown in Table 10. The highly scored PfAs were "health" "refreshing" "fun" and so on and these seem to be PfAs that are suitable for many people. In contrast, "preparing for a role" "writing blogs" are more user-specific PfAs and their average scores were low. From this result, we found that the effective PfAs are those that match the user's desire or situation. Identification of the most effective PfA considering user's preference and situation is future work.

8   Conclusions and Future Work

In this paper, we proposed a method of extracting purpose for motivation from social media texts and described how a service could use such purpose with the aim of motivating people to perform a target action. We evaluated our method and confirmed its effectiveness in extracting purposes and raising motivation. In the future, we will try to extract 5W1H (who, what, when, where, why, how) information from more experiences and select purpose and action to suit the user's situation. As shown in Fig. 1, detailed information about actions is also necessary along with motivation. We will collect information offered by existing services and establish an information navigation system that encourages people to perform a target action.

References

1. Fogg, B.J.: A Behavior Model for Persuasive Design. In: Proc. the 4th International Conference on Persuasive Technology, Article No. 40 (2009)
2. Bikhchandani, S., Hirshleifer, D., Welch, I.: Learning from the Behavior of Others: Conformity, Fads, and Informational Cascades. Journal of Economic Perspectives 12(3), 151–170 (1998)
3. Wakaki, H., Ariga, M., Nakata, K., Fujii, H., Sumita, K., Suzuki, M.: Building a Database of Purpose for Action from Word-of-mouth on the Web. In: Proc. Forum on Information Technology, pp. 15–22 (2011)
4. Zhu, D., Fukazawa, Y., Karapetsas, E., Ota, J.: Intuitive Topic Discovery by Incorporating Word-Pair's Connection into LDA. In: Proc. 2012 IEEE/WIC/AIM International Conferences on Web Intelligence and Intelligent Agent Technology, pp. 303–310 (2012)
5. Sakai, H., Masuyama, S.: Cause information extraction from financial articles concerning business performance. IEICE Trans. on Information and Systems E91-D(4), 959–968 (2008)
6. Sakaji, H., Sakai, H., Masuyama, S.: Automatic extraction of basis expressions that indicate economic trends. In: Washio, T., Suzuki, E., Ting, K.M., Inokuchi, A. (eds.) PAKDD 2008. LNCS (LNAI), vol. 5012, pp. 977–984. Springer, Heidelberg (2008)
7. Inui, T., Okumura, M.: Investigating the characteristics of causal relations in Japanese text. In: Proc. Workshop on Frontiers in Corpus Annotation II: Pie in the Sky, pp. 37–44 (2005)
8. Adomavicius, G., Tuzhilin, A.: Context-aware recommender systems. In: Recommender Systems Handbook, pp. 217–256 (2011)
9. The Japan Productivity Center: White Paper of Leisure 2013 (2013)
10. The Public Relations Department of the Cabinet Office: The opinion poll about lifelong learning (July 2012 investigation) (2012)
11. Imamura, K., Kikui, G., Yasuda, N.: Japanese dependency parsing using sequential labeling for semi-spoken language. In: Proc. ACL, pp. 225–228 (2007)
12. Kawakita, J.: A Scientific Exploration of Intellect ("Chi" no Tankengaku). Kodansha (1977)

Temporal Analysis of Sentiment in Tweets: A Case Study with FIFA Confederations Cup in Brazil

André Luiz Firmino Alves[1,2], Cláudio de Souza Baptista[1],
Anderson Almeida Firmino[1], Maxwell Guimarães de Oliveira[1],
and Hugo Feitosa de Figueirêdo[1]

[1] Information Systems Laboratory, Federal University of Campina Grande, Brazil
[2] Information Technology Coordination, State University of Paraíba, Brazil
andre@uepb.edu.br, {baptista,hugoff}@dsc.ufcg.edu.br,
anderson.firmino@ccc.ufcg.edu.br, maxwell@ufcg.edu.br

Abstract. The widespread of social communication media over the Web has made available a large volume of opinionated textual data. These media constitute a rich source for sentiment analysis and understanding of the opinions spontaneously expressed. The traditional techniques for sentiment analysis are based on POS Tagger and its usage ends up being too costly when considering the Portuguese language, due to the complex grammatical structure of this language. Faced with such problem, this paper presents a case study carried out in order to evaluate a technique developed for sentiment analysis based on Bayesian Classifiers. Although our technique could be applied to any language, the present study focused on tweets posted in Portuguese during the 2013 FIFA Confederations Cup. The achieved results represent one of the best applied to the Portuguese language.

Keywords: Analysis of Sentiment, Temporal Analysis, Tweets.

1 Introduction

Web 2.0 has propitiated a widespread of non-structured information by means of blogs, discussion forums, online product evaluation sites, microblogs and several social networks. This fact brought out new challenges and opportunities in the search and retrieval of information [2].

Microblogging is a communication tool very popular among Internet users [9]. The messages shared by microblogging users are not just about their personal lives, but also about opinions and information about products, people, facts and events in general. The sites providing microblogging services, such as Twitter, become rich sources for mining the user opinions. Since it is a rich source of real-time information, many entities (companies, government, etc.) have showed interest in knowing the opinions of people about services and products.

It is important to notice that opinions about several themes expressed by Web users are made in a spontaneous manner and in real time. The sentiment

H. Decker et al. (Eds.): DEXA 2014, Part I, LNCS 8644, pp. 81–88, 2014.

analysis appears providing the possibility of capturing opinions of the general public, in an automated way, about some theme. It is a recent research field that analyzes people's sentiments, evaluations, attitudes and emotions in the favor of the entities such as products, services, events, organizations and individuals [7]. Advanced techniques from machine learning, information retrieval and natural language processing (NLP) have been used to process large amounts of non-structured content generated by users, mainly in social media [11].

The analysis of opinionated comments expressed in social media requires too much effort to be carried out manually, mainly due to the volume of data, which requires an opinions' summarization. A common way of accomplishing the summarization is by means of the classification of the opinion of an object into categories: positive, negative and neutral. This kind of classification is referred in the literature as sentiment polarity or polarity classification [7].

Twitter is a rich source to understand people's opinions about many aspects of daily life. Performing sentiment analysis on tweets is not a trivial task due to the textual informality of the users. Algorithms that carry out sentiment analysis on tweets usually use NLP techniques, such as part-of-speech tagging. POS Tagging is used to detect subjective messages by identifying the grammatical classes of the words used in the text. POS Tagging tweets is not an easy task too if we consider the plenty of abbreviations used in the text, due to the character limitation of the messages, repeated letters in words to emphasize terms or even the absence of consonants. In texts written in Portuguese, these problems get worse due to the grammatical complexity, inherent of the language.

In this context, the present work exhibits a case study with sentiment analysis applied to tweets written in Portuguese and related to the FIFA's Confederations Cup, occurred in Brazil in 2013. The main objective of this study is to explore an approach that eliminates the need of POS Taggers in the identification of opinionated tweets. To identify the sentiment polarity, our approach uses two Bayesian classifiers. The first one is used to detect whether a tweet presents opinionated content, and the second one is used to classify the subjective polarity of the message as positive or negative.

2   Related Work

Sentiment analysis has been the most active research area in the field of NLP [3]. It has been used in many applications with several purposes, such as analysis of consumers' reviews about products or services, analysis of places or tourism regions by means of the tourists' comments, or analysis of subjects related to politics. Activities related to sentiment analysis comprise the detection of subjective or opinionated content, classification of the content polarity and summarization of the general sentiment of the evaluated entities. Several methods have already been proposed to classify the sentiment polarity of a text [7] [10] [11] and the main approaches used are based on machine learning techniques, semantic analysis techniques, statistical techniques and techniques based on lexical analysis or thesaurus.

The approaches to sentiment analysis that use machine learning implement classification algorithms such as Naive-Bayes, Support Vector Machine (SVM), Maximum Entropy, Decision Trees (C4.5), KNN (K-nearest neighbor) and Condition Random Field (CRF). One of the main limitations in the use of supervised learning is the need for labeled data for training and tests. In order to provide the collection of labeled data in an automated way, many works proposed the use of emoticons - characters that transmit emotions. About 87% of the tweets containing emoticons have the same sentiments represented in the text [6]. One of the main problems in using only emoticons in the collection of the data to train the classifiers is related to the recall metric, since emoticons are present in at most 10% of the tweets [4]. Pak & Paroubek [9] report good results on sentiment classification using the Naive-Bayes classifier and emoticons strategy to build the dataset to train a classifier and categorize tweets as positive or negative based on N-grams and in the grammatical classification of the words of the text by means of POS Tagger. Our work differs from the such work because it does not requires POS Tagger to identify and opinionated (subjective) content.

There is a lack of works in the literature that perform sentiment analysis using a corpus in Portuguese. Chaves et al. [1] and Tumitan & Becker [12] apply lexical analysis techniques based on thesauri, and Nascimento et al. [8] uses machine learning techniques. The first one presentes an algorithm based on both ontologies and a list of polarized adjectives that express sentiments to define the semantic orientation of the analyzed texts.

Faced with this scenario, our approach for sentiment analysis differs from the studied approaches mainly by the use of two classifiers, which eliminates the need of POS Tagger in the identification of an opinionated content. Thus, the first classifier detects whether a content is subjective or objective, and the second classifier identifies the polarity (positive or negative) of the content previously detected as opinionated.

3   Using Naive-Bayes for Classification of Tweets

The Naive-Bayes classifier uses techniques that work with the modeling of uncertainty by means of probabilities, considering the inputs independently. However, the Naive-Bayes presents optimum results even in problem classes that have highly dependent attributes [9]. This occurs due to the fact that the conditional independence of attributes is not a necessary condition for Naive Bayes optimality [5]. The Naive-Bayes algorithm basically brings the same mathematical foundations of the Bayes Theorem. Applying this theorem to the context of tweets classifiers, we have

$$P(c|t) \quad = \quad \frac{P(c)P(t|c)}{P(t)} \tag{1}$$

where $P(c)$ is the occurrence probability of the category, $P(t)$ is the occurrence probability of the tweet, $P(t|c)$ is the occurrence probability of the tweet considering the category occurred, and $P(c|t)$ is the probability of tweet, given that it occurred belonging to the category. The term $P(t|c)$ is computed taking into consideration the conditional probability of occurrence of each word that forms the tweet, since the category has occurred. This term could be written as:

$$P(t|c) \quad = \quad \prod_{1 \leq k \leq n} P(t_k|c) \tag{2}$$

where $P(t_k|c)$ is the probability of the term k occurring given that the category occurred, and n is the tweet length.

4   Case Study Design

We present in this section the overall design of our developed case study. It is subdivided into three subsections: selection of the corpus, which describes the used dataset; sentiment polarity classification of tweets, which addresses the usage of our technique; and the evaluation, which focuses on validation.

4.1   Selection of the Corpus

Since the text in microblogs are essentially informal, many challenges must be taken into account in order to perform the sentiment analysis on tweets: grammatical errors, slang, repeated characters, etc. Thus, it is necessary to deal with the text of a tweet in a specific manner. In general, some treatments for noise and lexical variation are: removal of special words like URLs, twitter user names or stopwords; usage of synonyms for decomposed terms; POS tagging's usage; recognition of entities; reduction of a term to its radical; and treatment of the composite terms forming HashTags, which are normally separated according to the capitalization of the letters.

Using Twitter's API, we collected approximately 300,000 tweets in Portuguese language concerning the theme of FIFA's Confederations Cup, which took place in Brazil, in 2013. We created a search engine that collected, every day, tweets containing at least one of the searched terms. Figure 1 presents a graphic containing the number of tweets obtained by the search terms. The data was collected between April 12th and August 12th in the year of 2013, approximately two months before the beginning and two months after the end of the competition, which occurred between July 15th and 30th, in 2013. The distinction of the collect period is important for a temporal analysis, enabling the perception of possible sentiment - opinion - of the Brazilian people with respect the theme of the cup in Brazil.

4.2   Sentiment Polarity Classification of the Tweets

In order to carry out this study, we needed to create a sentiment classifier. The classification process is presented by means of the diagram shown in Figure 2.

Fig. 1. Number of tweets obtained through query terms

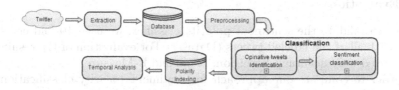

Fig. 2. Process for classification of tweets' Sentiment Polarity

Table 1. Number of Tweets Labelled (Training and Testing Sets)

Approach	Positive	Negative	Neutral	Total
Emoticons	1,468	492	-	1,960
Manual Labeling	326	321	463	1,110

After the data being collected and stored in a database, the tweets were submitted to a pre-processing, that included the treatment of HashTags and the removal of stopwords, special terms and user names.

A sentiment classifier capable of classifying the polarity of a tweet as positive or as negative was implemented. We opted by using the Naive-Bayes classification algorithm because it presents excellent results in classification problems [9]. This option required to have a previous-labeled training dataset. To obtain such dataset we defined a strategy based on two different approaches: Emoticons [9] and Manual Labeling. The first approach assumes that all words in the message have characters that express emotions, such as Happy ":-)", ":)", "=)", ":D" and Sad ":-(", ":(", "=(", ";(" emoticons, are also associated to the emotion of these characters. Thus, happy emoticons could be associated with positive polarity as well as sad emoticons could be associated with negative polarity. On the other hand, in the second approach, 1,500 collected tweets were randomly chosen and separated for manual labeling of the sentiment polarity. Ten volunteers were asked to give their opinions about the sentiments present in these tweets so that, using the majority vote of the discrepant opinions, only those that presented a dominant sentiment would be considered valid for labeling. Both methods for obtainment of sentiments labeling were used in the comparison and combination of results of the classifiers. Table 1 presents the number of tweets with sentiments labeled with basis on each approach.

Using distinct training sets, two binary classifiers were built: a classifier to check whether a tweet is subjective, that is, presents an opinion; and a polarity classifier to distinguish the sentiment as positive or negative. Once the sentiment classifier is trained, all tweets were analyzed and indexed with the opinion polarity obtained by the classifier. Finally, the collected sentiments were summarized by means of a temporal analysis, which enabled us to follow the general orientation of the sentiments expressed by the Brazilian people with respect to the subject of the Confederations Cup. Besides, a word frequency counter was used to detect the main terms cited in the tweets.

4.3   Evaluation

In order to validate the sentiment polarity classifier, we used 10-fold cross validation with all of the labeled tweets (Dataset). For evaluation of the results, the metrics used were accuracy, precision, recall and F-Measure.

In order to compare our approach for sentiment polarity classification of a tweet by using two binary classifiers aforementioned, we implemented another Bayesian classifier, used in other works [8] [9], which considers three possible classes (positive, negative, neutral). Thus, we use the same dataset and compare the achieved results.

5   Results and Discussion

Table 2 presents the results achieved with the developed classifiers. As it can be observed, the best results are obtained by our technique, which uses two binary classifiers, one to verify if a tweet is opinionated, and another one to determine the polarity of the sentiment contained in the tweet.

Table 3 presents a comparison of the datasets used to train the classifiers. We used the proposal of Pak & Paroubek [9] for automatic collection of data through emoticons to train the sentiment classifiers. The results achieved with the classifier by the training carried out with the data resulting from the automatic labeling were compared with the classifier trained with the data manually labeled. It can be noticed, that the classifier that uses the dataset obtained automatically presents the best results.

After building and validating the sentiments classifier and having used all the labeled tweets, the next step was to obtain the general semantic orientation of the sentiments expressed by the Brazilian people regarding the FIFA Confederations Cup. Figure 3a presents the result of the sentiments classifier applied to all of the collected tweets.

Table 2. Comparison of Developed Classifiers

Classifier	Dataset	Accuracy	Class	Precision	Recall	F-Measure
Subjective tweet classification + Polarity Classifier	Emoticons + Manual Labeling	0.777	Positive	0.91	0.742	0.817
			Negative	0.616	0.849	0.714
			Weighted Average	0.813	0.777	0.783
Polarity Classifier	Emoticons + Manual Labeling	0.610	Positive	0.804	0.607	0.692
			Negative	0.535	0.706	0.608
			Neutral	0.427	0.498	0.460
			Weighted Average	0.647	0.610	0.618

Table 3. Comparison of training classifier datasets

Classifier	Dataset	Accuracy	Class	Precision	Recall	F-Measure
Subjective tweet classification + Polarity Classifier	Emoticons + Manual Labeling	0.870	Positive	0.953	0.847	0.897
			Negative	0.748	0.916	0.824
			Weighted Average	0.885	0.870	0.873
	Manual Labeling	0.656	Positive	0.762	0.716	0.738
			Negative	0.469	0.529	0.497
			Weighted Average	0.668	0.656	0.661

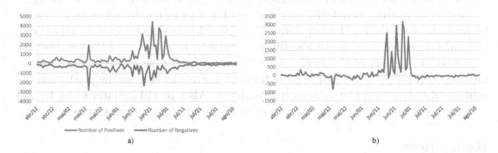

Fig. 3. Number of tweets obtained through query terms: a) Number of positive and negative tweets, and b) Positive polarity tweets during the competition

One way of obtaining the general semantic orientation of the sentiment expressed in the tweets is by subtracting the number of tweets with negative sentiments from the number of tweets with positive sentiments. Figure 3b presents the summary of the semantic orientation of the collected tweets.

The tool developed in the present work enables the decision maker to analyze the words more frequently used in a selected time interval, thus helping to identify possible dissatisfactions expressed in the messages or even complements about the evaluated objects. Figure 4a presents a cloud of words with the most frequent terms used in the tweets of 2013-05-14. Figure 4b presents a cloud of words obtained during the games played by the Brazilian team, in which the semantic orientation of the tweets was positive.

Fig. 4. Clouds of the most frequent terms in the generated tweets : a) negative polarity tweets in 05/14/2013, and b) positive polarity tweets during the competition

6   Conclusion and Further Work

We have presented a case study using an approach developed for detection of the polarity of the opinion in a message on Twitter, based on two Bayesian classifiers. The approach used discarded the necessity of using a POS Tagger for identification of opinionated tweets.

The results obtained by the sentiment classifier indicated an F-Measure of 0.783 and an accuracy of 74.2% for detection of sentiment polarity, which represent an excellent result for tweets in Portuguese, specially if we consider that the polarity of subjective content is not always consensual. For example, in annotations made by humans, consensus is hardly above 75% [10]. Moreover, a temporal analysis on the data was made aiming to identify the semantic orientation of the sentiments expressed by means of the tweets and the more cited terms in the opinionated messages. A future work will explore temporal series to help in the prediction of sentiments according to the detected tendency. Also, it will be interesting to investigate relationships across Twitter entities aiming to improve accuracy.

References

1. Chaves, M.S., de Freitas, L.A., Souza, M., Vieira, R.: PIRPO: An Algorithm to Deal with Polarity in Portuguese Online Reviews from the Accommodation Sector. In: Bouma, G., Ittoo, A., Métais, E., Wortmann, H. (eds.) NLDB 2012. LNCS, vol. 7337, pp. 296–301. Springer, Heidelberg (2012)
2. Eirinaki, M., Pisal, S., Singh, J.: Feature-based opinion mining and ranking. Journal of Computer and System Sciences 78(4), 1175–1184 (2012)
3. Feldman, R.: Techniques and applications for sentiment analysis. Comm. of the ACM 56(4), 82–89 (2013)
4. Gonçalves, P., Araújo, M., Benevenuto, F., Cha, M.: Comparing and combining sentiment analysis methods. In: Proceedings of the 1st ACM COSN, pp. 27–38. ACM Press, New York (2013)
5. Kuncheva, L.I.: On the optimality of naive bayes with dependent binary features. Pattern Recogn. Lett. 27(7), 830–837 (2006)
6. Li, Y.M., Li, T.Y.: Deriving Marketing Intelligence over Microblogs. In: Proceedings of the 44th HICSS, pp. 1–10. IEEE, Kauai (2011)
7. Liu, B.: Sentiment Analysis and Opinion Mining. Synthesis Lect. on Human Language Technologies, vol. 5. Morgan & Claypool Publishers (2012)
8. Nascimento, P., Aguas, R., Lima, D.D., Kong, X., Osiek, B., Xexéo, G., de Souza, J.: Análise de sentimento de tweets com foco em notícias. In: Proceedings of the CSBC 2012. SBC, Curitiba (2012)
9. Pak, A., Paroubek, P.: Twitter as a Corpus for Sentiment Analysis and Opinion Mining. In: Proceedings of the Seventh Conference on International Language Resources and Evaluation, LREC 2010 pp. 1320–1326 (2010)
10. Pang, B., Lee, L.: Opinion Mining and Sentiment Analysis. Foundations and Trends in Information Retrieval 2(2), 1–135 (2008)
11. Sharma, A., Dey, S.: A comparative study of feature selection and machine learning techniques for sentiment analysis. In: Proceedings of the 2012 ACM Research in Applied Computation Symposium, RACS 2012, p. 1. ACM Press (2012)
12. Tumitan, D., Becker, K.: Tracking Sentiment Evolution on User-Generated Content: A Case Study on the Brazilian Political Scene. In: SBBD 2013 pp. 1–6 (2013)

Searching for Local Twitter Users by Extracting Regional Terms

Takuji Tahara and Qiang Ma

Kyoto University, Kyoto, Japan 606-8501
{tahara@db.soc.,qiang@}i.kyoto-u.ac.jp

Abstract. Twitter users who tweet daily information on the local area are valuable information sources for location-based services. In this paper, we propose a novel method for searching for such local Twitter users on the basis of the linguistic features of their tweets. At first, by analyzing and comparing tweets describing different local areas, we extract regional terms of a target area from three aspects: the regional distributions of terms, the regional distribution of users whose tweets contain those terms, and the temporal distribution of tweets. Then, we construct a local area feature vector to represent the target area and compare it with a user vector to find local users. The experimental results validate our methods.

Keywords: Microblog, Twitter, Local information, Regional terms.

1 Introduction

Recently, the number of users of microblogging services has increased extremely. *Twitter*[1] is one of the most popular microblogging services. It enables its users to post a status update message in 140 characters, called a *tweet*. A lot of Twitter users post their daily experiences in real time. Some of these experiences contain useful daily information for local users, such as that *shop A* is more crowded than usual, there is a small special event, an incident, and so on. It is useful for local users to know this information in real time for their daily lives. The location and time of tweets are necessary data for obtaining such local information from Twitter. Each tweet contains its time information. However, it is not easy to obtain the location where the users tweeted. Although Twitter provides a geo-tag service, there are only 0.77%[2] tweets containing geo-tag information.

Sakaki et al. [1,2] showed that there are enough geo-tagged tweets about hot information such as the movement of earthquakes or celebrities, in which a lot of people are interested. However, in most cases, only local users are interested in daily information about the area they live in. Therefore, there are not enough geo-tagged tweets for such information. One considerable solution is to search for Twitter users who post local information and then obtain valuable information from these users' tweets. In this paper, we propose a method for searching for

---

[1] https://twitter.com
[2] http://semiocast.com

H. Decker et al. (Eds.): DEXA 2014, Part I, LNCS 8644, pp. 89–96, 2014.

local Twitter users by using regional terms extracted from tweets. Here, regional terms of a certain area denote the feature terms that can distinguish the local area and the Twitter users living there from its adjacent areas. Also, we call Twitter users who tweet daily information on a certain area *local users*.

The major contributions can be summarized as follows.

- We propose a novel notion of *localness* to extract regional terms by considering the regional distributions of terms, the regional distribution of users whose tweets containing those terms, and the temporal distribution of tweets. We treated terms with high localness as regional terms (Section 3.1). To the best of our knowledge, this is the first attempt at extracting regional terms by considering all three kinds of distributions.

- We propose a method for searching for local Twitter users by using the notion of localness. We first construct vectors of a local area and a candidate local user on the basis of regional terms. Second, we calculate the similarity between the vectors of the local area and the candidate user. We treat a user with high similarity as the local Twitter user of the target area (Section 3.2).

- We carry out experiments with real data to validate the proposed methods. The experimental results show that our method can be applied to discover local information from tweets in real time (Section 4).

2   Related Work

In addition to [1,2], there has been much research on event detection from Twitter in recent years.

Tsuchiya et al. [3] proposed a method of extracting information on railway troubles from tweets including the names of train lines by using SVM. Matsumura et al. [4] proposed a method of detecting events from tweets including place names by considering the number of tweets per unit of time. [3,4] can detect only events shown by tweets that include the names of places or lines. Most of those events are big-scale and interesting to people living in several areas.

Watanabe et al. [5] proposed a method of discovering events from tweets through Foursquare[3] and from tweetsincluding place names written on those tweets. Sugitani et al. [6] proposed a method of extracting events by using geo-tagged tweets in constant time. They cluster users who posted these tweets in nearby places and then discover events by extracting terms often appearing in one cluster. [5,6] detected relatively small-scale events. However, their methods detect few events in a small living area depending on the location data given by the users explicitly.

Several studies tried to estimate the living area of Twitter users by analyzing the contents of the users' tweets. Cheng et al. [7] proposed a method that uses a probability model based on the correlation of terms in tweets and a local area and a lattice-based neighborhood smoothing model. They tried to estimate users' location on the city level in the United States. Doumae et al. [8] proposed

---

[3] https://foursquare.com/

a method for making the topic of tweets of a living area by using LDA. They tried to estimate users' location on the prefecture level in Japan. We propose a method that estimates a smaller user living area than that in these studies.

Hasegawa et al. [9] made a co-occurence dictionary of a particular place and terms on the basis of the tf-idf method. By integrating similar dictionaries, they expanded the area and the period covered by the dictionary.

Nishimura et al. [10] proposed a method of estimating a user's location by using regional terms. They extract regional terms on the basis of the tf-idf method by treating tweets of one area as one sentence. However, we consider that tweets about a living area were posted by several people of the area several times.

3   Method of Searching for Local Twitter Users

3.1   Extracting Regional Terms

Regional terms are ones that users in a certain area more often post than other areas or terms that only users in the area often post. The step of extracting regional feature terms are : 1) Extracting candidate terms by morphological analysis of tweets, 2) Calculating the localness of candidate terms and treating high localness terms as regional terms.

Extraction of Candidate Terms from Tweets. We use only nouns and unknown words as candidates extracted from tweets by using MeCab[4], which is a morphological analysis tool for Japanese texts. However, as Li et al. [11] showed, because tweets are limited to 140 characters or less, there are grammatical errors or abbreviations of words. Terms extracted by conventional morphological analysis tools are not always the same as terms intended by authors. Therefore, we use the word-class N-gram method to extract the candidate terms.

Localness. The regional terms of a certain local area may appear more frequently in tweets of all users living in the area than the others or only in tweets of all users in the area. Tweets of an area are not written by one person at a time but by several people at different times. Therefore, we calculate the localness of a term by considering the following indications.

1. **Term Distribution:** The frequency of the term's appearance in the target local area by comparing with other areas
   We define the frequency of a term's appearance $tf(t_i, C_j)$ as the number of tweets containing the term t_i, and these tweets should be tweeted by the users in local area C_j. However, a term with high tf is not always a regional term. Most terms with high tf are often posted regardless of the areas. Regional terms should have higher tf in the target area than the others. Then, we modify the frequency in a relative manner as follows.

$$rtf(t_i, C_0) = \frac{tf(t_i, C_0)}{\frac{1}{|C|} \sum_j tf(t_i, C_j)} \tag{1}$$

---

[4] https://code.google.com/p/mecab/

where C_j is $localarea_j$, C_0 is the target local area, and $|C|$ is the total number of local areas.

2. **User Distribution:**
   - Whether users in other areas post the term
   Regional terms may be posted by users of the target local area and general terms by users of all the local areas. Then, we use formula (2) to consider whether users of the other areas post.

$$icf(t_i) = \frac{|C|}{cf(t_i)} \tag{2}$$

where $cf(t_i)$ is the number of local areas where users post a tweet including term t_i.
   - The number of users in the target area posting a term
   Regional terms should be representative ones in the target area. Terms with high rtf and icf posted by few users in the target area are not representable as regional terms. The representative feature based on the user distribution is formulated as follows.

$$uc(t_i, C_0) = \frac{u(t_i, C_0)}{|U_{C_0}|} \tag{3}$$

where $u(t_i, C_0)$ is the number of users of area C_0 posting a tweet including term t_i and $|U_{C_0}|$ is the number of users of area C_0.

3. **Temporal distribution:** How many days users in the target area post terms
   Terms with high rtf, icf, and uc may be regional terms. However, there are many cases in which these terms posted in a short period represent sudden incidents or events. To exclude such terms, we use formula (4) to consider the temporal features.

$$dc(t_i, C_0) = \frac{d(t_i, C_0)}{|D|} \tag{4}$$

where $dc(t_i, C_0)$ is how many days users in area C_0 posted a tweet including term t_i and $|D|$ is the number of total days.

Finally, we calculate the localness of term t_i in area C_0 as below.

$$loc(t_i, C_0) = rtf(t_i, C_0) * icf(t_i) * uc(t_i, C_0) * dc(t_i, C_0) \tag{5}$$

The terms with high localness (greater than a pre-specified threshold) are regarded as regional terms.

3.2 Searching for Local Twitter Users

Users often posting regional terms have high probabilities of being the local users of the target area. The steps of searching for local users are : 1) Constructing

Table 1. Number of Twitter users for each area

Area around Kyoto University	349	Yamashina-ku	315
Kita-ku	389	Shimogyo-ku	331
Kamigyo-ku	346	Minami-ku	335
Sakyo-ku	421	Ukyo-ku	350
Nakagyo-ku	399	Nishigyo-ku	322
Higashiyama-ku	316	Hushimi-ku	365

vectors of the target local area and a candidate local user on the basis of the regional feature terms (In this paper, the former is called a "local vector" and the latter a "user vector."), 2) Calculating Cosine similarity of the local vector and the user vector, 3) Selecting as a local user the user whose vector is similar to the local vector. Candidate users are collected on the basis of regional terms.

4   Experiments

We conducted evaluation experiments on the proposed method of extracting regional feature terms and searching for local users. In these experiments, we use the area around Kyoto University as the target local area and eleven districts of Kyoto City as the other areas. Experimental data are tweets posted by users of each area from October 1st to November 30th. The total number of tweets is 1,843,188, and the number of users of each area is about 350 - 400. We used Kyoto University students as users of the area around Kyoto University. Users of the other areas were collected by using conventional services of searching for Twitter users (Twipro[5] and Twitter Profile Search[6]).

4.1   Evaluation on Extracting Regional Feature Terms

We ranked regional terms by using the proposed method. To evaluate this method, normalized discounted cumulative gain (nDCG) was applied to the top k (k = 100, 500, 1000) results. We compared the proposed method with the following three methods (see also Formulas (1), (2), (3), and (4) in Section 3.1).

- Method B1 : Ranking terms by considering only $tf(t_i, C_0)$

$$loc_{B1}(t_i, C_0) = tf(t_i, C_0) \tag{6}$$

- Method B2 : Ranking terms by considering $tf(t_i, C_0)$ and $icf(t_i)$

$$loc_{B2}(t_i, C_0) = \log_2(1 + tf(t_i, C_0)) * \log_2(1 + icf(t_i)) \tag{7}$$

- Method B3 : Ranking terms by considering relative intra-document TF (RITF) introduced by Paik [12], $icf(t_i)$, $uc(t_i, C_0)$, and $dc(t_i, C_0)$

$$loc_{B3}(t_i, C_0) = ritf(t_i, C_0) * icf(t_i) * uc(t_i, C_0) * dc(t_i, C_0) \tag{8}$$

---

[5] http://twpro.jp/
[6] http://tps.lefthandle.net/

Table 2. nDCG score of top k of ranking with each method

k	Proposed method	Method B1	Method B2	Method B3
100	**0.884**	0.429	0.371	0.415
500	**0.836**	0.462	0.468	0.469
1000	**0.801**	0.472	0.523	0.490

$ritf(t_i, C_0)$ is the normalization of $tf(t_i, C_0)$, and calculated as below.

$$ritf(t_i, C_0) = \frac{\log_2(1 + tf(t_i, C_0))}{\log_2(1 + \frac{1}{|T_{C_0}|} \sum_j tf(t_j, C_0))} \qquad (9)$$

where $|T_{C_0}|$ is the number of vocabulary words taken out from tweets posted by users of C_0.

Table 2 reports the nDCG scores (k = 100, 500, 1000) of the ranking by each method. These results show that the nDCG scores of the proposed method were about 0.3 higher than those of the other methods. Therefore, the proposed method was effective in extracting regional feature terms.

4.2   Experiment on Searching for Local Twitter Users

In this experiment, we used 100 users per area as test data and the remaining users as training data. Also, we call users of the target area "positive users" and users of the other areas "negative users." As mentioned in 3.2, classification of local Twitter users was conducted basically on the basis of the cosine similarity of a local vector and a user vector. We use three methods of constructing a user vector and four methods of constructing local vector, respectively. Combining these methods, we tested twelve classification methods.

The methods of constructing a user vector $u : v_u = (w_{u1}, w_{u2}, ...w_{u1000})$ are the following three methods.

$$\text{Method } UV_b : w_{ui} = \begin{cases} 0 \text{ if } utf(h_i, U) = 0 \\ 1 \text{ else} \end{cases}$$

$$\text{Method } UV_f : w_{ui} = utf(h_i, u)$$

$$\text{Method } UV_{fd} : w_{ui} = utf(h_i, u) * \frac{udc(h_i, u)}{|D|}$$

where $utf(h_i, u)$ is the number of tweets including term h_i posted by user u_i and $udc(h_i, u)$ denotes how many days u posted tweets including h_i.

The methods of constructing a local vector are the following four methods.

- Method CV_{loc} : Features of the local vector are the top 1000 regional terms of the target area. These terms are extracted from training data by using the proposed method mentioned in 3.1. The weight of each feature is the localness score of the corresponding term calculated by *loc*.

Table 3. Recall, precision, and F-measure of each classification of users

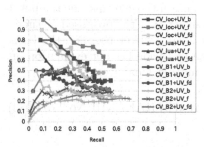

Fig. 1. PR curves of all classifications

Method	Recall	Precision	F-measure
$CV_{loc} + UV_b$	0.560	0.403	**0.469**
$CV_{lua} + UV_b$	0.550	0.320	0.404
$CV_{B1} + UV_b$	0.550	0.314	0.400
$CV_{B2} + UV_b$	0.540	0.195	0.286
$CV_{loc} + UV_f$	0.580	0.542	**0.560**
$CV_{lua} + UV_f$	0.510	0.395	0.445
$CV_{B1} + UV_f$	0.510	0.481	0.495
$CV_{B2} + UV_f$	0.570	0.291	0.385
$CV_{loc} + UV_{fd}$	0.450	0.441	**0.446**
$CV_{lua} + UV_{fd}$	0.630	0.251	0.359
$CV_{B1} + UV_{fd}$	0.50	0.379	0.431
$CV_{B2} + UV_{fd}$	0.690	0.220	0.334

– Method CV_{lua} : In the same way as method CV_{loc}, the features of the local vector are the top 1000 regional terms of the target area. The weight of each feature value is calculated by using formula (10) as follows.

$$w_{c_0 i} = \frac{1}{|L_{C_0}|} \sum_{j=1}^{|L_{C_0}|} w_{l_j i} \tag{10}$$

where $c_0 = (w_{c_0 1}, w_{c_0 2}, ... w_{c_0 n})$ is the local vector of the target area C_0, $v_l = (w_{l1}, w_{l2}, ... w_{ln})$ is the vector of a positive user in training data, and $|L_{C_0}|$ is the number of positive users in training data. To be brief, the local vector is the average of vectors of positive users in training data.

– Method CV_{B1} : Features of the local vector are the top 10,000 terms with high tf in tweets posted by positive users in training data. The weights of the local vector are made in the same way as method CV_{lua}.

– Method CV_{B2} : Features of the local vector are made by collecting the top 1000 terms with high tf of each positive user in training data. The weights of the local vector are made in the same way as method CV_{lua}.

Features of user vectors are the same as those of the local vector when using CV_{loc}, CV_{lua}, and CV_{B1}. When using CV_{B2}, however, features of user vectors are the top 1000 terms with high tf of each user. The thresholds used for classification methods using CV_{loc} are decided on the basis of pre-investigation and those for the other methods are those with the highest F-measure. Table 3 shows the recall, precision, and F-measure of each classification method. Figure 1 shows the PR curves of all the classification methods. These show that $CV_{loc} + UV_f$ achieved the best performance and classification methods using CV_{loc} achieved the best performance in any case of using UV_b, UV_f, and UV_{fd}.

5   Conclusion

In this paper, we proposed a method of searching for local Twitter users by using regional terms. For extracting regional terms, we proposed a novel notion of the localness of terms calculated by considering the regional distributions of terms, the regional distribution of users whose tweets contain those terms, the temporal distribution of tweets. The experiments results shows that the proposed methods of extracting regional terms and searching for local Twitter users on the basis of the proposed localness are effective. We plan to tackle further evaluation on extracting regional terms and experiments on another local area.

Acknowledgement. This work is partly supported by KAKENHI(No.25700033) and SCAT Reseach Funding.

References

1. Sakaki, T., Okazaki, M., Matsuo, Y.: Earthquake Shakes Twitter Users: Real-time Event Detection by Social Sensors. In: Proceedings of the 19th International Conference on World Wide Web, pp. 851–860 (2010)
2. Sakaki, T., Matsuo, Y.: Research on Real-time Event Detection from Social Media Prototype of Sighting Information Detection System. In: Proceedings of the 25th Annual Conference of the Japanese Society for Artificial Intelligence (2011) (in Japanese)
3. Tsuchiya, K., Toyoda, M., Kitsuregawa, M.: Extracting Details of Train Troubles from Microblogs. IEICE Technical Report 113(150), 175–180 (2013) (in Japanese)
4. Matsumura, T., Yasumura, M.: Proposal and prototyping of Twitter crowling and analysing system for town information. In: Proceedings of IPSJ interaction (2010) (in Japanese)
5. Watanabe, K., Ochi, M., Okabe, M., Onai, R.: Real-World Micro-Event Detection on Twitter. In: Proceedings of the 4th Rakuten R&D Symposium (2011) (in Japanese)
6. Sugitani, T., Shirakawa, M., Hara, T., Nishio, S.: Detecting Local Events by Analyzing Spatiotemporal Locality of Tweets. In: Proceedings of Advanced Information Networking and Applications Workshops, pp. 191–196 (2013)
7. Cheng, Z., Caverlee, J., Lee, K.: You Are Where You Tweet: A Content-based Approach to Geo-locating Twitter Users. In: Proceedings of the 19th ACM International Conference on Information and Knowledge Management, pp. 759–768 (2010)
8. Doumae, Y., Seki, Y.: Twitter User's Life Area Estimation Using Biased Topics Reflecting Area. IPSJ SIG Technical Report 2013(8), 1–6 (2013) (in Japanese)
9. Hasegawa, K., Ma, Q., Yoshikawa, M.: Trip Tweets Search by Considering Spatiotemporal Continuity of User Behavior. In: Liddle, S.W., Schewe, K.-D., Tjoa, A.M., Zhou, X. (eds.) DEXA 2012, Part II. LNCS, vol. 7447, pp. 141–155. Springer, Heidelberg (2012)
10. Nishimura, H., Suhara, Y., Susaki, S.: Twitter User's Residence Estimation Using Multi-class Classification with Area-specific Term Selection. IEICE Technical Report 112(367), 23–27 (2012) (in Japanese)
11. Li, C., Sun, A., Weng, J., He, Q.: Exploiting Hybrid Contexts for Tweet Segmentation. In: Proceedings of the 36th International ACM SIGIR Conference on Research and Development in Information Retrieval, pp. 523–532 (2013)
12. Paik, J.H.: A Novel TF-IDF Weighting Scheme for Effective Ranking. In: Proceedings of the 36th International ACM SIGIR Conference on Research and Development in Information Retrieval, pp. 343–352 (2013)

Location-Aware Tag Recommendations for Flickr

Ioanna Miliou[1] and Akrivi Vlachou[2]

[1] National Technical University of Athens, Greece
[2] Institute for the Management of Information Systems, R.C. "Athena", Greece
ioannamiliou@hotmail.com,avlachou@imis.athena-innovation.gr

Abstract. Flickr is one of the largest online image collections, where shared photos are typically annotated with tags. The tagging process bridges the gap between visual content and keyword search by providing a meaningful textual description of the tagged object. However, the task of tagging is cumbersome, therefore tag recommendation is commonly used to suggest relevant tags to the user. Apart from textual tagging based on keywords, an increasing trend of geotagging has been recently observed, as witnessed by the increased number of geotagged photos in Flickr. Even though there exist different methods for tag recommendation of photos, the gain of using spatial and textual information in order to recommend more meaningful tags to users has not been studied yet. In this paper, we propose novel location-aware tag recommendation methods and demonstrate the effectiveness of our proposed methods.

1 Introduction

Flickr allows users to upload photos, annotate the photos with tags, view photos uploaded by other users, comment on photos, create special interest groups etc. Currently, Flickr stores one of the largest online image collections with more than 8 billion photos (March 2013[1]) from more than 87 million users and more than 3.5 million new images uploaded daily. The tags are important for users to retrieve relevant photos among the huge amount of existing photos. Since multimedia data provide no textual information about their content, tags bridge the gap between visual content and keyword search by providing a meaningful description of the object. Thus, to make their photos searchable, users are willing to annotate their uploaded images with tags [2]. Nevertheless, tags reflect the perspective of the user that annotates the photo and therefore different users may use different tags for the same photo. This can be verified by the fact that photos of Flickr that depict the same subject may be described by a variety of tags. Tag recommendation [7] is commonly used to provide to the user relevant tags and enrich the semantic description of the photo.

Flickr motivates its users to geotag their uploaded photos[2]. Geotagging means to attach to a photo the location where it was taken. Photos taken by GPS-enabled cameras and mobile phones are geotagged automatically and location

---

[1] http://www.theverge.com/2013/3/20/4121574/flickr-chief-markus-spiering-talks-photos-and-marissa-mayer
[2] http://www.flickr.com/groups/geotagging/

H. Decker et al. (Eds.): DEXA 2014, Part I, LNCS 8644, pp. 97–104, 2014.

metadata, such as latitude and longitude, are automatically associated with the photos. Furthermore, photos may be also geotagged manually by the user when the photo is uploaded. Currently, there is an increasing trend in the number of geotagged photos in Flickr. Even though several recent studies [3] examine how relevant web objects can be retrieved based on both the spatial and textual information, the gain of using spatial information in order to recommend more meaningful tags to users has not been studied yet. Nevertheless, it is expected that nearby photos may depict similar objects, thus sharing common tags with higher probability. In this paper, we propose methods for tag recommendations based on both location and tag co-occurrence of the photos.

In details, we have created different data collections of geo-tagged photos of Flickr that are located in different cities and in Section 2 we describe our data collections. Then, Section 3 presents an overview of the location-aware tag recommendations system and describes the proposed location-aware tag recommendation methods. Moreover, we implemented a prototype system for location-aware tag recommendations over photos of Flickr and in Section 4 we demonstrate the effectiveness of location-aware tag recommendation through examples. Finally, in Section 5 we discuss related work and in Section 6 we provide some concluding remarks.

2   Data Collection

We have created three different data collections. Each of them contains 100.000 geotagged photos that are located in New York, Rome and London respectively. The collected photos are a random snapshot of the geotagged photos located in the aforementioned cities. For each city the boundary is defined by the bounding box provided at http://www.flickr.com/places/info/. The photos were collected between December 2012 and February 2013 and each photo has at least one tag describing it.

Our data collection of photos collected from Flickr located in New York contains 100.000 photos, with 1.502.454 tags in total, while the unique tags are 80.180. The photo collection of Rome has 897.185 tags in total and the unique tags are 41.843. Finally, the data collection of London has 1.428.047 tags in total and the unique tags are 110.231. By observing the statistical properties of the tags we conclude that the most popular tags should be excluded by our recommendation method because these tags are too generic to be helpful for recommendation. The popular tags include tags such as: NYC, New York, Rome, Italy, London, UK. Similar, the less popular tags with very small frequency (i.e., equal to 1) should be also excluded by our recommendation method, since these tags include words that are misspelled, complex phrases and very specific tags. For example consider the tags: drwho, loo, boring, SF, #noon, dv. Due to their low frequency it is expected that those tags can be useful only in very specific cases and thus are not suitable for recommending to other photos.

3   Recommendation Methods

In this section we describe our recommendation methods. The input of our methods is a photo p that is described by a location given by the owner of the photo and a set of tags $\{t_1, t_2, \dots\}$. The goal is to recommend to the use a set of relevant tags $\{t_1', t_2', \dots\}$ that could augment the description of p. Our methods rely on *tag co-occurrence*, i.e., the identification of tags frequently used together to annotate a photo. Furthermore, we enhance tag recommendation by taking explicitly into account the location of photos, in order to derive more meaningful co-occurring tags.

3.1   System Overview

Figure 1 gives a crisp overview of our location-aware tag recommendation system. Our system is built on an existing collection of photos that are geotagged, such as a subset of geotagged photos provided by Flickr. This information is necessary in order to identify frequently occurring tags, as well as to discover keywords that are used together as tags in many photos.

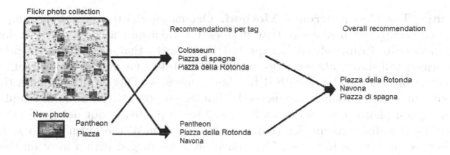

Fig. 1. System overview

We adopt a two-phase approach: in the first phase, a set of frequently co-occurring tags is discovered for each input tag $\{t_1, t_2, \dots\}$, while in the second phase, these sets of tags are combined to produce the final tag recommendation. In more details, for each given tag t_i a ranked list of n relevant tags to t_i is retrieved based on the tag co-occurrence and the distance between the given photo and the photos in which the tags co-occur. Each tag is associated with a score that expresses its relevance to given tag t_i. Then, in the second phase, the different lists of relevant tags are combined, by aggregating their partial scores, so that the k most relevant tags are recommended to the user.

Even though different aggregation functions are applicable, we employ a plain strategy of summing the partial scores. Thus, for each tag t_i', the overall score is defined as the sum of its scores in the ranked lists. Our goal is to produce more qualitative recommendations, by taking into account the location of the photo as well as the location of the existing tags.

3.2 Tag Recommendation Methods

We employ three different tag recommendation methods: (a) simple tag co-occurrence, (b) range tag co-occurrence, and (c) influence tag co-occurrence. The first method is location-independent and is used as a baseline, while the other two are novel, location-aware methods for tag recommendation.

Simple Tag Co-occurrence Method (Baseline). The simplest way to measure the relevance of an existing tag to a given tag is tag co-occurrence. Assuming that t_i is the given tag and t_j an existing tag, then we denote $\mathcal{P}_i$ (or $\mathcal{P}_j$) the sets of photos in which tag t_i (or t_j) appear. To compute the co-occurrence of tags t_i and t_j, we need a metric for set similarity. One commonly used metric to express the similarity based on co-occurrence is the Jaccard coefficient, which is defined as the size of the intersection of the two sets divided by the size of their union. Thus, for tags t_i and t_j, the Jaccard similarity is defined as:

$$Jaccard(t_i, t_j) = \frac{|\mathcal{P}_i \cap \mathcal{P}_j|}{|\mathcal{P}_i \cup \mathcal{P}_j|}.$$

Range Tag Co-occurrence Method. One major shortcoming of the simple tag co-occurrence method is that it does not take into account the location of the photo. Intuitively, it is expected that photos that are taken at nearby locations will share common tags, while photos taken far away from each other are less probable to be described by they same tags. This intuition guides the design of both location-aware methods that we propose. Given a radius r and a geo-tagged photo p, we define as $\mathcal{R}(p)$ the set of photos in our data collection that have a distance smaller than r to the location of the given photo p. In other words, photos in the set $\mathcal{R}(p)$ have been geo-tagged with a location that is within distance r from the location of the input photo p. Then, we define a novel measure that combines tag co-occurrence with location information:

$$Range(t_i, t_j) = \frac{|\mathcal{P}_i \cap \mathcal{P}_j \cap \mathcal{R}(p)|}{|\mathcal{P}_i \cup \mathcal{P}_j|}.$$

In this way, for tag co-occurrence, we take into account only the pairs of photos in which both tags appear and are geo-tagged withing a distance r. On the other hand, we divide with the total number of photos in which at least one of the tags appears, thus giving a penalty to tags that appear very often in photos that are distant to each other (i.e., outside the range r).

Influence Tag Co-occurrence Method. One drawback of range tag co-occurrence method is that a radius r needs to be defined as input, and it is not always straightforward how to set an appropriate value, without knowing the distribution of the locations of existing photos. Moreover, the defined range enforces a binary decision to whether a photo will be included or not in the tag co-occurrence computation, based on its distance being above or below the value r. For example, a very small value of radius may result in no photos with the given tag being located into the range, while on the other hand a large radius may result in most (or all) of the photos being located inside the range. Summarizing, the recommended tags are quite sensitive to the value of the radius, which is also hard to define appropriately.

To alleviate this drawback, we propose also a more robust and stable method than the plain range tag co-occurrence method. Given a radius r and a geo-tagged photo p, we define the *influence score* of two tags t_i and t_j as:

$$inflscore(t_i, t_j) = \sum_{p' \in \mathcal{P}_i \cap \mathcal{P}_j} 2^{\frac{-d(p',p)}{r}}$$

where $d(p', p)$ is the distance between the locations of p and p'. Then the relevance of a given tag t_i and an existing tag t_j is computed as:

$$Influence(t_i, t_j) = \frac{inflscore(t_i, t_j)}{|\mathcal{P}_i \cup \mathcal{P}_j|}.$$

The key idea behind the influence score is that tags that co-occur in nearby photos have a higher influence than tags that co-occur in distant photos. This is nicely captured in the above definition by the exponent, which gradually decreases the contribution of any photo p' the further it is located from p. Compared to the range tag co-occurrence method, this method does not enforce a binary decision on whether a photo will contribute or not to the score. Also, even though a radius r still needs to be defined, this practically has a smoothing effect on the influence score (rather than eliminating some photos), thus the score is not very sensitive to the value of r.

4 Empirical Study

In this section, we provide examples of the proposed recommendation methods of Section 3. To this end, we take into account also the conclusions drawn in Section 2. Therefore, to avoid tags that are too generic to be helpful for recommendation, we exclude from the recommendation tags that appear in more than 10% of the photos. Also, we remove from our photo collection photos that have more than 30 tags, as these tags cannot be considered to be representative for the photo. Moreover, photos that have only one tag cannot be used for tag recommendation that rely on co-occurrence of tags, therefore such photos are also removed from the photo collections.

In order to measure the distance between two photos, we convert the longitude and latitude of each photo to the Universal Transverse Mercator (UTM) projected coordinate system. Then, we apply the Euclidean distance in this transformed space.

Table 1. New York Harbor (Baseline recommends: "Newtown Creek", "Maspeth, New York", "DUGABO")

Radius	Photos	Range	Influence
500	1098	Frederic Bartholdi, nite, lens adapters	One New York Plaza, Statue of Liberty, Harbor
1000	3828	One New York Plaza, Harbor Statue of Liberty	One New York Plaza, Statue of Liberty, Harbor
1500	6117	One New York Plaza, Harbor, Statue of Liberty	Liberty Island, Statue of Liberty, Harbor
2000	8816	Harbor, One New York Plaza, Statue of Liberty	Liberty Island, Staten Island Ferry, Statue of Liberty

Table 2. Rome at Piazza della Rotonda (radius=100)

Query	Baseline	Range	Influence
Piazza	Navona, spagna, popolo	pantheon, Rotonda, della	pantheon, Navona, Rotonda
pantheon	colosseum, piazza di spagna, Piazza della Rotonda	Piazza della Rotonda, temple, Dome	Piazza della Rotonda, temple, Dome
Piazza and pantheon	Navona, spagna, popolo	Piazza della Rotonda, temple, Dome	Piazza della Rotonda, temple, Dome

In our first example we use the New York data collection. Assuming a user that uploads to Flickr a photo taken at the Battery Park $(40.703294, -74.017411)$ in the Lower Manhattan of New York. The user gives one tag to the photo namely "New York Harbor". Figure 2 shows our prototype system for this query. The recommendation results are shown in Table 1. In this example we study how the radius influences our two approaches, while the Baseline fails to recommend relevant tags. We notice that Range is more sensi-

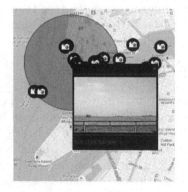

Fig. 2. Example of recommendation

tive to the radius than Influence. Table 1 shows also the number of photos that fall into the region of radius r. This explains the behavior of Range, as for small radius values there exist too few photos to make meaningful recommendations.

In the next example (Table 2) we study the case of a photo that is annotated by 2 tags before the tag recommendation. We use the Rome data collection and we assume that the photo is taken at Piazza della Rotonda in front of Pantheon (41.899134, 12.47681). We set the radius equal to 100 since in the historical center of Rome there are many nearby photos. Location-aware tag recommendation manages to give relevant tags also for generic terms like "Piazza". For "Piazza" and "pantheon" query, the Baseline returns the same results as "Piazza" because there is a higher co-occurrence between this tag and the others, while for the location-aware approaches the results are the same as "pantheon" because there are more photos with this tag nearby the given location.

Table 3. "Buckingham Palace" and "park"

	Baseline	Range	Influence
1	hyde	roadrace	the mall
2	Green Park	Piccadilly London	Green Park
3	the mall	Road Race Cycling	st james park'
4	Constitution Hill	the mall	Piccadilly London

Finally, we examine another example in which 2 tags are given ("Buckingham Palace" and "park"). This time we use the London data collection and the photo is located on the Birdcage Walk in front of the St. James's Park (51.501011, −0.133268). The radius is set to 500 and the results are depicted in Table 3. This example tries to illustrate a hard case, as one of the tags (i.e, "Buckingham Palace") is not directly related to the location and the other tag (i.e., "park") is quite generic. We notice that Range fails to return "St. James's Park" as a recommended tag, which is probably the most related term based on the location, but still both Range and Influence manage to recommend more relevant tags than the baseline.

5   Related Work

Automatic tag recommendation in social networks has emerged as an interesting research topic recently [8]. Especially in the case of Flickr, tag recommendation has been studied in [7,4]. In more details, [7] presents different tag recommendation strategies relying on relationships between tags defined by the global co-occurrence metrics. On the other hand, in [4] tag recommendation methods are studied that are personalized and use knowledge about the particular user's tagging behavior in the past. Nevertheless, none of the above methods takes into account the locations of photos. SpiritTagger [5] is a geo-aware tag suggestion tool for photos, but the proposed approach relies on the visual content (such as global color, texture, edge features) of the photo and on the global and local tag distribution. In contrast, our approach takes into account the tag co-occurrence and the distance between the given and the existing photos. An overview of the field of recommender systems can be found in [1]. The neighborhood is defined based on a graph and tags are propagated through existing edges. In [6] the authors also focus on geo-tagged photos and propose methods for placing photos uploaded to Flickr on the World map. These methods rely on the textual annotations provided by the users and predict the single location where the image was taken. This work is motivated by the fact that users spend considerable effort to describe photos [2,7] with tags and these tags relate to locations where they were taken.

6   Conclusions

Tag recommendation is a very important and challenging task, since it helps users to annotate their photos with more meaningful tags, which in turn enables retrieving relevant photos from large photos collections such as Flickr. Nowadays, more and more photos are geotagged, and therefore in this paper we investigate how to improve tag recommendation based on the spatial and textual information of the photos. To this end, we analyzed the tags of geo-tagged photos collected from Flickr and proposed two different location-aware tag recommendation methods.

Acknowledgments. The work of A. Vlachou was supported by the Action Supporting Postdoctoral Researchers of the Operational Program Education and Lifelong Learning (Actions Beneficiary: General Secretariat for Research and Technology), and is co-financed by the European Social Fund (ESF) and the Greek State.

References

1. Adomavicius, G., Tuzhilin, A.: Toward the next generation of recommender systems: A survey of the state-of-the-art and possible extensions. TKDE 17(6), 734–749 (2005)
2. Ames, M., Naaman, M.: Why we tag: motivations for annotation in mobile and online media. In: CHI, pp. 971–980 (2007)
3. Cong, G., Jensen, C.S., Wu, D.: Efficient retrieval of the top-k most relevant spatial web objects. PVLDB 2(1), 337–348 (2009)
4. Garg, N., Weber, I.: Personalized, interactive tag recommendation for flickr. In: RecSys, pp. 67–74 (2008)
5. Moxley, E., Kleban, J., Manjunath, B.S.: Spirittagger: a geo-aware tag suggestion tool mined from flickr. In: ACM MIR, pp. 24–30 (2008)
6. Serdyukov, P., Murdock, V., van Zwol, R.: Placing flickr photos on a map. In: SIGIR, pp. 484–491 (2009)
7. Sigurbjörnsson, B., van Zwol, R.: Flickr tag recommendation based on collective knowledge. In: WWW, pp. 327–336 (2008)
8. Song, Y., Zhang, L., Giles, C.L.: Automatic tag recommendation algorithms for social recommender systems. TWEB 5(1), 4 (2011)

Group-by and Aggregate Functions
in XML Keyword Search

Thuy Ngoc Le[1], Zhifeng Bao[2], Tok Wang Ling[1], and Gillian Dobbie[3]

[1] National University of Singapore
{ltngoc,lingtw}@comp.nus.edu.sg
[2] University of Tasmania & HITLab Australia
zhifeng.bao@utas.edu.au
[3] University of Auckland
gill@cs.auckland.ac.nz

Abstract. In this paper, we study how to support group-by and aggregate functions in XML keyword search. It goes beyond the simple keyword query, and raises several challenges including: (1) how to address the keyword ambiguity problem when interpreting a keyword query; (2) how to identify duplicated objects and relationships in order to guarantee the correctness of the results of aggregation functions; and (3) how to compute a keyword query with group-by and aggregate functions. We propose an approach to address the above challenges. As a result, our approach enables users to explore the data as much as possible with simple keyword queries. The experimental results on real datasets demonstrate that our approach can support keyword queries with group-by and aggregate functions which are not addressed by the LCA-based approaches while achieving a similar response time to that of LCA-based approaches.

1 Introduction

Like keyword search in Information Retrieval, its counterpart over XML data has grown from finding the matching semantics and retrieving basic matching results in the last decade [3,14,6,8,18,11,17], and now it is enabling many more opportunities for users to explore the data while keeping the query in the form of keywords with additional requirements, such as visualization, aggregation, query suggestion, etc. In this paper, we study how to support group-by and aggregate functions beyond the simple XML keyword search, which to our best knowledge, no such effort has been done yet. In this way, it alleviates users from learning complex structured query languages and the schema of the data. For example, consider the XML document in Figure 1, in which there exist two many-to-many relationship types between Lecturer and Course, and between Course and Student. Suppose a user wishes to know *the number of students registered for course Cloud*, ideally she can just pose a keyword query like {Cloud, count student}. It is even better if keyword queries can express *group-by* functions. For example, *finding the number of registered students for each course* can be expressed as {group-by course, count student}.

This motivates us to propose an approach for XML search which can support keyword queries with *group-by* and aggregate functions including *max, min, sum, avg,*

H. Decker et al. (Eds.): DEXA 2014, Part I, LNCS 8644, pp. 105–121, 2014.
© Springer International Publishing Switzerland 2014

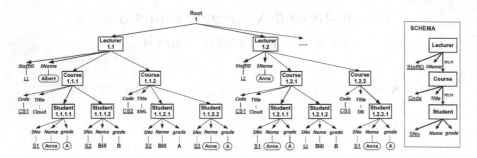

Fig. 1. An XML database

count (referred to as expressive keyword query) by just using a keyword based interface. As a result, our approach is able to provide a powerful and easy way to use a query
interface that fulfills a need not addressed by existing systems. Group-by and aggregate
functions are studied in XML structured queries such as [12,2] and in keyword search
over relational database (RDB) such as [10,13]. However, to the best of our knowledge,
there was no such work in XML keyword search.

Our approach has three challenges compared to the simple LCA-based approaches
for XML keyword search (i.e., approaches based on the LCA (Lowest Common
Ancestor) semantics) such as [3,14,6,8,18,11,17], which do not support group-by and
aggregate functions. Firstly, query keywords are usually ambiguous with different interpretations. Thereby, a query usually has different interpretations. In simple XML keyword search, an answer can be found without considering which query interpretation it
belongs to. On the contrary, in our approach, if all answers from different interpretations
are mixed altogether, the results for group-by and aggregate functions will be incorrect.
Secondly, an object and a relationship can be duplicated in an XML document because it
can appear multiple times due to many-to-many relationships. Such duplication causes
duplicated answers. Duplicated answers may overwhelm users but at least the answers
are still correct for simple XML keyword search. In contrast, duplicated answers cause
the wrong results for aggregate functions *count, sum, avg*. Thirdly, unlike simple XML
keyword search where all query keywords are considered equally and answers can be
returned independently, in our approach, query keywords are treated differently and the
answers need to be returned in a way that the group-by and aggregate functions can be
applied efficiently. Therefore, processing a keyword query with group-by and aggregate
functions is another challenge.

To overcome these challenges, we exploit the ORA-semantics (Object-Relationship-
Attribute-Semantics) introduced in our previous work [7,5,4]. The ORA-semantics includes the identification of nodes in XML data and schema. Once nodes in an XML
document are defined with the ORA-semantics, we can identify interpretations of a
keyword query. The ORA-semantics also helps determine many-to-many relationships
to detect duplication.

Contributions. In brief, we propose an approach for XML keyword search which can
support group-by and aggregate functions with the following contributions.

- Designing the syntax for an XML keyword query with group-by and aggregate
  functions (Section 2).

- Differentiating query interpretations due to keyword ambiguity in order not to mix together the results of all query interpretations (Section 3).
- Detecting duplication of objects and relationships to calculate aggregate functions correctly (Section 4).
- Processing XML keyword queries with group-by and aggregate functions including max, min, sum, avg, count efficiently (Section 5).
- Creating XPower, a system prototype for our approach. Experimental results on real datasets show that we can support most queries with group-by and aggregate functions which the existing LCA-based approaches cannot while achieving a similar response time to that of LCA-based approaches (Section 6).

2 Preliminary

2.1 Object-Relationship-Attribute (ORA)-Semantics

We defined the ORA-semantics as the identifications of nodes in XML data and schema. In XML schema, an internal node can be classified as object class, explicit relationship type, composite attribute and grouping node; and a leaf node can be classified as object identifier (OID), object attribute and relationship attribute. In XML data, a node can be object node or non-object node. Readers can find more information about the ORA-semantics in our previous works [7,5,4].

For example, the ORA-semantics of the XML schema and data in Figure 1 includes:
- Lecturer, Course and Student are object classes.
- StaffID, Code and SNo are OIDs of the above object classes.
- StaffID, SName, etc are attributes of object class Lecturer.
- There are two many-to-many relationship types between object classes: between Lecturer and Course, and between Course and Student.
- Grade is an attribute of the relationship type between Course and Student.
- Lecturer(1.1) is an object node; StaffID, L1, SName and Albert are non-object nodes of Lecturer(1.1).

Discovering ORA-semantics involves determining the identification of nodes in XML data and schema. If different matching nodes of a keyword have different identifications, then that keyword corresponds to different concepts of the ORA-semantics. In our previous work [7], the accuracy of discovery of the ORA-semantics in XML data and schema is high (e.g., greater than 99%, 93% and 95% for discovering object classes, OID and the overall process, respectively). Thus, for this paper, we assume the task of discovering the ORA-semantics has been done.

2.2 Expressive Keyword Query

This section describes the syntax of an expressive keyword query with group-by and aggregate functions including *max, min, count, sum, avg* supported by our approach. Intuitively, a group-by function is based on an object class or attribute. For example, {group-by course}, {group-by grade}. Thus, group-by must associate with an object class, or an attribute. On the other hand, aggregate functions *max, min, sum, avg* must associate with an attribute such as max grade, but not with an object class or a value because these functions are performed on the set of values of attributes. However, the aggregate function *count* can associate with all types of keyword: object class, attribute, and value because they all can be counted. For example, {count course}, {count StaffID}, {count A}. Based on these constraints, we define the syntax of an expressive keyword query in BackusNaur Form (BNF) as follows.

$\langle query \rangle ::= (\langle keyword \rangle[","])^*(\langle function \rangle[","])^*$
$\langle function \rangle ::= \langle group\_by\_fn \rangle \mid \langle aggregate\_fn \rangle$
$\langle group\_by\_fn \rangle ::= "group\_by"(\langle object\_class \rangle \mid \langle attribute \rangle)$
$\langle aggregate\_fn \rangle ::= \langle agg\_1 \rangle \mid \langle agg\_2 \rangle$
$\langle agg\_1 \rangle ::= ("max" \mid "min" \mid "sum" \mid "avg")(\langle attribute \rangle \mid \langle aggregate\_fn \rangle)$
$\langle agg\_2 \rangle ::= "count"(\langle keyword \rangle \mid \langle aggregate\_fn \rangle)$
$\langle keyword \rangle ::= \langle object\_class \rangle \mid \langle attribute \rangle \mid \langle value \rangle$

Since the parameters of aggregate function "count" are different from those of the other aggregate functions "max", "min", "sum", "avg", we use $\langle agg\_1 \rangle$ and $\langle agg\_2 \rangle$ to define them separately. With the above BNF, group-by and aggregate functions can be combined such as {group-by lecturer, group-by course, max grade, min grade} or nested such as {group-by lecturer, max count student}.

Terms related to an expressive keyword query. For ease of presentation, we use the following terms in this paper.

- *Group-by parameters* are query keywords following the term "group-by".
- *Aggregate functions* are the terms "max", "min", "sum", "avg" and "count".
- *Aggregate function parameters* are keywords following the aggregate functions.
- *Content keywords* are all query keywords except reserved words (i.e., the term "group-by" and the aggregate functions). Content keywords can be values, attributes, or object classes in the query.
- *Free keywords* are content keywords not as group-by parameter and not as aggregate function parameters.

3 Query Interpretation

The first challenge of our approach is that keywords are usually ambiguous with different interpretations as illustrated in Figure 2. Therefore, a query is also ambiguous with different interpretations, each of which corresponds to a way we choose the interpretation of keywords as described in the following concept.

Concept 1 (Query interpretation). *Given an expressive keyword query $Q = \{k_1, \ldots, k_n\}$, an interpretation of query Q is $\mathcal{I}_Q = \{i_1, \ldots, i_n\}$, where i_i is an interpretation of k_i.*

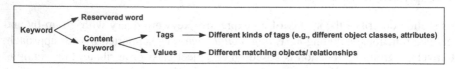

Fig. 2. Different possible interpretations of a keyword

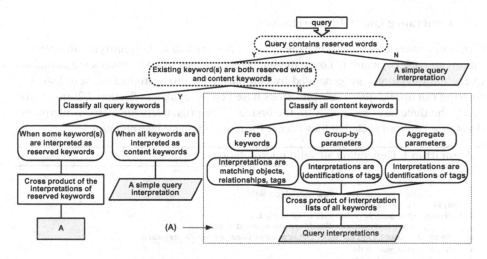

Fig. 3. Generating query interpretations

3.1 Impact of Query Ambiguity on the Correctness of the Results

In simple XML keyword search approaches such as [3,14,6,8,18,11,17], which do not support group-by and aggregate functions, an ambiguous keyword corresponds to a set of matching nodes, whose identifications are not considered. In these approaches, an answer can be found without considering which query interpretation it belongs to. In contrast, not differentiating query interpretations affects the correctness of group-by and aggregate functions as illustrated below.

Example 1. *Consider query* {Anna, count A} *issued to the XML data in Figure 1, in which one lecturer and two students have the same name* Anna. *They are object* <Lecturer:L2> *(w.r.t. object node* Lecturer (1.2)*), object* <Student:S1> *(w.r.t. object nodes* Student (1.1.1.1), Student (1.2.1.1), *and* Student (1.2.2.1)*), and object* <Student:S3> *(w.r.t. object node* Student (1.1.2.2)*). Both keywords* count *and* A *have only one interpretation in the XML data. Specifically,* count *is an aggregate function and* A *is a value of attribute* grade *of the relationship type between* student *and* course.

Intuitively, the query has three interpretations: (1) finding the number of grade A *of students taking courses taught by* Lecturer Anna, *(2) finding the number of grade* A *of* Student Anna *whose SNo is* S1, *and (3) similar to the second interpretation but for student* Student Anna *whose SNo is* S3. *If the query interpretations are not considered, the numbers of grade* A *corresponding to three interpretations are mixed and counted altogether instead of being counted separately. This makes the results of aggregate function* Count A *incorrect.*

Therefore, to calculate group-by and aggregate functions correctly, we need to process each query interpretation separately. To speed up the processing, we have an optimized technique, in which we do not process each query interpretation at the beginning, instead we process them together with group-by functions (discussed in Section 5.2).

3.2   Generating Query Interpretations

Generating all interpretations of a query contains three tasks: (1) identifying all interpretations of each keyword; (2) once the interpretation lists of all keywords are available, query interpretations are generated by computing the cross product of these lists; (3) filtering out invalid query interpretations based on the syntax in Section 2.2. Instead of doing the three tasks separately, in the first task, we proactively identify the interpretations of a keyword such that they do not form invalid query interpretations.

Algorithm 1. Generating all valid interpretations of a query

    Input: Query keywords $k_1, \ldots, k_n$
        The ORA-semantics
        The keyword-node lists and the node-object list
    Output: List of valid interpretations $qInt$
1 **Variable**: List of query keywords as reserved words L_{res}
2   A simple query interpretation I_s^q without reserved words
3 //`Task 1: identifying all interpretations of each keyword`
4 **for** *each query keyword k* **do**
5 | **if** *k is a reserved word* **then**
6 | └ Add k to L_{res}

7 //`Query does not contain reserved words`
8 **if** L_{res} *is empty* **then**
9 | // `It is a simple query without group-by or aggregate functions`
10 | Add all keywords to I_s^q // `All are content keywords`
11   └ return I_s^q // `We do not care about interpretations for this case`

12 //`Query contains reserved words`
13 **Variable**: The list of interpretations of a keyword k: L_k^i
14   The list of free keywords L_{fr}
15   The list of group-by parameters L_g
16   The list of aggregate functions and parameters $L_{a,f}$
17 **for** *each query keyword k in* L_{res} **do**
18 | $L_k^i \leftarrow$ content interpretation and reserved word interpretation
19 | $k.tags \leftarrow$ retrieve all identifications of k_i (k_i as tags) from ORA-semantics
20   └ $k.nodes \leftarrow$ retrieve all matching object nodes of k_i (k_i as values) from the keyword-node lists

21 $temp \leftarrow$ all cross product of the interpretation lists of keywords in L_{res}
22 **for** *each interpretation qI in temp* **do**
23 | // `All keywords are content keywords`
24 | **if** *All interpretations in qI are content interpretations* **then**
25 | | Add all query keywords to I_s^q
26 | └ Add I_s^q to $qInt$ // `a simple query interpretation`
27 | **else**
28 | | L_{fr}, L_g and $L_{a,f} \leftarrow$ parse all keywords based on L_{res}
29 | | **for** *each keyword k in* L_{fr} **do**
30 | | | $k.objects \leftarrow$ get objects of nodes for $k.nodes$ based on the node-object list
31 | | └ Add all $k.tags$ and $k.objects$ to L_k^i
32 | | **for** *each keyword k in* L_g **do**
33 | | └ Add all $k.tags$ to L_k^i
34 | | **for** *each keyword k in* $L_{a,f}$ **do**
35 | | | **if** *the aggregate function is "count"* **then**
36 | | | └ $k.tags \leftarrow$ get tags of nodes for $k.nodes$ based on the ORA-semantics
37 | └   └ Add all $k.tags$ to L_k^i

38 //`Task 2: generating all query interpretations`
39 $qInt \leftarrow$ add all cross product of the interpretation lists L_k^i's of all keywords

---

Identifying all interpretations of a keyword is not straightforward. Firstly, unlike a simple keyword query where interpretations of a keyword do not depend on those of the others, in our approach, to avoid invalid query interpretations, the interpretations of

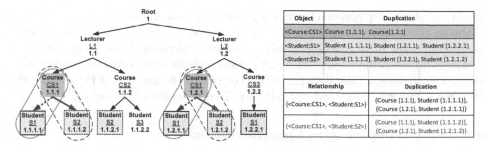

Fig. 4. Duplication of objects and relationships in the XML data in Figure 1

content keywords need to depend on those of the reserved words. Secondly, unlike simple keyword query where all keywords are treated equally, in our approach, different types of query keywords are treated differently and can provide different interpretations. Particularly, keywords as group-by and aggregate function parameters can only be interpreted as different tags, while keywords as free keywords can be interpreted as different matching objects[1], different matching relationships[2], or different tags as well.

Generating all valid query interpretations is illustrated in Figure 3 and is described in Algorithm 1. Since the reserved words may impact the interpretations of other keywords, we first need to identify whether a keyword is a reserved word, or a content keyword, or both. After that, the interpretations of content keywords depend on whether they are free keywords, group-by parameters or aggregate function parameters.

4   Duplication

Duplication of objects and relationships is another challenge of our approach. This section will discuss the impact of such duplication on the correctness of the results of aggregate functions and how to overcome it. For illustration, we use the XML data in Figure 1, in which duplicated objects and relationships are summarized in Figure 4, where we only show object nodes of the data for simplicity.

4.1   Duplicated Objects and Relationships

The duplication of objects is due to many-to-many $(m:n)$ or many-to-one $(m:1)$ relationships because in such relationship, the child object is duplicated each time it occurs in the relationship. For example, in the XML data in Figure 1, since the relationship between lecturer and course is $m:n$, a course can be taught by many lecturers such as <Course:CS1> are taught by <Lecturer:L1> and <Lecturer:L2>. Thus, the child object (<Course: CS1>) is shown as two object nodes Course (1.1.1) and Course(1.2.1).

$m:n$ and $m:1$ relationships cause not only duplicated objects, but also *duplicated relationships*. For example, as discussed above, in the XML data in Figure 1, because of

---

[1] An object matches keyword k when any of its object node matches k.

[2] A relationship matches keyword k when any of its involved objects matches k.

the $m : n$ relationship between lecturer and course, the child object (<Course:CS1>) is shown as two object nodes Course (1.1.1) and Course(1.2.1). Therefore, everything below these two object nodes is the same (duplicated), including the relationships between object (<Course:CS1>) and students such as the relationships between <Course:CS1> and <Student:S1> (the big dotted lines in Figure 4), and between <Course:CS1> and <Student:S2> (the big lines in Figure 4).

4.2 Impact of Duplication on Aggregate Functions

We show the impact of duplicated objects and relationships in the following examples.

Example 2 (Impacts of duplicated objects). *To count the number of students taught by lecturer Albert, a user can issue a query* {Albert, count student} *against the XML data in Figure 1. Without considering duplicated objects, the number of students is four. However, object node* Student (1.1.1.2) *and object node* Student (1.1.2.1) *refer to the same object* <Student:S2>. *Hence, only* three *students are taught by lecturer* Albert.

Example 3 (Impacts of duplicated relationships). *Recall query* {Anna, count A} *discussed in Example 1. We use the second interpretation, i.e., finding the number of grade A of* Student Anna *whose SNo is* S1, *to illustrate the impacts of duplicated relationships. In the XML data in Figure 1, the relationship between* <Student:S1> *and* <Course:CS1> *is duplicated twice (the big dotted lines in Figure 4). This makes attribute* grade *of this relationship duplicated. Without considering duplicated relationships, the number of grade A is* three. *In contrast, by keeping only one instance for each relationship, the answer is only* two.

Therefore, to perform aggregate functions *sum, avg* and *count* correctly, we must detect duplicated objects and relationships and keep only one instance for each of them. Duplication does not impact on the correctness of aggregate functions *max* and *min*.

4.3 Detecting Duplication

If there exists a $m : n$ or $m : 1$ relationship type between object classes A and B, then for all object classes (or relationship types) appearing as B or the descendants of B, the objects of those classes (or the relationships of those relationship types) may have duplication. Otherwise, with no $m : n$ or $m : 1$ relationship type, duplication does not happen. Therefore, to detect duplication, we first identify the possibility of duplication by checking $m : n$ and $m : 1$ relationship types. If there is no $m : n$ and no $m : 1$ relationship type, we can determine quickly the objects (or relationships) which are not duplicated. Identifying the possibility of duplication can be done with the ORA-semantics and is shown in Algorithm 2.

Once we determine that an object (or a relationship) is possibly duplicated, we determine whether two objects (of the same object class) are really duplicated by checking whether they have the same OID. A relationship is represented by a list of involved objects. Thus, two relationships (of the same relationship type) are duplicates if the two sets of objects involved by the two relationships are the same.

Algorithm 2. Detecting the possibility of duplication

Input: The aggregate function parameter p
       The ORA-semantics
1 $p.class \leftarrow$ get the object class of p (based on the ORA-semantics)
2 $p.ancestor \leftarrow$ get all ancestor object classes and itself of $p.class$ (based on the ORA-semantics)
3 $p.RelType \leftarrow$ get all relationship types with $p.ancestor$ involved in (based on the ORA-semantics)
4 $p.card \leftarrow$ get the cardinality of each relationship type in $p.RelType$ (based on the ORA-semantics)
5 **if** *existing* $m : n$ *or* $m : 1$ *relationship type in* $p.RelType$ *(determined in* $p.card$*)* **then**
6      $\lfloor$ $p.possibility \leftarrow$ TRUE

Detecting real duplication is integrated with calculating aggregate functions and will be described in Section 5.2. For each aggregate function parameter, if it is possibly duplicated, before applying an aggregate function on an object or a relationship instance, we check whether duplication occurs.

5 Indexing and Processing

Figure 5 describes the architecture of our approach, which consists of three indexes (presented in Section 5.1) and five processing components (discussed in Section 5.2). Like LCA-based approaches such as [3,14,6,8,18,11,17], our approach works on data-centric XML documents with no IDREFs and assumes updating does not frequently happen. Otherwise, adding or deleting one node can lead to change Dewey labels of all nodes in an XML document in those approaches.

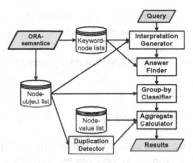

Fig. 5. The architecture

5.1 Labeling and Indexing

Labeling. Unlike conventional labeling schemes, where each node has a distinct label, we assign a *Dewey* label for only object nodes while non-object nodes use the same label with the object node they belong to as in Figure 1.

Indexing
A. Keyword-node lists. Each document keyword k has a list of matching object nodes.
B. Node-value list. We maintain a list of pairs of $\langle object\ node, values \rangle$ to retrieve values of an object node to operate group-by or aggregate functions.
C. Node-object list. We maintain a list of pairs of $\langle object\ node, object \rangle$ for two purposes. Firstly, it is used together with the keyword-node lists to find matching objects in order to generate query interpretations. Secondly, it is used to detect duplication.

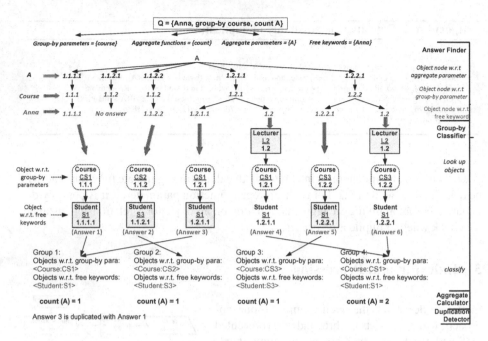

Fig. 6. Processing query $Q = \{$Anna, group-by course, count A$\}$

5.2 Processing

The processing of our approach can briefly be described as follows. As discussed in Section 3, we process each query interpretation separately in order not to mix together results of different query interpretations. *Interpretation Generator* is responsible for generating all valid interpretations of the input query. For each query interpretation, *Answer Finder* finds intermediate answers, each of which will be used as an operand of the aggregate function. *Group-by Classifier* classifies intermediate answers based on group-by parameters. *Aggregate Calculator* applies aggregate functions on intermediate answers. During the aggregate calculation, *Duplication Detector* detects *duplicated objects and relationships* in order to perform aggregate calculation correctly.

The above discussion is at conceptual level. At the lower level, we have an optimized technique for the process. We are aware that different query interpretations may produce the same sets of object nodes for group-by and aggregate function parameters. Therefore, instead of finding intermediate answers for each interpretation, we find all possible intermediate answers first, then classify them into suitable interpretations. Thereby, we can save costs of repeating the same thing. The optimized process is described in Algorithm 3[3].

---

[3] A query interpretation corresponds to a set of interpretations of free keywords, group-by parameters and aggregate function parameters. The input of Algorithm 3 is a query interpretation w.r.t. interpretations of group-by and aggregate function parameters, interpretations of free keywords will be handled after finding intermediate answers.

Interpretation Generator and *Duplication Detection* correspond to the discussion in Section 3 and Section 4 respectively. The following are details of the remaining three components. We use query {Anna, group-by course, count A} applied to the XML data in Figure 1 as a running example. Figure 6 shows the results produced by each component for the query.

Answer Finder. An intermediate answer contains a set of matching object nodes of content keywords which are aggregate function parameters, group-by parameters, and free keywords. We first find object nodes of aggregate function parameters (line 3 in Algorithm 3), then those of group-by parameters (line 6), and finally those of free keywords (line 11). This is because an intermediate answer corresponds to only one matching object node of an aggregate function parameter because it is used as only one operand in an aggregate function.

An object node of a *group-by parameter* must have an ancestor-descendant relationship with that of the aggregate function parameter because group-by functions are based on the relationship among objects. In XML, these relationships are represented through ancestor-descendant relationships (edges).

All nodes in an intermediate answer must be meaningfully related. For this purpose, we agree with the argument in [15] that the object class of the LCA (Lowest Common Ancestor) of nodes in an answer must belong to the LCA of the object classes of these nodes. We use this property to find object nodes of *free keywords* in an answer.

In the running example (in Figure 6), the aggregate function parameter A has five matching nodes. For each of them, we find the corresponding object nodes of the group-by parameter course. For each pair of object nodes of the aggregate function parameter and the group-by parameter, we find the corresponding object nodes of the free keyword Anna.

Scope. Like LCA-based approaches, our approach does not consider the cases where two objects are connected through several relationships which do not form an ancestor-descendant chain. This constraint is to find group-by parameter.

Group-by Classifier. We classify an intermediate answer into a group based on objects w.r.t. the group-by parameters and the free keywords. Thus, we first retrieve the corresponding objects from the set of nodes of an answer. Then, we compare that set of objects of an answer with that of groups (lines 14-19). In the running example, for the first answer, <Course:CS1> and <Student:S1> are objects corresponding to the group-by parameter and the free keyword respectively. So, it is classified into Group 1 which has the same set of objects.

Aggregate Calculator. Aggregate Calculator calculates aggregate functions on objects and relationships (or their attributes and values) of the aggregate function parameters in each group. During the calculation, we check the duplication of objects and relationships. We first check the possibility of duplication based on schema (line 25, referred to Algorithm 2). If this possibility is true, we will check whether the considered object or relationship is duplicated or not (line 26) before processing the aggregate functions (line 27). In the running example, two answers in Group 1 are duplicated. Thus, the result of Count A in this group is only one.

Algorithm 3. Processing an expressive keyword query

Input: Free keyword $k_1, \ldots, k_n$
The group-by parameters $g_1, \ldots, g_q$
The aggregate function parameters $a_1, \ldots, a_p$
The aggregate functions $f_1, \ldots, f_p$
The ORA-semantics
Indexes: the keyword-node lists, the node-value list, the node-object list

```
1   //Answer Finder
2   for each aggregate function parameter aᵢ do
3       for each matching object node a_node ∈ matchNode(aᵢ) do
4           //find the corresponding object nodes matching group-by parameters
5           for each group-by parameter gᵢ do
6               L_{gᵢ_node} ← the list of object nodes u such that u ∈ matchNode(gᵢ) and u is an ancestor or
                  a descendant of a_node
7           for each set of object nodes {g₁_node, . . . , g_q_node}, gᵢ_node ∈ L_{gᵢ_node} do
8               highest ← the highest object node among gᵢ_node's and a_node
9               //find the corresponding object nodes matching free keywords
10              for each free keyword kᵢ do
11                  L_{kᵢ_node} ← findObjectNodesFreeKeyword(kᵢ, highest)

12              //Group-by Classifier
13              L_group ← {group| group ← {g₁_node, .., g_q_node, k₁_node, .., k_n_node}, where
                  kᵢ_node ∈ L_{kᵢ_node}}
14              for each group ∈ L_group do
15                  unit ← {a_node, group}
16                  if exist groupᵢ matches group then
17                      classify unit into groupᵢ //Classify unit

18                  else
19                      create new class group_{i+1}
20                      Updated objects w.r.t. group-by parameters and free keywords for group_{i+1}
21                      classify unit into group_{i+1} //Classify unit into new class

22                  //Aggregate Calculator
23                  if aᵢ is a relationship or a relationship attribute (based on the ORA-semantics) then
24                      rel(a_node) ← get the relationship w.r.t. a_node

25                  else
26                      obj(a_node) ← get the object w.r.t. a_node

27                  if aᵢ.possibility is TRUE then
28                      if DetectDuplication (obj(a_node)/rel(a_node)) is FALSE then
29                          ApplyAggregateFunction (fᵢ, aᵢ, a_node)
```

Complexity. Since the number of aggregate function parameters, group-by parameters and free keywords are few, the *For* loops in line 1, line 4 and line 9 do not have much affect on the complexity. Since all the lists of matching object nodes are sorted by the pre-order of labels of matching object nodes, the finding of matching object nodes of group-by parameters and free keywords can be obtained efficiently. Thereby, the complexity of finding all intermediate answers depends on the number of matching object nodes of aggregate function parameters. Thus, for Answer Finder, for each matching object of an aggregate function parameter, the costs are $\log(G)$ and $\log(K)$ for finding object nodes of a group-by parameter and of a free keyword in an answer respectively, where G and K are the length of their lists of matching object nodes respectively. In Group-by Classifier, the cost for classifying in the worst case is $\log(Gr)$ where Gr is the total number of groups (sorted). For Aggregate Calculator, the cost in the worst case is $O \times \log(O)$ where O is the maximum number of objects in a group.

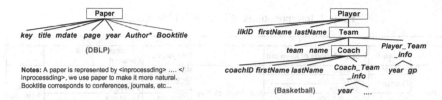

Fig. 7. A part of schema of DBLP and Basketball used in experiments

Table 1. Queries for tested datasets

DBLP	User search intention	Query	XPower suports	Ambi-guity	Dupli-cation
QD1	Count the papers of Yi Chen	Yi Chen, count paper	Yes	Yes	No
QD2	Count the co-authors of Yi Chen	Yi Chen, count author	No	N.A	N.A
QD3	How many years Yi Chen have published papers and in how many conferences	Yi Chen, count booktitle, count year	Yes	Yes	Yes
QD4	Count the papers of Yi Chen in each conference for each year	Yi Chen, group-by year, group-by booktitle, count paper	Yes	Yes	Yes
QD5	How many conferences has Diamond published papers	Diamond, count booktitle	Yes	Yes	Yes
QD6	Find the latest year Diamond published paper in IEEE-TIT	Diamond, IEEE-TIT, max year	Yes	Yes	No
QD7	Count the papers of Brown published in each conference	Brown, group-by booktitle, count paper	Yes	Yes	No
QD8	Count the conferences Brown has published papers	Brown, count booktitle	Yes	Yes	Yes
Basketball					
QB1	How many players and coaches in team Celtics	Celtics, count player, count coach	Yes	No	Yes
QB2	How many teams Michael have worked for	Michael, count team	Yes	Yes	Yes
QB3	How many players Thomas have worked with	Thomas, count player	No	N.A	N.A
QB4	How many players Johnson have worked with	Johnson, count player	Partial	Yes	Yes
QB5	Find the latest year Edwards has worked for team Hawks	Edwards, Hawks, max year	Yes	Yes	No
QB6	When did player Edwards start to work	Edwards, min year	Yes	Yes	No
QB7	Count players in each team which Michael has worked for	Michael, group-by team, count player	Partial	Yes	No
QB8	How many players and coaches of each team	group-by team, count player, count coach	Yes	No	Yes

6 Experiment

We have implemented a framework, called XPower for evaluation on several aspects: enhancement, impacts of query interpretation and duplication, and efficiency. We compare XPower with XKSearch [14] because it is one of the most popular XML keyword search approaches. Like other XML keyword search approaches, XKSearch does not support group-by and aggregate functions, and does not consider the effects of duplication and query interpretation on search results. Thus, we compare with XKSearch on only efficiency. The experiments were performed on an Intel(R) Core(TM)i7 CPU 3.4GHz with 8GB of RAM. We used the subsets of two real datasets: Basketball (45MB)[4] and DBLP (570MB)[5,6]. A part of the schema of each dataset is given in Figure 7.

---

[4] http://www.databasebasketball.com/

[5] http://dblp.uni-trier.de/xml/

[6] In the updated DBLP dataset, authors of the same name can be distinguished. We make use this in differentiating query interpretations related to authors.

Table 2. Interpretations of keywords in tested queries

		DBLP			Basketball	
	Keyword	Interpretations of keyword		Keyword	Interpretations of keyword	
value	Yi Chen	6 authors	Match values only	Celtics	1 team	No keyword in multiple groups: value, tag, reserved word
	Diamon	many authors		Hawks	1 team	
		many titles		Edwards	4 players	
	Brown	many authors		Thomas	15 players	
		many titles		Michael	2 coaches and 13 players	
	IEEE-TIT	one conference		Johnson	5 coaches and 14 players	
tag	author	object class/ attribute	Match both tags and values of Title (in class Paper)	team	object class	
	paper	object class		coach	object class	
	year	attribute (in class paper)		player	object class	
	booktitle	object class/ attribute		year	relationship attribute	
reserved word	group-by	not in DBLP		group-by	not in document	
	count	in title of paper	both reserved words and values	count	not in document	
	max	in title and author		max, min	not in document	

6.1   Enhancement Evaluation

Table 1 shows eight queries for each dataset used in the experiments and Table 2 provides interpretations of keywords in those queries. Table 1 also shows whether XPower was accurate on the tested queries. As can be seen, XPower can return answers for seven out of eight queries for each Basketball and DBLP dataset. This is because XPower handles group-by functions based on relationships, which are represented as edges in XML. However, in QD2, Yi Chen is an author, and it does not have any direct relationship with another author. In other words, there is no ancestor-descendant relationship between authors. Therefore, XPower cannot provide any answer for this query. This is similar to QB3 of Basketball. For QB4 and QB7 in Basketball, Michael and Johnson can be both players and coaches. If they are interpreted as players, XPower cannot provide an answer for the same reason as QB3. If they are interpreted as coaches, XPower can provide answers.

6.2   Impact of Query Interpretation due to Keyword Ambiguity

Table 3 shows three different results for each query in Basketball in three different scenarios: (1) XPower considering both query interpretation and duplication, (2) only considering query interpretation but not duplication, and (3) only considering duplication but not differentiating query interpretation. Because of space constraints, we only showed the results and explanations for Basketball although DBLP has similar results. We also describe whether duplication and keyword ambiguity impact on the results of each query in Table 1. As can be seen, keyword ambiguity impacts on the correctness of the results of all queries in DBLP, and five out of seven queries in Basketball (not considering queries with no answer). This verifies the importance of differentiating query interpretations. Otherwise, the results of all query interpretations are mixed together.

6.3   Impact of Duplication

As we can see in Table 1, duplication impacts on the correctness of the results for four out of eight queries in both DBLP and Basketball. This is fewer than those affected by

Table 3. Results of queries of Baketball dataset

	XPower results	Results if not filter duplication	Reasons for duplication	Results if not differentiate queryinterpretation	Explain
QB1	count player = 215 count coach = 13	count player = **2795** count coach = 13	Team Celtics has been coached by 13 coaches, thus its players are duplicated 13 times. Coaches are not duplicated.	same results with XPower	
QB2	**15 answers** for 15 persons (2 coaches, 13 players), each has a number of teams they have worked for. Sum of these numbers are **69**.	count team = **298**	Michael as players: a player can work with the same team (duplicated) under different coaches.	**1 answer:** count team = 69	
QB4	No answer for Johnson as players. Johnson as coaches: **5 answers** for 5 coaches, each has a number of players. Total number is **136**.	count player = **219**	A player can works for more than 1 team (duplicated) in different years under the same coach Johnson	**1 answer:** count player = 136	mix the results of all interpretations
QB5	**2 answers** for 2 players Edwards in team Hawks, with max year 1997, 2004 resp.	same results with XPower	duplication does not affect aggregate function max	**1 answer:** max year = 2004	
QB6	**4 answers** w.r.t. min year for 4 players Edwards: 1993, 1995, 1981, 1977 resp.	same results with XPower	duplication does not affect aggregate function min	**1 answer:** min year = 1977	
QB7	**6 answers:** Michael as players: No answer. Michael as coach 1: 3 teams (count players = 153, 256, 82 resp.) Michael as coach 2: 3 teams (count	same results with XPower	Although players are duplicated in documents, they are not duplicated under the pair of 1 coach and 1 team	**4 answers:** 4 teams (count player = 153, 236, 512, 164 resp.).	
QB8	Provide the number of players and those of coaches for each team	count player: diff count team: same	If a team is duplicated, all of its players are duplicated.	same results with XPower	

ambiguity but this number is still significant. This agrees our arguments about the importance of detecting duplication. Otherwise, the results of aggregate functions would not be correct. The number of queries affected by duplication is fewer than that of ambiguity because in Basketball, coaches are not duplicated, only teams and players can be duplicated. In DBLP, papers are not duplicated either. Therefore, there is no impact on the functions count coach in Basketball and count paper in DBLP. Moreover, duplication does not affect *max* and *min* functions as in QB5 and QB6.

6.4   Efficiency Evaluation

Figure 8 shows the response time of XPower (XP as abbreviation) and XKSearch (XK as abbreviation) for queries tested except the ones (QB3 and QD2) XPower does not provide any answer. Since XKSearch does not support group-by and aggregate functions, we dropped reserved words of tested queries when running XKSearch. Although XPower has the overhead of doing group-by and aggregate functions, the response time of queries are similar to those of XKSearch. This is because XPower does not find all SLCAs because many SLCAs do not correspond to any intermediate answer. For queries with complicated group-by and aggregate functions (e.g., QB1, QB7, QB8, QD3, QD4 and QD7), the overhead of processing those functions makes XPower run slightly slower than XKSearch. The response time of XPower is dominated by that of Answer Finder. Aggregate Calculator costs more than Group-by Classifier because it needs to detect duplication.

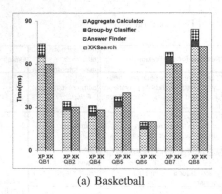
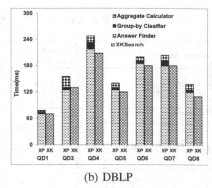

(a) Basketball   (b) DBLP

Fig. 8. Efficiency comparison of XPower and XKSearch on Basketball and DBLP (Dropping reversed words of tested queries when running XKSearch)

7   Related Work

XML keyword search. Most approaches for XML keyword search are based on the concept of LCA (Lowest Common Ancestor) first proposed in XRANK [3]. Later, many approaches extend the concept of LCA to filter less relevant answers. XKSearch [14] defines Smallest LCAs (SLCAs) to be the LCAs that do not contain other LCAs. Meaningful LCA (MLCA) [8] incorporates SLCA into XQuery. VLCA [6] and ELCA [18] introduces the concept of valuable/exclusive LCA to improve the effectiveness of SLCA. XReal [1] proposes an IR-style approach for ranking results. MaxMatch [9] investigates an axiomatic framework that includes the properties of monotonicity and consistency. MESSIAH [11] handles the cases of missing values in optional attributes. Although these works can improve effectiveness of the search, they do not support group-by and aggregate functions.

Group-by and aggregate functions. Group-by and aggregate functions are studied in XML structured queries such as [12,2] and in keyword search over relational database (RDB) such as [10,13]. However, there is no such work in XML keyword search. Arguably, XML can be shredded into RDB, and then we can apply the techniques of RDB for XML. However, [16] proves that the relational approaches are not as efficient as the native approaches (XML is used directly) in most cases.

8   Conclusion and Future Work

We proposed an approach to support queries with group-by and aggregate functions including sum, max, min, avg, count to query a data-centric XML document with a simple keyword interface. We processed query interpretations separately in order not to mix together the results of different query interpretations. To perform aggregate functions correctly, we detected duplication of objects and relationships. Otherwise, the results of aggregate functions may be wrong. Experimental results in real datasets showed the enhancement of our approach, the importance of detecting duplication and differentiating query interpretations on the correctness of aggregate functions. These results also showed the optimized techniques enable our approach to be almost as efficient as LCA-based approaches although it has some overhead. In the future, we will handle the problem by different techniques, including transferring XML keyword queries to structured queries such as XQuery queries. Moreover, we will solve the case where object nodes of group-by and aggregate function parameters do not have ancestor-descendant relationships.

References

1. Bao, Z., Ling, T.W., Chen, B., Lu, J.: Efficient XML keyword search with relevance oriented ranking. In: ICDE (2009)
2. Gokhale, C., Gupta, N., Kumar, P., Lakshmanan, L.V.S., Ng, R., Prakash, B.A.: Complex group-by queries for XML. In: ICDE (2007)
3. Guo, L., Shao, F., Botev, C., Shanmugasundaram, J.: XRANK: Ranked keyword search over XML documents. In: SIGMOD (2003)
4. Le, T.N., Ling, T.W., Jagadish, H.V., Lu, J.: Object semantics for XML keyword search. In: Bhowmick, S.S., Dyreson, C.E., Jensen, C.S., Lee, M.L., Muliantara, A., Thalheim, B. (eds.) DASFAA 2014, Part II. LNCS, vol. 8422, pp. 311–327. Springer, Heidelberg (2014)
5. Le, T.N., Wu, H., Ling, T.W., Li, L., Lu, J.: From structure-based to semantics-based: Towards effective XML keyword search. In: Ng, W., Storey, V.C., Trujillo, J.C. (eds.) ER 2013. LNCS, vol. 8217, pp. 356–371. Springer, Heidelberg (2013)
6. Li, G., Feng, J., Wang, J., Zhou, L.: Effective keyword search for valuable LCAs over XML documents. In: CIKM (2007)
7. Li, L., Le, T.N., Wu, H., Ling, T.W., Bressan, S.: Discovering semantics from data-centric XML. In: Decker, H., Lhotská, L., Link, S., Basl, J., Tjoa, A.M. (eds.) DEXA 2013, Part I. LNCS, vol. 8055, pp. 88–102. Springer, Heidelberg (2013)
8. Li, Y., Yu, C., Jagadish, H.V.: Schema-free XQuery. In: VLDB (2004)
9. Liu, Z., Chen, Y.: Reasoning and identifying relevant matches for XML keyword search. In: PVLDB (2008)
10. Tata, S., Lohman, G.M.: SQAK: doing more with keywords. In: SIGMOD (2008)
11. Truong, B.Q., Bhowmick, S.S., Dyreson, C.E., Sun, A.: MESSIAH: missing element-conscious SLCA nodes search in XML data. In: SIGMOD (2013)
12. Wu, H., Ling, T.W., Xu, L., Bao, Z.: Performing grouping and aggregate functions in XML queries. In: WWW (2009)
13. Wu, P., Sismanis, Y., Reinwald, B.: Towards keyword-driven analytical processing. In: SIGMOD (2007)
14. Xu, Y., Papakonstantinou, Y.: Efficient keyword search for smallest LCAs in XML databases. In: SIGMOD (2005)
15. Zeng, Y., Bao, Z., Jagadish, H.V., Ling, T.W., Li, G.: Breaking out of the mismatch trap. In: ICDE (2014)
16. Zhang, C., Naughton, J., DeWitt, D., Luo, Q., Lohman, G.: On supporting containment queries in relational database management systems. In: SIGMOD (2001)
17. Zhou, J., Bao, Z., Wang, W., Ling, T.W., Chen, Z., Lin, X., Guo, J.: Fast SLCA and ELCA computation for XML keyword queries based on set intersection. In: ICDE (2012)
18. Zhou, R., Liu, C., Li, J.: Fast ELCA computation for keyword queries on XML data. In: EDBT (2010)

Efficient XML Keyword Search
Based on DAG-Compression

Stefan Böttcher, Rita Hartel, and Jonathan Rabe

Institute for Computer Science, University of Paderborn,
Fürstenallee 11, 33102 Paderborn, Germany

Abstract. In contrast to XML query languages as e.g. XPath which require
knowledge on the query language as well as on the document structure, keyword
search is open to anybody. As the size of XML sources grows rapidly, the need for
efficient search indices on XML data that support keyword search increases. In
this paper, we present an approach of XML keyword search which is based on the
DAG of the XML data, where repeated substructures are considered only once,
and therefore, have to be searched only once. As our performance evaluation
shows, this DAG-based extension of the set intersection search algorithm[14,15],
can lead to search times that are on large documents more than twice as fast as
the search times of the XML-based approach. Additionally, we utilize a smaller
index, i.e., we consume less main memory to compute the results.

Keywords: Keyword Search, XML, XML compression, DAG.

1 Introduction

1.1 Motivation

The majority of the data within the internet is available nowadays in form of semi-
structured data (i.e. HTML, XML or 'XML dialects'). When searching for certain information in huge document collections, the user typically (1) has no knowledge of
the structure of the document collection itself and (2) is a non-expert user without any
technical knowledge of XML or XML query languages. These requirements are met by
XML keyword search where the user specifies the searched information in form of a list
of keywords (i.e., neither knowledge of the document structure nor of any specific query
language is required) and document fragments are returned that contain each keyword
of the specified keyword list. For this purpose, the need for efficient keyword search
approaches on XML data is high.

1.2 Contributions

Our paper presents *IDCluster*, an approach to efficient keyword search within XML data
that is based on a DAG representation of the XML data, where repeated substructures
exist only once and therefore have to be searched only once. IDCluster combines the
following features and advantages:

H. Decker et al. (Eds.): DEXA 2014, Part I, LNCS 8644, pp. 122–137, 2014.
© Springer International Publishing Switzerland 2014

- Before building the index, IDCLuster removes redundant sub-trees and splits the document into a list of so-called *redundancy components*, such that similar sub-trees have to be indexed and searched only once.
- For keyword search queries where parts of the results are contained partially or completely within repeated sub-trees, the DAG-based keyword search outperforms the XML keyword search by a factor of more than two on large documents, whereas it is comparably fast for keyword search queries where all results occur in sub-trees that exist only once within the document.

To the best of our knowledge, IDCluster is the first approach showing these advantages.

1.3   Paper Organization

The rest of the paper is organized as follows. Section 2 introduces the underlying data model and presents the ideas of the set intersection keyword search algorithm [14][15] on which our approach is based. Section 3 presents our adaptation of the ideas of Zhou et al. to build the index based on the document's DAG instead of the document's tree in order to avoid repeated search within identical structures. Section 4 contains the performance evaluations, Section 5 discusses the advantages of our approach to already existing approaches, and finally, Section 6 concludes the paper with a short summary.

2   Preliminaries

2.1   Data Model

We model XML trees as conventional labeled ordered trees. Each node represents an element or an attribute, while each edge represents a direct nesting relationship between two nodes. We store a list of all keywords k_n contained in the element name or attribute name respectively, or in any text value for each node n. Thereby, we tokenize each text label into keywords at its white-space characters. E.g., a node with label "name" and text value "Tom Hanks" is split into the three keywords *name*, *Tom* and *Hanks*. The keywords in k_n are called **directly contained** keywords of n. Keywords that are directly contained in a descendant of n are called **indirectly contained** keywords of n. Keywords that are either directly or indirectly contained in n are called **contained** keywords of n. Furthermore, we assign each node its pre-order traversal number assigned as ID. Fig. 1 shows a sample XML tree. The ID of each node is written top left next to it. The element or attribute name is given inside the node's ellipse, while the text values are given below.

2.2   Query Semantics

In the last decade, different search goals for XML keyword search queries have been proposed. These search goals form subsets of the common ancestor (CA) nodes. For a query Q, the set of common ancestors $CA(Q)$ contains all nodes that contain every keyword of Q. For the given query $Q_{ex} = \{USA, English\}$ and the example tree of Fig. 1, the common ancestors are $C(Q_{ex}) = \{1, 2, 4, 5, 11, 12\}$. The two most widely

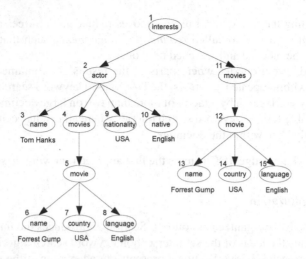

Fig. 1. Example of an XML tree

adopted CA subsets are the smallest lowest common ancestor (SLCA) and the exclusive lowest common ancestor (ELCA). All nodes in $CA(Q)$ that do not have descendants in $CA(Q)$ are in $SLCA(Q)$, e.g. $SLCA(Q_{ex}) = \{5, 12\}$. A node n is in $ELCA(Q)$, when n contains each keyword of Q outside of each subtree of a descending CA node of n, e.g., $ELCA(Q_{ex}) = \{2, 5, 12\}$. Node 2 is in $ELCA(Q_{ex})$, because when we would remove node 5 (which is CA) and its descendants, node 2 still contains both keywords: *USA* in node 9 and *English* in node 10. Note that the given definitions imply $SLCA(Q) \subseteq ELCA(Q) \subseteq CA(Q)$.

2.3   IDLists

This paper's algorithm is an improvement of the set intersection keyword search algorithm FwdSLCA and its modifications as proposed by Zhou et al. [14,15], i.e., an improvement of the currently best known approach to XML keyword search.

Zhou et al. use IDLists as an index to efficiently perform keyword search. An IDList is an inverted list of nodes that contain (directly or indirectly) a certain keyword. For each node n the IDList provides three values: The *ID* of n, the position of n's parent inside the IDList named $PIDPos$, and the number N_{Desc} of nodes within n's subtree directly containing the keyword. IDLists are sorted by ID. Fig. 2 shows the IDLists L_{USA} and $L_{English}$ of the keywords *USA* and *English* respectively. IDLists can be easily generated by a single pass through the document.

2.4   Search

The general idea is to use set intersection on the IDLists of the query keywords to find CA nodes. The found CA nodes are then checked against the SLCA or ELCA semantics respectively to calculate the result set.

L_{USA}	Pos	0	1	2	3	4	5	6	7	8
	ID	1	2	4	5	7	9	11	12	14
	PIDPos	-1	0	1	2	3	1	0	6	7
	N_{Desc}	3	2	1	1	1	1	1	1	1

$L_{English}$	Pos	0	1	2	3	4	5	6	7	8
	ID	1	2	4	5	8	10	11	12	15
	PIDPos	-1	0	1	2	3	1	0	6	7
	N_{Desc}	3	2	1	1	1	1	1	1	1

Fig. 2. IDLists for the keywords *USA* and *English*

The basic SLCA search algorithm of [14,15], *FwdSLCA*, uses the method *fwdGetCA* to efficiently calculate all CA nodes in ascending order. This is done by maintaining pointers to the current position C_i in each IDList L_i, selecting the highest ID of all C_i, and binary searching for this ID in the remaining IDLists. Due to the ascending order, a CA node n is SLCA if and only if the next found CA node is not a child of n. This can be checked using the IDList's $PIDPos$.

For checking the ELCA semantics, the algorithm *FwdELCA* uses an additional stack for storing two arrays for each visited node n: One holds the N_{Desc} values of n, the other holds the added N_{Desc} values of n's CA children. As soon as every CA child of n has been found, the differences of those two arrays indicate whether n is ELCA.

Beside *FwdSLCA*, two more SLCA search algorithms based on set intersection are proposed in [14,15]: The algorithm *BwdSLCA* finds CA nodes in reversed order, and as a result can efficiently skip ancestors of CA nodes (note that ancestors of CA nodes by definition never can be SLCA). The algorithm *BwdSLCA+* additionally improves the binary search required for CA calculation by shrinking the search space. The shrinkage is based on the parent position information given by the $PIDPos$ value.

For *FwdELCA*, one alternative is proposed: The algorithm *BwdELCA* finds CA nodes in reversed order and shrinks the binary search space like *BwdSLCA+*. The ancestor skipping introduced in *BwdSLCA* cannot be implemented for ELCA search since inner nodes can be an ELCA too.

3   Our Approach

The algorithm described in section 2.4 is not redundancy-sensitive, i.e., whenever there are repeated occurrences of the same sub-tree, these occurrences are searched repeatedly. However, the goal of our approach is to follow the idea of DAG-based compression approaches and to exploit structural redundancies in order to perform faster keyword search. This is done by splitting the original XML tree into disjoint **redundancy components**. A redundancy component is a subgraph of the original tree which occurs more than once within the tree. Note that splitting the XML tree into disjoint redundancy components can be performed bottom-up by DAG-compression linear in time of the tree. We then search each redundancy component only once and combine the results to get the complete result set, i.e., the same result as if we had performed the search based on the tree and not based on the DAG.

3.1   Index

Our index is called IDCluster. It contains the IDLists for each redundancy component and the redundancy component pointer map (RCPM) which is used for combining the results of redundancy components. The IDCluster is generated in two passes through the document.

The first pass is an extended DAG compression where nodes are considered as being identical when (a) they directly contain the same keywords and (b) all children are identical. When a node n' is found which is identical to a node n found earlier, node n' is deleted and the edge from its parent $parent(n')$ to n' is replaced by a new edge from $parent(n')$ to n. This new edge is called an offset edge and contains the difference between the IDs of n' and n as an additional integer value. The information contained in offset edges will later be used to recalculate the original node ID of n'. Furthermore, for each node, we store the $OccurrenceCount$ which indicates the number of identical occurrences of this node. Fig. 3 shows the XML tree after the first traversal. The nodes with a white background have an $OccurrenceCount$ of 1, while nodes with a grey background have an $OccurenceCount$ of 2. E.g., node 5 is identical to node 12 in the original tree. Hence node 12 is deleted and implicitly represented by node 5 and an offset edge on its path with the offset +7. Node 5 has an $OccurrenceCount$ of 2, which indicates that node 5 now represents two nodes of the original tree.

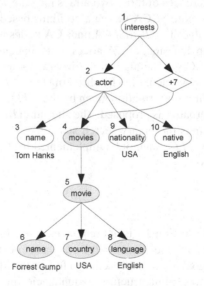

Fig. 3. DAG compressed XML tree after the first traversal

In the second pass, the $OccurenceCount$ values are used for selecting redundancy components: Each connected component consisting only of nodes with the same $OccurenceCount$ is selected as a redundancy component. (The redundancy component furthermore includes additional dummy nodes which represent nested redundancy components and are introduced below). Identifying redundancy components and constructing their IDLists can be done easily in a single document traversal utilizing the *OccurenceCount*. For each redundancy component, a distinct set of IDLists is created. Each IDList entry for nodes belonging to a redundancy component rc is stored in the set of IDLists for rc. The IDLists also include additional entries that represent nested redundancy components. These additional entries, called dummy nodes, have the same ID as the root node of the represented nested redundancy component rc_{nested} and are only added to IDLists of keywords contained in rc_{nested}. Fig. 4 shows the IDLists created for the keywords *USA* and *English*. Dummy nodes are the entries at the positions 2 and 4 in L_{USA}^{rc0} and positions 2 and 4 in $L_{English}^{rc0}$.

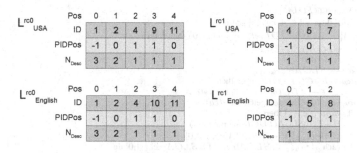

Fig. 4. Created IDLists for the keywords *USA* and *English* as part of an IDCluster

The dummy nodes are added to the RCPM. The key of each entry in the RCPM is the ID of the respective dummy node. Each entry contains the identifier of the redundancy component which the dummy node is representing and an offset. The offset is given by the offset edge between rc and rc_{nested}, or +0 if the edge between rc and rc_{nested} is not an offset edge. Fig. 5 shows the RCPM of the IDCluster. Note that only one RCPM is required, no matter how many keywords are indexed.

Key	RC	Offset
4	rc1	+0
11	rc1	+7

Fig. 5. Created redundancy component pointer map (RCPM) as part of an IDCluster

3.2 Search

SLCA Computation. The main idea for SLCA computation is to utilize the base algorithms for searching each redundancy component individually. Due to the added dummy nodes, any SLCA which is found in a redundancy component and is not a dummy node, is also SLCA in the original tree. SLCA, which are dummy nodes, indicate that more SLCAs are inside the redundancy component, which the dummy node is representing.

This way, all redundancy components containing SLCA can be searched starting at the root redundancy component. The offset values stored in the RCPM are used for recalculating the original ID of SLCA contained in nested redundancy components.

```
 1: DagFwdSLCA(IDCluster,0)

 2: procedure DagFwdSLCA(IDCluster,rc_cur)
 3:     IDLists ⟵ IDCluster.getIDLists(rc_cur)
 4:     while ¬ EoL(IDLists) do
 5:         v ⟵ FwdGetCA(IDLists)
 6:         if v ≠ null then
 7:             if u ≠ null ∧ v.parent ≠ u then
 8:                 SLCA[rc_cur].add(u)
 9:             u ⟵ v
10:             ADVANCE(IDLists)
11:     if u ≠ null then SLCA[rc_cur].add(u)
12:     if SLCA[rc_cur][0] = IDLists[0].getID(0) then return
13:     size ⟵ SLCA[rc_cur].size
14:     for i ⟵ 0, size do
15:         slca ⟵ SLCA[rc_cur][i]
16:         if slca ∈ IDCluster.RCPM then
17:             rc_nes ⟵ IDCluster.getPointer(slca)
18:             os ⟵ IDCluster.getOffset(slca)
19:             if ¬done[rc_nes] then
20:                 DagFwdSLCA(IDCluster,rc_nes)
21:             SLCA[rc_cur][i] ⟵ SLCA[rc_nes][0] + os
22:             for each slca_nes ∈ SLCA[rc_nes] \ SLCA[rc_nes][0] do
23:                 SLCA[rc_cur].add(slca_nes + os)
24:     done[rc_cur] ⟵ true
25: end procedure

26: function EoL(IDLists)
27:     for each IDList ∈ IDLists do
28:         if IDList.C_i ≥ IDList.length then
29:             return true
30:     return false
31: end function

32: procedure ADVANCE(IDLists)
33:     for each IDList ∈ IDLists do
34:         IDList.C_i ⟵ IDList.C_i+1
35: end procedure
```

Fig. 6. Alternative SLCA search algorithm for redundancy components

The basic search algorithm for IDCluster is an extension of FwdSLCA [15] and is depicted in Fig. 6. The algorithm starts searching in the redundancy component with the ID 0, which by definition is the redundancy component containing the document's root node. In lines 4-11 the procedure DagFwdSLCA performs an SLCA search similar to FwdSLCA on one redundancy component. In lines 12-23 the results from this redundancy component are processed. To check whether an SLCA is a dummy node, the SLCA is looked up in the RCPM in line 16. If the SLCA is a dummy node, more SLCA can be found in the nested redundancy component. Therefore, the nested redundancy component is searched recursively in line 20 if it was not searched before. The results from the nested redundancy component are now modified by the offset and added to the SLCA list in lines 21-23. Note that in line 21 the dummy node is deleted from the

SLCA list by replacing it with an actual SLCA contained in the nested redundancy component. The only exception to this proceeding is when the SLCA list contains only the root node of the current redundancy component. This is a special case since an RCPM lookup would return dummy node information (the dummy node and the root node of the nested redundancy component have the same ID) and cause an infinite loop.

If we consider our example, a search for the keywords *USA* and *English* starts with the call of the main algorithm for the root redundancy component rc_0. In lines 12-23, node 4 and node 11 are calculated as the SLCA results for this redundancy component. In the first iteration in lines 14-23, node 4 is identified as a dummy node (line 16). As there are no results yet for the nested redundancy component, rc_1 is searched recursively now. In rc_1, the only SLCA result is node 5. Since node 5 is not found in the RCPM (line 16), the SLCA result list for rc_1 remains unchanged and the recursion terminates. Back in the parenting recursion for rc_0, inside the SLCA result list, the dummy node 4 is replaced with the first result of the nested redundancy component, increased by the offset, 5+0=5 in line 21. Since no more SLCA results exist in rc_1, the loop in lines 22-23 is skipped. In the next iteration of the outer loop, node 11 is identified as a dummy node in line 16. The nested redundancy component is once again rc_1, for which at this point results already exist. Therefore, rc_1 is not searched again and the dummy node is replaced by the SLCA result from rc_1 increased by the offset given by the RCPM, 5+7=12. At this point, the outer loop terminates, and with it, the algorithm terminates, while the SLCA result list for rc_0 contains nodes 5 and 12 as the final result.

One advantage of this approach is that the algorithm given by Zhou et al. is integrated as unmodified module. This means our approach will benefit from any improvements made to the base algorithm, like the parent skipping introduced in BwdSLCA or the improved binary search introduced in BwdSLCA+.

ELCA Computation. ELCA search can be implemented in a similar manner, which is shown in Figure 7. Lines 3-13 are similar to the algorithm given by Zhou et al. [15]. The following lines are adopted from the DagFwdSLCA algorithm. Note that a redundancy component can contain multiple ELCA, even when the root is ELCA. So, if the first ELCA is the root, the first ELCA is just skipped in line 16 instead of aborting the whole function call as in DagFwdSLCA.

In a search for *USA* and *English*, the ELCA nodes found in rc_0 are 2, 4 and 11. Node 2 is not in the RCPM and therefore a final ELCA. Node 4 is a dummy node and forces a search in rc_1. The ELCA list in rc_1 contains node 5 only. Therefore, node 4 in the ELCA list of rc_0 is replaced with 5 plus the offset 0. Node 11 is also a dummy node pointing to rc_1. Since rc_1 was already searched, node 11 in the ELCA list of rc_0 is replaced with 5 plus the offset 7. Therefore, the final ELCA results are 2, 5, and 12.

4 Evaluation

To test the performance of this paper's algorithms, comprehensive experiments were run. The experiments focus on the comparison between the base algorithms and their respective DAG variants introduced in this paper. An evaluation of the base algorithms showing their superior performance in comparison to other classes of keyword search algorithms can be found in the original papers [14][15].

```
 1: DAGFWDELCA(IDCluster,0)

 2: procedure DAGFWDELCA(IDCluster,rc_cur)
 3:     IDLists ⟵ IDCluster.getIDLists(rc_cur)
 4:     while ¬ EoL(IDLists) do
 5:         v ⟵ FWDGETCA(IDLists)
 6:         if v ≠ null then
 7:             while ¬S.empty ∧ S.top ≠ v.parent do
 8:                 PROCESSSTACKENTRY
 9:             S.push(v)
10:         else break
11:         ADVANCE(IDLists)
12:     while ¬S.empty do
13:         PROCESSSTACKENTRY
14:     start ⟵ 0
15:     if ELCA[rc_cur][0] = IDLists[0].getID(0) then
16:         start ⟵ 1
17:     size ⟵ ELCA[rc_cur].size
18:     for i ⟵ start, size do
19:         elca ⟵ ELCA[rc_cur][i]
20:         if elca ∈ IDCluster.RCPM then
21:             rc_nes ⟵ IDCluster.getPointer(elca)
22:             os ⟵ IDCluster.getOffset(elca)
23:             if ¬done[rc_nes] then
24:                 DAGFWDELCA(IDCluster,rc_nes)
25:             ELCA[rc_cur][i] ⟵ ELCA[rc_nes][0] + os
26:             for each elca_nes ∈ ELCA[rc_nes] \ ELCA[rc_nes][0] do
27:                 ELCA[rc_cur].add(elca_nes + os)
28:     done[rc_cur] ⟵ true
29: end procedure
```

Fig. 7. Alternative ELCA search algorithm for redundancy components

4.1 Setup

All experiments were run on a Xeon E5-2670 with 256GB memory and Linux OS. The algorithms were implemented in Java 1.6.0_24 and executed using the OpenJDK 64-Bit Server VM. The time results are the averages of 1,000 runs with warm cache.

The XML version of the music database discogs.com[1] is used as testdata. It contains 4.2 million records of music releases of a size of 12.6GB. To evaluate the effects of different database sizes, smaller file sizes are created by successivly removing the second half of the set of all remaining records. Thereby, additional databases having the sizes of 0.8GB/1.6GB/3.3GB/6.5GB are created.

For the evaluation, 3 categories of queries are proposed:

- Category 1: Queries consisting of nodes that will not be compressed in a DAG. DAG-based algorithms cannot exploit these kinds of queries. Since the DAG-based algorithms still need time for verifying the absence of RC-Pointers (nodes with entries in the RCPM), they should have worse performance than the base algorithms.
- Category 2: Queries consisting of nodes that will be compressed in a DAG, but having common ancestors (CA) which still cannot be compressed. This means that all results will be in the first redundancy component. DAG-based algorithms can exploit the fact that the IDLists for the first redundancy can be shorter than the

---

[1] http://www.discogs.com/data/

Table 1. Properties of queries

query	CA	S_{ca}	ELCA	S_{elca}	SLCA	S_{slca}
Q1	14818739	0%	7697608	0%	7697603	0%
Q2	3560569	0%	3560568	0%	3560567	0%
Q3	1299279	0%	1299278	0%	1299277	0%
Q4	824063	0%	823952	0%	823840	0%
Q5	616919	0%	616907	0%	616905	0%
Q6	207865	0%	207864	0%	207863	0%
Q7	3438507	78%	709713	98%	703949	99%
Q8	2389299	78%	486093	98%	481211	99%
Q9	1307891	73%	328419	99%	328386	99%

Table 2. Properties of keywords

keyword	path	S_{path}	nodes	S_{nodes}
image	14862218	0%	7716896	0%
uri	22521466	0%	7699158	0%
release	4788936	9%	4447144	3%
identifiers	2845185	6%	1422641	12%
vinyl	9088129	72%	2343810	97%
electronic	5305014	66%	1773128	99%
12"	3976817	77%	812587	97%
uk	1753528	50%	872902	95%
description	36872585	76%	12774491	92%
rpm	3460698	78%	715987	98%
45	2524740	76%	531299	95%
7"	3371487	77%	689983	97%

IDLists for the base algorithms. On the other hand, the absence of all RC-Pointers still has to be verified. These advantages and disadvantages might cancel each other depending on the situation.

- Category 3: Queries with results that can be compressed in a DAG. This means that there has to be at least a second redundancy component containing all keywords. DAG-based algorithms should have a better performance than the base algorithms for queries from this category.

Queries of different lengths for all categories are selected randomly using the 200 most frequent keywords. Table 1 shows the properties of these queries. The columns CA, ELCA and SLCA show the total number of CA-, ELCA- or SLCA-nodes respectively. The columns S_{ca}, S_{elca} and S_{slca} show the savings by DAG compression, e.g. a total number of 100 CA and 80% CA savings imply that 20 CA are left in the XML tree after DAG compression. The properties of the used keywords can be found in Table 2. The column **nodes** shows the number of nodes directly containing the respective keyword; the column **path** shows the number of nodes directly or indirectly containing the keyword. The columns S_{nodes} and S_{path} accordingly show the compression savings.

4.2   Experiment I: Category

In the first experiment, the performance of the base algorithm FwdSLCA is compared to the DAG-based algorithm DagFwdSLCA. Database size and query length are fixed, while the category is altered. The results are shown in Fig. 8.

The results confirm that in Category 1, the performance of the DAG-based algorithm is a bit worse than the performance of the base algorithm. This is as expected, since there are no redundancies which can be exploited by a DAG-based algorithm, but the DAG-based algorithm has a certain overhead for verifying that no further redundancy components needs to be searched, resulting in a worse performance. In Category 2, the performance of both algorithms is very similar with the base algorithm being slightly faster. The better relative performance of the DAG-based algorithm can be traced back to the IDLists being shorter than the IDLists used in the base algorithm. Finally, in Category 3, the DAG-based algorithm is more than twice as fast as the base algorithm.

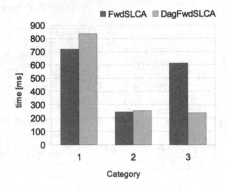

Fig. 8. Comparison with different categories. (database size: 12.6GB; query length: 3).

4.3   Experiment II: Query Length

In this experiment, the length of the queries is modified. Fig. 9 shows the results for categories 1 and 3 with a fixed database size.

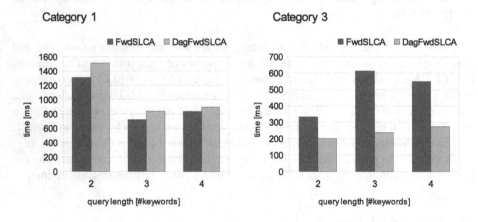

Fig. 9. Comparison with different query lengths. (database size: 12.6GB).

The general tendency is the same: The base algorithm is a bit better for Category 1 queries, while the DAG-based algorithms are better for Category 3 queries. The Category 1 results suggest that the gap between both algorithms gets smaller the more keywords are used. This is plausible, since the overhead for the DAG-based algorithms depends on the amount of results. Adding more keywords to a query can reduce the amount of results, but never increase it (see Table 1).

4.4   Experiment III: Database Size

The third experiment examines the effects of database size and is shown in Fig. 10 for Category 1 and Category 3 with a fixed query length.

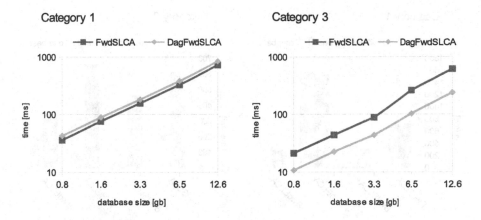

Fig. 10. Comparison with different database sizes. (query length: 3).

The exponential growth of the database size leads to an exponential growth in the search time for both algorithms. Minor changes in the proportions between both algorithms can be traced back to minor changes in the keyword frequencies and compression savings. Therefore, the database size seems not to have a direct impact on the performance ratio between the base algorithm and the DAG-based algorithm.

4.5 Experiment IV: Algorithm

In the last experiment, algorithms FwdSLCA, BwdSLCA+, FwdELCA and BwdELCA as proposed by Zhou et al. are compared to our DAG-based variants. Database size and query length are fixed.

For Category 1, the DAG-based algorithms always have a small overhead independent of the algorithm. In Category 3, the relative difference is smaller for backward algorithms, but still significant. Backward search performs generally better than forward search of the same type. The only exceptions are the DAG-based variants of FwdSLCA and of BwdSLCA+ in Category 3. The backward algorithm is actually slower. This result suggests that a major part of the speedup in BwdSLCA+ is generated by parent skipping. Due to the DAG compression many of the cases in which parent skipping provides benefits are already optimized.

4.6 Index Size

The size of the IDCluster differs from the size of IDLists in two aspects. On the one hand, additional space is required for storing the RCPM. On the other hand, less space is required for storing the IDLists due to DAG compression.

The RCPM can be stored in different ways which affect the required memory space and the time performance. In this evaluation, the RCPM is stored as an array containing the redundancy component identifier and the offset. The node ID is implicitly represented by the position in the array. The size of the array has to be big enough to contain all node IDs. This way of storing the RCPM is optimized for time performance.

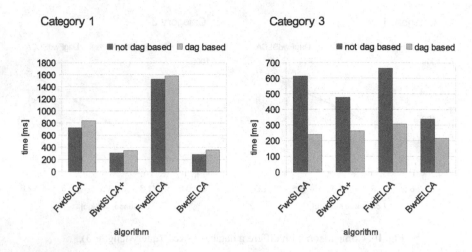

Fig. 11. Comparison with different algorithms. (database size: 12.6GB, query length: 3).

The DAG compression strongly depends on the XML database used. The additionally created dummy nodes also have to be considered.

In a typical use case, both effects, additional memory for RCPM and reduced memory due to DAG compression, are likely to cancel each other.

In the Discogs database, the total amount of nodes in the IDLists index is 3.9 billion. Considering the 2 (3) integers per node required for performing an SLCA (ELCA) search and an integer size of 4 bytes, the total IDLists index sums up to 28.7GB (43.0GB). The amount of nodes in an IDCluster is only 3.0 billion for the same database. Storing these nodes sums up to 22.5GB (33.7GB) for SLCA (ELCA) search. However storing the RCPM in an array as described above (with redundancy component identifier and offset both as 4 bytes integer) for all 656 million distinct nodes requires an additional 4.9GB. So, the total memory required for the IDCluster is 27.3GB (38.6GB) for SLCA (ELCA) search.

5   Related Work

There exist several approaches that address the problem of keyword search in XML. These approaches can be roughly divided into two categories: approaches that enhance the quality of the search results by considering the semantics of the queries on the one hand, and approaches that enhance the performance of the computation of the set of query results on the other hand.

Within the first category, XSEarch [3] presents a query semantics that returns only those XML fragments, the result nodes of which are *meaningfully related*. Meaningful related means that the result nodes intuitively belong to the same entity. In order to check this, they examine whether a pair of result nodes has two different ancestor nodes that have the same label (e.g., two nodes with label "author", such that the first keyword belongs to author1 and the second one to author2).

[5] not only focusses on an efficient, stack-based algorithm for keyword search based on inverted element lists of the node's DeweyIDs, but also aims to rank the search results in such a way, that the user gets the (probably) most interesting results prior to the other results. SUITS [16] is a heuristics-based approach, and the approach presented in [10] uses probabilistic scoring to rank the query results. In order to enhance the usability, [8] and [7] propose an approach on how to group the query results by category.

Within the second category (efficient result computation) most approaches are based on finding a set of SLCA (or ELCA) nodes for all matches of a given keyword list.

Early approaches were computing the LCA for a set of given keywords on the fly. [11] proposes the meet-operator that computes the LCA for a pair of nodes that match two query strings without requiring additional knowledge on the document structure.

In contrast, recent approaches try to enhance the query performance by using a pre-computed index.

[4] proposes an extension of XML-QL by keyword search. In order to speed-up the keyword search, it computes the so-called "inverted file" for the XML document – a set of inverted element lists – and stores the contents within a relational database.

[9] presents two approaches to compute the Meaningful Lowest Common Ancestor (MLCA), a concept similar to the SLCA. Its first approach allows computing the MLCA with the help of standard XQuery operations, whereas its second approach is a more efficient approach that utilizes a stack-based algorithm for structural joins.

Similar to XRANK [5] is the stack-based approach presented in [6]. In contrast to the previous stack-based approaches, the authors propose to use a combination of preorder position, postorder position, and depth of the node instead of the DeweyID.

XKSearch [13] is an indexed-based approach to compute the LCA. They store inverted element lists consisting of DeweyIDs of the nodes. They start searching for the results at node n of the shortest keyword list, and they check for the other keyword lists whether the node l (or r) being the next node to the left (or right) of n has a smaller distance to n. Then, they use n and the nearest node (l or r) to compute the LCA.

[12] presents an anchor-based approach to compute the SLCA. From the set of current nodes of each relevant keyword list, they search the so-called anchor, i.e., that node that is closest to all current nodes. As soon as an anchor is identified, they try to exchange each node n_i of each other keyword list L_i by the next node next(n_i) of L_i, in order to check, whether next(n_i) is closer to the anchor than n_i and whether next(n_i) defines a new anchor. Finally, the set of anchor nodes form the set of LCA candidates that do not have another LCA candidate child is then reduced to the set of SLCA nodes.

JDeweyJoin [2] returns the top-k most relevant results. This approach computes the results bottom-up by computing a kind of join on the list of DeweyIDs of the nodes in the inverted element list. Whenever it finds a prefix that is contained in all relevant element lists, the node with this prefix as ID is a result candidate. In addition, it uses a weight function to sort the list entries in such a way, that it can stop the computation after k results, returning the top-k most relevant results.

[14] and [15] belong to the intersection-based approaches. They present a more efficient, but more space-consuming approach. The elements of their inverted element lists do not only contain the nodes that have the keyword as label, but also contain all ancestor-nodes of these nodes, and for each node, the inverted element lists contain the

ID of the parent node. Therefore, they can compute the SLCAs by intersecting the inverted element lists of the keywords and by finally removing each result candidate, the descendant of which is another result candidate.

Like the contributions of the second category, our paper focusses on efficient result computation. It follows the idea of the intersection-based approaches. However, different from all other contributions and similar to a prior approach [1], instead of computing an XML-index, we compute a DAG-Index. This helps to compute several keyword search results in parallel, and thereby speeds-up the SLCA computation. To the best of our knowledge, DAG-Index is the first approach that improves keyword search by using XML compression before computing the search index.

6 Conclusions

We have presented IDCLuster, an indexing and search technique that shares common sub-trees in order to index and to search redundant data only once.

As our performance evaluation shows, using the DAG-based index of IDCluster, the intersection-based keyword search algorithms can be significantly improved, i.e., gain a speed-up up to a factor of more than 2.

Therefore, we consider the idea to cluster repeated data collections to be a significant contribution to all applications that have to search in large data collections with high amounts of copied, backed-up or redundant data.

Acknowledgments. We would like to thank Wei Wang for helpful discussions and comments during our project cooperation.

This work has been supported by DAAD under project ID 56266786.

References

1. Böttcher, S., Brandenburg, M., Hartel, R.: DAG-Index: A compressed index for XML Keyword Search. (Poster) 9th International Conference on Web Information Systems and Technologies, WEBIST 2013 (2013)
2. Chen, L.J., Papakonstantinou, Y.: Supporting top-k keyword search in XML databases. In: 2010 IEEE 26th International Conference on Data Engineering (ICDE), pp. 689–700. IEEE (2010)
3. Cohen, S., Mamou, J., Kanza, Y., Sagiv, Y.: XSEarch: a semantic search engine for XML. In: Proceedings of the 29th International Conference on Very Large Data Bases, VLDB 2003, vol. 29, pp. 45–56. VLDB Endowment (2003), http://dl.acm.org/citation.cfm?id=1315451.1315457
4. Florescu, D., Kossmann, D., Manolescu, I.: Integrating keyword search into XML query processing. Computer Networks 33(1), 119–135 (2000)
5. Guo, L., Shao, F., Botev, C., Shanmugasundaram, J.: XRANK: ranked keyword search over XML documents. In: Proceedings of the 2003 ACM SIGMOD International Conference on Management of Data, pp. 16–27. ACM (2003)
6. Hristidis, V., Koudas, N., Papakonstantinou, Y., Srivastava, D.: Keyword proximity search in XML trees. IEEE Transactions on Knowledge and Data Engineering 18(4), 525–539 (2006)

7. Koutrika, G., Zadeh, Z.M., Garcia-Molina, H.: Data clouds: summarizing keyword search results over structured data. In: Proceedings of the 12th International Conference on Extending
   Database Technology: Advances in Database Technology, pp. 391–402. ACM (2009)
8. Li, J., Liu, C., Zhou, R., Wang, W.: Suggestion of promising result types for XML keyword
   search. In: Proceedings of the 13th International Conference on Extending Database Technology, pp. 561–572. ACM (2010)
9. Li, Y., Yu, C., Jagadish, H.: Schema-free XQuery. In: Proceedings of the Thirtieth International Conference on Very Large Data Bases, vol. 30, pp. 72–83. VLDB Endowment (2004)
10. Petkova, D., Croft, W.B., Diao, Y.: Refining keyword queries for XML retrieval by combining
    content and structure. In: Boughanem, M., Berrut, C., Mothe, J., Soule-Dupuy, C. (eds.)
    ECIR 2009. LNCS, vol. 5478, pp. 662–669. Springer, Heidelberg (2009)
11. Schmidt, A., Kersten, M., Windhouwer, M.: Querying XML documents made easy: Nearest
    concept queries. In: Proceedings of the 17th International Conference on Data Engineering,
    pp. 321–329. IEEE (2001)
12. Sun, C., Chan, C.Y., Goenka, A.K.: Multiway SLCA-based keyword search in XML data.
    In: Proceedings of the 16th International Conference on World Wide Web (2007)
13. Xu, Y., Papakonstantinou, Y.: Efficient keyword search for smallest LCAs in XML databases.
    In: Proceedings of the 2005 ACM SIGMOD International Conference on Management of
    Data, pp. 527–538. ACM (2005)
14. Zhou, J., Bao, Z., Wang, W., Ling, T.W., Chen, Z., Lin, X., Guo, J.: Fast SLCA and ELCA
    computation for XML keyword queries based on set intersection. In: 2012 IEEE 28th Inter-
    national Conference on Data Engineering (ICDE), pp. 905–916. IEEE (2012)
15. Zhou, J., Bao, Z., Wang, W., Zhao, J., Meng, X.: Efficient query processing for XML key-
    word queries based on the IDList index. The VLDB Journal, 1–26 (2013)
16. Zhou, X., Zenz, G., Demidova, E., Nejdl, W.: SUITS: Constructing Structured Queries from
    Keywords. LS3 Research center (2008)

Finding Missing Answers due to Object Duplication in XML Keyword Search

Thuy Ngoc Le, Zhong Zeng, and Tok Wang Ling

National University of Singapore
{ltngoc,zengzh,lingtw}@comp.nus.edu.sg

Abstract. XML documents often have duplicated objects, with a view to maintaining tree structure. Once object duplication occurs, two nodes may have the same object as the child. However, this child object is not discovered by the typical LCA (Lowest Common Ancestor) based approaches in XML keyword search. This may lead to the problem of missing answers in those approaches. To solve this problem, we propose a new approach, in which we model an XML document as a so-called XML IDREF graph so that all instances of the same object are linked. Thereby, the missing answers can be found by following these links. Moreover, to improve the efficiency of the search over XML IDREF graph, we exploit the hierarchical structure of the XML IDREF graph so that we can generalize the efficient techniques of the LCA-based approaches for searching over XML IDREF graph. The experimental results show that our approach outperforms the existing approaches in term of both effectiveness and efficiency.

1 Introduction

Since XML has become a generally accepted standard for data exchange over the Internet, many applications use XML to represent the data. Therefore, keyword search over XML documents has attracted a lot of interests. The popular approach for XML keyword search is the LCA (Lowest Common Ancestor) semantics [4], which was inspired by the hierarchical structure of XML. Following this, many extensions of the LCA semantics such as SLCA [17], MLCA [13], ELCA [19] and VLCA [10] have been proposed to improve the effectiveness of the search. However, since these approaches only search up to find common ancestors, they may suffer from the problem of missing answers as discussed below.

1.1 The Problem of Missing Answers due to Object Duplication

XML permits nodes to be related through parent-child relationships. However, if the relationship type between two object classes is many-to-many without using IDREF, an object can occur at multiple places in an XML document because it is duplicated for each occurrence in the relationship. We refer such duplication as *object duplication*.

Example 1. *Consider an XML document in Figure 1 where the relationship type between student and course is many-to-many ($m : n$) and students are listed as children of courses. When a student takes two courses, this student is repeated under*

H. Decker et al. (Eds.): DEXA 2014, Part I, LNCS 8644, pp. 138–155, 2014.

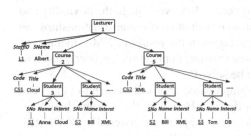

Fig. 1. XML data tree

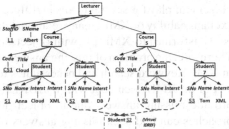

Fig. 2. XML IDREF graph w.r.t. XML data tree in Figure 1

both courses. For example, both courses <Course:CS1>[1] and <Course:CS2> are taken by <Student:S2>, which is repeated as the two groups of nodes, starting at node 4 and node 6 under the two courses. This causes the duplication of object <Student:S2>.

When object duplication happens, two nodes may have the same object as the child. However, this *common child object* is not discovered by the LCA-based approaches because they only search up from matching nodes for common ancestors, but never search down to find common information appearing as descendants of matching nodes. We call this incident as the problem of *missing answers due to object duplication* which leads to loss of useful information as illustrated below.

Example 2. *Consider keyword query* {CS1, CS2} *issued against the XML data in Figure 1, where the keywords match object identifier of two courses (node 2 and node 5). The LCA-based approaches return only* <Lecturer:L1> *(node 1) as an answer. However, as discussed in Example 1, object* <Student:S2> *is the common student taking both matching courses and thus it should also be an answer. Intuitively, the two courses are not only taught by the same lecturer (*<Lecturer:L1>*), but also taken by the same student (*<Student:S2>*). As we can see, common information related to query keywords appearing as both ancestors and descendants are meaningful to users.*

Object duplication can be eliminated by ID/IDREF. However, to maintain the tree structure for ease of understanding, readability and retrieval, XML designers may duplicate objects instead of using ID/IDREF. In practice, object duplication is a common scheme for maintaining a view of tree structure. For example, suppose 300 students take course A. Among them, 200 students also take another course B. Then, if students are listed as children of courses, these 200 students are duplicated under both courses. Many real XML datasets, including IMDb[2] and NBA[3] (used in experiments of XML research works such as [16,15]), contain object duplication. In IMDb, an actor or actress can play in many movies, and a company can produce several movies. In NBA, a

---

[1] <Course:CS> denotes an object which belongs to object class Course and has object identifier CS1.

[2] http://www.imdb.com/interfaces

[3] http://www.nba.com

player can play for several teams in different years. Moreover, due to the flexibility and exchangeability of XML, many relational datasets with many-to-many relationships can be transformed to XML [3] with object duplication in the resulting XML documents. Therefore, the problem of missing answers due to object duplication frequently happens in XML keyword search and necessitates to be solved.

For an XML document with ID/IDREF, graph-based approaches such as [11,7] can provide missing answers due to object duplication. However, those graph-based approaches can find such missing answers only if all objects are covered by ID/IDREF mechanism. Otherwise, those graph-based approaches do not recognize instances of the same object appearing in different places in an XML document. In such cases, they cannot find missing answers either.

1.2 Our Approach and Contributions

In this paper, we propose an approach for keyword search over a data-centric XML document which can find missing answers due to object duplication. The input XML document in our approach can contain both objects under ID/IDREF mechanism and duplicated objects. For the latter, we propose a virtual object node to connect all instances of the same object via virtual IDREFs. The resulting model is called *XML IDREF graph*. "Virtual" here means we do not modify XML documents and ID/IDREF links are virtually created with the sole goal of finding missing answers.

A challenge appears when we have to deal with an XML IDREF *graph*, not a *tree* anymore. Searching over an arbitrary graph-structured data has been known to be equivalent to the group Steiner tree problem, which is NP-Hard [2]. In contrast, keyword search on XML tree is much more efficient thanks to the hierarchical structure of XML tree. This is because the search in an XML tree can be reduced to find LCAs of matching nodes, which can be efficiently computed based on node labels.

We discover that XML IDREF graph is a special graph. Particularly, it is an XML tree (with parent-child (PC) edges) plus a portion of *IDREF edges*. An IDREF edge is an edge from a referring node to a referred node. Although these nodes refer to the same object, we can treat them as having a parent-child relationship, in which the parent is the referring node and the child is the referred node. This shows that XML IDREF graph still has hierarchy, which enables us to generalize efficient techniques of LCA-based approaches (based on the hierarchy) for searching over our proposed XML IDREF graph. Thereby, we do not have to traverse the XML IDREF graph to process a keyword query.

Contribution. In brief, we make the following contributions.

- We argue that LCA-based approaches, which only search up to find common ancestors, may miss meaningful answers due to object duplication. To find such missing answers, we model an XML document as an XML IDREF graph, in which all instances of the same object are connected by a virtual object node.
- We discover the hierarchical structure of an XML IDREF graph which distinguishes it from an arbitrary graph. Based on this hierarchical structure, we can generalize techniques of the LCA-based approaches for an efficient search.

- The experimental results show that our approach outperforms both the graph-based and LCA-based approaches in term of both effectiveness and efficiency.

Roadmap. The rest of the paper is organized as follows. We introduce data model and answer model in Section 2. Our approach is described in Section 3. The experiment and evaluation are provided in Section 4. We review related works in Section 5. Finally, we conclude the paper in Section 6.

2 Data and Answer Model

2.1 Data Model

In XML, an object can be referred to either by duplicating it under the referrer or by using ID/IDREF. The former causes object duplication whereas the latter does not. With ID/IDREF, an object has only one instance, and other objects refer to it via ID/IDREF. Without ID/IDREF, an object can be represented as many different instances. We propose virtual ID/IDREF mechanism, in which we assign a virtual object node as a hub to connect all instances of the same object by using virtual IDREF edges. The resulting model is called an XML IDREF graph which is defined as followed.

Definition 1 (XML IDREF graph). *An XML IDREF graph $G(V, E)$ is a directed, labeled graph where V and E are nodes and edges of the graph.*
- *$V = V_R \sqcup V_V$ where V_R and V_V are real and virtual nodes respectively. A real node is an object node in XML document. A virtual node is a virtual object node to connect all instances of the same object in XML document.*
- *$E = E_R \sqcup E_V$ where E_R and E_V are real edges and virtual edges respectively. A real edge (can be a real PC edge or real IDREF edge) is an edge between two real nodes. A virtual edge is the edge links an instance of a duplicated object (real node) to a virtual object node.*

For example, Figure 2 shows an XML IDREF graph with two virtual edges from node 4 and node 6 to a virtual object node (node 8) because node 4 and node 6 are instances of the same object <Student:S2>.

XML permits some objects under ID/IDREF mechanism and some other objects with duplication co-exist in an XML document. In this case, the resulting XML IDREF graph has two types of IDREF: real and virtual. Thus, an XML IDREF graph may have three types of edges: PC edge, real IDREF edge and virtual IDREF edge.

The hierarchical structure of XML IDREF graph. We observe that an XML IDREF graph still has hierarchy with parent-child (PC) relationships represented as containment edges (PC edges) or referenced edges. This is because nodes in a referenced edge can be considered as having PC relationship, in which the parent is the referring node and the child is the referred node.

Importance of the hierarchical structure of XML IDREF graph. Once we discover the hierarchical structure of an XML IDREF graph, we can inherit the efficient search techniques of LCA-based approaches which based on the hierarchical structure of XML tree. Thereby, we do not have to traverse the XML IDREF graph to process a keyword query as graph-based search does. This brings a huge improvement on efficiency. Without the property of the hierarchy, generally, in graph-based search, matching nodes will be expanded to all directions until they can connect to one another. In theory, there can be exponentially many answers under the Steiner tree based semantics: $O(2^m)$ where m is the number of edges in the graph. The graph-based search has been well known to be equivalent to the group Steiner tree problem, which is *NP-Hard* [2].

Generating XML IDREF graph. An object instance in XML is usually represented by a group of nodes, rooted at the object class tagged node, followed by a set of attributes and their associated values to describe its properties. In this paper, we refer to the root of this group node as an *object node* and the other nodes as *non-object nodes*. Hereafter, in unambiguous contexts, we use object node as the representative for a whole object instance, and nodes are object nodes by default. For example, matching node means matching object nodes. Among non-object nodes, *object identifier* (OID) can uniquely identify an object. To generate an XML IDREF graph from an XML document, we need to detect object instances of the same object. Since an object is identified by object class and OID, we assume that two object instances (object nodes as their representatives) are of the same object if they belong to the same *object class* and have the same *OID*.

We assume that the data is consistent and we work on a single XML document. Data integration, data uncertainty, and heterogeneous data are out of the scope of this work. Object classes and OIDs can be discovered from XML schema and data by our previous works [12], which achieve high accuracy (greater than 98% for object classes and greater than 93% for OIDs). Therefore, we assume the task of discovering object class and OID has been done. Interested readers can find more details in [12].

2.2   Answer Model

Consider a n-keyword query $Q = \{k_1, \ldots, k_n\}$. An answer to Q contains three kinds of nodes: *matching nodes*, *center nodes* and *connecting nodes*. A matching node contains keyword(s). A center node connects all matching nodes through some intermediate nodes (called connecting nodes). Based on the hierarchical structure of XML IDREF graph, there exist ancestor-descendant relationships among nodes in an XML IDREF graph. Therefore, we can define an answer to Q as follows:

Definition 2. *Given a keyword query $Q = \{k_1, \ldots, k_n\}$ to an XML IDREF graph G, an answer to Q is a triplet $\langle c, \mathbb{K}, \mathbb{I} \rangle$, where $c, \mathbb{K}$, and $\mathbb{I}$ are called the* center node, *the set of* matching nodes *and the set of* connecting nodes *(or intermediate nodes) respectively.* $\mathbb{K} = \bigcup_1^n u_i$ *where u_i contains k_i. An answer satisfies the following properties:*

 – *(P1: Connective) For every i, c is an ancestor of u_i or for every i, c is a descendant of u_i, i.e., c is either a common ancestor or a common descendant of u_i's.*
 – *(P2: Informative) For any answer $\langle c', \mathbb{K}', \mathbb{I}' \rangle$ where $\mathbb{K}' = \bigcup_1^n u_i'$ and u_i' contains k_i:*
   • *if c and c' are both common ancestors of u_i's, and of u_i''s respectively, and c' is a descendant of c, then $\forall i\ u_i \notin \mathbb{K}'$.*
   • *if c and c' are both common descendant of u_i's, and of u_i''s respectively, and c' is a ancestor of c, then $\forall i\ u_i \notin \mathbb{K}'$.*
 – *(P3: Minimal) It is unable to remove any node in an answer such that it still satisfies properties P1 and P2.*

Among nodes in an answer, the center node is the most important one because it connects matching nodes through connecting nodes. It corresponds to both common ancestors and common descendants (Figure 3(a)). Intuitively, common ancestors are similar the LCA semantics while common descendants provide the missing answers. We do not return the subgraph in Figure 3(b) because it may provide meaningless answers. In other words, a center node has only incoming edges (common descendant), or only outgoing edges (common ancestor), but not both.

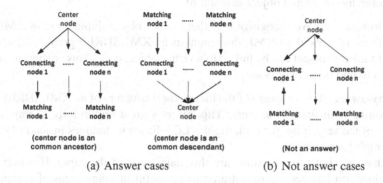

(a) Answer cases   (b) Not answer cases

Fig. 3. Illustration for answers

The second Property P2 of Definition 2 is to avoid overlapping information in answers. Each answer needs to contribute new information by having its own set of matching nodes, i.e., matching nodes of an answer cannot also be matching nodes of other answers where the latter is an ancestors/descendants of the former one. This property is similar to the constraint in the ELCA [19] semantics.

3 Our Approach

Our approach takes a data-centric XML document as the input, models it as an XML IDREF graph, and returns answers as defined in Definition 2. In the input XML document, objects under IDREF mechanism and objects with duplication can co-exist.

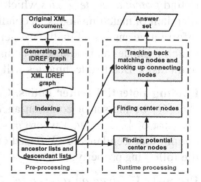

Fig. 4. The process of our approach

The process of our approach, as shown in Figure 4, comprises of two major components for pre-processing and runtime processing. For pre-processing, there are two main tasks, namely generating XML IDREF graph (discussed in Section 2.1), and indexing (will be discussed in Section 3.2). For runtime processing, there are three main tasks, each of which corresponds to a property of an answer in Definition 2. Particularly, task 1 is to find potential center nodes, task 2 is to find real center nodes, and task 3 is to track back matching nodes and look up connecting nodes. These steps will be discussed in Section 3.3. Before discussing detailed techniques, we provide the overview of the approach at the conceptual level in Section 3.1.

3.1 Overview of the Approach

A. Features of Our Approach. Our approach has three main features: duplication-aware, hierarchy-aware and object-orientation.

Duplication-aware. We recognize that there exists object duplication in XML documents. Thus, we model an XML document as an XML IDREF graph so that all instances of the same object can be linked to a virtual object node. This enables us follows these links to find *missing answers*.

Hierarchy-aware. We are aware of the hierarchical structure of an XML IDREF graph and exploiting it in finding answers. This offers a great opportunity to improve the efficiency of the search by generalizing the LCA-based techniques instead of doing a general graph-based search.

Note that the above two features are the main focus of the paper. However, since object orientation has been demonstrated to be useful in many areas of computation (including database field), our approach also relies upon object orientation.

Object-orientation. All nodes of the same object instance can be grouped. Among these nodes, object node is the most important one and should be chosen as the representative of the group. Instead of working with all these nodes, we only work with the representative of each group, i.e., the object node, and associate all non-object nodes to the corresponding object node. This largely reduces search space and improves the efficiency of the search.

B. Basic Ideas of Runtime Processing. Our runtime processing has three main tasks corresponding to three properties of an answer based on Definition 2. Firstly, we need to find *potential center nodes* which can be either a common ancestor or a common descendant of matching nodes. Secondly, we have to find *center nodes* which can provide informative answers. Finally, we get *full answers* w.r.t. a center node by tracking back matching nodes and looking up connecting nodes. Follows are theories behind the three main tasks. Detailed techniques will be provided in Section. 3.3.

Finding potential center nodes. Consider a keyword query $Q = \{k_1, \ldots, k_n\}$. Let $Anc(Q)$ be the set of common ancestors of Q, i.e., $\forall u \in Anc(Q)$, u is a common ancestor of $\{u_1, \ldots, u_n\}$ where u_i contains k_i. Similarly, let $Dec(Q)$ denote the set of common descendants of Q. Based on *Property P1* of Definition 2, obviously, we have the following property:

Property 1. *Given an answer $\langle c, \mathbb{K}, \mathbb{I} \rangle$ for a keyword query Q, the* center node $c \in Anc(Q) \cup Des(Q)$.

For a set of nodes, common descendants of these nodes can only be object nodes (real or virtual) which is referred by some other node(s) by IDREF links. We call them *referred object node*. Let $Ref(k)$ is the set of referred object nodes w.r.t. k, each of which is a descendant of some node containing k. For example, in Figure 2, $Ref(Cloud) = \{8\}$. Let $Ref(Q)$ be the set of referred object nodes w.r.t. Q. We have the following property about common descendants.

Property 2. *Given a keyword query Q, $Des(Q) = Ref(Q)$.*

By Property 1 and Property 2, $Anc(Q)$ and $Ref(Q)$ can provide potential center nodes. Since $Anc(Q) = \bigcap_1^n Anc(k_i)$, and $Ref(Q) = \bigcap_1^n Ref(k_i)$, in order to find $Anc(Q)$ and $Ref(Q)$, we use computation of *set intersection*. The computation of set intersection has been used to find SLCA and ELCA in [18] and has been shown to be more efficient than the traditional computation based on common prefix of labels when dealing with XML tree.

Finding center nodes. Among potential center nodes, we identify *real center nodes* by checking *Property P2* of Definition 2, which infers that an answer should have its own matching nodes from its ancestor/descendant answers to be informative. Let $Des(c, k)$ denote the set of descendants of node c which contains keyword k. $Center\_Des(c)$ denotes center nodes which are descendants of c. $Content\_Des(c, k)$ denotes the set of matching nodes w.r.t. all nodes in $Center\_Des(c)$. We have the following property.

Property 3. *Given a keyword query $Q = \{k_1, \ldots, k_n\}$ and $c \in Anc(Q)$, if $Des(c, k) - Content\_Des(c, k) \neq \emptyset \ \forall i = 1..n$, then c is a real center node.*

We use bottom up for checking common ancestors. For a *common ancestor c*, after removing matching nodes of descendant answers out of $Des(c, k)$'s, if c still has its own matching nodes, then it is a real center node. This is similar for checking whether a *common descendant* is a center node. However, the process is top down.

Tracking back matching nodes and looking up connecting nodes. To return a full answer $\langle c, \mathbb{K}, \mathbb{I} \rangle$ to users, after having a center node c, we need to get the corresponding set $\mathbb{K}$ of matching nodes and set $\mathbb{I}$ of connecting nodes such that they satisfy *Property P3* of Definition 2. We follow the below property.

Property 4. *Given an answer $\langle c, \mathbb{K}, \mathbb{I} \rangle$, if $\mathbb{I}$ is the set of nodes on the paths from the center node c to matching nodes in $\mathbb{K}$, then that answer is minimal.*

3.2   Labling and Indexing

A. Labeling. Different from conventional labeling schemes, where each node has a distinct label, we only label *object nodes*. All *non-object nodes* are assigned the same label with their corresponding object nodes. This is the feature of object-orientation of our approach. By this labeling scheme, a keyword matching a non-object node is considered as matching the corresponding object node and the number of labels is largely reduced. This brings huge benefits for the efficiency because the search space is reduced. We use *number* instead of Dewey for labeling because in XML IDREF graph, a node can have multiple parents. The other reason is that computation on number is faster than on Dewey since as each component of the Dewey label needs to be accessed

Table 1. The ancestor lists for keywords Cloud and XML

	L^a_{Cloud} (k = Cloud)			L^a_{XML} (k = XML)				
Matching node and its ancestors	u	1	2	3	u	1	2	3	5	7
Parent of u	Par(u)	0	1	2	Par(u)	0	1	2	1	5
Descendants of u directly containing k	Des(u,k)	2,3	2,3	3	Des(u,k)	3,5,7	3	3	3,7	7
Referer of u containing k if u is IDREF node	IDREF(u,k)	∅	∅	∅	IDREF(u,k)	∅	∅	∅	∅	∅

and computed. Labels are compatible with the *document order* of the XML document. Besides labeling real nodes in XML document, we also label *virtual nodes*. Each virtual node is also assigned a label which succeeds labels of real nodes. For example, in Figure 2, real object nodes are labeled from 1 to 7, and the virtual node is labeled 8.

B. Indexing. Apart from traditional inverted list where each keyword corresponds to a set of matching nodes, to capture both ancestor-descendant relationships and ID/IDREFs in XML IDREF graph, and to facilitate the search, it is necessary to have complex techniques on indexing. A keyword k corresponds to an ancestor list L^a_k and a descendant list L^d_k to facilitate the computation of finding *common ancestors* and finding *common descendants* respectively.

Ancestor list. Each entry in an ancestor list L^a_k is a quadruple which includes:
 – u: an object node that matches k or its ancestors. This will be used in finding common ancestors.
 – $Par(u)$: the parent of object node u. If u is the root, then $Par(u) = -1$. If the parent of u is the root, then $Par(u) = 0$.
 – $Des(u, k)$: the set of descendant object nodes (and itself) of u, which directly contains k. This is used for Property P2.
 – $IDREF(u, k)$: the referrer of u w.r.t. k if u is an referred object node. Otherwise, $IDREF(u, k) = \emptyset$. For example, in Figure 2, $IDREF(node\ 8, Cloud)$ is $node\ 4$. This is used in presenting output when the common ancestor is a referred object node.

For example, Table 1 shows the ancestor lists for keywords Cloud and XML in the XML IDREF graph in Figure 2, where each entry corresponds to a column in the tables. For instance, the first entry of L^a_{Cloud} corresponds to node 1; the parent of node 1 is node 0 (the root); node 2, 3 are descendants of node 1 containing Cloud; and node 1 is not an referred object node.

Descendant referred object node list. Similar to an ancestor list, each entry in a descendant referred object node list L^d_k is a quadruple which includes:
 – u: an *referred object node* which is a descendant of an matching object node. This will be used in finding common descendants. Note that a common descendant can only be an referred object node.
 – $Child(u)$: set of referred object nodes which are children of u.
 – $Match(u, k)$: the set of ancestor object nodes of u, which directly contains k. This is used for Property P2.
 – $Path(u, k)$: paths from each node in $Match(u, k)$ to u. This will be used in presenting output.

Table 2. The descendant referred object node lists for keywords Cloud and XML

	L^d_{Cloud} (k = Cloud)		L^d_{XML} (k = XML)	
IDREF node	u	8	u	8
Children of **u**	Child(u)	∅	Child(u)	∅
Matching node	Match(u,k)	2	Match(u,k)	5
Path from **Match(u,k)** to **u**	Path(u,k)	2,4,8	Path(u,k)	5,6,8

For example, Table 2 shows the descendant referred object node lists for keywords Cloud and XML in the XML IDREF graph in Figure 2, where each entry corresponds to a column in the tables. Among object nodes (nodes 1,2,3) matching keyword Cloud, only node 2 has referred object node 8 as descendant. Nodes on the path from node 2 to node 8 are 2-4-8.

3.3 Runtime Processing

Given a keyword query $Q = \{k_1, \ldots, k_n\}$ to an XML IDREF graph, there are three steps for finding answers to Q, which are (1) finding potential center nodes (common ancestors and common descendant referred object nodes), (2) finding center nodes and (3) generating full answers. This section presents detailed techniques on these steps.

Step 1: finding potential center nodes. Based on Property 1 and Property 2, $Anc(Q)$ and $Ref(Q)$ are potential center nodes, where $Anc(Q)$ and $Ref(Q)$ are the set of common ancestors and the set of common descendant referred object nodes respectively. For each keyword k, we retrieve $Anc(k)$ and $Ref(k)$ from the first field, i.e., the field containing u of the ancestor list L^a_k and the descendant referred object list L^d_k respectively. Let $Anc(Q)$ and $Ref(Q)$ be the set intersection of $Anc(k)$'s and $Ref(k)$'s for all keywords k respectively. The computation of set intersection can leverage any efficient existing algorithms for set intersection. In this work, we use a simple yet efficient set intersection for m ordered sets $L_1, \ldots, L_m$ by scanning both lists in parallel, which requires $\sum_i(|L_i|)$ operations in worst case.

Algorithm 1 is to find common ancestors. For each common ancestor, we get the corresponding information (line

Algorithm 1: Finding potential center nodes

Input: Ancestor lists L^a_i of keyword k_i, $\forall i = 1..n$
Output: L^a_c: list of common ancestors

```
1  L^a_c ← ∅
2  //SetIntersection(L_1, ..., L_n)
3  for each cursor C_i do
4      C_i ← 1
5  index ← 1
6  for each element e in L_1 do
7      cur ← L_1[e]
8      next ← L_1[e + 1]
9      //Search e in the other lists
10     for each inverted list L_i from L_2 to L_n do
11         while L_i[C_i] < next do
12             if cur = L_i[C_i] then
13                 break;
14             C_i++;
15         break;
16     //update if e is a common one
17     if L_i[C_i] < next then
18         L^a_c[index].u ← e
19         L^a_c[index].Par(u) ← Par(e)
20         for each keyword k do
21             Add(Des(u, k)) to
                 L^a_c[index].Des(u, Q)
22             Add(IDREF(u, k)) to
                 L^a_c[index].IDREF(u, Q)
```

Table 3. Common ancestors of query {Cloud, XML}

L^a_c

Common ancestor	u	1	2	3
Parent of **u**	Par(u)	0	1	2
Descendants of **u** directly containing query keywords	Des(u, Q)	{2}, {5}	{2}, ∅	{3}, {3}
Referer of **u** if **u** is IDREF node	IDREF(u, Q)	∅	∅	∅

17-22). The result is illustrated in Table 3, where $Des(u, Q)$ and $IDREF(u, Q)$ contains n components of $Des(u, k)$ and $IDREF(u, k)$ respectively for all keywords k. This figure shows the common ancestor list for query $Q = \{$Cloud, XML$\}$. From L^a_{Cloud} and L^a_{XML}, we get $Anc(Cloud) = \{1, 2, 3\}$ and $Anc(XML) = \{1, 2, 3, 5, 7\}$. So, $Anc(Q) = \{1, 2, 3\}$. For common ancestor 1, $Des(1, Q) = \{2\}, \{5\}$ means $Des(1, Cloud) = \{2\}$ and $Des(1, XML) = \{5\}$.

Finding common descendant is similar. However, the differences are the information we get for each potential center node. Particularly, from line 17 to line 22, for each common descendant referred object node u, we will update $u, Child(u), Match(u, k)$ and $Path(u, k)$.

Step 2: finding real center nodes. Among potential center nodes, we need to find real center nodes by checking whether they have their own matching nodes. Based on Property 3, $Des(c, k_i) - Content\_Des(c, k), \forall i = 1..n$ is checked bottom up. Initially, each common ancestor c, we can get n sets $Des(c, k_i)$, $i = 1..n$ from the ancestor lists. Among common ancestors $Anc(Q)$, we start from those having no descendant in $Anc(Q)$ (bottom up). They are center nodes. For the parent c' of each new center node c, we update $Des(c', k_i), \forall i = 1..n$ by removing $Des(c, k_i)$ out of $Des(c', k_i)$, $\forall i = 1..n$. Finally, if $Des(c', k_i) \neq \emptyset$, $i = 1..n$, then c' is center node. This is similar for finding common descendant refered object nodes. The list of common ancestors is sorted so that an ancestor occurs before its children. Therefore, to find center nodes from the list of common ancestors, we start from the end of the list of common ancestors because the descendant is then considered first. A considered common ancestor will be filtered out of the list if it is not an center node.

The progress is given in Algorithm 2. If $Des(c, k_i) - Content\_Des(c, k) \neq \emptyset$, $\forall i = 1..n$, then it is a center node. Otherwise, it is filtered out (line 2, 3). For a returned center nodes u, we update the set of exclusive matching descendants of all common ancestors which are ancestors of u (line 5, 6, 7), for each of which, if the set exclusive matching descendants w.r.t. any keyword is empty, we remove it from the set of common ancestors (line 8, 9). Thereby, after finding a center node, we proactively filter out a lot of its ancestors if they are not center nodes.

Algorithm 2: Finding real center nodes

Input: L^a_c: the list of common ancestors
1 **for** *each element e from the end of L^a_c* **do**
2   **for** *each keyword k* **do**
3   **if** $Des(u, k)$ *is* $\emptyset$ **then**
4   L^a_c.Remove(e)

5   //Update the sets of exclusive matching descendants of COAs
6   **for** *each parent node e'.u of e.u in L^a_c* **do**
7      $v \leftarrow e'.u$ **for** *each keyword k* **do**
8        $Des(v, k)$.Remove($Des(u, k)$)
9   **if** $Des(v, k)$ *is* $\emptyset$ **then**
10   L^a_c.Remove(v)

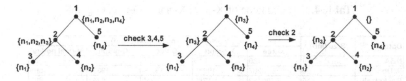

Fig. 5. Illustration of checking center nodes

The way to get the list of common descendants is similar to that of common ancestors but we start from the beginning of the list of descendant referred object nodes because an descendant occurs before its parents.

For example, Figure 5 illustrates the way our approach checks whether a potential center node is a real center node or not. First of all, our approach checks nodes 3, 4 and 5 in the most left figure because they have no descendant. After this checking step, descendants containing keywords of node 2 and node 1 is updated. Specifically, we remove n_1 and n_2 for node 2 because they are contained in node 3 and node 4 (descendants of node 1), and similarly remove n_1, n_2, and n_4 for node 1. We continue checking until the root is reached.

Step 3: tracking back matching nodes and looking up connecting nodes. Once we have found a center node c, we generate the answer $\langle c, \mathbb{K}, \mathbb{I} \rangle$ accordingly by identifying $\mathbb{K}$ and $\mathbb{I}$. The $Des(u, Q)$ and $Match(u, Q)$ fields allow us to produce the matching nodes $\mathbb{K}$ for answers with simple extension and without affecting space and time complexity. If c is a common descendant, $\mathbb{I}$ can be retrieved by the field $Path(u, k)$. If c is a common ancestor, to find the path from c to a matching node u, we need to find backward from u to c by using Par field in the ancestor lists. Finally, to return the full answers, we need another index from a label to the content of whole matching node.

4 Experiment

This section studies how the features of our approach, including hierarchy-aware, duplication-aware and object-orientation, impact on the performance. We will show the impacts of each feature as well as the impacts of all features on the effectiveness and efficiency. The experiments were performed on an Intel(R) Core(TM)i7 CPU 3.4GHz with 8GB of RAM.

4.1 Experimental Settings

Datasets. We used three real datasets including **NBA**[4], **IMDb**[5], and **Basketball**[6]. In IMDb, an actor or actress can play in many movies, and a company can produce several movies. In NBA and Basketball, a player can play for several teams in different years.

---

[4] http://www.nba.com
[5] http://www.imdb.com/interfaces
[6] http://www.databasebasketball.com/stats_download.htm

Table 4. Size and node of X-tree, X-graph, O-tree, O-graph

Dataset	Size (MB)	Number of nodes (thousand)				
		X-tree	X-graph	O-tree	O-graph	Virtual nodes
NBA	2.4	180	215	50	85	35
IMDb	90	48267	49835	18254	19822	1568
Basketball	56	30482	31285	5042	5845	803

We pre-processed them and used the subsets with the sizes 2.4MB, 90MB and 56MB for NBA, IMDb and Basketball respectively.

Discovering object classes and OIDs of datasets. We first apply [12] to automatically discover object classes and OIDs of datasets. We then manually adjust the results to get 100% of accuracy for the results of discovery. This is to make sure that the discovery step does not affect our results.

Modeling datasets. Each dataset corresponds to four models:
- An X-tree: an XML tree with object duplication and without ID/IDREF.
- An X-graph: an XML IDREF graph obtained from an X-tree.
- An O-tree (XML object tree): obtained from an X-tree by labelling only object nodes and assigning all non-object nodes the same label with the corresponding object nodes.
- An O-graph (XML object graph): obtained from an X-graph in the same manner with obtaining an O-tree from an X-tree.

Size and node of datasets. The size of the three datasets and the number of nodes of the X-tree, X-graph, O-tree and O-graph are given in Table 4. The numbers of nodes of an X-graph, O-graph is the sum of those of the corresponding X-tree, O-tree respectively and the number of virtual nodes.

Queries. We randomly generated 183 queries from value keywords of the three real datasets. To avoid meaningless queries, we retained 110 queries and filtered out 73 generated queries which are not meaningful at all (e.g., queries only containing articles and preposition). The remaining queries include 15, 57 and 38 queries for NBA, IMDb and Basketball datasets, respectively.

Compared approaches. Since our approach can supports both XML documents with and without IDREFs, we compare the performance of our approach with both tree-based approaches (Set-intersection [18]), and XRich [8] and a graph-based approach (BLINKS [5]).

Running compared approaches. Since Set-intersection and XRich work on XML tree, we can run them with X-tree and O-tree. Since BLINK works with XML graph, we ran it on X-graph and O-graph. Our approach is also a graph-based approach, thus we can run it on X-graph and O-graph. Since X-tree, X-graph, O-tree and O-graph can be derived from one another with a necessary minor cost and they represent the same information, it is fair to use them together for comparison.

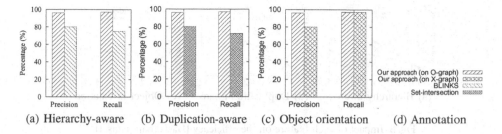

(a) Hierarchy-aware   (b) Duplication-aware   (c) Object orientation   (d) Annotation

Fig. 6. Impact of each feature on the effectiveness [Basketball dataset]

Methodology. Hierarchy-aware, duplication-aware and object orientation are three features which impact the effectiveness and the efficiency of our approach. Among compared approaches, BLINKS does not have concepts of object and hierarchy, Set-intersection does not have the concept of object and duplication, and XRich has all the three similar features, but works on XML tree only. Thus, we have the following methodology to show the impact of each feature and of all features.

Impact of hierarchy-aware. To show the impact of the hierarchical structure on graph search, we compared our approach with BLINKS, a non-hierarchy graph-based approach. To separate with the impact of object orientation, we also operated BLINKS at object level, i.e., ran BLINKS on O-graph.

Impact of duplication-aware. For duplication-aware, we compared our approach with Set-intersection, an unaware duplication approach. To separate with the impact of object orientation, we also operated BLINKS at object level, i.e., ran BLINKS on O-tree.

Impact of object orientation. To show the impact of object orientation, we ran our approach at object level (i.e., O-graph) as well as at node level (i.e., on X-graph).

Impact of all three features. To show the impact of all features, we ran compared algorithms on the data they initially designed for. Particularly, we ran BLINKS on X-graph, Set-intersection on X-tree, and our approach on O-graph. Besides, we also compare with XRich (on O-tree) because XRich has three similar features.

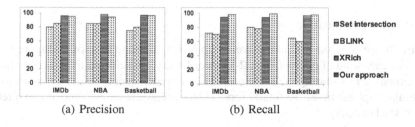

(a) Precision   (b) Recall

Fig. 7. Impact of all features on the effectiveness

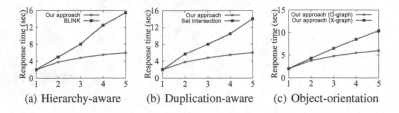

(a) Hierarchy-aware   (b) Duplication-aware   (c) Object-orientation

Fig. 8. Impact of each feature on the efficiency [Basketball dataset]

4.2 Effectiveness Evaluation

Metrics. To evaluate the effectiveness, we used standard *Precision* ($\mathcal{P}$) and *Recall* ($\mathcal{R}$) metrics. We randomly selected a subset (20 queries) of 110 generated queries for effectiveness evaluation. To compute precision and recall, we conducted surveys on the above 20 queries and the test datasets. We asked 25 researchers of our database labs to interpret 20 queries. Interpretations from at least 18 out of 25 researchers are manually reformulated into schema-aware XQuery queries and the results of these XQuery queries are used as the ground truth.

Impact of each feature. Figure 6 shows the impacts of each feature on the effectiveness. As can be seen, each feature can help to improve effectiveness. Follows are the reasons of improvement w.r.t. each feature. The hierarchical structure enables us to avoid meaningless answers caused by unrelated matching nodes. BLINKS uses the distinct root semantics which is similar to the LCA semantics, thus its recall is affected by the problems of not returning common descendants. Duplication-aware help return missed answers. Object orientation improves precision because it enables us to avoid meaningless answer caused by returning only non-object nodes.

Impact of all features. Figure 7 shows the impacts of all features on the effectiveness. As discussed above, all these features have impacts on the effectiveness. Among them, duplication-aware has the highest impact. Thus, the more features an approach possesses, the higher precision and recall are. Particularly, our approach has highest precision and recall because our approach has all three features. BLINKS has higher performance than Set-intersection because BLINKS works with graph, a duplication-aware data. Compare to XRich, our approach has slightly higher recall while the precision is similar.

4.3 Efficiency Evaluation

Metrics. To measure the efficiency, we compared the running time of finding returned nodes. For each kind of queries, e.g., 2-keyword query, we selected five queries among 110 retained queries sharing the same properties. For each query, we ran ten times to get the average response time. We finally reported the average response time of five queries for each kind of query.

Impact of each feature. Figure 8 shows the impact of each features on the efficiency. As can be seen, all features improve efficiency, among which, the hierarchy has the

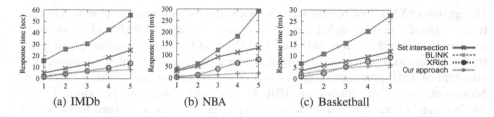

(a) IMDb   (b) NBA   (c) Basketball

Fig. 9. Impact of all features on the efficiency (varying number of query keywords)

most impact. The reasons for the improvement are follows. The hierarchy enables us to avoid NP-Hard problem of the general graph search and it just extends visited nodes to two directions: ancestor and descendant rather than to all directions. Duplication-aware and object orientation both help reduce the search space.

Impact of all features. The response time of algorithms is shown in Figure 9, in which we varied the number of query keywords. As discussed above, all features impact the efficiency. Thus, the more features an algorithm has, the more efficient it is. Our approach obtains the highest efficiency because it has all three features. Among the others, BLINKS is the least efficient because it is affected by structure which does not have hierarchy. Compared to XRich which based on tree-structure, our approach can get almost the similar response time, even better when the number of keywords is increased.

5   Related Work

LCA-based XML keyword search. XRANK [4] proposes a stack based algorithm to efficiently compute LCAs. XKSearch [17] defines Smallest LCAs (SLCAs) to be the LCAs that do not contain other LCAs. Meaningful LCA (MLCA) [13] incorporates SLCA into XQuery. VLCA [10] and ELCA [19] introduces the concept of valuable/exclusive LCA to improve the effectiveness of SLCA. MaxMatch [14] investigates an axiomatic framework that includes the properties of monotonicity and consistency. Although extensive works have been done on improving the effectiveness of LCA-based approaches, these works commonly still suffers from the problem of missing answers because of undetected object duplication in XML document. Moreover, these works only can work with XML documents with no IDREF. Recently, XRich [8] takes common descendants into account of answers. However, its input XML document must not contain IDREF or n-ary relationships ($n \geq 3$).

We generalize the technique of set intersection [18] in processing queries. However, there are several differences. First, that work considers XML tree where a node has only one parent whereas ours can deal with the case where a node has multiple parents by employing more complex indexes and searching techniques for XML IDREF graph. Secondly, that work operates at node level whereas we operate at object level, which enables us to improve both efficiency (by reducing search space) and effectiveness (by avoiding meaningless answers) of the search. Most importantly, that work only search up to find common ancestors but miss common descendants because it cannot detect object duplication. In contrast, we have techniques to find missing answers.

Graph-based XML keyword search. BANKS [1] uses backward search to find Steiner tree in labeled, directed graph. Later, Bidirectional [6] improves BANKS by using bidirectional (backward and forward) search. EASE [11] introduces a unified graph index to handle keyword search on heterogeneous data. Without the exploiting hierarchical structure of XML graph, the graph search in general suffers from NP-Hard Problem. Moreover, since XML graph (with IDREF) can contain object duplication, but these works cannot detect object duplication. Thus they may also miss answers. To the best of our knowledge, only [9] can provide the such missing answers. Nevertheless, this work transfers XML to a graph which is similar to relational database and follows Steiner tree semantics. Thus, it suffers from the inefficiency and may return meaningless answers because matching nodes may not be (or weakly) related.

6   Conclusion

We introduced an approach to handle the problem of missing answers due to object duplication for keyword search in a data-centric XML document. We model the input XML document as an XML IDREF graph where all instances of the same object are connected via a virtual object node (duplication-aware). We only work with object nodes and associate non-object nodes to the corresponding object nodes (object-orientation). More importantly, we discover the hierarchical structure of XML IDREF graph to inherit LCA-based techniques for an efficient search (hierarchy-aware). The experiments showed the impact of each and all feature(s) (duplication-aware, hierarchy-aware, object-orientation) to the efficiency and effectiveness, which made our approach outperforms the compared approaches in term of both efficiency and effectiveness.

References

1. Bhalotia, G., Hulgeri, A., Nakhe, C., Chakrabarti, S., Sudarshan, S.: Keyword searching and browsing in databases using BANKS. In: ICDE (2002)
2. Dreyfus, S.E., Wagner, R.A.: The steiner problem in graphs. Networks (1971)
3. Fong, J., Wong, H.K., Cheng, Z.: Converting relational database into XML documents with DOM. Information & Software Technology (2003)
4. Guo, L., Shao, F., Botev, C., Shanmugasundaram, J.: XRANK: Ranked keyword search over XML documents. In: SIGMOD (2003)
5. He, H., Wang, H., Yang, J., Yu, P.S.: BLINKS: ranked keyword searches on graphs. In: SIGMOD (2007)
6. Kacholia, V., Pandit, S., Chakrabarti, S., Sudarshan, S., Hrishikesh Karambelkar, R.D.: Bidirectional expansion for keyword search on graph databases. In: VLDB (2005)
7. Kargar, M., An, A.: Keyword search in graphs: finding r-cliques. PVLDB (2011)
8. Le, T.N., Ling, T.W., Jagadish, H.V., Lu, J.: Object semantics for XML keyword search. In: Bhowmick, S.S., Dyreson, C.E., Jensen, C.S., Lee, M.L., Muliantara, A., Thalheim, B. (eds.) DASFAA 2014, Part II. LNCS, vol. 8422, pp. 311–327. Springer, Heidelberg (2014)
9. Le, T.N., Wu, H., Ling, T.W., Li, L., Lu, J.: From structure-based to semantics-based: Towards effective XML keyword search. In: Ng, W., Storey, V.C., Trujillo, J.C. (eds.) ER 2013. LNCS, vol. 8217, pp. 356–371. Springer, Heidelberg (2013)
10. Li, G., Feng, J., Wang, J., Zhou, L.: Effective keyword search for valuable LCAs over XML documents. In: CIKM (2007)

11. Li, G., Ooi, B.C., Feng, J., Wang, J., Zhou, L.: EASE: Efficient and adaptive keyword search on unstructured, semi-structured and structured data. In: SIGMOD (2008)
12. Li, L., Le, T.N., Wu, H., Ling, T.W., Bressan, S.: Discovering semantics from data-centric XML. In: Decker, H., Lhotská, L., Link, S., Basl, J., Tjoa, A.M. (eds.) DEXA 2013, Part I. LNCS, vol. 8055, pp. 88–102. Springer, Heidelberg (2013)
13. Li, Y., Yu, C., Jagadish, H.V.: Schema-free XQuery. In: VLDB (2004)
14. Liu, Z., Chen, Y.: Reasoning and identifying relevant matches for XML keyword search. PVLDB (2008)
15. Tao, Y., Papadopoulos, S., Sheng, C., Stefanidis, K.: Nearest keyword search in XML documents. In: SIGMOD (2011)
16. Termehchy, A., Winslett, M.: EXTRUCT: using deep structural information in XML keyword search. PVLDB (2010)
17. Xu, Y., Papakonstantinou, Y.: Efficient keyword search for smallest LCAs in XML databases. In: SIGMOD (2005)
18. Zhou, J., Bao, Z., Wang, W., Ling, T.W., Chen, Z., Lin, X., Guo, J.: Fast SLCA and ELCA computation for XML keyword queries based on set intersection. In: ICDE (2012)
19. Zhou, R., Liu, C., Li, J.: Fast ELCA computation for keyword queries on XML data. In: EDBT (2010)

A Study on External Memory Scan-Based Skyline Algorithms

Nikos Bikakis[1,2], Dimitris Sacharidis[2], and Timos Sellis[3]

[1] National Technical University of Athens, Greece
[2] IMIS, "Athena" R.C., Greece
[3] RMIT University, Australia

Abstract. Skyline queries return the set of non-dominated tuples, where a tuple is dominated if there exists another with better values on all attributes. In the past few years the problem has been studied extensively, and a great number of external memory algorithms have been proposed. We thoroughly study the most important scan-based methods, which perform a number of passes over the database in order to extract the skyline. Although these algorithms are specifically designed to operate in external memory, there are many implementation details which are neglected, as well as several design choices resulting in different flavors for these basic methods. We perform an extensive experimental evaluation using real and synthetic data. We conclude that specific design choices can have a significant impact on performance. We also demonstrate that, contrary to common belief, simpler skyline algorithm can be much faster than methods based on pre-processing.

Keywords: Experimental evaluation, experimental survey, disk-based algorithm.

1 Introduction

The *skyline query*, or skyline operator as it was introduced in [3], has in the past few years received great attention in the data management community. Given a database of objects, the skyline query returns those objects which are not dominated. An object *dominates* another, if it has better values on all attributes, and strictly better value on at least one. Finding the skyline is also known as the Pareto-optimal set, or maximal vectors problem in multi-objective optimization research, where it has been studied extensively in the past, but only for in-memory computations. For example the well-known divide and conquer algorithm of [7] has complexity $O(N \log^{d-2} N)$, for $d \geq 2$, where N is the number of objects, and d their dimensionality; the algorithm is optimal for $d = 3$.

The interest in external memory algorithms has sparked after the seminal work in [3]. The most efficient method in terms of worst-case Input/Output (I/O) operations is the algorithm in [15], which requires in the worst case $O\left((N/B) \log_{M/B}^{d-2}(N/B)\right)$ I/Os, where M is the memory size and B the block (minimum unit of transfer in an I/O operation) size in terms of objects. However,

H. Decker et al. (Eds.): DEXA 2014, Part I, LNCS 8644, pp. 156–170, 2014.
© Springer International Publishing Switzerland 2014

in practice, other external-memory algorithms proposed over the past years can be faster.

This work studies in detail an important class of practical algorithms, the *scan-based skyline algorithms*. An algorithm of this class performs multiple passes over an input file, where the input file in the first pass is the database, and in a subsequent pass it is the output of the previous pass. The algorithm terminates when the output file remains empty after a pass concludes. Generally speaking, during each pass, the algorithm maintains in main memory a small *window* of incomparable objects, which it uses to remove dominated objects from the input file. Any object not dominated is written to the output file.

Although the studied algorithms are specifically designed to operate in external memory, little attention has been given to important implementation details regarding memory management. For example, all algorithms assume that the unit of transfer during an I/O operation is the object, whereas in a real system is the block, i.e., a set of objects. Our work addresses such shortcomings by introducing a more realistic I/O model that better captures performance in a real system. Furthermore, by thoroughly studying the core computational challenge in these algorithms, which is the management of the objects within the window, we introduce several novel potentially interesting policies.

Summarizing, the contributions of our study are the following:

- Based on a standard external memory model [1], we appropriately adapt four popular scan-based algorithms, addressing in detail neglected implementation details regarding memory management.
- We focus on the core processing of scan-based algorithms, the management of objects maintained in the in-memory window. In particular, we introduce various policies for two tasks: traversing the window and evicting objects from the window. Both tasks can have significant consequences in the number of required I/Os and in the CPU time.
- We experimentally evaluate concrete disk-based implementations, rather than simulations, of all studied algorithms and derive useful conclusions for synthetic and real datasets. In particular, we demonstrate that, in many cases and contrary to common belief, algorithms that pre-process (typically, sort) the database are not faster.
- We perform an extensive study of our proposed policies, and reach the conclusion that in some settings (dimensionality and dataset distribution) these policies can reduce the number of dominance checks by more than 50%.

2 Preliminaries

2.1 Definitions

Let $\mathcal{O}$ be a set of *d-dimensional objects*. Each object $o \in \mathcal{O}$ is represented by its *attributes* $o = (o^1, o^2, \ldots, o^d)$. The *domain* of each attribute, is the positive real numbers set $\mathbb{R}^+$. Without loss of generality, we assume that an object o_1 is *better* than another object o_2 on an attribute j, iff $o_1^j < o_2^j$. An object o_1

dominates another object o_2, denoted by $o_1 \succ o_2$, iff (1) $\forall i \in [1, d]$, $o_1^i \leqslant o_2^i$ and (2) $\exists j \in [1, d]$, $o_1^j < o_2^j$. The *skyline* of an object set $\mathcal{O}$, denoted as $SL(\mathcal{O})$, is the set of objects in $\mathcal{O}$ that are not dominated by any other object of $\mathcal{O}$. Formally, $SL(\mathcal{O}) = \{o_i \in \mathcal{O} \mid \nexists o_k \in \mathcal{O} : o_k \succ o_i\}$.

2.2   External Memory I/O Model

This section describes an external memory model, similar to that of [1]. The unit of transfer between the main memory and the external memory (i.e., the disk) is a single *block*.[1] Any external memory algorithm, like the skyline methods, read/write blocks from/to disk *files*. We assume that files are stored contiguously on disk, and therefore a new block is written always at the end of a file.

We denote as $N = |\mathcal{O}|$ the size of the database, i.e., N is the total number of objects to be processed. We measure the fixed size B of a block in terms of objects (tuples). Similarly, main memory can fit M objects, with the requirements that $M < N$ (and often much smaller) to justify the need for external memory algorithms, and $M > 2B$ to support basic in-memory operations.

We next discuss Input/Output (I/O) operations. We assume no input or output buffers, so that blocks from the disk are transferred directly to (resp. from) the disk from (resp. to) the main memory. Equivalently, the input/output buffers share the same memory of size M with the algorithm.

We categorize I/O operations in two ways. Naturally, a *read* transfers data from the disk, whereas a *write* transfers data to the disk. The second categorization is based on the number of blocks that are transferred. Note that a read (resp. write) operation transfers at least one block and at most $\lfloor \frac{M}{B} \rfloor$ blocks into main memory (resp. disk). We also remark that in disks, the *seek time*, i.e., the time it takes for the head to reach the exact position on the ever spinning disk where data is to be read or written, is a crucial parameter in disk performance. Reading or writing k consecutive blocks on the disk is much faster than reading or writing k blocks in arbitrary positions on the disk. The reason is that only one seek is required in the first case, compared to the k seeks for the second. Therefore, we distinguish between *sequential* and *random* I/Os. A random I/O incorporates the seek time, whereas a sequential I/O does not. For example, when a procedure reads k blocks sequentially from the disk, we say that it incurs 1 random read and $k - 1$ sequential reads.

3   A Model for Scan-Based Skyline Algorithms

3.1   Design Choices

All skyline algorithms maintain a set of objects, termed *window*, which consists of possible skyline objects, actual skyline objects, or some arbitrary objects in general. A common procedure found in all algorithms is the following. Given some candidate object not in the window, *traverse* the window and determine if

---

[1] [1] assumes that P blocks can be transferred concurrently; in this work we set $P = 1$.

the candidate object is dominated by a window object, and, if not, additionally determine the window objects that it dominates. Upon completion of the traversal and if the candidate is not dominated, the skyline algorithm may choose to insert it into the window, possible *evicting* some window objects.

In the aforementioned general procedure, we identify and focus on two distinct design choices. The first is the *traversal policy* that determines the order in which window objects are considered and thus dominance checks are made. This design choice directly affects the number of dominance checks performed and thus the running time of the algorithm. An ideal (but unrealistic) traversal policy would require only one dominance check in the case that the candidate is dominated, i.e., visit only a dominating window object, and/or visit only those window objects which the candidate dominates.

The second design choice is the *eviction policy* that determines which window object(s) to remove so as to make room for the candidate object. This choice essentially determines the dominance power of the window, and can thus indirectly influence both the number of future dominance checks and the number of future I/O operations.

We define four window traversal policies. The *sequential traversal policy* (*sqT*), where window objects are traversed *sequentially*, i.e., in the order they are stored. This policy is the one adopted by all existing algorithms. The *random traversal policy* (*rdT*), where window objects are traversed in *random* order. This policy is used to gauge the effect of others. The *entropy-based traversal policy* (*enT*), where window objects are traversed in ascending order of their *entropy* (i.e., $\sum_{i=1}^{d} \ln(o^i+1)$) values. Intuitively, an object with a low entropy value has greater dominance potential as it dominates a large volume of the space.

In addition to these traversal policies, we define various ranking schemes for objects, which will be discussed later. These schemes attempt to capture the dominance potential of an object, with higher ranks suggesting greater potential. Particularly, we consider the following traversal policies. The *ranked-based traversal policy* (*rkT*), where window objects are traversed in descending order based on their *rank* values. Moreover, we consider three hybrid random-, rank-based traversal policies. The *highest-random traversal policy* (*hgRdT*), where the k objects with the highest rank are traversed first, in descending order of their rank; then, the random traversal policy is adopted. The *lowest-random traversal policy* (*lwRdT*), where the k objects with the lowest rank are compared first, before continuing with a random traversal. Finally, the *recent-random traversal policy* (*rcRdT*), where the k most recently read objects are compared first, before continuing with a random traversal.

Moreover, we define three eviction policies. The *append eviction policy* (*apE*), where the *last* inserted object is removed. This is the policy adopted by the majority of existing algorithms. The *entropy-based eviction policy* (*enE*), where the object with the *highest entropy* value is removed. Finally, the *ranked-based eviction policy* (*rkE*), where the object with the *lowest rank* value is removed. In case of ties in entropy or rank values, the most recent object is evicted.

We next discuss ranking schemes used in the ranked-based traversal and eviction policies. Each window object is assigned a rank value, initially set to zero. Intuitively, the rank serves to identify "promising" objects with high dominance power, i.e., objects that dominate a great number of other objects. Then, the skyline algorithm can exploit this information in order to reduce the required dominance checks by starting the window traversal from promising objects, and/or evict non-promising objects.

We define three ranking schemes. $r0R$: the rank of an object o at a time instance t, is equal to the number of objects that have been dominated by o until t. In other words, this ranking scheme counts the number of objects dominated by o. $r1R$: this ranking is similar to $r0R$. However, it also considers the number of objects that have been dominated by the objects that o dominates. Let $rank(o)$ denote the rank of an object o. Assume that object o_1 dominates o_2, Then, the rank of o_1 after dominating o_2 is equal to $rank(o_1) + rank(o_2) + 1$. $r2R$: this ranking assigns two values for each object o, its $r1R$ value, as well as the number of times o is compared with another object and none of them is dominated (i.e., the number of incomparable dominance checks). The $r1R$ value is primarily considered to rank window objects, while the number of incomparable check is only considered to solve ties; the more incomparable checks an object has, the lower its rank.

3.2   Algorithm Adaptations for the I/O Model

BNL. The *Block Nested Loop* (BNL) [3] algorithm is one of the first external memory algorithms for skyline computation. All computations in BNL occur during the window traversal. Therefore, BNL uses a window as big as the memory allows. In particular, let W denote the number of objects stored in the window, and let O_b denote the number of objects scheduled for writing to disk (i.e., in the output buffer). The remaining memory of size $I_b = M - W - O_b$ serves as the input buffer, to retrieve objects from the disk. Note that the size of the I/O buffers I_b and O_b vary during the execution of BNL, subject to the restriction that the size of the input buffer is always at least one disk block, i.e, $I_b \geq B$, and that the output buffer never exceeds a disk block, i.e., $O_b \leq B$; we discuss later how BNL enforces this requirements.

We next describe memory management in the BNL algorithm. BNL performs a number of passes, where in each an input file is read. For the first pass, the input file is the database, whereas the input file in subsequent passes is created at the previous pass. BNL terminates when the input file is empty. During a pass, the input file is read in *chunks*, i.e., sets of blocks. In particular, each read operation transfers into main memory exactly $\lfloor \frac{I_b}{B} \rfloor$ blocks from disk, incurring thus 1 random and $\lfloor \frac{I_b}{B} \rfloor - 1$ sequential I/Os. On the other hand, whenever the output buffer fills, i.e., $O_b = B$, a write operation transfers into disk exactly 1 block and incurs 1 random I/O.

We now discuss what happens when a chunk of objects is transfered into the input buffer within the main memory. For each object o in the input buffer, BNL traverses the window, adopting the sequential traversal policy (sqT). Then, BNL performs a two-way dominance check between o and a window object w. If o is dominated by w, o is discarded and the traversal stops. Otherwise, if o dominates w, object w is simply removed from the window.

At the end of the traversal, if o has not been discarded, it is appended in the window. If W becomes greater than $M - O_b - B$, BNL needs to move an object from the window to the output buffer to make sure that enough space exists for the input buffer. In particular, BNL applies the append eviction policy (apE), and selects the last inserted object, which is o, to move into the output buffer. If after this eviction, the output buffer contains $O_b = B$ objects, its contents are written to the file, which will become the input file of the next pass.

A final issue is how BNL identifies an object o to be a skyline object, BNL must make sure that o is dominance checked with all surviving objects in the input file. When this can be guaranteed, o is removed from the window and returned as a result. This process is implemented through a *timestamp mechanism*; details can be found in [3].

SFS. The *Sort Filter Skyline* (SFS) [4] algorithm is similar to BNL with one significant exception: the database is first sorted by an external sort procedure according to a monotonic scoring function. SFS can use any function defined in Section 3.1.

Similar to BNL, the SFS algorithm employs the sequential window traversal policy (sqT) and the append eviction policy (apE). There exist, however, two differences with respect to BNL. Due to the sorting, dominance checks during window traversal are one-way. That is an object o is only checked for dominance by a window object w. In addition, the skyline identification in SFS is simpler than BNL. At the end of each pass, all window objects are guaranteed to be results and are thus removed and returned.

LESS. The *Linear Elimination Sort for Skyline* (LESS) [5] algorithm improves on the basic idea of SFS, by performing dominance checks during the external sort procedure. Recall that standard external sort performs a number of passes over the input data. The so-called zero pass (or sort pass) brings into main memory M objects, sorts them in-memory and writes them to disk. Then, the k-th (merge) pass of external sort, reads into main memory blocks from up to $\lfloor M/B \rfloor - 1$ files created in the previous pass, merges the objects and writes the result to disk.

LESS changes the external sort procedure in two ways. First, during the zero pass, LESS maintains a window of size W_0 objects as an elimination filter to prune objects during sorting. Thus the remaining memory $M - W_0$ is used for the in-memory sorting. The window is initially populated after reading the first $M - W_0$ objects by selecting those with the lowest entropy scores. Then for each object o read from the disk and before sorting them in-memory, LESS performs a window traversal. In particular, LESS employs the sequential traversal policy (sqT) performing a one-way dominance check, i.e., it only checks if o is dominated. Upon comparing all input objects with the window, the object with the lowest entropy o_h is identified. Then, another sequential window traversal (sqT) begins, this time checking if o_h dominates the objects in the window. If o_h survives, it is appended in the window, evicting the object with the highest entropy score, i.e., the entropy-based eviction policy (enE) is enforced.

The second change in the external sort procedure is during its last pass, where LESS maintains a window of size W objects. In this pass, as well as any subsequent skyline processing passes, LESS operates exactly like SFS. That is the sequential traversal policy (sqT) is used, one-way dominance checks are made, and window objects are removed according to the append eviction policy (epE).

RAND. In the Randomized multi-pass streaming (RAND) algorithm [13], each pass in RAND consists of three phases, where each scans the input file of the previous pass. Therefore, each pass essentially corresponds to three reads of the input file. In the first phase, the input file is read and a window of maximum size $W = M - B$ is populated with randomly sampled input objects (using reservoir sampling).

In the second phase, the input file is again read one block at a time, while the window of W objects remain in memory. For each input object o, the algorithm traverses the window in sequential order (sqT), performing one-way dominance checks. If a window object w is dominated by o, w is replaced by o. Note that, at the end of this phase, all window objects are skyline objects, and can be returned. However, they are not removed from memory.

In the third phase, for each input object o, RAND performs another sequential traversal of the window (sqT), this time performing an inverse one-way dominance check. If o is dominated by a window object w, or if o and w correspond to the same object, RAND discards o. Otherwise it is written on a file on the disk, serving as the input file for the next pass. At the end of this phase, the memory is cleaned.

4   Related Work

External memory skyline algorithms can be classified into three categories: (1) scan-based, (2) index-based, and (3) partitioning-based algorithms.

The *scan-based* approaches perform multiple passes over the dataset and use a small window of candidate objects, which is used to prune dominated objects. The algorithms of this category can be further classified into two approaches: with and without pre-processing. Algorithms of the first category, directly process the set of objects, in the order in which they are stored, or produced (e.g., in the case of pipelining multiple operators). The BNL [3] and RAND [13] algorithms, detailed in Section 3.1, lie in this category. On the other hand, methods in the second category perform an external sort of the objects before, or parallel to the skyline computation. The SFS [4] and LESS [5], also detailed in Section 3.1, belong to this category. Other algorithm, include Sort and Limit Skyline algorithm (SaLSa) [2], which is similar to SFS and additionally introduces a condition for early terminating the input file scan, and Skyline Operator on Anti- correlated Distributions (SOAD) [14], which is also similar to SFS but uses different sorting functions for different sets of attributes.

In *index-based* approaches, various types of indices are used to guide the search for skyline points and prune large parts of the space. The most well-known and efficient method is the Branch and Bound Skyline (BBS) [12] algorithm. BBS employs an R-tree, and is shown to be I/O optimal with respect to this index. Similarly, the Nearest Neighbor algorithm (NN) [6] also uses an R-tree performing multiple nearest neighbor searches to identify skyline objects. A bitmap structure is used by Bitmap [16] algorithm to encode the input data. In the Index [16] algorithm, several B-trees are used to index the data, one per dimension. Other methods, e.g., [9,10], employ a space-filling curve, such as the Z-order curve, and use a single-dimensional index. The Lattice Skyline (LS) algorithm [11] builds a specialized data structure for low-cardinality domains.

In the *partitioning-based* approaches, algorithms divide the initial space into several partitions. The first algorithm in this category, D&C [3] computes the skyline objects adopting the divide-and-conquer paradigm. A similar approach with stronger theoretical guarantees is presented in [15]. Recently, partitioning-based skyline algorithms which also consider the notion of incomparability are proposed in [17,8]. OSP [17] attempts to reduce the number of checks between incomparable points by recursively partition the skyline points. BSkyTree [8] enhances [17] by considering both the notions of dominance and incomparability while partitioning the space.

5   Experimental Analysis

5.1   Setting

Datasets. Our experimental evaluation involves both synthetic and real datasets. To construct synthetic datasets, we consider the three standard distribution types broadly used in the skyline literature. In particular, the distributions are: anti-correlated (ANT), correlated (CORR), and independent (IND). The synthetic datasets are created using the generator developed by the authors of [3].

We also perform experiments on three real datasets. *NBA* dataset consists of 17,264 objects, containing statistics of basketball players. For each player we consider 5 statistics (i.e., points, rebound, assist, steal blocks). *House* is 6-dimensional dataset consists of 127,931 objects. Each object, represents the money spent in one year by an American family for six different types of expenditures (e.g., gas, electricity, water, heating, etc.). Finally, *Colour* is a 9-dimensional dataset, which contains 68,040 objects, representing the first three moments of the RGB color distribution of an image.

Implementation. All algorithms, described in Section 3.1, were written in C++, compiled with gcc, and experiments were performed on a 2.6GHz CPU. In order to accurately convey the effect of I/O operations, we disable the operating system caching, and perform direct and synchronous I/O's.

The size of each object is set equal to 100 bytes, as was the case in the experimental evaluation of the works that introduced the algorithms under investigation. Finally, the size of block is set to 2048 bytes; hence each block contains 20 object.

Metrics. To gauge efficiency of all algorithms, we measure: (1) the number of disk I/O operations, which are distinguished into four categories, read, write operations, performed during the pre-processing phase (i.e., sorting) if any, and read, write operations performed during the main computational phase; (2) the number of dominance checks; (3) the time spent solely on CPU processing denoted as CPU Time and measured in seconds; (4) the total execution time, denoted as Total Time and measured in seconds; In all cases the reported time values are the averages of 5 executions.

5.2 Algorithms Comparison

Table 1 lists the parameters and the range of values examined. In each experiment, we vary a single parameter and set the remaining to their default

Table 1. Parameters

Description	Parameter	Values
Number of Objects	N	50k, 100K, **500K**, 1M, 5M
Number of Attributes	d	3, **5**, 7, 9, 15
Memory Size	$M/N(\%)$	0.15%, 0.5% **1%**, 5%, 10%

(bold) values. SFS and LESS sort according to the entropy function. During pass zero in LESS, the window is set to one block.

Varying the Number of Objects. In this experiment, we vary the number of objects from 50K up to 5M and measure the total time, number of I/O's and dominance checks, and CPU time, in Figures 1–4.

The important conclusions from Figure 1 are two. First, RAND and BNL outperform the other methods in anti-correlated datasets. This is explained as follows. Note that the CPU time mainly captures the time spent for the following task: dominance checks, data sorting in case of LESS/SFS, and skyline identification, in case of BNL. From Figure 4 we can conclude that BNL spends a lot of CPU time in skyline identification. BNL requires the same or more CPU time than RAND, while BNL performs fewer dominance checks than RAND. This is more clear in the case of independent and correlated datasets where the cost for dominance checks is lower compared to the anti-correlated dataset. In these datasets, the BNL CPU time increased sharply as the cardinality increases.

The second conclusion is that, in independent and correlated datasets, the performance of BNL quickly degrades as the cardinality increases. This is due to the increase of the window size, which in turn makes window maintenance and skyline identification more difficult.

Figure 2 shows the I/O operations performed by the algorithms. We observe that BNL outperforms the other methods in almost all settings. Particularly, in the correlated dataset, LESS is very close to BNL. Also, we can observe that, in general, the percentage of write operations in LESS and SFS is much higher than in BNL and RAND. We should remark that, the write operations are generally more expensive compared to the read operations. Finally, for LESS and SFS, we can observe that the larger amount of I/O operations are performed during the sorting phase.

Regarding the number of dominance checks, shown in Figure 3, LESS and SFS perform the fewest, while RAND the most, in all cases. Figure 4 shows the CPU time spent by the methods. SFS spends more CPU time than LESS even though they perform similar number of dominance checks; this is because SFS sorts a larger number of object than LESS. Finally, as previously mentioned, BNL spends considerable CPU time for skyline identification.

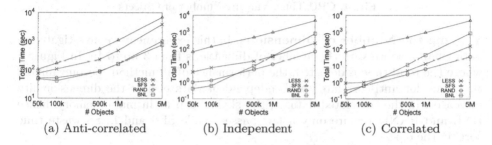

(a) Anti-correlated   (b) Independent   (c) Correlated

Fig. 1. Total Time: Varying Number of Objects

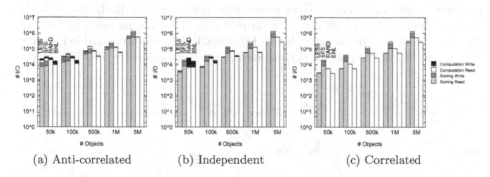

(a) Anti-correlated   (b) Independent   (c) Correlated

Fig. 2. I/O Operations: Varying Number of Objects

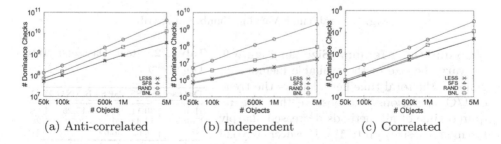

(a) Anti-correlated   (b) Independent   (c) Correlated

Fig. 3. Dominance Checks: Varying Number of Objects

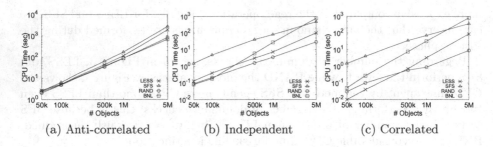

Fig. 4. CPU Time: Varying Number of Objects

Varying the Number of Dimensions. In this experiment we investigate the performance as we vary the number of dimensions from 3 up to 15. In Figure 5 where the total time is depicted, the performance of all methods become almost the same for anti-correlated and independent datasets, as the dimensionality increases. In the correlated dataset, the skyline can fit in main memory, hence BNL and RAND require only a few passes, while SFS and LESS waste time sorting the data.

Regarding I/O's (Figure 6), BNL outperforms all other methods in all cases, while LESS is the second best method. Similarly, as in Figure 2, LESS and SFS performs noticeable more write operations compared to BNL and RAND. Figure 7 shows that LESS and SFS outperforms the other method, performing the same number of dominance checks. Finally, CPU time is presented in Figure 8, where once again the cost for skyline identification is noticeable for BNL.

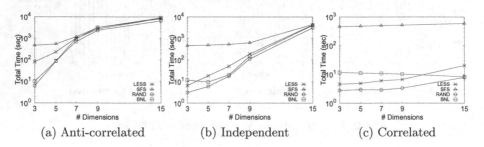

Fig. 5. Total Time: Varying Number of Attributes

Varying the Memory Size. In Figure 9, we vary the size of the available memory. In general, the total time here, follows the trend of I/O operations. We observe that the required time of all methods decreased sharply for memory sizes up to 1%. However, beyond this point, the time is almost stable as the

Table 2. Real Datasets: Total Time (sec)

Dataset	LESS	SFS	RAND	BNL
House	30.11	178.21	15.25	4.98
Colour	14.43	90.73	3.70	1.28
NBA	9.45	26.68	0.71	0.41

memory size increases, with the exception of BNL, where the time slightly increases (due to the skyline identification cost w.r.t. window size).

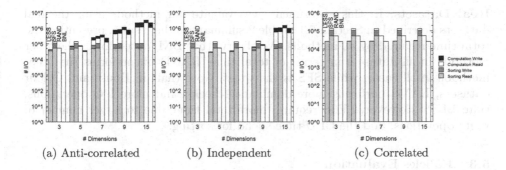

(a) Anti-correlated   (b) Independent   (c) Correlated

Fig. 6. I/O Operations: Varying Number of Attributes

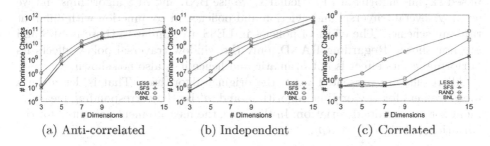

(a) Anti-correlated   (b) Independent   (c) Correlated

Fig. 7. Dominance Checks: Varying Number of Attributes

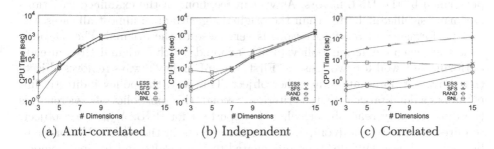

(a) Anti-correlated   (b) Independent   (c) Correlated

Fig. 8. CPU Time: Varying Number of Attributes

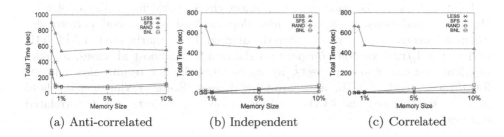

(a) Anti-correlated   (b) Independent   (c) Correlated

Fig. 9. Total Time: Varying Memory Size

Real Datasets. In this experiment, we evaluate our methods using the real datasets described in Section 5.1. Table 2 summarizes the results, presenting the total time required by all methods. We observe that BNL outperforms the other methods in all datasets in terms of total time. RAND outperforms the other methods in all cases, while SFS is the worst. Note that, in House and Colour datasets, RAND performs more dominance checks, and more I/O operations, than LESS. However, LESS requires more total time, due to larger number of write operations, and the CPU time spend for sorting.

5.3   Policies Evaluation

In this experiment, we study the effect of different window policies in scan-based skyline algorithms. Particularly, we use BNL and SFS algorithms and we employ several traversal and eviction and policies, in conjunction with different ranking schemes. The effect of policies in LESS are similar to those in SFS and are not shown. Regarding RAND, only the window traversal policy affects its performance; its effect is not dramatic and hence it is also not shown.

All results are presented w.r.t. the original algorithms. That is, let m be a measurement for the original algorithm, and m' be the corresponding measurement for an examined variation. In this case, the measurement presented for the variation is $1 + (m' - m)/m$.

BNL. We first study the performance of BNL under the 10 most important policy and ranking scheme combinations. Figure 10 shows the I/O operations performed by the BNL flavors. As we can see, none of the examined variations performs significant better than the original algorithm. In almost all cases, the I/O performance of most variations is very close to the original. The reason is that the append eviction policy (apE), adopted by the original BNL already performs very well for two reasons. First, the apE policy always removes objects that have not dominated any other object. This way, the policy indirectly implements a dominance-oriented criterion. Second, the apE policy always removes the most recently read object, which is important for BNL. A just read object, requires the most time (compared to other objects in the window) in order to be identified as a skyline, thus propagated to the results and freeing memory. Hence, by keeping "older" objects we increase the probability of freeing memory in the near future. Still it is possible to marginally decrease the number of I/Os.

Figure 11 shows the number of dominance checks performed. We can observe that, in several cases, the variants that adopt rank-based traversal, perform significant fewer dominance checks than the original. Particularly, the rkT/rkE/r1R and rkT/rkE/r2R variants outperform the others in almost all cases, in independent and correlated datasets, by up to 50%. Similar results also hold for low dimensionalities in the anti-correlated dataset. However, this does not hold in more dimensions, due to the explosion of skyline objects in anti-correlated datasets.

SFS. Here, as in the previous experiment, we examine the performance of SFS algorithm adopting several policies. Similar to BNL, none of SFS variants

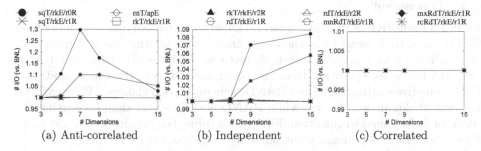

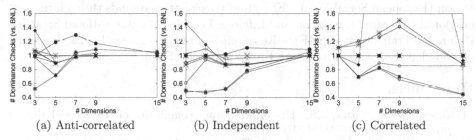

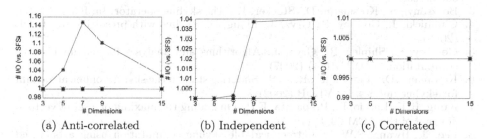

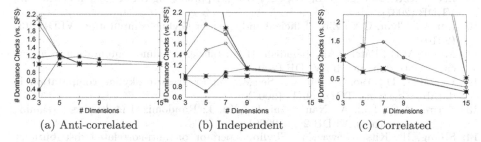

perform noticeable fewer I/O operations (Figure 12). Regarding the dominance checks (Figure 13), in anti-correlated and independent datasets, most of variants have similar performance to the original algorithm. Only for correlated datasets, ranked-based policies exhibit significant performance gains.

5.4 Discussion

In an I/O-sensitive setting, i.e., when I/O operations cost significantly more than CPU cycles, BNL seems to be the ideal choice, as it performs less I/O operations than all other methods in almost all settings. Additionally, BNL and RAND perform less write operation than the other methods. On the other hand, in a CPU-sensitive setting, LESS and RAND seem to be good choices. LESS performs the fewest dominance checks, while RAND doesn't spend time for sorting the data, or for skyline identification. Finally, regarding the policies tested, the rank-based ones show significant gains but only in CPU-sensitive settings.

Acknowledgements. This research has been co-financed by the European Union (European Social Fund – ESF) and Greek national funds through the Operational Program "Education and Lifelong Learning" of the National Strategic Reference Framework (NSRF) – Research Funding Program: Thales. Investing in knowledge society through the European Social Fund.

References

1. Aggarwal, A., Vitter, J.S.: The input/output complexity of sorting and related problems. Commun. ACM 31(9) (1988)
2. Bartolini, I., Ciaccia, P., Patella, M.: Efficient sort-based skyline evaluation. TODS 33(4) (2008)
3. Börzsönyi, S., Kossmann, D., Stocker, K.: The skyline operator. In: ICDE (2001)
4. Chomicki, J., Godfrey, P., Gryz, J., Liang, D.: Skyline with presorting. In: ICDE (2003)
5. Godfrey, P., Shipley, R., Gryz, J.: Algorithms and analyses for maximal vector computation. VLDBJ 16(1) (2007)
6. Kossmann, D., Ramsak, F., Rost, S.: Shooting stars in the sky: An online algorithm for skyline queries. In: VLDB (2002)
7. Kung, H.T., Luccio, F., Preparata, F.P.: On finding the maxima of a set of vectors. Journal of the ACM 22(4) (1975)
8. Lee, J., Hwang, S.W.: Bskytree: scalable skyline computation using a balanced pivot selection. In: EDBT (2010)
9. Lee, K.C.K., Zheng, B., Li, H., Lee, W.C.: Approaching the skyline in z order. In: VLDB (2007)
10. Liu, B., Chan, C.Y.: Zinc: Efficient indexing for skyline computation. VLDB 4(3) (2010)
11. Morse, M.D., Patel, J.M., Jagadish, H.V.: Efficient skyline computation over low-cardinality domains. In: VLDB (2007)
12. Papadias, D., Tao, Y., Fu, G., Seeger, B.: Progressive skyline computation in database systems. TODS 30(1) (2005)
13. Sarma, A.D., Lall, A., Nanongkai, D., Xu, J.: Randomized multi-pass streaming skyline algorithms. VLDB 2(1) (2009)
14. Shang, H., Kitsuregawa, M.: Skyline operator on anti-correlated distributions, vol. 6 (2013)
15. Sheng, C., Tao, Y.: Worst-case i/o-efficient skyline algorithms. TODS 37(4) (2012)
16. Tan, K.L., Eng, P.K., Ooi, B.C.: Efficient progressive skyline computation. In: VLDB (2001)
17. Zhang, S., Mamoulis, N., Cheung, D.W.: Scalable skyline computation using object-based space partitioning. In: SIGMOD (2009)

On Continuous Spatial Skyline Queries
over a Line Segment

Wei Heng Tai[1], En Tzu Wang[2], and Arbee L.P. Chen[3,*]

[1] Cloud Computing Center for Mobile Applications,
Industrial Technology Research Institute, Hsinchu, Taiwan
Henry-0516@hotmail.com
[2] Computational Intelligence Technology Center,
Industrial Technology Research Institute, Hsinchu, Taiwan
m9221009@em92.ndhu.edu.tw
[3] Department of Computer Science, National Chengchi University, Taipei, Taiwan
alpchen@cs.nccu.edu.tw

Abstract. Traditional skyline queries consider data points with static attributes, e.g., the distance to a beach of a hotel. However, some attributes considered for a skyline query can be dynamic, e.g., the distance to a beach of a moving vehicle. Consider a scenario as follows. A person in a moving vehicle issues a skyline query to find restaurants, taking into account the static attributes of the restaurants as well as the distance to them. This scenario motives us to address a novel skyline query considering both static and dynamic attributes of data points. Given a data set D (e.g., restaurants) with a set of static attributes in a two-dimensional space, a query line segment l (e.g., the route for driving), and a distance constraint of r (for choosing a restaurant), we want to find out the skylines along l considering the static and dynamic attributes of the data points, satisfying the location constraint of r. We propose two methods to solve the problem. In the first method, we find some special location points which partition l into sub-segments and also make the skylines in the adjacent sub-segments different. In the second method, we apply some properties to identify data points which need not be considered for computing skylines. Moreover, to reduce the number of sub-segments, we propose an approximate method to compute skylines and define a similarity function to measure the similarity between the exact and approximate results. A series of experiments are performed to evaluate the exact methods and the results show that the second method is more efficient. We also perform experiments to compare the exact and approximate results and it shows that there is a trade-off between the reduction of the number of sub-segments and the accuracy of the results.

Keywords: Skyline, dynamic attributes, location data, query line segment.

1   Introduction

Location-based services (LBS) have been widely studied in recent years. Most of the applications on LBS are developed for offering users location information. However, since a

---

* Corresponding author.

H. Decker et al. (Eds.): DEXA 2014, Part I, LNCS 8644, pp. 171–187, 2014.
© Springer International Publishing Switzerland 2014

lot of objects located in our surroundings, offering them which one is better? The skyline query can help to answer this question as it plays an important role in multi-critiria decision making. Given two objects $O_1 = (x_1, x_2, x_3,..., x_d)$ and $O_2 = (y_1, y_2, y_3,..., y_d)$, if $x_i \leq y_i$ for all $1 \leq i \leq d$, and there exists a dimension j such that $x_j < y_j$, we say O_1 dominates O_2. The skyline contains the objects which are not dominated by any other objects. As shown in Figure 1, there are eight objects (hotels) on a two-dimensional space, and each object has two attributes: the price and distance to the beach. Suppose that we prefer the hotels which have lower prices and are closer to the beach. Then, O_3 is better than O_2, O_4, O_5, and O_6 in terms of the price and distance. The skyline result contains three objects (O_1, O_3, and O_7) which are not dominated by the other objects.

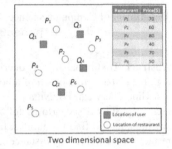

Fig. 1. An example of a skyline   **Fig. 2.** An example of a spatial skyline query

The traditional skyline problem considers that all dimensions of a data point are static. For example, the distance to the beach of a hotel is static. Some people start to consider the dynamic attributes in the recent years. There are many variations related to the traditional skyline problem, such as the dynamic skyline [2], the reverse dynamic skyline [3], and the spatial skyline query [6]. The spatial skyline query (SSQ) is one of the variations, which considers the dynamic attribute. The spatial skyline query is described as follows. Given a set of data points D and a set of query points Q, SSQ retrieves the points from D, not dominated by any other points in D, by considering a set of attributes. For each data point, these attributes are its distances to the query points in Q. The domination is determined with respect to both spatial and nonspatial attributes of D. We show an example of the spatial skyline query in Figure 2. Suppose there are six restaurants (P_1 to P_6) in a two-dimensional space. The users (Q_1 to Q_4) want to find a restaurant in terms of distance and price. We can see that P_6 dominates P_5 for each user because of the lower price and shorter distances.

Sometimes users may issue the skyline query on a moving trajectory instead of a fixed location. Although the skyline query returns the non-dominated objects for users, users need not want the skyline points which are too far to them. For example, a user gets a skyline list of restaurants and a specific restaurant in the list with very good services but locates 10KM away. We assume that the user has a preference radius which is the longest tolerant distance for them. Consider a scenario as follows. The route of a moving vehicle is a series of road segments. A user on the route may issue a skyline query to find restaurants based on static attributes such as price and dynamic attributes such as distance. Accordingly, a new skyline query is addressed in this paper. Since the skyline query over a road segment can be computed individually, we define the problem we are solving in this paper as follows. Given a data set D, a query line segment l (to describe a part of the route), and a distance of r (to describe the acceptable distance for users' access) in a two-dimensional space, the skyline query retrieves a full partition of a sequence of the line segment l, such that each sub-segment of a sequence contains the corresponding skyline points within the distance constraint of r. The result is as shown in Figure 3. We recommend the data point e between p_1 and p_2, the data points d and e between p_2 and p_3, and the data point d between p_3 and p_4 for users.

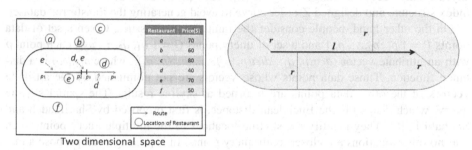

Fig. 3. A continuous spatial skyline query   **Fig. 4.** The candidate region of a line segment

It is more challenging to process the skyline query on a route because it has high computation cost. In fact, we need not re-compute the skyline result continuously. In this paper, we propose two methods to solve the continuous spatial skyline query. In the basic method, we find some intersection points which may change the skyline results and compute the results on these points. In the second method, we use some properties to, 1) prune some data points which must not be the skyline points, and 2) reduce some skyline computations on certain intersection points. In order to reduce the number of sub-segments, we also propose an approximate method. We set a threshold to avoid probing into the small MBRs in the R-tree traversing. Three synthetic datasets and two real datasets are used to test our methods and the experiment results demonstrate that the second method is better than the basic method.

The remainder of this paper is organized as follows. The related works of the skyline query are reviewed in Section 2. Section 3 introduces the terminologies used in this paper and formalizes the problem to be solved. We detail our methods in Section 4. The approximate method is introduced in Section 5. The experiment results are presented and analyzed in Section 6, and finally, Section 7 concludes this work.

2   Related Works

The skyline problem has a long history. The first work is proposed by Borzsonyi et al. [1]. They study the skyline computation problem in the context of databases and proposed a SQL syntax for the skyline query. The traditional skyline problem considers that all dimensions of a data point are static. Many works start to consider the skyline problem with dynamic attribute. The variant skyline queries regarding dynamic attributes are developed. The Papadias et al. proposed a BBS search technique to reduce the search space and briefly how to use it to solve the dynamic skyline query in [2]. Dellis and Seeger use the semantics of the dynamic skyline to introduce the dynamic skyline queries in [3]. Given a point q, the reverse skyline query returns the data points whose corresponding dynamic skylines contain q. Sacharidis et al. propose a caching mechanism in [4]. For a new query, they used the information of past queries to speed up the dynamic skyline processing. Wang et al. first solve the dynamic skyline query consider range query in [5]. They proposed a method based on grid index structure and designed Z-order curve to avoid generating the transferred dataset.

On the other hand, people consider the multiple query issues. Given a set of data points $P = \{ p_1, p_2,..., p_m \}$ and a set of query points $Q = \{ q_1, q_2,..., q_k \}$, each point p with an attribute vector $\langle dist(p, q_1), dist(p, q_2), ..., dist(p, q_k) \rangle$, where $dist(\cdot,\cdot)$ is a distance function. Those data points whose vectors are not dominated by the attribute vectors of the other data points are returned as skyline results. The spatial skyline query, which $dist(\cdot,\cdot)$ is the Euclidean distance, is first proposed by Sharifzadeh and Shahabi in [6]. They identify the skyline locations to the multiple query points such that no other locations are closer to all query points. In [7], Geng et al. propose a method which combines the spatial information with non-spatial information to obtain the skyline results. In [8], Qianlu et al. propose an efficient algorithm to solve general spatial skyline which can provide a minimal set of them that contain optimal solutions of any distance with different types. Deng et al. propose the multi-source skyline query in [9] and consider the distance measure as the distance of the shortest path in road network. Li et al. solve the multi-source skyline query with multi-dimensional space in [10]. [11] and [12] solve the metric skyline query, where $dist(\cdot,\cdot)$ is a metric distance.

Some people consider the continuous skyline query. Huang, Z et al. propose continuous skyline query processing on moving object [13]. They process the possible skyline changing from one time to another and, making the skyline result updated and available continuously. Jang et al. propose a method in [14] which processes the continuous skyline in road network. They define the shortest range of targets by pre-computing and it can get the skyline result efficiently.

As mentioned, many works consider the skyline problem and combine it with continuous query. However, the preference radius is also important for users. They need not know the skyline points which are too far to them. In this paper, we consider a preference radius and want to find a full partition of the line segment l, such that each sub-segment of l contains the skyline points with a dynamic spatial attribute and static attributes.

3   Preliminaries

In this section, the problem of continuous spatial skylines regarding a query line segment is formally defined. In addition, terminology used in this paper is also introduced here. To simplify the explanation in this paper, we focus on a line segment in the following discussion.

3.1   Problem Formulation

Suppose that a set of data points $D = \{d_1, d_2, ..., d_n\}$ are located at the coordinate of (x_i, y_i) in a two-dimensional space, $\forall i = 1$ to n, each of which contains a set of static attributes $M = \{m_1, m_2, ..., m_k\}$. Given a query line segment l, each location point of l has a corresponding distance to each data point in D, regarded as the dynamic attribute of the data point in D. Since users may not be interested in the data points located too far away from them (from l), we use a distance of r to issue a restriction of locations. The problem to be solved in this paper is addressed as follows. Given a data set D, a query line segment l, and a distance of r, the continuous spatial skyline query regarding l retrieves a full partition of l, such that each sub-segment in the partition contains a corresponding skyline result taking into account both of the static and dynamic attributes and also satisfying the location restriction of r.

3.2   Terminology

Our approaches are based on the *R-tree* index structure. We assume that the R-tree index of the coordinates of data points in D is constructed in advance. Moreover, since a distance of r is used to restrict the locations of data points, when the query is issued, we can quickly find out the candidate data points which have chances of being skyline results. Only the data points with minimum distances to l less than or equal to r need to be taken into account, which form a special range query as Figure 4 illustrates. By processing this special range query using the original R-tree index structure, the original R-tree can be reduced to a *meta-R-tree* of D by pruning the minimal bounding rectangles (MBRs) not overlapping the range query and also pruning the leaf nodes not satisfying the range query. This region formed by the special range query is named *Candidate Region (CR)*.

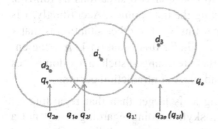

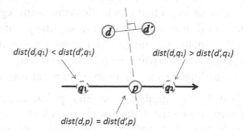

Fig. 5. The entering and leaving points   **Fig. 6.** The example of a swap point

For each data point in CR, we take its location as a center to generate a circle with a radius of r. This circle must intersect l because the minimum distance from its center to l is smaller than or equal to r. As shown in Figure 5, there are a set of data points ($D = d_1$, d_2, and d_3) in CR and a line segment l with the *start* and *end points* equal to q_s and q_e, respectively, i.e. $l = \overline{q_s q_e}$. The circumferences of the circles with centers equal to d_1, d_2, and d_3 partition l into sub-segments. The intersection points of the circumferences of D are q_{1e}, q_{2l}, q_{1l}, and q_{3e}. As can be seen, d_1, d_2, and d_3 cover (which means *valid* in) $\overline{q_{1e} q_{1l}}$, $\overline{q_{2e} q_{2l}}$, and q_{3e}, respectively. These points e.g., q_{1e} and q_{1l}, are named the *entering point* and *leaving point* with respect to d_1. Notice that, while traversing the query line segment l, we use the *partial list* to continuously maintain the valid data points.

If all location points in a sub-segment s are closer to a data point d than the other data point d', then we say that $order(d) < order(d')$ with respect to s. Here, the dynamic attribute regarding distance is abstracted to the concept of ordering. For any two data points d and d', their *perpendicular bisector* may intersect l. This intersection point is defined a *swap point*, which has a special property descripted as follows. As shown in Figure 6, the intersection point between $l = \overline{q_1 q_2}$ and the bisector of d and d' is p. According to the property of bisectors, all location points in $\overline{q_1 p}$ are closer to d than d' and vice versa. Therefore, $order(d)$ is smaller than $order(d')$ with respect to $\overline{q_1 p}$ and $order(d')$ is smaller than $order(d)$ with respect to $\overline{p q_2}$, which means that the ordering of d and d' changes while traversing p. The union of all of the entering, leaving, and swap points is defined *Check Set* (denoted CS) and only the location points in CS may affect the skyline results.

Lemma 1: Given a query line segment l and a location point p in l which divides l into two sub-segments s_1 and s_2. If p is not in CS, the skyline regarding s_1 and that regarding s_2 must be identical.

Proof: We assume that p is not in CS but the skyline regarding s_1 and that regarding s_2 are different. Case 1: The number of skyline points regarding s_1 is larger than that regarding s_2. There must exist a data point x such that x is a skyline point regarding s_1 but not a skyline point regarding s_2. We need to consider two situations as follows. One situation is that p is the leaving point regarding the data point x. Accordingly, x is not valid in s_2, making it not a skyline point regarding s_2. The other situation is that x is dominated by another data point y regarding s_2. Under this situation 1) p is the entering point regarding y or 2) $order(x)$ and $order(y)$ are changed such that y dominates x regarding s_2 (p is the swap point regarding x and y). A contradiction occurs.

Case 2: The number of skyline points regarding s_2 is larger than that regarding s_1. There must exist a data point x such that x is a skyline point regarding s_2 but not a skyline point regarding s_1. We need to consider two situations as follows. One situation is that p is the entering point regarding the data point x. Accordingly, x is valid in s_2, making it a skyline point regarding s_2. The other situation is that x is not dominated by any skyline points regarding s_1. Under this situation 1) p is the leaving point of a specific data point y which dominates x regarding s_1 or 2) $order(x)$ and $order(y)$ are changed such that y cannot dominate x regarding s_2. A contradiction occurs.

Case 3: The number of skyline points regarding s_1 and that regarding s_2 are equal. There must exist at least two data points x and y such that x is a skyline point regarding s_1 but not a skyline point regarding s_2 and y is a skyline point regarding s_2 but not a skyline point regarding s_1. It means a data point exists to dominate y regarding s_1 and a data point exists to dominate x regarding s_2. Since the number of skyline points regarding s_1 and that regarding s_2 are equal, $order(x)$ and $order(y)$ are changed (p is the swap point regarding x and y). According to these cases discussed above, p is in the check set. A contradiction occurs. Therefore, if p is not in CS, the skyline regarding s_1 and that regarding s_2 must be identical.   ■

4   Approaches to Finding Skyline Results

Our approaches to finding the skyline result regarding a query line segment are described in this section.

4.1   The Basic Method

In the basic method, we first find the entering points, leaving points, and the swap points from the candidate region and then, add these points to Check Set which may change the global skyline result. For each point in Check Set, we compute the distance from each intersection point to itself and take it as the dynamic attribute. After we get static attributes and a dynamic attribute, we can compute the global skyline points which are not dominated by other points, and add the sub-segments to result list. We verify the global skyline points on the intersection points where the entering point and the leaving point are the same. When we finished, we merge the sub-segments with same skyline. Finally, we return the result list for user.

The example is shown in Figure 7. We process the R-tree structure and keep the six data points (a, b, c, d, e, f) which are in the candidate region. Let these data points be the center of circle and find the entering points and leaving points, such as a-plus is the entering point of a and a-minus is the leaving point of a, as shown in Figure 7(a). The entering point and leaving point of the data point f is the same. We put f into the verify list and verify it after we find the global skyline on other intersection points. We also compute the perpendicular bisector of each data point and find the swap points. The swap point of a and b on l are denoted by (a, b) in Figure 7(b). Then, we add these points to check set list where the data points may change the global skyline points. For each intersection point, we compute the distance from each data point to itself and take the distance order as a dynamic attribute. Consider the static attributes (Table 1) and the dynamic attribute. We compute the skyline on the points which are in the check set list. Before merging the sub-segments, we verify the global skyline point from verify list and f doesn't dominate the skyline points. The result is shown in Figure 7(c).

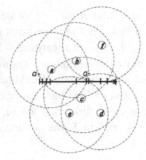

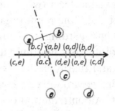

(a) Finding the entering points and leaving points.

(b) Finding the swap points

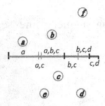

(c) The result of basic method

Fig. 7. The example of basic method

Table 1. Attributes of the data points

Data Points	Attribute 1	Attribute 2
a	40	40
b	50	50
c	60	70
d	70	60
e	65	65
f	65	80

4.2　The Pruning Method

We do many computations for finding the intersection points in the basic method. In order to avoid lots of computation cost, we have some lemmas and properties to help us prune some data points.

Lemma 2: Given two data points d, $d' \in CR$, and a query line l. There exist arbitrary two points p and p' on l such that the line is divided into three intervals. If the distance order is $order(d) < order(d')$ at p and p', the distance order is $order(d) < order(d')$ between p and p'.

Proof: Consider the following two cases between p and p'.

Case 1: There are no swap point of d and d' between p and p' on l: We compute that the distance order are $order(d) < order(d')$ at p and p'. The distance order only change when passing the swap point of d and d'. Because there are no swap points of d and d' at the interval, the distance order between p and p' are fixed. Hence, the distance order between p and p' is $order(d) < order(d')$.

Case 2: There are a swap point of d and d' between p and p' on l: We compute that the distance order are $order(d) < order(d')$ at p and p'. The distance order only change when passing the swap point of d and d'. Because there are a swap point of d and d' at the interval, the distance order must change. However, the distance order change from $order(d) < order(d')$ to $order(d) > order(d')$ and doesn't change the order afterward.

It is impossible the distance order at p and p' are $order(d) < order(d')$ but the order is $order(d) > order(d')$ between them. Hence, this case does not happen. ∎

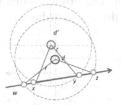

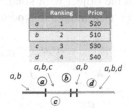

	Ranking	Price
a	1	$20
b	2	$10
c	3	$30
d	4	$40

Fig. 8. The example of Lemma 3   **Fig. 9.** The example of Lemma 4

Lemma 3: Given two data points $d, d' \in CR$, a radius r, and a query line l. Assume the circle centered at d intersects l at two points w and z, and the circle centered at d' intersects l at two points x and y. If the interval $wz \supset$ the interval xy, then $\forall p \in$ interval xy, $dist(p,d) < dist(p,d')$.

Proof: We take two points x and y. We compute $yd < r = yd'$ and $xd < r = xd'$ by property of circle. By Lemma 2, if the distance order at x and y are $order(d) < order(d')$, the distance order between x and y are also $order(d) < order(d')$. Hence, the interval of $wz \supset$ the interval of xy, $\forall p \in$ interval of xy such that $dist(p,d) < dist(p,d')$. ∎

Property 1: Given two data points $d, d' \in CR$, a radius r, and a query line l. Assume the circle centered at d and l intersect at two points, the two points form an interval $i1$. Also, the circle centered at d' and l intersect at two points, the two points form an interval $i2$. We say the data point d' must not be the skyline result if $i1$ contains $i2$ and all the static attributes of d are better than d'.

Proof: Assume data point d' is one of the skyline points in certain sub-segments. By Lemma 3, the order(d) < order(d') in $i2$, Otherwise, all the static attributes of the object d are better than the object d'. Hence, the data point d dominates the data point d'. The data point d' is not the skyline result. →←   ∎

Given the data points $d, d' \in CR$, a query line l, and a radius r. We find the effect interval of each data point on l. The partial list stores the data points which need to be considered on l. Lemma 4 shows the property of the skyline subspace.

Lemma 4: If data points d and d' are static skyline points in the partial list, they must be one of the global skyline points.

Figure 9 is an example. Suppose data points a and b are static skyline points. The result is shown these two points are also the global skyline because no objects dominate the two points even considering the dynamic spatial attribute.

Given the data points$\in CR$, a query line l. We can find all the swap points of each pair of the data point on l. The distance order list stores the order of all data points which are in the candidate region. Lemma 5 also shows the property of the skyline subspace.

Lemma 5: If a data point is ordered the first in the distance order list, the data point must be one of the global skyline points.

Figure 10 is an example. We compute the distance order on each sub-segment. The result shows the object with smallest order must be the one of the skyline points because no objects dominate this point even considering the static attributes.

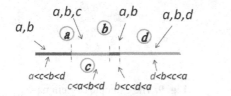 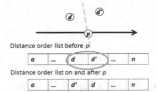

Fig. 10. The example of Lemma 5   **Fig. 11.** The example of Lemma 6

Lemma 6: Given the two data points d, d' ∈ CR, and a query line l. We can find a swap point of d and d' on l. If we check the swap point, the order(d) and the order(d') are adjacent in the distance order list.

Proof: Consider following two cases.

Case 1: There are no swap points which are composed of d or d' have the same location with (d,d'). Assume we check the swap points of d and d' on l, two points are not adjacent in the distance order list. We can find at least a point p locate between d and d' in distance order list and the distance order is order(d) < order(p) < order(d'). Because we check the swap points of d and d', the distance order of other data points is fixed. The order is order(d') < order(p) < order(d) after the swap point. No swap points which are composed of d or d' have the same location with (d,d'). It is impossible occur the distance order of d and p is change from order(d) < order(p) to order(d) > order(p). →←. Hence, d and d' are adjacent in the distance order list if we check the swap point of d and d' on l.

Case 2: Exist the swap points which are composed of d or d' have the same location with (d,d'). Assume exist a point p such that (p,d), (p,d') have the same location with (d,d'), and the order is order(d) ≤ order(p) ≤ order(d') before these swap points. When we check these swap points, the check order must be (d,p), (d,d'), and (p,d') according to the original location of p, d, and d'. Before we check the (d,d'), we first check the (d,p). The order is changed from order(d) ≤ order(p) ≤ order(d') to order(p) ≤ order(d) ≤ order(d'). Hence, d and d' are adjacent in the distance order list if we check the swap point of d and d' on l.   ∎

When we find the entering points, leaving points, and swap points, we compute the skyline points on the check set where the objects may cause the skyline change. By Lemmas 4 and 5, we observe the static skyline point must be the one of the answer and the data point is first in the distance order also must be one of the answer. By Lemma 6, when computing global skyline points on swap points, we can take the data points which are composed the swap point as a bundle. The distance order of data points is changed only in the bundle. Now, we have a property 2 and it can help us reduce some skyline computation from the check set.

Property 2: Given two data points d, $d' \in CR$, a query line l. The swap point of d and d' on l divides l into two intervals $i1$ and $i2$. The global skyline points of $i2$ are same as the global skyline points of $i1$ if it satisfies one of following cases.

Case 1: d and d' are the static skyline points.

Case 2: d and d' are not the global skyline points of $i1$.

Proof: Assume we compute the skyline on the swap point of d and d'. In the case 1, the data points d and d' are static skyline points. The attributes of other data points are static. By Lemma 4, the static skyline must be the one of global skyline result. By Lemma 6 the $order(d)$ and the $order(d')$ are adjacent in the distance order list. When we swap the $order(d)$ and $order(d')$, it doesn't affect the order of other data points. Hence the global skyline of $i2$ is same as $i1$. In the case 2, the data points d and d' on l are not the global skyline of $i1$. By Lemmas 4 and 5, the data points d and d' are not the static skyline points, and the $order(d)$ and $order(d')$ are not first in the distance order list. It exist some data points such that these data points dominate d and d'. By Lemma 6 the order of d and d' are adjacent in the distance order list. We only swap the distance order of d and d' on the swap point, so these data points still dominate the data points d and d'. Hence the global skyline of $i2$ is same as $i1$.   ∎

In the pruning method, we find the entering points and leaving points of data points which locate in the candidate region. And we reduce some data points which are not contributive for result by Property 1. Given the d, $d' \in CR$, if all the static attributes of d dominate the static attribute of d' and the effect interval of d contains the effect interval of d', we can prune the data point d' because it is a redundant data point, as shown in Figure 12(a). The effect interval of b contains the effect interval of e and the static attributes of e are dominated by b, we prune the data point e by Property 1. We find the swap points of all remainder data points. Then, add these intersection points into check set list and compute the global skyline from it. For swap points of the two data points on l, we have property 2 to reduce some skyline computations. We don't need to re-compute the skyline points if the data points satisfy 1) they are the static skyline points or 2) they are not the global skyline points of previous interval. According Property 2, we can reduce some skyline computation cost. For the dynamic attribute, we don't need to compute the distance on each intersection point. We just compute the distance order on start point, and swap the order by Lemma 6. The example is shown in Figure 12 (b) and (c). When we compute the skyline on the swap point of the data point b and c, they are not the previous skyline points. The skyline points are same as previous by Property 2 Case 2. When we compute the skyline on the swap point of the data point c and d, they are the static skyline points. The skyline points are also same as previous by Property 2 Case 1. After we find the global skyline points, we verify it on the intersection points where the entering points and the leaving points are the same. When we finished, we merge the sub-segments which is same as previous. Finally, we return the result list for user.

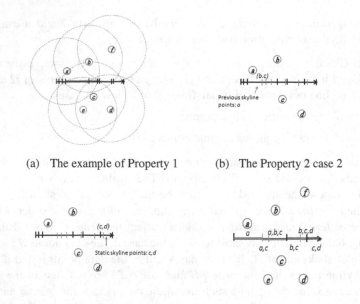

(a)   The example of Property 1   (b)   The Property 2 case 2

(c)The example of Property 2 case 1   (d) The result of pruning method

Fig. 12. The example of pruning method

5   Extension

Since many sub-segments can be produced, we propose an approximate method which can reduce the number of sub-segments. Moreover, we define a similar function to measure the similarity between the exact results and the approximate results, and we compare and analysis it by several experiments.

5.1   The Approximate Approach

Based on R-tree structure, we process the MBR before we execute the pruning method. At first we keep the MBRs which locate in the candidate region. We define the threshold as the radius divide by the approximate factor. For each MBR, if the diagonal of MBR is smaller than threshold, we take the internal node of MBR as a new data point, and take the new static attributes which are the smallest attribute of data points among the rectangle. The example is shown in Figure 13. We keep the MBRs ($R8$, $R9$, $R12$ and $R13$) which locate in the candidate region, and process each MBR whose diagonal is smaller than threshold. The $R8$, $R9$ and $R12$ satisfy the condition, we take the internal node of itself as new node. The $R13$ don't satisfy the condition, we probe the MBR and get objects g and h. According our processing, the number of objects is from eight to five. We expect through process these MBRs may reduce the number of data points efficiently, moreover reduce the number of sub-segments.

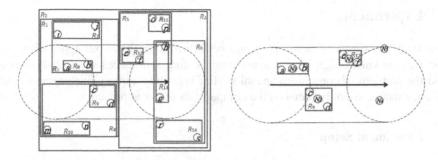

(a) The original R-tree structure.   (b) The meta-R-tree.

Fig. 13. An example of the approximate method

Table 2. The table of new data points

Data points	Set	Attribute 1	Attribute 2
N1	{ a, b}	Min(a1,b1)	Min(a2,b2)
N2	{ c, d}	Min(c1,d1)	Min(c2,d2)
N3	{ e, f}	Min(e1,f1)	Min(e2,f2)
N4	{g}	Min(g1)	Min(g2)
N5	{h}	Min(h1)	Min(h2)

5.2   The Similar Function

In order to analysis our approximate method. We define a similar function which can measure the distance between the exact result and approximate result. We modify the well know distance: Jaccard distance. The Jaccard distance defined as follow.

$$j(A,B) = \frac{|A \cup B| - |A \cap B|}{|A \cup B|} = \frac{|A \Delta B|}{|A \cup B|}$$

We modify the distance function to match our problem. The similar distance function as follow

$$d(x, y) = \frac{1}{L} \sum_i l_i * \left(\frac{|s(x_{l_i}) \Delta s(y_{l_i})|}{|s(x_{l_i}) \cup s(y_{l_i})|} \right)$$

Table 3. The notation of similar function

Notation	Description
d(x,y)	The similar distance between query line x and line y.
L	The total length of query line.
li	The length of i-th sub-segment.
x_{li}	The i-th sub-segment at line x
$s(x_{li})$	The skyline set of x_{li}

6   Experiments

In this section, a series of experiments are performed to evaluate our approaches. To the best of our knowledge, since there are no existing approaches specifically focusing on the problem of continuous spatial skyline regarding a query line segment with preference radius, we only focus on the comparisons of our approaches.

6.1   Experiment Setup

We have two real datasets. One dataset is from website (http://www.we8there.com) which records the comment of user on restaurant or hotel. We fetch the data for restaurants in California and the distribution as shown in Figure 16(a). The data size is about 400, the three static attributes are the price, the service, and the ranking of restaurant, and the coordinate of each object is range from ([0, 100K], [0, 100K]). The other dataset is from website (http://www.census.gov/geo/www/tiger) and distribution as shown in Figure 16(b). We fetch the location in Los Angles. The data size is about 350K, the static is 2-dimension by independent generator like [10], and the coordinate of each object is range from ([0, 3200], [0, 1800]).

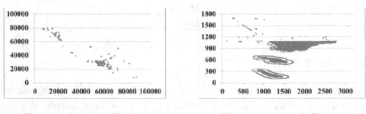

(a) The real dataset 1(California)   (b) The real dataset 2 (Los Angle)

Fig. 14. The data distributions of the real datasets

We use the length l of query line and preference radius r as our factor. On the synthetic datasets, we set the l and r with a range of [100, 200] and [50, 150], respectively. Following the experiments parameters used in [15], the length l is set from 0.05% ~ 0.15% of side. Therefore, we set 0.15 as the default value of l when considering r as the experiment parameter. We also set r = 100 as the default value while changing l. On real data one, we set the l and r with range of [10K, 20K] and [5K, 15K], set the 15K as the default value of l and set the 10K as the default value of r. On real data two, we set 150 as the default value of l and set 100 as the default value of r. All of the algorithm are implemented in C++ and performed on a PC with Intel Core 2 Quad 2.66GHz CPU, 8GB main memory, and under the windows 7 64bits operating system.

6.2   Experiment Result

The running time of the three methods is shown in Figure 15. As can be seen, the running time of the basic method is worse than the pruning methods. We compare the number of points for Property 1 as shown in Figure 16 and for Property 2 as shown in Figure 17. As can be seen, the Property 1 can prune about 50% points no matter what we vary l or r. It can prune many intersection points because more points satisfy Property 2 Case 2. And according our observation, many points are not located in the pre-skyline list. It reduce many skyline computation cost.

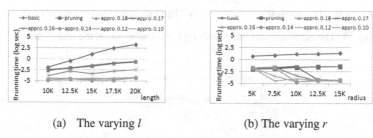

(a) The varying *l*   (b) The varying *r*

Fig. 15. The running time comparisons of the methods on the real dataset 1

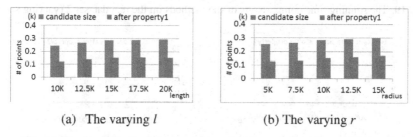

(a) The varying *l*   (b) The varying *r*

Fig. 16. The number of point comparisons for property 1 on the real dataset 1

We compute the accuracy between pruning and approximate method as shown in Figure 18(a) and compare the number of sub-segments as shown in Figure 18(b). The high accuracy causes many sub-segments.

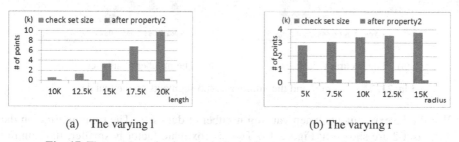

(a) The varying l   (b) The varying r

Fig. 17. The number of point comparisons for property 2 on the real dataset

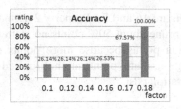
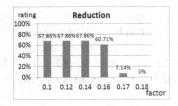

(a)   The accuracy   (b)   The number of sub-segments

Fig. 18. The accuracy and the number of sub-segments on the real dataset 1

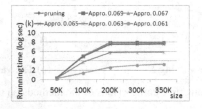

Fig. 19. The running time comparisons of the methods on the real dataset 2 (varying size)

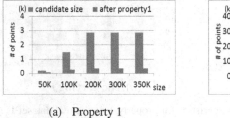

(a)   Property 1   (b)   Property 2

Fig. 20. The number of point comparisons of the property on the real dataset 2

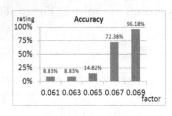
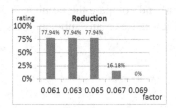

(a)   The accuracy   (b)   The number of sub-segments

Fig. 21. The accuracy and the number of sub-segments on the real dataset 2

We do the experiments when varying number of data size. The running time on the real dataset 2 are shown in Figure 19. The approximate factor is smaller, the running time is shorter. In Figure 20, it still has good performance on the Property 1 and Property 2 with varying number of data size. The accuracy and the number of sub-segments result are also same as previous with varying number of data size. The accuracy and difference of number of sub-segments are shown in Figure 21.

7   Conclusion

In this paper, we introduce the continuous spatial skyline query considering line segments. We propose two methods to solve this problem. In the basic method, for some intersection points which may cause the skyline changed, we compute the distance between each data point to those points to get the dynamic attribute. Then, we find the skyline with static attributes and a dycnamic attribute considering the points in *CS*. In the pruning method, we prune some data points which could not be the skyline by Property 1 and reduce skyline computation by Property 2. To avoid generating many sub-segments, we propose an approximate method which can reduce the number of data points in *CR*. We do the experiments and analyze their performances. The experiments demonstrate that the pruning method is better than the basic one. We use approximate factor to get the approximate result and it shows that there is a trade-off between the reduction of sub-segments and the accuracy of the results.

References

[1] Börzsönyi, S., Kossmann, D., Stocker, K.: The Skyline Operator. In: International Conference on Data Engineering (ICDE), pp. 421–430 (2001)

[2] Papadias, D., Tao, Y., Fu, G., Seeger, B.: Progressive skyline computation in database systems. ACM Trans. Database Syst (TODS) 30(1), 41–82 (2005)

[3] Dellis, E., Seeger, B.: Efficient Computation of Reverse Skyline Queries. In: International Conference on Very Large Data Bases (VLDB), pp. 291–302 (2007)

[4] Sacharidis, D., Bouros, P., Sellis, T.K.: Caching Dynamic Skyline Queries. In: Ludäscher, B., Mamoulis, N. (eds.) SSDBM 2008. LNCS, vol. 5069, pp. 455–472. Springer, Heidelberg (2008)

[5] Wang, W.-C., Wang, E.T., Chen, A.L.P.: Dynamic Skylines Considering Range Queries. In: Yu, J.X., Kim, M.H., Unland, R. (eds.) DASFAA 2011, Part II. LNCS, vol. 6588, pp. 235–250. Springer, Heidelberg (2011)

[6] Sharifzadeh, M., Shahabi, C.: The Spatial Skyline Queries. In: International Conference on Very Large Data Bases (VLDB), pp. 751–762 (2006)

[7] Geng, M., Arefin, M.S., Morimoto, Y.: A Spatial Skyline Query for a Group of Users Having Different Positions. In: International Conference on Networking and Computing (ICNC), pp. 137–142 (2012)

[8] Lin, Q., Zhang, Y., Zhang, W., Li, A.: General Spatial Skyline Operator. In: Lee, S.-g., Peng, Z., Zhou, X., Moon, Y.-S., Unland, R., Yoo, J. (eds.) DASFAA 2012, Part I. LNCS, vol. 7238, pp. 494–508. Springer, Heidelberg (2012)

[9] Deng, K., Zhou, X., Shen, H.T.: Multi-source Skyline Query Processing in Road Networks. In: International Conference on Data Engineering (ICDE), pp. 796–805 (2007)

[10] Li, C., He, W., Chen, H.: Multi-Source Skyline Queries Processing in Multi-Dimensional Space. In: Zaki, M.J., Yu, J.X., Ravindran, B., Pudi, V. (eds.) PAKDD 2010, Part I. LNCS, vol. 6118, pp. 471–479. Springer, Heidelberg (2010)

[11] Chen, L., Lian, X.: Efficient Processing of Metric Skyline Queries. IEEE Trans. Knowl. Data Eng. (TKDE) 21(3), 351–365 (2009)

[12] Fuhry, D., Jin, R., Zhang, D.: Efficient skyline computation in metric space. In: International Conference on Extending Database Technology (EDBT), pp. 1042–1051 (2009)

[13] Huang, Z., Lu, H., Ooi, B.C., Tung, A.K.H.: Continuous Skyline Queries for Moving Objects. IEEE Trans. Knowl. Data Eng. (TKDE) 18(12), 1645–1658 (2006)

[14] Jang, S., Yoo, J.: Processing Continuous Skyline Queries in Road Networks. In: International Symposium on Computer Science and its Applications (CSA), pp. 353–356 (2008)

[15] Gao, Y., Zheng, B.: Continuous obstructed nearest neighbor queries in spatial databases. In: Special Interest Group on Management of Data (SIGMOD), pp. 577–590 (2009)

Effective Skyline Query Execution in Wireless Broadcast Environments

Chuan-Ming Liu and Kai-An Yu

Department of Computer Science and Information Engineering
National Taipei University of Technology, Taipei, Taiwan
{cmliu,t100598045}@ntut.edu.tw

Abstract. Using data broadcasting to provide services has attracted much attention for years and many types of location based services (LBSs) data broadcasting have been studied. The skyline query is one of the LBSs and can find the results that meet one of the conditions given in the query. In this paper, we consider the skyline query in wireless broadcast environments and each data object has two kinds of attributes: spatial and non-spatial attributes. Different queries at different locations thus will have different skyline sets. Two indexing structures based on R*-trees will be studied: the original R*-tree and AR*-tree which indexes the data objects mainly using spatial attributes and the non-spatial attribute is augmented to the R*-tree. We then propose effective approaches that allow the clients to execute the skyline query at any location. Last, the proposed broadcasting protocols are evaluated by performing the experiments in terms of the latency, tuning time, and memory usage.

Keywords: Data Broadcasting, Skyline Query, Latency, Tuning Time, Memory Usage, Location-based Services.

1 Introduction

Emerging technologies on wireless communication, positioning systems, and networking make it possible for people to access information ubiquitously. In such a wireless mobile environment, the bandwidth of the downlink is much larger than the bandwidth of the uplink [5]. Under such an asymmetric environment, the traditional one-to-one client/server communication structure is not an appropriate method to provide services because of the bottleneck of the uplink. *Data broadcasting* thus provides an alternative and effective way to disseminate information to a large amount of mobile clients.

Using data broadcasting to provide information services has been studied extensively in recent years, including the location-based services (LBSs). LBS is one of the important information services and provides the information related to the positions of the mobile clients. The *skyline query* is one of the interesting and useful query types in location-based services. For instance, a client may want to find the nearest and cheapest hotel. In many cases, the client can not find the hotel which meets the conditions simultaneously. Instead, the client may have some hotels that are good enough and meet one of the conditions. Such a query is the skyline query.

There are many applications for the skyline query and many methods for processing the skyline query have been proposed [10]. These methods work on the static data

H. Decker et al. (Eds.): DEXA 2014, Part I, LNCS 8644, pp. 188–195, 2014.

and all the values of the attributes of the data should be known before the query process. In wireless mobile environments, the query point may move. If the location of the query point changes, all the distances need to be recomputed before computing the skyline, thus being time-consuming. Hence, employing the existing methods for skyline query in wireless mobile environments may not be effective. In this paper, we consider the skyline query in wireless broadcast environments where the locations of the query points are different and propose effective skyline query protocols.

Two cost measures are usually considered when using data broadcasting to provide information services: *latency* and *tuning time*. The latency is the time elapsed between issuing and termination of the query and the tuning time is the amount of time spent on listening to the channel. These two cost measures can represent the service time and power consumption of mobile clients respectively. A broadcast cycle is one instance of all the broadcast data(with/without index). Besides, we also consider the *memory usage* of the query process since the memory is usually one of the limited resources in mobile clients. In order to simultaneously optimize the latency, tuning time, and memory usage at the client, when designing the data broadcasting protocols, we investigate how a server schedules the broadcast and what the query process is with the corresponding broadcast at the client side. Our broadcast schedule will include an R*-tree as the index on the data for effectively reducing the tuning time [3]. Many types of queries have been studied on broadcast R-trees in recent years [4,7,9].

Although data broadcasting has been studied for years and the skyline query is useful, there are few studies about the skyline query in wireless broadcast environments. To our knowledge, only [12] considers the skyline query in wireless broadcast environments and studies for the static values of attributes. The data object we consider has two kinds of attributes: one is the *spatial* attribute and the other is the *non-spatial* attribute. The spatial attributes are used to compute the distance to the query point which will be an attribute to be considered with for the skyline query. The distance depends on the location of the query. This differentiates our working problem from the other methods [10]. Two ways for indexing data objects based on R*-tree are considered in this paper. One is the original R*-tree. By referring to [2,6], the other uses the R*-tree with the spatial attributes and augments the non-spatial attribute to that R*-tree.

In the rest of this paper, we first review the related work and introduce the preliminaries in Section 2. In Section 3, we present and discuss the broadcast schedules. Scheduling the index tree for broadcast involves determining the order by which the index nodes are sent out. A mobile client tunes into the broadcast and operates independently according to the broadcast schedule. An algorithm for executing a skyline query on the corresponding broadcast schedules is proposed and analyzed in Section 4. The experimental evaluation is discussed in Section 5. Our experimental results show that the proposed skyline query protocol can effectively derive the skyline on the broadcast data and achieves a short latency and smaller tuning time with less memory usage on the mobile client. Section 6 concludes this paper.

2   Preliminaries

In this work, we consider the skyline query where the spatial attributes are the coordinates on the plane and there is only one non-spatial attribute. For the spatial part, we consider the distance to be as shorter as better. For the non-spatial attribute, the smaller value is preferred in the query. One can define the values of the attributes and distance to be the minimum or maximum depending on the applications. In order to manage the multidimensional data set, we consider the R*-tree for indexing the data objects. An index node o_i of an R-tree uses the *minimum bounding rectangle* (MBR) as its index which surrounds the MBRs of its children and contains the information of its children, including the MBRs of the children. Given a query point q, the proposed skyline query process needs three types of distance metrics: *mindist*, *minmaxdist*, and *maxdist*, with respect to a node o_i in the index tree as below.

- $mindist(q, o_i)$ is the minimum distance from q to o_i's MBR;
- $minmaxdist(q, o_i)$ is the minimum distance of the maximum distances from q to each face of o_i's MBR;
- $maxdist(q, o_i)$ is the maximum distance from q to o_i's MBR.

If o_i is a point data, then $mindist(q, o_i) = minmaxdist(q, o_i)$. $mindist(q, o_i)$ is the minimum possible distance to a child of o_i and $minmaxdist(q, o_i)$ is the upper bound of the distance from the query point to the closest object within the MBR of o_i.

The authors of [10] reviewed many existing algorithms for skyline queries in database. The divide-and-conquer approach divides the whole dataset into several partitions so that each partition fits in memory. The final skyline is obtained by merging the partial skyline. The Block Nested Loop (BNL) method in [1] is a direct approach to compute the skyline. Block nested loop builds on the concept by scanning all the objects and keeping a list of candidate skyline points in memory. The index approach [11] organizes objects into lists according to the smallest attribute of each object. The query process starts with objects having the smallest values in all the list and repeats by considering two cases: (i) computing the skyline for the new objects and (ii) among the computed skyline objects, adding the ones not dominated by any existing skyline objects. The NN skyline query in [8] uses the results of an NN search on an R-tree to partition the data recursively. Then the partition regions are used to prune the irrelevant objects for efficiency. The branch-and-bound approach, BBS [10], is similar to the previous NN method algorithms and adopts the branch-and-bound paradigm on the R-tree for skyline query process. During the process, BBS uses mindist to prune the irrelevant objects when visiting each node.

3   Broadcast Schedules

We consider two ways for indexing data objects based on R*-trees. One is the original R*-tree. The other uses the R*-tree with the spatial information and augments the non-spatial attributes to the data and index nodes. The latter index structure is thus named as *AR*-tree* (Augmented R*-tree). For a data object p, we let (p_x, p_y) denote the location coordinate and p_z the non-spatial attribute, respectively. With the coordinates, a mobile client can compute the distance from the query point to a data object. Figures 1 (a) and (b) show an R*-tree and AR*-tree for the same data set respectively. In general, the node structure for these two index trees will be similar. Each index node will have the node ID and the information about the children, including the MBR of each child.

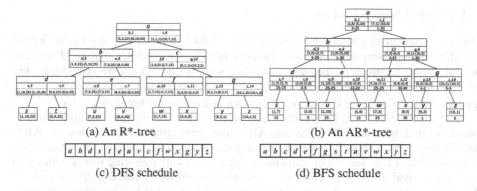

(a) An R*-tree   (b) An AR*-tree

(c) DFS schedule   (d) BFS schedule

Fig. 1. An R*-tree (a) and AR*-tree (b) for a data set and the broadcast schedules of (c) the R*-tree with DFS and (d) the AR*-tree with BFS

In wireless broadcast environment, the server simply broadcasts the data objects with the index according to some rules. We let a node in the index tree correspond to a packet in the broadcast for simplicity. Two broadcast schedules are proposed by two kinds of tree traversals, breadth-first search(BFS) and depth-first search(DFS). Figures 1 (c) and (d) show the broadcast schedules for the R*-tree with DFS and for the AR*-tree with BFS, respectively.

4   The Proposed Skyline Search Process

To effectively execute the query, the proposed client query process will refer to the following properties about the dominance of two MBRs by observations. Note that a data object (point) is a degenerate MBR. For two nodes o_i and o_j in the index tree with their corresponding MBRs in the spatial attributes in the index tree, we consider the relation of the different types of distances to the query point q for spatial attributes and then use the distances and other non-spatial attributes to derive the dominance of these two nodes. We list all the properties without proof due to the space limitation.

Given two nodes o_i and o_j in the index tree with their corresponding MBRs m_i and m_j respectively, o_i dominates o_j if and only if one of the following cases holds:

Property 1
   1. maxdist(o_i, q) < mindist(o_j, q) and
   2. *For non-spatial attribute, maximum of o_i's is smaller than the minimum of o_j's.*
Property 2
   1. minmaxdist(o_i, q) < mindist(o_j, q) and
   2. *For non-spatial attribute, maximum of o_i's is smaller than the minimum of o_j's.*
Property 3
   1. maxdist(o_i, q) < mindist(o_j, q) and
   2. *For non-spatial attribute, minimum of o_i's is smaller than the minimum of o_j's.*

4.1 Skyline Search Process

The proposed query process will use Property 2 and 3 to effectively derive the skyline result for some query q. In the query process, two data structures, *C-list* and *R-list*, are maintained to facilitate the process. *C-list* keeps the nodes that will be visited later and *R-List* stores the nodes which are in the current resulting skyline set during the search process. All the nodes in *C-List* and *R-List* are ordered by the broadcast order.

The query process SKYLINE-PROCESS starts with receiving the root node and can work on both broadcast R*-trees and AR*-trees, respectively. We now use the AR*-tree in Fig. 1 (b) to show how the algorithm performs with BFS data broadcast schedule shown in Fig. 1 (d). The client starts the query process with receiving root a. Since root a is an internal node, each child of a will be checked for candidacy (line 1 and 2). Suppose child node b is first considered. Since *C-List* and *R-List* are empty, no node dominates b or be dominated by b (line 3- 6). So, node b is inserted into *C-List* (line 7). Similarly, node c is also inserted into *C-List*. Then line 9 determines that node b is the next node to be received since b is broadcast earlier than c. The process continues (line 11) to receive node b.

SKYLINE-PROCESS(v)

```
1    if (v is an internal node)
2        for each child u of v
3            if u is not dominated by any w ∈ C-List ∪ R-list
4                for each w ∈ C-List ∪ R-list
5                    if w is dominated by u
6                        remove w from C-List or R-list
7                insert u to C-List
8    else /* v is a leaf node */
9        insert the v to R-list
         /* the node closest to the currently examined node in the broadcast*/
10   w = FINDNEXT(C-List);
11   SKYLINE-PROCESS(w)
```

When examining node b, the child nodes d and e are considered. Node d is inserted into *C-List*. Node e is then ignored since minmaxdist(d) is less than mindist(e) and node d's maximum non-spatial attribute value is less than node e's minimum non-spatial attribute value. So node d dominates node e. Similarly, node g is inserted into *C-List* and node f is discarded. Afterwards, node d and g will be received and examined and nodes s, t, and y will be inserted into *C-List*. When node s is received, since s is a leaf, it is inserted into *R-List* (line 8- 9). The process stops when *C-List* is empty and *R-List* contains the resulting skyline, nodes s, t, and y.

5   Simulation Experimental Results

In the simulation experiment study, we use synthetic data and the amount of the data is from 10,000 to 100,000 with an increment of 10,000. The cost measures include the tuning time, latency, and memory usage. We use the node-based metric where the number of nodes is counted in the experiments as the metric unit and one node corresponds to one broadcast packet. The packet size ranges from 88 to 888 bytes and the default size is 246 bytes. Different packet sizes correspond to different fanouts in the index trees. The fanout of an index tree is the maximum number children of an index node in the index tree. Each data object is generated uniformly within the unit square $[0,1] \times [0,1]$ and non-spatial attribute is scaled from 0.01 to 0.60 randomly. The data are broadcast in BFS and DFS fashions with AR*-tree and R*-tree, thus four broadcast schedules: AR*-DFS, AR*-BFS, R*-DFS, and R*-BFS. The data reported are the average of 2,000 different queries.

5.1 Tuning Time

Fig. 2 shows the performance of tuning time. Both AR*-BFS and R*-BFS are better than AR*-DFS and R*-DFS, respectively. With the BFS broadcast schedules, since the query process can get the index of MBRs early and prune irrelevant nodes. From the other angle, in Figure 2(a), the BFS and DFS broadcast schedules based on AR*-trees perform better than the BFS and DFS ones using R*-trees respectively. To explore this, we measure the *distance range* of each node v. Given a query point q, the distance range of a node v is the difference of $maxdist(v, q)$ and $mindist(v, q)$. Figure 3(a) presents the average distance ranges for the AR*-trees and R*-trees on different data sets with different packet sizes. In general, the distance range for an AR*-tree on a given data set is much less than the one for an R*-tree.

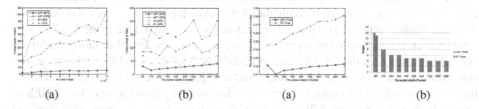

(a)   (b)   (a)   (b)

Fig. 2. Comparison on the tuning time for (a) datasets of different sizes and (b) different packet sizes

Fig. 3. The distance ranges (a) and heights (b) of the AR*-tree and R*-tree with 100,000 data objects

Figure 2(b) presents the tuning time for the four broadcast schedules in terms of KBs with different packet sizes and have the similar trend as Figure 2(a). The broadcast schedule using BFS is better than the one using DFS and the one based on AR*-trees is better in terms of tuning time. Figure 3(b) shows the height of AR*-tree and R*-tree with different packet size for 100,000 data objects. For the index trees with same height, the smaller packet size leads to a fewer tuning time. This is because the index trees with smaller packet size is more compact than the ones with larger packet size. For the broadcast schedule using BFS, the fluctuation of the curve is not obvious because less nodes are received during the query process.

5.2 Latency

Surprisingly, in our experimental results, the schedules based on BFS are better than the ones based on DFS in terms of latency in general. Figure 4(a) shows the trend. Using the BFS broadcast schedules, the client can decide the skyline right after receiving the last member of the resulting skyline by pruning the candidates in *C-list*. Thus, in general, the schedules based on BFS have a shorter latency than the ones based on DFS in average. As for the broadcast schedules using DFS, AR*-DFS is usually better than R*-DFS. Figure 4(b) shows the latency for the four broadcast schedules with different packet size and the results follow the trend we discuss above. AR*-DFS is usually better than R*-DFS and the curves of AR*-BFS and R*-BFS are tangled. More latency can be expected for the broadcast schedules using BFS. As shown in Figure 4(b), when the packet size is small (say 88 or 176), the latency for the broadcast schedules using DFS is better than or almost equal to the latency using BFS.

5.3 Memory Usage

As the above discussion, the broadcast schedules using BFS may have less tuning and fewer latency. However, the query process with the BFS broadcast schedules needs more memory to store the possible candidates and Figure 5 shows the trend. AR*-BFS and R*-BFS need to store more index nodes since the BFS broadcast schedules broadcast the index nodes first and then the data leaves. For the BFS broadcast schedules, the irrelevant index nodes will be pruned only by index nodes first. Due to the MBRs of the index nodes, the number of index nodes to be pruned may be few, thus more candidates should be stored. For the DFS broadcast schedules, AR*-DFS and R*-DFS can access some data nodes (leaves) at earlier stage and may prune more irrelevant nodes. Hence, the broadcast schedules using DFS may use less memory than the ones using BFS as shown in Figure 5(a). For different packet sizes, using the same index structure and broadcast schedule may have the similar memory usage. Figure 5(b) demonstrates this trend.

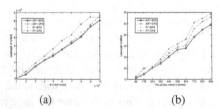
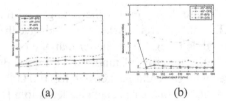

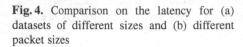

| (a) | (b) | (a) | (b) |

Fig. 4. Comparison on the latency for (a) datasets of different sizes and (b) different packet sizes

Fig. 5. Comparison on the memory usage for (a) datasets of different sizes and (b) different packet sizes

6   Conclusions

In this paper, we propose a skyline query protocol in wireless broadcast environments which allows the clients to have the skyline queries at different locations. At the server side, we use R*-tree and AR*-tree to index the data objects respectively and consider the broadcast schedules using BFS and DFS order. At the client side, an effective skyline query process for the corresponding broadcast is proposed using the relations among mindist, maxdist, and minmaxdist of two MBRs. The process can effectively ignore the irrelevant nodes during the query process, thus achieving fewer tuning time, shorter latency, and less memory usage. The simulation experiments are extensively performed for further study and discussion. As the results show, the broadcast schedules using BFS can have fewer tuning time and shorter latency than the ones using DFS but need more memory usage. Using AR*-tree as the index outperforms using R*-tree as the index due to smaller distance range.

Acknowledgement. This work is partially supported by NSC under the Grant 102-2221-E-027-088-.

References

1. Borzsony, S., Kossmann, D., Stocker, K.: The skyline operator. In: Proceedings of the 17th International Conference on Data Engineering, pp. 421–430 (2001)
2. Cormen, T.H., Leiserson, C.E., Rivest, R.L.: Introduction to Algorithms. McGraw-Hill, New York (1990)
3. Gaede, V., Günther, O.: Multidimensional access methods. ACM Computing Surveys 30(2), 170–231 (1998)
4. Hambrusch, S., Liu, C.-M., Aref, W.G., Prabhakar, S.: Efficient query execution on broad-casted index tree structures. Data and Knowledge Engineering 60(3), 511–529 (2007)
5. Imieliński, T., Viswanathan, S., Badrinath, B.R.: Data on air: Organization and access. IEEE Transactions on Knowledge and Data Engineering 9(3), 353–372 (1997)
6. Jun, E.R., Cho, B.K., Chung, Y.D., Liu, L.: On processing location based top-k queries in the wireless broadcasting system. In: Proceedings of the 2010 ACM Symposium on Applied Computing, pp. 585–591 (2010)
7. Jung, H.R., Chung, Y.D., Liu, L.: Processing generalized k-nearest neighbor queries on a wireless broadcast stream. Information Sciences 188, 64–79 (2012)
8. Kossmann, D., Ramsak, F., Rost, S.: Shooting stars in the sky: An online algorithm for sky-line queries. In: Proceedings of the 28th International Conference on Very Large Data Bases, pp. 275–286 (2002)
9. Liu, C.-M., Fu, S.-Y.: Effective protocols for knn search on broadcast multi-dimensional index trees. Information Systems 33(1), 18–35 (2008)
10. Papadias, D., Tao, Y., Fu, G., Seeger, B.: Progressive skyline computation in database sys-tems. ACM Transactions on Database Systems 30(1), 41–82 (2005)
11. Tan, K.-L., Eng, P.-K., Ooi, B.C.: Efficient progressive skyline computation. In: Proceedings of the 27th International Conference on Very Large Data Bases, pp. 301–310 (2001)
12. Wang, C.-J., Ku, W.-S.: Efficient evaluation of skyline queries in wireless data broadcast en-vironments. In: Proceedings of the 20th International Conference on Advances in Geographic Information Systems, pp. 442–445 (2012)

A Memory-Efficient Tree Edit Distance Algorithm

Mateusz Pawlik and Nikolaus Augsten

University of Salzburg, Austria
{mateusz.pawlik,nikolaus.augsten}@sbg.ac.at

Abstract. Hierarchical data are often modelled as trees. An interesting query identifies pairs of similar trees. The standard approach to tree similarity is the tree edit distance, which has successfully been applied in a wide range of applications. In terms of runtime, the state-of-the-art for the tree edit distance is the RTED algorithm, which is guaranteed to be fast independently of the tree shape. Unfortunately, this algorithm uses twice the memory of the other, slower algorithms. The memory is quadratic in the tree size and is a bottleneck for the tree edit distance computation.

In this paper we present a new, memory efficient algorithm for the tree edit distance. Our algorithm runs at least as fast as RTED, but requires only half the memory. This is achieved by systematically releasing memory early during the first step of the algorithm, which computes a decomposition strategy and is the main memory bottleneck. We show the correctness of our approach and prove an upper bound for the memory usage. Our empirical evaluation confirms the low memory requirements and shows that in practice our algorithm performs better than the analytic guarantees suggest.

1   Introduction

Data with hierarchical dependencies are often modeled as trees. Tree data appear in many applications, ranging from hierarchical data formats like JSON or XML to merger trees in astrophysics [20]. An interesting query computes the similarity between two trees. The standard measure for tree similarity is the tree edit distance, which is defined as the minimum-cost sequence of node edit operations that transforms one tree into another. The tree edit distance has been successfully applied in many applications [1, 2, 14, 18], and has received considerable attention from the database community [3, 5–8, 11–13, 16, 17].

The fastest algorithms for the tree edit distance (TED) decompose the input trees into smaller subtrees and use dynamic programming to build the overall solution from the subtree solutions. The key difference between the different TED algorithms is the decomposition strategy, which has a major impact on the runtime. Early attempts to compute TED [9, 15, 23] use a hard-coded strategy, which disregards or only partially considers the shape of the input trees. This may lead to very poor strategies and asymptotic runtime differences of up to

H. Decker et al. (Eds.): DEXA 2014, Part I, LNCS 8644, pp. 196–210, 2014.

a polynomial degree. The most recent development is the Robust Tree Edit Distance (RTED) algorithm [19], which operates in two steps (cf. Figure 1(a)). In the first step, a decomposition strategy is computed. The strategy adapts to the input trees and is shown to be optimal in the class of LRH strategies, which contains all previously proposed strategies. The actual distance computation is done in the second step, which executes the strategy.

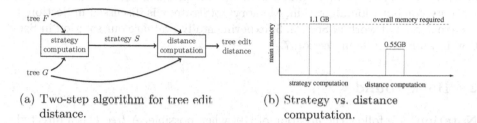

(a) Two-step algorithm for tree edit distance.

(b) Strategy vs. distance computation.

Fig. 1. Strategy computation requires more memory than actual distance computation

In terms of runtime, the overhead for the strategy computation in RTED is small compared to the gain due to the better strategy. Unfortunately, this does not hold for the main memory consumption. Figure 1(b) shows the memory usage for two example trees (perfect binary trees) of 8191 nodes: the strategy computation requires 1.1GB of RAM, while the execution of the strategy (i.e., the actual distance computation) requires only 0.55GB. Thus, for large instances, the strategy computation is the bottleneck and the fallback is a hard-coded strategy. This is undesirable since the gain of a good strategy grows with the instance size.

This paper solves the memory problem of the strategy computation. We present MemoryOptStrategy, a new algorithm for computing an optimal decomposition strategy. We achieve this by computing the strategy bottom-up using dynamic programming and releasing part of the memoization tables early. We prove an upper bound for the memory requirements which is 25% below the best case for the distance computation. Thus, the strategy computation is no longer the main memory bottleneck. In our extensive experimental evaluation on various tree shapes, which require very different strategies, the memory for the strategy computation never exceeds the memory for the distance computation. With respect to the strategy algorithm proposed for RTED [19], we reduce the memory by at least 50%. For some tree shapes our algorithm even runs in linear space, while the RTED strategy algorithm always requires quadratic space. Summarizing, the contributions of this paper are the following:

- *Memory efficiency.* We substantially reduce the memory requirements w.r.t. previous strategy computation algorithms. We develop a new MemOptStrategy algorithm which computes the strategy by traversing the trees bottom-up and systematically releasing rows early. We show the correctness of our approach and prove an upper bound for the memory usage.

- *Shape and size heuristics.* Our MemOptStrategy algorithm requires at most 50% of the memory that the RTED strategy algorithm needs. We observe that the worst case happens for very specific tree shapes only. We devise heuristics which deal with a number of frequent worst-case instances and thus make our solution even more effective in practice.

The paper is structured as follows. Section 2 sets the stage for the discussion of strategy algorithms. In Section 3 we define the problem. Our memory-efficient strategy computation algorithm, MemoryOptStrategy is discussed in Section 4. We treat related work in Section 5, experimentally evaluate our solution in Section 6, and conclude in Section 7.

2   Background

Notation. We follow the notation of [19] when possible. A *tree* F is a directed, acyclic, connected graph with nodes $N(F)$ and edges $E(F) \subseteq N(F) \times N(F)$, where each node has at most one incoming edge. In an edge (v, w), node v is the *parent* and w is the *child*, $p(w) = v$. A node with no parent is a *root* node, a node without children is a *leaf*. Children of the same node are *siblings*. Each node has a *label*, which is not necessarily unique within the tree. The nodes of a tree F are strictly and totally ordered such that (a) $v > w$ for any edge $(v, w) \in E(F)$, and (b) for any two siblings f, g, if $f < g$, then $f' < g$ for all descendants f' of f, and $f < g'$ for all descendants g' of g. The tree traversal that visits all nodes in ascending order is the *postorder* traversal.

F_v is the *subtree rooted in node* v of F iff F_v is a tree, $N(F_v) = \{x : x = v$ or x is a descendant of v in $F\}$, and $E(F_v) \subseteq E(F)$. A *path* in F is a subtree of F in which each node has at most one child.

We use the following short notation: $|F| = |N(F)|$ is the size of tree F, we write $v \in F$ for $v \in N(F)$.

Example 1. The nodes of tree F in Figure 2 are $N(F) = \{v_1, v_2, v_3, v_4, v_5, v_6, v_7, v_8, v_9, v_{10}, v_{11}, v_{12}, v_{13}\}$, the edges are $E(F) = \{(v_{13}, v_4), (v_{13}, v_{10}), (v_{13}, v_{12}), (v_4, v_1), (v_4, v_3), (v_3, v_2), (v_{10}, v_5), (v_{10}, v_9), (v_9, v_8), (v_8, v_6), (v_8, v_7), (v_{12}, v_{11})\}$, the node labels are shown in italics in the figure. The root of F is $r(F) = v_{13}$, and $|F| = 13$. F_{v_8} with nodes $N(F_{v_9}) = \{v_6, v_7, v_8\}$ and edges $E(F_{v_9}) = \{(v_8, v_6), (v_8, v_7)\}$ is a subtree of F. The postorder traversal visits the nodes of F in the following order: $v_1, v_2, v_3, v_4, v_5, v_6, v_7, v_8, v_9, v_{10}, v_{11}, v_{12}, v_{13}$.

Strategy Computation. The tree edit distance is the minimum cost sequence of node edit operations that transforms tree F into G. The standard edit operations are: delete a node, insert a node, rename the label of a node. The state-of-the-art algorithms compute the tree edit distance by implementing a well-known recursive solution [21] using dynamic programming [9, 19, 21, 23].

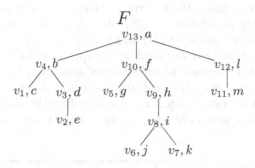

Fig. 2. Example tree

They devise so called *decomposition strategies* to decompose the input trees into smaller subtree pairs for which the distance is computed first. The results of smaller subproblems are used to compute the distances of bigger problems. Pawlik and Augsten [19] introduce *path strategies* as a subset of all possible decomposition strategies. Together with the *general tree edit distance* algorithm, path strategies generalize all previous approaches. We revisit the concept of path strategies and the computation of the optimal strategy in RTED.

In the first step of RTED (cf. Figure 1(a)) a path strategy is computed, which maps each pair of subtrees (F_v, G_w) of two input trees F and G to a root-leaf path in either F_v or G_w[1]. The resulting strategy is optimal in the sense that it minimizes the number of subproblems (i.e., the number of distances) that must be computed.

The RTEDStrategy algorithm [19] (Algorithm 1) has the following outline. The strategy is computed in a dynamic programming fashion. The algorithm uses six memoization arrays, called *cost arrays*: L_v, R_v, H_v of size $|F||G|$, and L_w, R_w, H_w of size $|G|$. Each cell in a cost array stores the cost of computing the distance for a specific subtree pair and a specific path type. The names of the cost arrays indicate the path type, which may be left (L), right (R), or heavy (H) in the class of LRH strategies considered in [19]. All state-of-the-art strategies fall into the LRH class. The algorithm computes the final result in a bottom-up manner starting with the leaf nodes. The core of the algorithm are two nested loops (lines 3 and 4). They iterate over the nodes of the input trees in postorder, which ensures that a child node is processed before its parent. Inside the innermost loop, first the minimal cost of computing the distance for the particular subtree pair (F_v, G_w) is calculated based on the previously computed distances for the children of nodes v and w (line 7). Second, the path corresponding to that cost is send to the output (line 8). Finally, the distances for the parent nodes $p(v)$ and $p(w)$ are updated (lines 9 and 10).

---

[1] RTED considers LRH strategies which allow left, right, and heavy paths. For example, the path in F from v_{13} to v_7 in Figure 2 is a heavy path, i.e., a parent is always connected to its child that roots the largest subtree (the rightmost child in case of ties). The path from v_{13} to v_1 is left, and the path from v_{13} to v_{11} is right.

Since all the operations inside the innermost loop are done in constant time, the time complexity of the algorithm is $O(|F||G|)$. The space complexity is given by the three cost arrays of the size $|F||G|$ and is $O(|F||G|)$.

In this paper we show that the strategy computation in RTED may require twice the memory of the actual distance computation step. We remove that bottleneck by reducing the size of the quadratic cost arrays L_v, R_v, and H_v. We show that our new MemOptStrategy algorithm uses significantly less memory than RTEDStrategy, and never uses more memory then the distance computation.

Algorithm 1. RTEDStrategy(F, G)

---

1   L_v, R_v, H_v : arrays of size $|F||G|$
2   L_w, R_w, H_w : arrays of size $|G|$
3   **for** $v \in F$ *in postorder* **do**
4 | **for** $w \in G$ *in postorder* **do**
5 | | **if** v *is leaf* **then** $L_v[v, w] \leftarrow R_v[v, w] \leftarrow H_v[v, w] \leftarrow 0$
6 | | **if** w *is leaf* **then** $L_w[w] \leftarrow R_w[w] \leftarrow H_w[w] \leftarrow 0$
7 | | compute the minimal cost of distance computation for (F_v, G_w)
8 | | output the path corresponding to the minimal cost for (F_v, G_w)
9 | | **if** v *is not root* **then** update values for $L_v[p(v), w], R_v[p(v), w], H_v[p(v), w]$
10 | | **if** w *is not root* **then** update values for $L_w[p(w)], R_w[p(w)], H_w[p(w)]$

---

3   Problem Definition

As outlined above, the path strategy introduced by Pawlik and Augsten [19] generalizes all state-of-the-art algorithms for computing the tree edit distance. The RTED algorithm picks the optimal strategy. Unfortunately, the computation of the optimal strategy requires more space than executing the strategy, i.e., computing the actual tree edit distance. This limits the maximum size of the tree instances which can be processed in given memory.

In this paper we analyse the memory requirements of strategy and distance computation in the RTED algorithm. We devise a new technique to significantly reduce the memory usage of the strategy computation. We develop the new MemOptStrategy algorithm, which reduces the required memory by at least 50% and thus never requires more memory than the distance computation.

4   Memory Efficient Tree Edit Distance

The main memory requirement is a bottleneck in the tree edit distance computation. The strategy computation in RTED exceeds the memory needed for executing the strategy. Our MemOptStrategy algorithm (Algorithm 2) reduces the memory usage by at least 50% and in practice never uses more memory than it is required for executing the strategy. We achieve that by decreasing the maximal size of the data structures used for strategy computation.

4.1   Memory Analysis in RTED

The asymptotic space complexity is quadratic for both, computing the strategy and executing it (cf. Figure 1(a)). However, due to different constants and depending on the strategy, the strategy computation may use twice the memory of distance computation. This is undesirable since the strategy computation becomes a memory bottleneck of the overall algorithm.

We analyse the memory usage of RTED. The strategy computation is dominated by three cost arrays L_v, R_v, H_v which take $3|F||G|$ space in total. Each time a strategy path is computed, it is send to the output. However, due to the fact that the entire strategy is required in the decomposition step, it is materialized in a quadratic array of the size $|F||G|$ [19]. Thus, the strategy computation requires $4|F||G|$ memory.

The memory of distance computation depends on the strategy. The tree edit distance is computed by so called *single-path functions* which process a single path from the strategy. Different path types require different amount of memory. Independent of the strategy, a distance matrix of size $|F||G|$ is used.

a) If only left and/or right paths are used in the strategy, the single path functions require only one array of size $|F||G|$. Thus $2|F||G|$ of space is needed in total.

b) If a heavy path is involved in the strategy, one array of size $|F||G|$ and one array of size $|G|^2$ are used. Thus the distance computation needs $2|F||G| + |G|^2$ space in total.

We observe that the distance computation uses always less memory than the strategy. Thus, the strategy computation is the limiting factor for the maximum tree size that can be computed in given memory.

The goal is to reduce the memory requirements of the strategy computation algorithm to meet the minimum requirements of the distance computation. We observe that some of the rows in the cost arrays are not needed any more after they are read. We develop an early deallocation technique to minimize the size of the cost arrays.

4.2   MemOptStrategy Algorithm

The strategy algorithm in RTED uses three cost arrays of quadratic size (L_v, R_v, H_v), where each row corresponds to a node $v \in F$. The outermost loop (line 3) iterates over the nodes in F in postorder. In each iteration only two rows of the cost arrays are accessed: the row of node v is read and the row of its parent $p(v)$ is updated. Once the row of node v is read it will not be accessed later and thus can be deallocated. In our analysis we count the maximum number of concurrently needed rows.

We observe that a row which corresponds to a leaf node stores only zeros. Thus we store a single read-only row for all leaf nodes in all cost arrays and call it *leafRow* (cf. line 4).

We define a set of four operations that we need on the rows of a cost array:

- read(v) / read(*leafRow*) - read the row of node v / read *leafRow*,
- allocate(v) - allocate a row of node v,
- update(v) - update values in the row of node v,
- deallocate(v) - deallocate the row of node v.

The same operations are synchronously applied to the rows of all the three cost arrays (L_v, R_v, H_v). To simplify the discussion, we consider a single cost array and sum up in the end.

We present the new MemOptStrategy algorithm, Algorithm 2, which dynamically allocates and deallocates rows in the cost arrays. The lines with the grey background implement our technique. At each iteration step, i.e., for each node $v \in F$ in postorder, MemOptStrategy performs the following steps on a cost array:

> if v is not root of F
>   if row for $p(v)$ does not exist: allocate($p(v)$) (line 5)
>   read(v)   (line 8)
>   update($p(v)$)   (line 10)
>   if v is not a leaf: deallocate(v) (line 12)

Algorithm 2. MemOptStrategy(F, G)

---

1 L_v, R_v, H_v : arrays of size $|F|$

2 $L_w, R_w, H_w,$ *leafRow* : arrays of size $|G|$

3 **for** $v \in F$ *in postorder* **do**

4   **if** v *is leaf* **then** $L_v[v] \leftarrow R_v[v] \leftarrow H_v[v] \leftarrow$ *leafRow*

5   **if** *row for* $p(v)$ *is not allocated* **then** allocate a row for $p(v)$ in L_v, R_v, H_v

6   **for** $w \in G$ *in postorder* **do**

7   **if** w *is leaf* **then** $L_w[w] \leftarrow R_w[w] \leftarrow H_w[w] \leftarrow 0$

8   compute the minimal cost of distance computation for (F_v, G_w)

9   output the path corresponding to the minimal cost for (F_v, G_w)

10   **if** v *is not root* **then** update values for $L_v[p(v), w], R_v[p(v), w], H_v[p(v), w]$

11   **if** w *is not root* **then** update values for $L_w[p(w)], R_w[p(w)], H_w[p(w)]$

12   **if** v *is not leaf* **then** deallocate row of v in L_v, R_v, H_v

---

Depending on the position of a node in the tree (root, leaf, leftmost child) a specific sequence of operations must be performed. Figure 3 shows the possible sequences and assigns them to the respective node types.

$S0 = \langle\rangle$
$S1 = \langle$allocate$(p(v))$, read($leafRow$),
         update$(p(v))\rangle$
$S2 = \langle$allocate$(p(v))$, read(v),
         update$(p(v))$, deallocate$(v)\rangle$
$S3 = \langle$read(v), update$(p(v))$, deallocate$(v)\rangle$
$S4 = \langle$read($leafRow$), update$(p(v))\rangle$

	leaf	non-leaf
leftmost child	S1	S2
not leftmost child	S4	S3
root		S0

Fig. 3. Rows operations depending on the node position

Example 2. We study the number of concurrently needed rows for the example tree in Figure 4. The table in the figure shows the operations performed for every node of the example tree and the rows stored in memory after each step. The subscript numbers of the nodes represent their postorder position. We iterate over the nodes in postorder, i.e., $v_1, v_2, v_3, v_4, v_5, v_6$. Depending on the node type, different operations are performed. The nodes trigger the following operation sequences: $v_1 - S1$, $v_2 - S1$, $v_3 - S3$, $v_4 - S2$, $v_5 - S4$, $v_6 - S0$.

The cost arrays in RTEDStrategy for the example tree have six rows (one for each node). From the table in Figure 4 we see that a maximum of only three rows ($leafRow$, rows for v_3 and v_4) need to be stored at the same time.

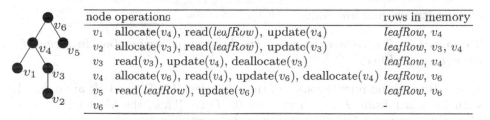

node operations		rows in memory
v_1	allocate(v_4), read($leafRow$), update(v_4)	$leafRow$, v_4
v_2	allocate(v_3), read($leafRow$), update(v_3)	$leafRow$, v_3, v_4
v_3	read(v_3), update(v_4), deallocate(v_3)	$leafRow$, v_4
v_4	allocate(v_6), read(v_4), update(v_6), deallocate(v_4)	$leafRow$, v_6
v_5	read($leafRow$), update(v_6)	$leafRow$, v_6
v_6	-	-

Fig. 4. Rows operations on an example tree

We now consider the general case and count the number of rows that must be kept in memory concurrently. Each of the operation sequences allocates and/or deallocates rows. With $s(v) \in \{S0, S1, S2, S3, S4\}$ we denote the operation sequence required for node v, with $d(S)$, $S \in \{S0, S1, S2, S3, S4\}$, we denote the difference between the number of allocated and deallocated rows by a particular operation sequence S, i.e., $d(S0) = 0$, $d(S1) = 1$, $d(S2) = 0$, $d(S3) = -1$, $d(S4) = 0$. We count the maximum number of concurrent rows during the postorder traversal of tree F with the following formula.

$$cnr(F) = 1 + \max_{v \in F}\{ \sum_{w \in F, w < v} d(s(w))\}$$

The *leafRow* is always kept in memory. In addition, each node contributes with $d(S)$ concurrently needed rows, where S is the operation sequence required by the node. For the tree in Figure 4 we have $cnr(F) = 1 + \max\{1, 1 + 1, 1 + 1 - 1, 1 + 1 - 1 + 0, 1 + 1 - 1 + 0 + 0, 1 + 1 - 1 + 0 + 0 + 0\} = 3$.

We observe that only operation sequence $S1$ adds more rows than it deallocates, i.e., $d(S1) > 0$. An upper bound on the number of concurrently needed rows is given by the number of nodes that require operation sequence $S1$. The only type of node that falls into this category is a leaf node which is the leftmost child of its parent.

Lemma 1. *The maximum number of concurrently needed rows of each cost array is at most $\frac{|F|}{2}$.*

Proof. The maximum number of the concurrently needed rows for a tree F is the number of its leftmost-child leaf nodes, which is bound by $\lfloor \frac{|F|}{2} \rfloor$: Each such node, in order to be a leftmost-child leaf node, must have a different parent. An example of a tree with $\lfloor \frac{|F|}{2} \rfloor$ leftmost-child leaf nodes is the right branch tree (the right branch tree is a symmetric tree to the left branch tree in Figure 5(a)). This results in $\lfloor \frac{|F|}{2} \rfloor$ different parent nodes for $\lfloor \frac{|F|}{2} \rfloor$ leftmost-child leaf nodes. If $|F|$ is odd, then $\lfloor \frac{|F|}{2} \rfloor + \lfloor \frac{|F|}{2} \rfloor = |F| - 1$ and there is one node that we did not count. This node can become a leftmost-child leaf node of either a node which is a leftmost-child leaf or of a node that has a leftmost-child leaf. Then, the old leftmost-child leaf becomes the parent or a right sibling of the new leftmost-child leaf, respectively. If $|F|$ is even, $\lfloor \frac{|F|}{2} \rfloor + \lfloor \frac{|F|}{2} \rfloor = |F|$. Thus, the maximum number of the leftmost-child leaf nodes is $\lfloor \frac{|F|}{2} \rfloor$. □

Theorem 1. *The memory required for the strategy computation in MemOpt-Strategy is $1.5|F||G|$.*

Proof. By expiring rows the sizes of the cost arrays L_v, R_v, and H_v are reduced with Lemma 1 from $|F||G|$ to at most $0.5|F||G|$. Thus, the MemOptStrategy algorithm requires $1.5|F||G|$ memory in the worst case. □

4.3   Shape and Size Heuristics

We further devise two heuristic techniques that are easy to apply yet effective in reducing the memory.

Shape. The MemOptStrategy algorithm iterates over the nodes in postorder. One can think of a symmetric algorithm which iterates over the nodes in right-to-left postorder[2]. Then the maximum number of concurrently needed rows is equal to the maximum number of the rightmost-child leaves in a tree (instead of leftmost-child leaves). For the right branch tree, which is the worst case for postorder, the right-to-left postorder algorithm needs only two rows. Thus, the direction of the tree traversal matters. Let $\#l(F)$ and $\#r(F)$ be the number

---

[2] In right-to-left postorder, children are traversed from right to left and before their parents. For example, traversing nodes of the tree in Figure 4 gives the following sequence: $v_5, v_2, v_3, v_1, v_4, v_6$.

of leftmost-child and rightmost-child leaf nodes in F, respectively. The *shape heuristic* says: If $\#l(F) < \#r(F)$, then use postorder, otherwise use right-to-left postorder to compute the strategy.

Size. The length of the rows in the cost array is the size of the right-hand input tree, thus the array size can be reduced by switching the input parameters. The *size heuristic* is as follows: If $|G| \cdot \min\{\#l(F), \#r(F)\} < |F| \cdot \min\{\#l(G), \#r(G)\}$, then the number of rows should depend on F and the length of the rows on G.

5   Related Work

Tree Edit Distance Algorithms. The fastest algorithms for the tree edit distance are dynamic programming implementations of a recursive solution which decomposes the input trees to smaller subtrees and subforests. The runtime complexity is given by the number of subproblems that must be solved. Tai [21] proposes and algorithm that runs in $O(n^6)$ time and space, where n is the number of nodes. Zhang and Shasha [23] improve the complexity to $O(n^4)$ time and $O(n^2)$ space. Klein [15] achieves $O(n^3 \log n)$ runtime and space. Demaine et al. [9] develop an algorithm which requires $O(n^3)$ time and $O(n^2)$ space and show that their solution is worst-case optimal. The most recent development is the RTED algorithm by Pawlik and Augsten [19]. They observe that each of the previous algorithms is efficient only for specific tree shapes and runs into its worst case otherwise. They point out that the runtime difference due to a poor algorithm choice may be of a polynomial degree. They devise the general framework for computing the tree edit distance shown in Figure 1(a).

Path Strategies. Each of the tree edit distance algorithms uses a specific set of paths to decompose trees and order the computation of subproblems in the dynamic programming solution. The algorithms by Zhang and Shasha [23], Klein [15], and Demaine et al. [9] use hard-coded strategies which do not require a strategy computation step, but the resulting algorithms are efficient only for specific tree shapes. Dulucq and Touzet [10] compute a decomposition strategy in the first step, then use the strategy to compute the tree edit distance. However, they only consider strategies that decompose a single tree, and the overall algorithm runs in $O(n^3 \log n)$ time and space. More efficient strategies that decompose both trees were introduced by Pawlik and Augsten [19]. Their entire solution requires $O(n^3)$ time and $O(n^2)$ space. The resulting strategy is shown to be optimal in the class of LRH strategies, which cover all previous solutions. However, the strategy computation may need twice as much memory as the distance computation. Our memory-efficient strategy computation algorithm, MemOptStrategy, improves over the strategy computation in RTED [19] and reduces the memory by at least 50%. We never need more memory for the strategy computation than for the tree edit distance computation.

Approximations. More efficient approximation algorithms for the tree edit distance have been proposed. Zhang [22] proposes an upper bound, *constrained tree edit distance*, which is solved in $O(n^2)$ time and space. Guha et al. [13] develop a lower bound algorithm by computing the string edit distance between the preorder and postorder sequences of node labels in $O(n^2)$ time and space. Augsten et al. [4] decompose the trees into pq-grams and propose a lower bound algorithm which requires $O(n \log n)$ time and $O(n)$ space. Finis et al. [11] develop the RWS-Diff algorithm (Random Walk Similarity Diff) to compute an approximation of the tree edit distance and an edit mapping in $O(n \log n)$ time.

6 Experiments

In this section we experimentally evaluate our MemOptStrategy algorithm and compare it to RTED [19]. The experiments on real-world and synthetic data confirm our analytical results. We show that computing the strategy with MemOptStrategy is as efficient as in RTED and requires significantly less memory. The memory requirements of MemOptStrategy are below the memory used to execute the strategy and compute the actual tree edit distance.

Set-up. All algorithms were implemented as single-thread applications in Java 1.7. We run the experiments on a 4-core Intel i7 3.70GHz desktop computer with 8GB RAM memory. We measure main memory and runtime for each of the two steps in the process: the strategy and the distance computation (cf. Figure 1(b)). The output of the strategy computation is a strategy which can be stored in an array of size $|F||G|$ for a pair of trees, F and G. Our memory measurements for the two strategy algorithms also include the space required to store the computed strategy, which we materialize in main memory due to efficiency reasons.

The Datasets. We test the algorithms on both synthetic and real world data. We generate synthetic trees of three different shapes: left branch (LB), zigzag (ZZ), and full binary (FB) (Figure 5). We also generate random trees (Random) varying in depth and fanout (with a maximum depth of 15 and a maximum fanout of 6).

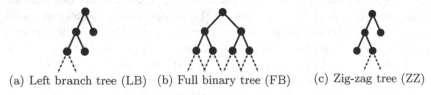

(a) Left branch tree (LB)   (b) Full binary tree (FB)   (c) Zig-zag tree (ZZ)

Fig. 5. Shapes of the synthetic trees

We use three real world datasets with different characteristics. SwissProt[3] is an XML protein sequence database with 50000 medium sized and flat trees (average depth 3.8, maximum depth 4, average fanout 1.8, maximum fanout 346, average size 187). TreeBank[4] is an XML representation of natural language syntax trees with 56385 small and deep trees (average depth 10.4, maximum

---

[3] http://www.expasy.ch/sprot/
[4] http://www.cis.upenn.edu/~treebank/

depth 35, average fanout 1.5, maximum fanout 51, average size 68). TreeFam[5] stores 16138 phylogenetic trees of animal genes (average depth 14, maximum depth 158, average fanout 2, maximum fanout 3, average size 95).

Memory measurement. We first measure the memory requirements for strategy computation and compare it to the memory needed for executing the strategy. We measure only the heap memory allocated by the Java Virtual Machine. The non-heap memory is independent of the tree size and varies between 3.5MB and 5MB in all our tests.

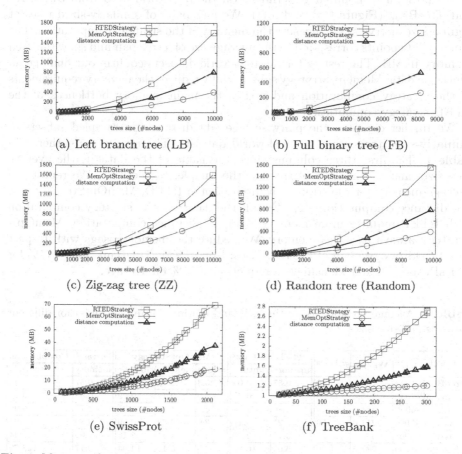

(a) Left branch tree (LB)

(b) Full binary tree (FB)

(c) Zig-zag tree (ZZ)

(d) Random tree (Random)

(e) SwissProt

(f) TreeBank

Fig. 6. Memory of strategy computation compared to executing the strategy on synthetic and real-world datasets

Figures 6(a)-6(d) show the memory requirements for different tree shapes and sizes. We use three datasets containing synthetic trees of a specific shape (LB, ZZ, FB) and a dataset with randomly generated trees (Random). Each synthetic dataset contains trees with up to 10000 nodes. The graphs show the memory

---

[5] http://www.treefam.org/

usage in MB of (a) the RTEDStrategy algorithm, (b) our MemOptStrategy algorithm, and (c) the distance computation step. In all tested scenarios Mem-OptStrategy requires significantly less memory than RTED. The RTEDStrategy algorithm uses much more (up to 200%) memory than is needed for executing the strategy thus forming a main memory bottleneck. On the contrary, MemOpt-Strategy always uses less memory than the distance computation. In particular, we observe that most of the memory allocated by MemOptStrategy is used to materialize the output (i.e., the computed strategy) in main memory.

We performed a similar experiment on the real-world data from SwissProt and TreeBank (Figure 6(e) and 6(f)). We pick pairs of similarly-sized trees at regular size intervals. We measure the memory of the strategy computation. The plotted data points correspond to the average size of a tree pair and show the used memory in MB. The results for the real-world datasets confirm our findings on synthetic data: MemOptStrategy substantially reduces the memory requirements of the strategy computation and breaks the main memory bottleneck of the RTED strategy computation.

We further compute the pairwise tree edit distance on sampled subsets of similarly-sized trees from the real-world datasets. The results are gathered in Table 1. The first three columns show the name of the dataset, the average tree size, and the number of trees in the sample, respectively. We report the average memory for the strategy computation in RTED, MemOptStrategy, and the distance computation step. For all the datasets, the strategy computation with RTED requires more memory than the distance computation; MemOpt-Strategy always uses less memory. We reduce the memory usage with respect to RTEDStrategy in all of the test cases: from 7% for TreeBank-100 to 71% for SwissProt-2000, and we achieve a reduction of 51% on average.

Table 1. Average memory usage (in MB) and runtime (in milliseconds) for different datasets and tree sizes

Dataset	Avg.size	#Trees	RTEDStrategy		MemOptStrategy			Distance Computation	
			memory	runtime	#rows	memory	runtime	memory	runtime
TreeBank-100	98.7	50	0.57	0.02	5.3	0.53	0.03	0.57	1.43
TreeBank-200	195.4	50	1.10	0.07	6.0	0.72	0.09	0.88	6.27
TreeBank-400	371.3	10	2.86	2.14	6.3	1.30	2.31	1.77	26.99
SwissProt-200	211.0	50	1.21	0.15	3.0	0.70	0.09	0.89	4.92
SwissProt-400	395.0	50	3.12	2.25	3.0	1.31	2.16	1.92	15.44
SwissProt-1000	987.5	20	16.64	16.95	3.0	5.13	14.94	9.17	123.79
SwissProt-2000	1960.1	10	63.20	64.85	3.0	17.80	55.11	33.10	502.88
TreeFam-200	197.7	50	1.16	0.15	9.6	0.74	0.15	0.88	9.86
TreeFam-400	402.6	50	3.33	2.52	12.3	1.46	2.37	2.05	55.48
TreeFam-1000	981.9	20	16.73	18.33	14.1	5.58	17.33	9.27	453.51

Number of rows. We perform an experiment on our real-world datasets and count the average number of concurrently needed rows in the cost arrays during the strategy computation with MemOptStrategy. This shows the memory reduction on the cost arrays only and does not consider the materialized strategy. The results are shown in column 6 of Table 1. For RTEDStrategy the number of rows is equal to the number of nodes of the left-hand input tree (column 2 - Avg. size). For MemOptStrategy the number of rows varies from 5% of the tree size for TreeBank-100 to 0.15% for SwissProt-2000, and we obtain a reduction of 97.2% in average. This confirms our analytical results and shows the effectiveness of our approach.

Runtime. We measure the average runtime of computing and executing the strategies. The results for different datasets are shown in the runtime columns of Table 1. They show that MemOptStrategy in most cases is slightly faster than RTEDStrategy. This is due to low memory allocation. Compared to the distance computation, the strategy requires only a small fraction of the overall time. Figure 7 shows how the strategy computation scales with the tree size for the SwissProt dataset. Compared to the distance computation, the strategy computation requires 6% of the overall time on average.

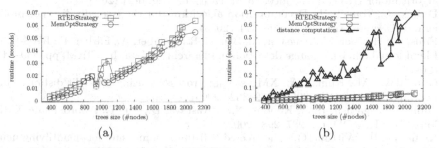

Fig. 7. Runtime difference between RTEDStrategy and MemOptStrategy algorithms (a) compared to the overall time of the distance computation (b)

7   Conclusion

In this paper we developed a new strategy computation algorithm for the tree edit distance. The strategy computation is a main memory bottleneck of the state-of-the-art solution, RTED [19]. The memory required for the strategy computation can be twice the memory needed for the actual tree edit distance computation, thus forming a bottleneck. Our MemOptStrategy algorithm reduces the memory requirements by at least 50% compared to RTED and never uses more memory than the distance computation. Our extensive empirical evaluation on synthetic and real world datasets confirmed our theoretical results.

Acknowledgements. This work was partially supported by the SyRA project of the Free University of Bozen-Bolzano, Italy.

References

1. Akutsu, T.: Tree edit distance problems: Algorithms and applications to bioinformatics. IEICE Trans. on Inf. Syst. 93-D(2), 208–218 (2010)
2. Aoki, K.F., Yamaguchi, A., Okuno, Y., Akutsu, T., Ueda, N., Kanehisa, M., Mamitsuka, H.: Efficient tree-matching methods for accurate carbohydrate database queries. Genome Informatics 14, 134–143 (2003)
3. Augsten, N., Barbosa, D., Böhlen, M., Palpanas, T.: Efficient top-k approximate subtree matching in small memory. IEEE TKDE 23(8), 1123–1137 (2011)
4. Augsten, N., Böhlen, M.H., Gamper, J.: The pq-gram distance between ordered labeled trees. ACM TODS 35(1) (2010)
5. Chawathe, S.S.: Comparing hierarchical data in external memory. In: VLDB, pp. 90–101 (1999)
6. Cobena, G., Abiteboul, S., Marian, A.: Detecting changes in XML documents. In: ICDE, pp. 41–52 (2002)
7. Cohen, S.: Indexing for subtree similarity-search using edit distance. In: SIGMOD, pp. 49–60 (2013)
8. Dalamagas, T., Cheng, T., Winkel, K.-J., Sellis, T.K.: A methodology for clustering XML documents by structure. Inf. Syst. 31(3), 187–228 (2006)
9. Demaine, E.D., Mozes, S., Rossman, B., Weimann, O.: An optimal decomposition algorithm for tree edit distance. ACM Trans. on Alg. 6(1) (2009)
10. Dulucq, S., Touzet, H.: Decomposition algorithms for the tree edit distance problem. J. Discrete Alg. 3(2-4), 448–471 (2005)
11. Finis, J.P., Raiber, M., Augsten, N., Brunel, R., Kemper, A., Färber, F.: RWS-Diff: Flexible and efficient change detection in hierarchical data. In: CIKM, pp. 339–348 (2013)
12. Garofalakis, M., Kumar, A.: XML stream processing using tree-edit distance embeddings. ACM TODS 30(1), 279–332 (2005)
13. Guha, S., Jagadish, H.V., Koudas, N., Srivastava, D., Yu, T.: Approximate XML joins. In: SIGMOD, pp. 287–298 (2002)
14. Heumann, H., Wittum, G.: The tree-edit-distance, a measure for quantifying neuronal morphology. BMC Neuroscience 10(suppl. 1), P89 (2009)
15. Klein, P.N.: Computing the edit-distance between unrooted ordered trees. In: Bilardi, G., Pietracaprina, A., Italiano, G.F., Pucci, G. (eds.) ESA 1998. LNCS, vol. 1461, pp. 91–102. Springer, Heidelberg (1998)
16. Korn, F., Saha, B., Srivastava, D., Ying, S.: On repairing structural problems in semi-structured data. Proceedings of the VLDB Endowment 6(9) (2013)
17. Lee, K.-H., Choy, Y.-C., Cho, S.-B.: An efficient algorithm to compute differences between structured documents. IEEE TKDE 16(8), 965–979 (2004)
18. Lin, Z., Wang, H., McClean, S.: Measuring tree similarity for natural language processing based information retrieval. In: Hopfe, C.J., Rezgui, Y., Métais, E., Preece, A., Li, H. (eds.) NLDB 2010. LNCS, vol. 6177, pp. 13–23. Springer, Heidelberg (2010)
19. Pawlik, M., Augsten, N.: RTED: A robust algorithm for the tree edit distance. Proceedings of the VLDB Endowment, 334–345 (2011)
20. Springel, V., White, S.D.M., Jenkins, A., Frenk, C.S., Yoshida, N., Gao, L., Navarro, J., Thacker, R., Croton, D., Helly, J., Peacock, J.A., Cole, S., Thomas, P., Couchman, H., Evrard, A., Colberg, J., Pearce, F.: Simulations of the formation, evolution and clustering of galaxies and quasars. Nature 435 (2005)
21. Tai, K.-C.: The tree-to-tree correction problem. J. ACM 26(3), 422–433 (1979)
22. Zhang, K.: Algorithms for the constrained editing distance between ordered labeled trees and related problems. Pattern Recognition 28(3), 463–474 (1995)
23. Zhang, K., Shasha, D.: Simple fast algorithms for the editing distance between trees and related problems. SIAM J. Comput. 18(6), 1245–1262 (1989)

Efficient Evacuation Planning for Large Cities

Ajay Gupta and Nandlal L. Sarda

Indian Institute of Technology Bombay, Mumbai, India
{ajay,nls}@cse.iitb.ac.in

Abstract. Given a large city represented by nodes and arcs, with a subset of these nodes having people required to be evacuated in a disaster situation, and a subset of destination nodes where the people may be taken for safety, the evacuation route planner finds routes in this network to evacuate the people in minimum possible time. Evacuation route planning in pre-known disasters such as hurricanes is of utmost importance for civic authorities. Computing such evacuation routes in a large graph with road infrastructure constraints such as road capacities and travelling times can be challenging. The Capacity Constrained Route Planner (CCRP) is a well studied heuristic algorithm proposed by Lu, George and Shekhar for solving this problem. It uses shortest path computations as a basis for finding and scheduling possible evacuation paths in the graph. However, the algorithm fails to scale to very large graphs. Other algorithms based on CCRP like CCRP++ and Incremental Data Structure based CCRP have been proposed to scale them to larger graphs. In this paper, we analyze these algorithms from performance perspective and suggest a faster algorithm by extending the CCRP algorithm that avoids recomputing partial paths. We have carried out experiments on various graph structures which show a vast improvement in runtime without impacting the optimality of CCRP results. For instance, for the city of Oldenburg with 382285 people, 6105 nodes and 7034 edges, our algorithm produced an evacuation plan in 2.6 seconds as compared to 123.8 seconds by CCRP and 9.3 seconds by CCRP++.

1 Introduction

Evacuation route planning in transportation networks[1] or buildings[2] forms a critical part of disaster management and defence systems. Mass media communication was used in earlier days to communicate the possible extent of disaster to affected population. No planned approach to evacuate the population under threat was possible due to lack of information and evacuation strategies. The capacity constraints of the transportation network were not considered leading to chaos at the time of evacuation, thereby increasing the evacuation time and possibly increasing the threat to people during the evacuation. Therefore, efficient tools which can generate effective evacuation plans are required which can help avoid unwanted casualties. In situations involving threat to life, it is extremely important to reduce the evacuation time, considering the constraints of the network.

H. Decker et al. (Eds.): DEXA 2014, Part I, LNCS 8644, pp. 211–225, 2014.

An evacuation planning algorithm assumes the availability of an accurate graph structure and the location of the people to be evacuated. With this information, we can plan optimized evacuation routes for all the people. Minimizing the average evacuation time, minimizing the evacuation egress time and maximizing the flow of evacuees have been the focus areas. Research work in this area has been mainly done using the following two approaches: (1) Linear Programming approaches[3,4] which use Time Expanded Graphs and Integer Programming to produce optimal evacuation plans with minimum evacuation time. (2) Heuristic approaches such as Capacity Constrained Route Planner (CCRP)[5] which model capacity as time series and use shortest path heuristic to generate evacuation plans. Linear Programming approach gives optimized results in terms of evacuation time, but it can be used for only small networks due to the increased memory requirements and the high runtime. CCRP is a greedy algorithm which solves the scalability and high runtime issues with Linear Programming approach but it can deviate from the optimized results to a small extent. However, even with the improvements, the CCRP algorithm doesn't scale to larger networks. A variation to the CCRP, called CCRP++[6] was proposed recently which aims to solve the scalability issue with CCRP. However, the CCRP++ algorithm deviates from the CCRP algorithm to solve the problem as described in Section 2.3. In this paper, we have proposed a new algorithm and data structure related refinements to CCRP. The algorithm executes as per the CCRP algorithm, but provides a significant improvement in runtime as compared to CCRP as well as CCRP++ by reusing data from previous iterations of CCRP.

The remainder of this paper is organized as follows: in section 2, we discuss in brief the evacuation problem and the CCRP, CCRP++ and IDS based CCRP algorithms. In section 3, we describe our algorithm and explain the reasons for it being able to perform better than CCRP. Section 4 consists of the experiments conducted to demonstrate the improvements of our algorithm. In section 5, we conclude our work and look at the future work in this area.

2   Problem Statement and Literature Review

2.1   Evacuation Planning

The area for which evacuation is to be performed is represented as a graph/network G of nodes N and arcs/edges E. The graph can be directed or undirected. A directed graph allows flow along an edge in single direction. An undirected graph assumes flow along both directions of the edge. In this paper, we assume the graphs to be undirected. The algorithms can be easily worked out for directed graphs. A subset S of N is the collection of source nodes where people who need to be evacuated are present. Another subset D is the collection of destination nodes where the people may be taken for safety. Each node in the graph has a maximum capacity which indicates the maximum number of people that can be at that node at an instance of time. The source nodes have an initial occupancy. It is assumed that the destination nodes have an infinite capacity. Each edge has a travel time, which is the number of time units taken to reach from one end to the other. The edges also have a intake capacity, which is the maximum number of people that can be admitted per unit time. Such a graph is the input to the evacuation planner.

The evacuation planner gives as output an evacuation plan giving details of the route to be taken by people at various source nodes to reach a destination node. The people from the same source are divided into different groups. A path is scheduled for each group from source to a destination. The evacuation plan for each group includes details of the entire path such as the time at which a group arrives at a node, the time at which it leaves that node and the edge taken to reach the next node along the path. The evacuation is planned taking into consideration the capacity of the nodes and edges at different time intervals. The objectives of an evacuation planning algorithm are to minimize the evacuation egress time. Egress time is the time at which the last person reaches a destination node, assuming evacuation starts at time 0. At the same time, the algorithm should reduce the cost of computation for generating the evacuation plan.

A model of a transportation network with 10 nodes is shown in Fig. 1. Each node has an id, initial occupancy and maximum capacity. The graph consists of 2 source nodes with 20 evacuees at node N1 and 30 at node N2. N8 and N10 are the destination nodes. The occupancy and the maximum capacity for destination nodes are shown as -1 since we have assumed that destination nodes have infinite capacity.

2.2   CCRP Algorithm

CCRP[5] is an algorithm based on the greedy approach. It uses a generalized shortest path search algorithm taking into account the route capacity constraints. It models capacity as a time series because available capacity of each node and edge may vary at different time instances during the evacuation. This algorithm can divide evacuees from each source into multiple groups and assign

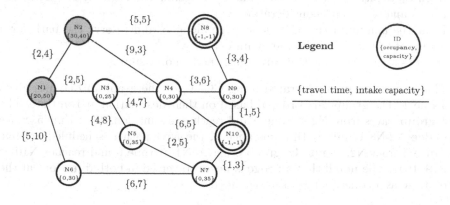

Fig. 1. Network Model

Table 1. Evacuation Plan using CCRP

Group Id	Source	No. of Evacuees	Route	Time Stamps
1	N2	5	N2–N8	0–5
2	N2	5	N2–N8	1–6
3	N2	5	N2–N8	2–7
4	N2	5	N2–N8	3–8
5	N2	5	N2–N8	4–9
6	N1	3	N1–N3–N5–N7–N10	0–2–6–8–9
7	N2	5	N2–N8	5–10
8	N1	2	N1–N3–N5–N7–N10	0–2–6–8–10
9	N1	1	N1–N3–N5–N7–N10	1–3–7–9–10
10	N1	4	N1–N2–N8	0–2–11
11	N1	1	N1–N2–N8	1–3–11
12	N1	3	N1–N3–N5–N7–N10	1–3–7–9–11
13	N1	1	N1–N3–N4–N9–N10	1–3–7–10–11
14	N1	3	N1–N2–N8	1–7–12
15	N1	3	N1–N2–N8	2–7–12

a route and time schedule to each group of evacuees based on an order that is prioritized by each group's destination arrival time. The capacity of the nodes and edges are changed based on the time at which these groups use the respective nodes and edges along the route. A brief description of the steps involved in the algorithm is given below. It uses priority queue implementation of Dijkstra's shortest path algorithm for finding shortest path from a source to a destination.

1. The source nodes S in the graph are all connected to a supersource node S_0 by an edge of infinite capacity and travel time 0. This step ensures that the shortest path algorithm need not be executed for all source nodes in each iteration.
2. S_0 is added to the priority queue. The nodes in priority queue are ordered based on the distance calculated from the supersource node.
3. While there are evacuees in any of the source nodes, find the shortest path R from S_0 to any of the destination nodes D taking the capacity of the various nodes and edges into consideration.
4. Find the minimum capacity node/edge along R and reserve the path for a group of size equal to the minimum capacity.
5. If there are evacuees left at any source node, go to step 2.

The evacuation plan generated using CCRP for the graph in Fig. 1 is given in Table 1. The groups are ordered based on their destination arrival time. The first group leaves from N2 at time 0 and reaches destination N8 at time 5. Since the edge N2-N8 is full at time instance 0, the next path is scheduled at time instance 1 from N2. Similarly, group 6 leaves N1 at time 0 and reaches N10 at time 9. It may be noted that an edge is never traversed in both directions at the same time as a consequence of using shortest paths.

The CCRP algorithm chooses a route based on the earliest arrival time. The complexity of CCRP is $O(P * N * \log N)$ where $N * \log N$ is the complexity of Dijkstra's priority queue implementation and P is the number of people. Analyzing the algorithm, we see that in each iteration, CCRP executes Dijkstra's shortest path algorithm from all the source nodes until a destination node is found. However, only one of the paths (the one that reached a destination) from a source node towards the destination is used in one iteration. The expansions done from other source nodes in other directions need to be performed despite the fact that it will not be used in current iteration and will again be performed in subsequent iteration. As a result, CCRP works well for small and medium sized networks, but for large graphs with many source nodes, it doesn't scale well due to these unwanted expansions.

2.3 CCRP++ Algorithm

CCRP++[6] also makes use of the earliest arrival time heuristic to produce an evacuation plan. It overcomes the problem of unwanted expansion of paths in CCRP. For each source node, it calculates the shortest path and reserves the path if capacities along the nodes and edges in the path at the scheduled time are available. The algorithm makes use of two priority queues: PreRQ is an auxiliary priority queue that is used for ordering sources based on shortest destination arrival time. These shortest paths are not reserved, i.e., people are not scheduled to travel along these shortest paths. RQ is a reserve priority queue that is used to hold paths with reserved capacities. The algorithm steps are described below.

1. Calculate the shortest path from all the source nodes in S and insert them in PreRQ. Pop the smallest destination arrival time path from PreRQ and insert it into RQ.
2. While there are evacuees at any source
   (a) If the shortest path q in PreRQ has arrival time less than the arrival time of the entry p at head of RQ, recheck the arrival time of q. We need to recheck the arrival time because there could be other paths which were reserved after paths in PreRQ were calculated. As a result, the edges or nodes along the path in q may not be available. The check ensures that there is capacity of at least 1 unit along path q before q can be scheduled. If the arrival time of q is greater than p, go to step 2(c).
   (b) If the arrival time of q is the same as before, insert it into RQ and go to step 2. If the arrival time of q is not the same as before, update the arrival time of the route, insert it into PreRQ and go to step 2.
   (c) Find the minimum capacity m along path p and schedule a group of size m along p.
   (d) Recalculate the shortest path from the source of path p and insert it into the preRQ. Go to step 2.

We see that the CCRP++ algorithm calculates only a single path from a source to a destination in each iteration. However, the paths are calculated without knowing whether the arrival time will remain the same during reservation due to possibility of reservation by other paths along the edges of this path. Hence, it works on the premise that more paths will be disjoint and will not need to be recalculated, leading to efficient execution. Rechecking the route before reserving it becomes an overhead in CCRP++.

2.4   Incremental Data Structure(IDS) Based CCRP Algorithm

Kim et al.[7] have proposed changes to the CCRP algorithm to scale CCRP to large graphs. After a destination node is found by CCRP algorithm, the IDS based CCRP algorithm removes the shortest path found. This path removal can result in orphan branches, i.e., there could be parts in the expanded tree with no parent. These need to be removed by marking the nodes in these orphan branches as unvisited. After the path is removed, there is also a possibility that the certain sources get blocked due to the presence of common nodes in the path removed and the paths expanded from other sources. The algorithm proposes to use a sorted list for each node containing the earliest arrival time from the adjacent nodes. After removing the path, the first entry in the sorted list of the nodes along this path are removed. We have not implemented and tested this algorithm since it forms a certain special case of our algorithm described in the section below.

3   CCRP* : Extending CCRP for Performance

3.1   Reusing Expansion for Alternate Paths

Our algorithm aims to reduce the overheads involved in CCRP and CCRP++ and to further optimize the IDS based approach by utilizing the expansion generated from previous runs of CCRP and performing minimal changes to calculate a route in next iteration. Dijkstra's algorithm used in CCRP starts from a source node and until a destination node is found, it continues the search. This search leads to the formation of a tree structure where each node n has only one parent node from which the shortest path to n is available.

Consider Fig. 2a. It shows a path obtained by applying Dijkstra's algorithm from source s_0. The shortest path found is s_0--n_0--...--n_1--n_2--...--d. Assume that the edge e has the minimum capacity along this path. Also assume that we start a group from s_0 at time t_0 and reach n_1 at time t_1. The time to travel edge e is t_e and the time to reach n_2 via n_1 is t_2. We make the following important observations:

1. Since the edge n_1--n_2 is having minimum capacity, the path from s_0 to n_1 will have an additional available capacity of at least 1 unit during the same time interval, i.e. at least 1 more person will be able to leave s_0 at time t_0

and reach n_1 at time t_1 along the same route. The edge e at time t_1 has zero intake capacity left. The edge e can only be used at time $t_1 + 1$. Hence, a delay of one time unit will be required before a person can reach node n_2 via the same path, i.e., we can reach n_2 at time $t_2 + 1$. As a result, all nodes below n_2 will have travel time increased by one unit if the same path to those nodes is taken.

2. Suppose that in the graph, there was an edge e_2 between nodes n_3 and n_2. Also, assume that the time taken to reach node n_2 via n_3 was equal to t_2. Since a node can have only one parent node, the algorithm could have chosen n_1 to be the parent instead of n_3. Now, after reserving the path from s_0--d, the time to reach n_2 increases by one unit if we are to take the path via node n_1. However, there is a shorter path to n_2 via node n_3. If the CCRP algorithm was executed, it would have taken the route via n_3. This is true for any node below the minimum capacity edge.

From the above two observations, we can conclude that after a path is reserved, only the travel time to a node below the *minimum capacity edge* in the tree formed by Dijkstra's algorithm *may* change. Hence, we need to ensure that the correct travel time to these nodes is calculated in subsequent iteration. CCRP* algorithm overcomes this issue by setting the travel time to all nodes below node n_2 to infinity and by inserting adjacent scanned nodes into priority queue. Hence, as seen in Fig. 2b and Fig. 2c, CCRP will start the execution in next iteration from node S_0 whereas CCRP* algorithm will start the next iteration from n_1 and n_3.

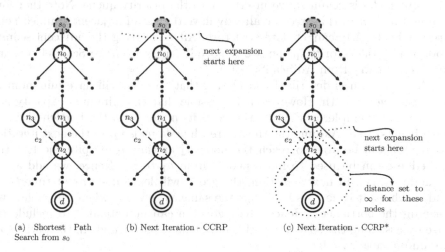

(a) Shortest Path Search from s_0

(b) Next Iteration - CCRP

(c) Next Iteration - CCRP*

Fig. 2. Shortest Path Search from Source Node s_0

3.2 CCRP* Algorithm

In Dijkstra's algorithm, for a graph with non negative edges, the distance to a scanned(visited) node does not change. Since the travel time to the set of nodes below the minimum capacity edge (nodes belonging to the tree rooted at n_2 in Fig. 2c) can change in CCRP, we need to reset such nodes and recalculate the travel time to these nodes from their adjacent nodes. The adjacent nodes can be scanned or unscanned. If a node is unscanned, it may or may not be in the priority queue. If it is in the priority queue, it doesnt need to be added. If it is not in the priority queue, it means the travel time to it is infinity and hence should not be added to priority queue. Hence, for the nodes below the minimum capacity edge, we only add the adjacent scanned nodes to the priority queue. This ensures that the minimum travel time to nodes below the minimum capacity edge in subsequent iteration can be calculated. To calculate the updated travel time to the target node (node n_2 in Fig. 2) of the minimum capacity edge, we also need to add the parent of this node (node n_1) to the priority queue. If all the evacuees from a source have been assigned a path, then the tree formed from this source is no longer required. Hence, the travel time to the nodes in this tree needs to be set to infinity so that these nodes can be used in other paths from other source nodes. The steps to implement these ideas are presented in Algorithm 1.

In the CCRP* algorithm (Algorithm 1), the AdjScanSet for a node is a set consisting of scanned adjacent nodes of node n which need to be added to priority queue before proceeding to the next iteration. It ensures correct calculation of travel time to the node n in future iterations. Lines 7 to 11 are used to reset a part of the expansion tree. This reset procedure involves setting the distance to all nodes in this tree to infinity and removing them from the priority queue if present. We also scan the adjacent nodes v of each of these nodes. If any of the adjacent nodes is scanned, we insert it into the priority queue. Note that for a scanned node n in this tree, we already have the list of adjacent scanned nodes present in the AdjScanSet data structure. After inserting the adjacent scanned node v into the priority queue, we remove its entry from AdjScanSet of n and remove n's entry from AdjScanSet of v.

The algorithm is described considering that an edge will have minimum capacity along the path. However, it is possible that the minimum capacity is at a node. For example, in Fig. 2a, the node n_2 might form the bottleneck in the path from s_0 to d. Assume the time to reach node n_2 is t_2. In the next iteration, no person will be able to reach the node n_2 at time t_2 despite the fact that the edge e can be used to move people from n_1 to n_2. Hence, we add a wait at node n_1, i.e., in next iteration, the group will leave node n_1 at time $t_1 + 1$ and reach n_2 at time $t_2 + 1$. This is the same as the case where the edge e was forming the bottleneck. Hence, the algorithm can be extended to include this case by assuming the edge above this minimum capacity node in the path as the minimum capacity edge.

Algorithm 1. CCRP*

Input: A set of nodes N and edges E with capacity constraints and travel time
A set of source nodes S including initial evacuee occupancy O and a set
of destination nodes D

Output: Evacuation plan including route schedules of evacuees on each route

---

1 Insert all the source nodes into the priority queue PQ
2 **while** *any source node $s \in S$ has evacuees* **do**
3 | Using priority queue implementation of Dijkstra's shortest path algorithm,
   | find the shortest path p from any source node in PQ to any destination
   | node taking into account the node and edge capacity constraints.
4 | During the shortest path search, while scanning an adjacent node v of a
   | node u, if v is found to be scanned and if u and v do not have parent-child
   | relationship, we insert v in AdjScanSet of u and insert u in AdjScanSet of v.
5 | Find the minimum capacity c_{min} and the corresponding minimum capacity
   | edge e from path p
6 | Reserve p for a group of size c_{min} from p's source s
7 | **if** *all the evacuees from s are evacuated* **then**
8 | | Reset the entire tree rooted at s (AdjScanSet used in this step to add
   | | nodes to priority queue)
9 | **else**
10 | | Reset the tree rooted at the target node n of edge c in p and insert the
   | | parent of n in PQ
11 | **end**
12 **end**

3.3   Complexity Analysis

The worst case complexity of our algorithm remains the same as that of CCRP, i.e., $O(P * N * \log N)$. This is because the minimum capacity nodes/edges may always occur extremely close to the source nodes resulting in almost all the nodes in the tree getting reset. However, for large graphs with many sources and randomly distributed capacities of the nodes and edges, the algorithm will perform significantly less work as compared to CCRP. This is because only a part of the tree from one of these sources will be affected as shown in Fig. 2c. There can be variation in the results obtained in CCRP and our algorithm because the ordering of the nodes in priority queue in the two algorithms can be different. This is because our algorithm preserves the queue from earlier iteration whereas the CCRP constructs the entire queue again in each iteration. As compared to CCRP++, our algorithm doesn't always need to start a search for shortest path from a source node. Also, unlike CCRP++, only a part of the tree is cut. Hence, the branches from the same source in other directions are available in the tree for subsequent evaluations. CCRP++ algorithm finds the shortest path from each source separately. Hence, a node in the graph may be a part of two trees from two different source nodes. As a result, CCRP++ will expand the tree below this common node from both sources. Since our algorithm inserts all source nodes into priority queue before the shortest path evaluation, the expansion of tree below a common node will be done only once.

4   Experimental Results

To conduct experiments, we have used a machine with Windows 8, 2.7 GHz Intel Core I5 processor and 8 GB RAM. We have implemented the three algorithms CCRP, CCRP++ and CCRP* using Java. We have used the priority queue data structure for Dijkstra's algorithm from the standard Java library. Each node structure contains entire path up to that node from the source. We have conducted experiments on graphs of buildings, road network of the City of Oldenburg and the road network of California.

We have evaluated the following parameters for the different algorithms.

– **Egress Time:** It is the time taken by the last evacuee to reach a destination node, assuming evacuation starts at time 0.

Table 2. Test Results for Building 101

Evacuees	Nodes		Edges		Sources		Destinations		Avg. Edge Cap.
323	57		62		37		3		25

Algorithm	Egress Time	Avg. Evacuation. Time	Groups	Routes	Max. Waiting Time	Avg. Waiting Time	Runtime (secs)	SpeedUp	Avg. Nodes/Path
CCRP	55	22	185	16	5	0.81	0.181	--	12.12
CCRP++	56	19.1	159	19	19	1.19	0.209	0.86	12.12
CCRP*	55	21.8	184	16	5	0.81	0.137	1.32	12.12

Table 3. Randomly Distributed Sources and Destinations

(a) Test Results for City of Oldenburg

Evacuees	Nodes		Edges		Sources		Destinations		Avg. Edge Cap.
382295	6105		7034		939		1183		999

Algorithm	Egress Time	Avg. Evacuation. Time	Groups	Routes	Max. Waiting Time	Avg. Waiting Time	Runtime (secs)	SpeedUp	Avg. Nodes/Path
CCRP	6064	220.55	54921	960	181	5.73	123.79	--	3.71
CCRP++	6064	220.55	54930	960	181	5.71	9.349	13.24	3.71
CCRP*	6064	220.55	54921	960	181	5.72	2.641	46.87	3.71

(b) Test Results for California

Evacuees	Nodes		Edges		Sources		Destinations		Avg. Edge Cap.
321693	21048		21693		1839		2312		999

Algorithm	Egress Time	Avg. Evacuation. Time	Groups	Routes	Max. Waiting Time	Avg. Waiting Time	Runtime (secs)	SpeedUp	Avg. Nodes/Path
CCRP	3068	21.637	79544	3900	43	1.28	1052.86	--	6.84
CCRP++	3068	21.637	79524	3896	58	1.29	47.48	22.17	6.84
CCRP*	3068	21.635	79542	3901	43	1.28	3.61	291.65	6.84

- **Average Evacuation Time:** It is the average of the amount of time taken by each evacuee to reach a destination node. If T is the egress time, average evacuation time can be calculated as

$$\frac{\sum_{t=0}^{T}(\text{Number of people reaching a destination at time } t) * t}{\text{Total number of people}}$$

- **Routes:** It is the number of distinct paths which were generated. Hence, at least one group is scheduled by the evacuation planning algorithm along each of these paths.
- **Waiting Time:** It is the amount of time for which people had to wait at a node. We have evaluated the maximum waiting time and the average waiting time at a node.
- **Speedup:** Speedup for an algorithm A is the ratio of the time taken by CCRP to the time taken by A to generate an evacuation plan for a graph G.

4.1 Tests for Small Graphs

In order to verify the correctness of our implementations, we executed the CCRP, CCRP++ and CCRP* algorithms on the graph in Fig. 1 and on two small building networks; Building 101 as provided in [2] and CSE building at IIT Bombay. The results for the example in Fig. 1 were the same as those of CCRP, though in different order. The results for the Building 101 is shown in Table 2. The egress times generated by the three algorithms are almost the same as CCRP which verifies the correctness of our implementations. For all the small graphs, CCRP* gave results similar to CCRP for other factors too like the paths taken, number of routes, number of groups, number of nodes per path and the waiting time. This points to the correct implementation and working of our algorithm. The speedup obtained as compared to CCRP is not much. This can be due to the extra processing done for reusing data from previous iteration which may not be worth doing for smaller graphs.

Random Distribution of Sources and Destinations. Using the data available from [8] for different regions, we conducted tests comparing the CCRP, CCRP++ and CCRP*. We first randomly distributed the source and destination nodes and assigned random capacities to the nodes and edges. The results for the city of Oldenburg and California are given in Table 3a and Table 3b. We see that CCRP* algorithm has generated evacuation plans much faster than both CCRP and CCRP++. The results generated have the same egress time, average evacuation time and average waiting time for all three algorithms.

4.2 Tests for Large Graphs

Variation of Location of Destinations and Number of Source Nodes. We conducted tests using the California state data[8] to compare the three algorithms by varying the number of source nodes and the distribution of sources and destination nodes for the network. We varied the number of sources and the location of destinations keeping approximately the same number of evacuees. The evacuees were distributed uniformly at random over the entire state. The following four cases were considered:

- **Case 1:** Destinations on eastern border and larger number of source nodes.
- **Case 2:** Destinations on eastern border and smaller number of source nodes.
- **Case 3:** Destinations are a few clustered set of nodes distributed across the state and larger number of source nodes.
- **Case 4:** Destinations are a few clustered set of nodes distributed across the state and smaller number of source nodes.

Case 1 and Case 2 have all destinations on the eastern border of the state. The destinations here are chosen to evaluate performance when the entire region needs to be evacuated. For Case 3 and Case 4, the destinations are a few clustered set of nodes distributed across the state. Here, we assumed that there are shelter locations within the region of evacuation. The results are shown in Table 4a, Table 4b, Table 5a and Table 5b.

We can see in Table 4a and Table 4b that the amount of time taken by CCRP* algorithm has increased, despite the decrease in the number of source nodes. This might be due to the fact that larger trees are produced with less source nodes and hence the number of nodes to be reset before starting the next iteration will be more. The same trend can be seen in Table 5a and Table 5b. All three algorithms have improved performance when the destinations are spread across the state. The gains of CCRP* over CCRP++ are more when the paths contain more nodes (Table 4a and Table 5a). CCRP++ precomputes routes assuming that the nodes in those routes will be disjoint. With greater number of nodes per path, the probability that nodes in the routes will be disjoint could get reduced, affecting the performance of CCRP++.

Table 4. Test Results for California Data Set with Destination Nodes on Eastern Border: Variation in Number of Source Nodes

(a) More Source Nodes(Case 1)

Evacuees	Nodes		Edges		Sources		Destinations		Avg. Edge Cap.
321693	21048		21693		1839		2312		136

Algorithm	Egress Time	Avg. Evacuation. Time	Groups	Routes	Max. Waiting Time	Avg. Waiting Time	Runtime (secs)	SpeedUp	Avg. Nodes/Path
CCRP	5758	136.65	16520	1856	4	0.039	1180.7	--	164.68
CCRP++	5758	136.37	16492	1858	4	0.039	831.98	1.41	164.59
CCRP*	5758	136.63	16518	1857	4	0.039	119.7	9.86	164.67

(b) Less Source Nodes(Case 2)

Evacuees	Nodes		Edges		Sources		Destinations		Avg. Edge Cap.
319985	21048		21693		756		2312		136

Algorithm	Egress Time	Avg. Evacuation. Time	Groups	Routes	Max. Waiting Time	Avg. Waiting Time	Runtime (secs)	SpeedUp	Avg. Nodes/Path
CCRP	5310	145.27	17910	772	9	0.07	1186.97	--	157.07
CCRP++	5310	145.05	17879	773	9	0.07	660.61	1.79	157.05
CCRP*	5310	145.24	17904	771	9	0.07	140.652	8.439	157.07

Table 5. Test Results for California Data Set with Clustered Destination Nodes: Variation in Number of Source Nodes

(a) More Source Nodes(Case 3)

Evacuees	Nodes	Edges	Sources	Destinations	Avg. Edge Cap.
330762	21048	21693	1915	2478	136

Algorithm	Egress Time	Avg. Evacuation. Time	Groups	Routes	Max. Waiting Time	Avg. Waiting Time	Runtime (secs)	SpeedUp	Avg. Nodes/Path
CCRP	3538	52.17	13545	1929	4	0.04	559.4	--	78.88
CCRP++	3538	52.08	13535	1929	4	0.04	102.23	5.47	78.86
CCRP*	3538	54.47	13544	1929	4	0.04	28.91	19.34	78.89

(b) Less Source Nodes(Case 4)

Evacuees	Nodes	Edges	Sources	Destinations	Avg. Edge Cap.
312540	21048	21693	764	2478	136

Algorithm	Egress Time	Avg. Evacuation. Time	Groups	Routes	Max. Waiting Time	Avg. Waiting Time	Runtime (secs)	SpeedUp	Avg. Nodes/Path
CCRP	3236	55.58	13587	777	10	0.07	536.81	--	78.06
CCRP++	3236	55.58	13587	776	10	0.07	96.67	5.55	78.02
CCRP*	3236	55.61	13615	776	10	0.07	31.18	17.21	78.07

Table 6. Test Results for California data set : Variation in Number of Evacuees

(a) 749165 Evacuees

Evacuees	Nodes	Edges	Sources	Destinations	Avg. Edge Cap.
749165	21048	21693	1839	2312	136

Algorithm	Egress Time	Avg. Evacuation. Time	Groups	Routes	Max. Waiting Time	Avg. Waiting Time	Runtime (secs)	SpeedUp	Avg. Nodes/Path
CCRP	5467	186.05	50552	2006	10	0.1	3486.3	--	165.51
CCRP++	5467	183.02	49937	2019	50	0.1	2824.87	1.23	165.47
CCRP*	5467	186.15	50584	2005	10	0.1	256.62	13.58	165.48

(b) 1107673 Evacuees

Evacuees	Nodes	Edges	Sources	Destinations	Avg. Edge Cap.
1107673	21048	21693	1839	2312	136

Algorithm	Egress Time	Avg. Evacuation. Time	Groups	Routes	Max. Waiting Time	Avg. Waiting Time	Runtime (secs)	SpeedUp	Avg. Nodes/Path
CCRP	5759	204.85	81905	2095	25	0.16	5567	--	165.9
CCRP++	5759	199.63	79986	2296	122	0.16	5339	1.04	165.29
CCRP*	5759	204.36	81726	2100	26	0.16	373.7	14.9	165.88

Variation in Number of Evacuees. We also wished to see the effect of the number of evacuees on the different algorithms. We used a graph for California with destinations same as those in Case 1. Table 4a, Table 6a and Table 6b show the effect of the variation in number of evacuees. The increased number of evacuees have a much larger effect on the performance of both CCRP++ and CCRP as compared to the CCRP* algorithm. We can also see that the number of routes used do not increase as drastically as the number of groups which have increased.

Table 7. Test Results for California Data Set: Variation of Edge Capacities

(a) Average Edge Capacity : 37

Evacuees	Nodes	Edges	Sources	Destinations	Avg. Edge Cap.
330762	21048	21693	1915	2478	37

Algorithm	Egress Time	Avg. Evacuation. Time	Groups	Routes	Max. Waiting Time	Avg. Waiting Time	Runtime (secs)	SpeedUp	Avg. Nodes/Path
CCRP	3541	161.89	43288	1986	13	0.15	3494.161	--	76.41
CCRP++	3541	160.41	42939	1986	14	0.15	278.161	12.56	76.29
CCRP*	3541	161.84	43279	1987	13	0.14	56.8	61.52	76.4

(b) Average Edge Capacity : 10

Evacuees	Nodes	Edges	Sources	Destinations	Avg. Edge Cap.
330762	21048	21693	1915	2478	10

Algorithm	Egress Time	Avg. Evacuation. Time	Groups	Routes	Max. Waiting Time	Avg. Waiting Time	Runtime (secs)	SpeedUp	Avg. Nodes/Path
CCRP	3851	594.42	141457	2639	135	0.55	5667.59	--	81.02
CCRP++	3784	550.94	133182	3083	257	0.62	2621.33	2.162	80.13
CCRP*	3851	591.94	141065	2642	134	0.55	164.91	34.36	80.91

Variation of Average Edge Capacities. We varied the edge capacity for the Case 3 graph. Table 5a, Table 7a and Table 7b show the results of this test. We can see that apart from increasing number of groups and the increasing runtime for generating the evacuation plan, the number of distinct routes generated, the average evacuation time and the average waiting time increases significantly with the decrease in the edge capacities. The egress time does not increase drastically with decrease in edge capacities. This is understandable because with reduced edge capacities, the number of distinct routes taken have increased, leading to more utilization of the network capacities. The reduced evacuation time of CCRP++ in Table 7b seems to be a one-off case peculiar to this graph and source nodes.

5   Conclusions

In this paper, we have presented CCRP* algorithm, a modified version of the evacuation planning algorithm CCRP to scale it to much larger graphs and generate results significantly quicker than both CCRP and CCRP++. By reusing the expanded tree generated in previous iterations by CCRP algorithm and by doing minimum changes to this tree before the start of next iteration, the amount of work required for calculation of routes is significantly reduced. We conducted experiments by varying many different parameters of the input graph to get useful insights of CCRP, CCRP++ and CCRP* algorithm. The results showed improvements in speedup of approximately 10 to 20 times as compared to CCRP and approximately 5 to 15 times as compared to CCRP++. CCRP* algorithm performs well when the paths have bigger length and when the distribution of people is more widespread over the transportation network. In future, we would like to evaluate the evacuation planning problem to check the sensitivity of such evacuation plans to variations in parameters such as the spatial distribution of population at various source nodes, location of sources and destinations, structure of input graph, etc.

References

1. Geisberger, R.: Advanced route planning in transportation networks. PhD thesis
2. Chalmet, L., Francis, R., Saunders, P.: Network models for building evacuation. Fire Technology 18(1), 90–113 (1982)
3. Bhushan, A., Sarda, N.: Modeling of building evacuation using ladders. Fire Safety Journal 55, 126–138 (2013)
4. Kennington, J., Helgason, R.: Algorithms for network programming. A Wiley-Interscience publication. John Wiley & Sons Australia, Limited (1980)
5. Lu, Q., Huang, Y., Shekhar, S.: Evacuation planning: A capacity constrained routing approach. In: Chen, H., Miranda, R., Zeng, D.D., Demchak, C., Schroeder, J., Madhusudan, T. (eds.) ISI 2003. LNCS, vol. 2665, pp. 111–125. Springer, Heidelberg (2003)
6. Yin, D.: A scalable heuristic for evacuation planning in large road network. In: Proceedings of the Second International Workshop on Computational Transportation Science, IWCTS 2009, pp. 19–24. ACM, New York (2009)
7. Kim, S., George, B., Shekhar, S.: Evacuation route planning: scalable heuristics. In: Proceedings of the 15th Annual ACM International Symposium on Advances in Geographic Information Systems, GIS 2007, pp. 20:1–20:8. ACM, New York (2007)
8. Li, F., Cheng, D., Hadjieleftheriou, M., Kollios, G., Teng, S.H.: On trip planning queries in spatial databases. In: Bauzer Medeiros, C., Egenhofer, M., Bertino, E. (eds.) SSTD 2005. LNCS, vol. 3633, pp. 273–290. Springer, Heidelberg (2005)

Benchmarking Database Systems
for Graph Pattern Matching*

Nataliia Pobiedina[1], Stefan Rümmele[2], Sebastian Skritek[2], and Hannes Werthner[1]

[1] TU Vienna, Institute of Software Technology and Interactive Systems, Austria
{pobiedina,werthner}@ec.tuwien.ac.at
[2] TU Vienna, Institute of Information Systems, Austria
{ruemmele,skritek}@dbai.tuwien.ac.at

Abstract. In graph pattern matching the task is to find inside a given graph some specific smaller graph, called pattern. One way of solving this problem is to express it in the query language of a database system. We express graph pattern matching in four different query languages and benchmark corresponding database systems to evaluate their performance on this task. The considered systems and languages are the relational database PostgreSQL with SQL, the RDF database Jena TDB with SPARQL, the graph database Neo4j with Cypher, and the deductive database Clingo with ASP.

1 Introduction

Graphs are one of the most generic data structures and therefore used to model data in various application areas. Their importance has increased, especially because of the social web and big data applications that need to store and analyze huge amounts of highly interconnected data. One well-known task when dealing with graphs is the so-called *graph pattern matching*. Thereby the goal is to find inside a given graph a smaller subgraph, called *pattern*. This allows to explore complex relationships within graphs as well as to study and to predict their evolution over time [1]. Indeed, graph pattern matching has lots of applications, for example in software engineering [2], in social networks [3–5], in bioinformatics [3, 4] and in crime investigation & prevention [5, 6].

Solving graph pattern matching tasks can be done in two ways. The first way is to use specialized algorithms. A comparison of various specialized algorithms for graph pattern matching has been done recently [4]. The second way is to express this problem in the query language of a database system. In the database area the comparison of different systems is an important topic and has a long tradition. Therefore, there exist already studies comparing databases in the context of graph queries [7–9]. Thereby the authors compare the performance of various databases for different query types, like adjacency queries, reachability queries, and summarization queries. But we want to point out that these works do not study graph pattern matching queries, which are computationally harder. To the best of our knowledge, there is no work on the comparison of database systems for this type of queries.

---

* N. Pobiedina is supported by the Vienna PhD School of Informatics; S. Rümmele & S. Skritek are supported by the Vienna Science and Technology Fund (WWTF), project ICT12-15.

H. Decker et al. (Eds.): DEXA 2014, Part I, LNCS 8644, pp. 226–241, 2014.

Additionally, most papers that compare database systems with respect to graph problems, evaluate relational and graph database systems. A family of database systems that is less known in this area is the family of *deductive database systems*. These systems are widely used in the area of artificial intelligence and knowledge representation & reasoning. They are especially tailored for combinatorial problems of high computational complexity. Since graph pattern matching is an NP-complete problem, they lend themselves as a candidate tool for the problem at hand. Another family of database systems that are suitable for the task at hand are RDF-based systems. These databases are tailored to deal with triples that can be seen as labeled edges. Hence, graph problems can be naturally expressed as queries for these systems.

To close the mentioned gaps, we conduct an in-depth comparison of the viability of relational databases, graph databases, deductive databases, and RDF-based systems for solving the graph pattern matching problem. The results of this work are the following:

- We build a benchmark set including both synthetic and real data. The synthetic data is created using two different graph models while the real-world datasets include a citation network and a global terrorist organization collaboration network.
- We create sample graph patterns for the synthetic and real-world datasets. Again, part of these patterns are generated randomly. The second set of patterns is created using frequent graph pattern mining. This means we select specific patterns which are guaranteed to occur multiple times in our benchmark set.
- We express the graph pattern matching problem in the query languages of the four database systems we use. These are SQL for relational databases, Cypher for graph databases, ASP for deductive databases, and SPARQL for RDF-based systems.
- We conduct an experimental comparison within a uniform framework using PostgreSQL as an example of relational database system, Jena TDB representing RDF-based systems, Neo4j as a representative for graph databases systems, and Clingo as a deductive database. Based on our experimental results we draw conclusions and offer some general guidance for choosing the right database system.

2 Preliminaries

In this work we deal with *undirected, simple graphs*, that means graphs without self-loops and with not more than one edge between two vertices. We denote a (labeled) graph G by a triple $G = (V, E, \lambda)$, where V denotes the set of *vertices*, E is the set of *edges*, and λ is *labeling function* which maps vertices and/or edges to a set of labels, e.g. the natural numbers $\mathbb{N}$. An edge is a subset of V of cardinality two.

Let $G = (V_G, E_G, \lambda_G)$ and $P = (V_P, E_P, \lambda_P)$ be two graphs. An *embedding* of P into G is an injective function $f : V_P \to V_G$ such that for all $x, y \in V_P$:

1. $\{x, y\} \in E_P$ implies that $\{f(x), f(y)\} \in E_G$;
2. $\lambda_P(x) = \lambda_G(f(x))$; and
3. $\lambda_P(\{x, y\}) = \lambda_G(\{f(x), f(y)\})$.

This means if two vertices are connected in P then their images are connected as well. Note that there is no requirement for the images to be disconnected if the original vertices are. This requirement would lead to the notion of *subgraph isomorphism*.

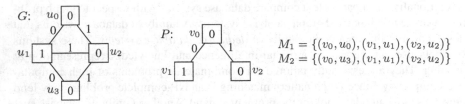

Fig. 1. Examples data graph G, pattern graph P, and resulting embeddings M_1, M_2

Instead, the problem of *graph pattern matching* is defined as follows: Given two graphs, G and P, where P is the smaller one, called *pattern* or pattern graph, the task is to compute all embeddings of P into G. Figure 1 shows an example of a graph G, a pattern graph P and all possible embeddings of P into G. In practice, we may stop the pattern matching after the first k embeddings are found.

The problem of deciding whether an embedding exists is NP-complete, since a special case of this problem is to decide if a graph contains a clique of certain size, which was shown to be NP-complete [10]. However, if the pattern graph is restricted to a class of graphs of bounded treewidth, then deciding if an embedding exists is fixed-parameter tractable with respect to the size of P, i.e., exponential with respect to the size of P but polynomial with respect to the size of G [11].

3   Related Work

Related work includes theoretical foundations of graph pattern matching and benchmarks comparing different databases.

Pattern Matching. Pattern match queries have gained a lot of attention recently with different extensions being introduced [12]. For example, Zou et al. [3] introduce a distance pattern match query which extends embeddings so that edges of the pattern graph are matched to paths of bounded length in the data graph. The authors demonstrate how such distance pattern match queries can be used in the analysis of friendship, author collaboration and biological networks. Fan et al. [5] introduce a graph pattern match query with regular expressions as edge constraints. Lee et al. [4] compare several state-of-the-art algorithms for solving graph pattern matching. They compare performance and scalability of algorithms such as VF2, QuickSI, GraphQL, GADDI and SPath. Each of these algorithms uses a different data structure to store graphs.

We investigate how available database systems perform with regard to graph pattern matching. To the best of our knowledge, there is no experimental work on this issue.

Benchmarking Database Systems for Graph Analysis. Social web and big data applications deal with highly interconnected data. Modeling this type of data in a relational database causes a high number of many-to-many relations. That is why a number of the so-called NoSQL databases have been developed [13]. Among them, graph databases are especially interesting since they often offer a proper query language. Nevertheless, since these databases are young compared to relational databases, and their query optimizers are not mature enough, it is an open question whether it is worth switching from a relational database to a dedicated graph database. Angles [14] outlines four types of queries on graphs:

- adjacency queries, e.g., list all neighbors of a vertex;
- reachability queries, e.g., compute the shortest path between two nodes;
- pattern matching queries, which are the focus of this paper; and
- summarization queries, e.g., aggregate vertex or edge labels.

There are already several works comparing the performance of current database systems for graph analysis. Dominguez et al. [15] study several graph databases (Neo4j, Jena, HypergraphDB and DEX) as to their performance for reachability and summarization queries. They find that DEX and Neo4j are the most efficient databases.

The other works include not only graph databases, but also relational database systems. Vicknair et al. [8] compare Neo4j, a graph database, and MySQL, a relational database, for a data provenance project. They conclude that Neo4j performs better for adjacency and reachability queries, but MySQL is considerably better for summarization queries. In the comparison they also take into account subjective measures, like maturity, level of support, ease of programming, flexibility and security. They conclude that, due to the lack of security and low maturity in Neo4j, a relational database is preferable. It is worth noting that they used Neo4j v1.0 which did not have a well developed query language and was much less mature than MySQL v5.1.42.

Holzschuher and Peinl [7] show that Neo4j v1.8 is much more efficient for graph traversals than MySQL. Angles et al. [9] extend the list of considered databases and include two graph databases (Neo4j and DEX), one RDF-based database (RDF-3X), and two relational databases (Virtuoso and PostgreSQL) in their benchmark. They show that DEX and Neo4j are the best performing database systems for adjacency and reachability queries. However, none of these works consider graph pattern matching queries.

4   Benchmark for Graph Pattern Matching

Our benchmark consists of three main components: database systems, datasets and query sets. The first component includes the choice of the systems, their setup, the used data representation and encodings of pattern graphs in a specific query language. The second component consists of data graphs that are synthetically generated according to established graph models or chosen from real-world data. The last component contains the construction of pattern graphs which are then transformed into queries according to the first component and used on the datasets from the second component.

4.1   Database Systems

The database systems, which we compare, are PostgreSQL, Neo4j, Clingo, and Jena TDB. These systems have in common that they are open source and known to perform well in their respective area. But they differ considerably in the way they store data and the algorithms used to execute queries. However, all four systems allow to execute the four mentioned types of graph queries. We present in this section the data schema in each of these systems as well as the query statements in four different query languages. The used data schemas are general purpose schemas for graph representation and are not specifically tailored to graph pattern matching.

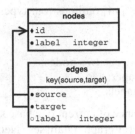 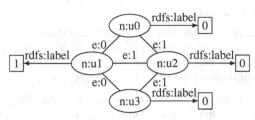

(a) Entity relationship diagram of a graph in PostgreSQL

(b) RDF representation for the example data graph in Figure 1

Fig. 2. Data schema for PostreSQL and Jena TDB

PostgreSQL (SQL). PostgreSQL is an open-source *object-relational database system* with SQL being the query language. We use PostgreSQL server v9.1.9. The data schema we use for a graph consists of two tables: nodes and edges (see Figure 2a). The primary key of the table "nodes" is the attribute "id" and it contains an index over the attribute "label". The attribute "label" is also indexed in the table "edges". The attributes "source" and "target" constitute the primary key of this table. Since we are dealing with undirected graphs, the edge table contains two entries per edge. The SQL query we use to find all embeddings of the pattern graph in Figure 1 is the following:

```
select v0.id, v1.id, v2.id
from nodes v0, nodes v1, nodes v2, edges e0, edges e1
where v0.label=0 and v1.label=1 and v2.label=0 and
      v0.id<>v2.id and
      e0.source=v0.id and e0.target=v1.id and e0.label=0 and
      e1.source=v0.id and e1.target=v2.id and e1.label=1;
```

Listing 1.1. SQL query for the pattern graph from Figure 1

As we see, we need to do many joins for the tables "nodes" and "edges" corresponding to the amounts of vertices and edges in the pattern graph. The way the query is written, we leave it up to the database query optimizer to define the join order.

It is possible to use a denormalized data schema when we have only one table which contains all information about vertices and edges. One can also manually optimize the query. However, this is not the scope of the paper. The same data schema has been used in previous benchmarking works [8, 9]. Also, our focus is at how the database engine can optimize the query execution. Such setting corresponds to a typical production scenario when an average user is not an expert.

Jena TDB (SPARQL). In RDF (Resource Description Framework), entities comprise of triples (*subject*, *predicate*, *object*). The interpretation is that the subject applies a predicate to the object. RDF-based data can be also regarded as a graph where the subject and the object are nodes, and predicate corresponds to a link between them. There exist several RDF-based database systems. We choose Jena which is an open-source framework for building RDF based applications and provides several methods to store and interact with the data. For our purposes, the SPARQL server Fuseki provides the database server that is used for query answering. We use Fuseki 1.0.1, with the included Jena TDB for the persistent storage of the data. There are two main reasons for choosing TDB over SDB. Firstly, TDB was repeatedly reported to provide better performance. Secondly, especially for the comparison with PostgresSQL, it is convenient to use the native triple store TDB instead of SDB that is backed up by an SQL database.

We encode the graph in RDF by representing each node of the graph as a resource (i.e., we generate an IRI for every node). The edge relation is described as properties of these resources that state to which other nodes a node is connected. Since the choice of the node that appears in the subject position of the RDF triple implies an order on the edges, we create two triples for each edge where we switch the subject and object positions. In the presence of edge labels, we introduce a separate property for each label. As a result, the structure of the RDF data resembles the original graph (cf. Figure 2b).

The following query looks for embeddings of the graph pattern from Figure 1:

```
SELECT ?X0 ?X1 ?X2
WHERE {?X0 e:0 ?X1 . ?X0 e:1 ?X2 .
  ?X0 a t:node . ?X0 rdfs:label ``0'' .
  ?X1 a t:node . ?X1 rdfs:label ``1'' .
  ?X2 a t:node . ?X2 rdfs:label ``0'' .
  FILTER ( (?X0 ≠ ?X1) && (?X0 ≠ ?X2) && (?X1 ≠ ?X2))}
```

Listing 1.2. SPARQL query for the pattern in Figure 1 (omitting prefix definitions)

Neo4j (Cypher). Neo4j is a *graph database system* with Cypher being its own query language. Neo4j does not rely on a relational data layout, but a network model storage that natively stores nodes (vertices), relationships (edges) and attributes (vertex and edge labels). Neo4j has a dual free/commercial license model. We use Neo4j v1.9 which introduced considerable enhancements and optimization to the query language Cypher.

It is possible to access and to modify data in Neo4j either with Cypher queries or directly via a Java API. Additionally, Neo4j can be embedded into the application or accessed via REST API. Experimental results show that Cypher via REST API performs slower than Cypher in embedded mode [7]. It is clear that queries using an embedded instance are faster than those accessing the database over the network. However, since we deploy the relational and RDF databases as servers and send queries from a client application, we also use REST API to send Cypher queries to the Neo4j server. Moreover, such a configuration models most real-world scenarios more accurately. The data schema of a graph in Neo4j corresponds to the representation in Figure 1. Vertex and edge labels are indexed with Lucene.

There are several ways to express a pattern match query with Cypher. The most straightforward way is to start with matching one vertex from the pattern graph, and match all edges in one "MATCH" statement of the Cypher query. We cannot specify all vertices as the starting nodes in Cypher, since it results in a completely different set of answers. This shortcoming is unfortunate since the user is left with the task to choose the most appropriate starting node.

As an alternative, it is possible to write nested queries in Cypher. This allows to match the pattern graph one edge at a time and transfer the intermediate results to the next level:

```
START v0 = node:my_nodes(label='0') MATCH v0-[e0]-v1
WHERE v1.label=1 and e0.label=0
WITH v0, v1 MATCH v0-[e1]-v2
WHERE v2.label=0 and id(v0)<>id(v2) and e1.label=1
RETURN id(v0), id(v1), id(v2);
```

Listing 1.3. Nested Cypher query for the pattern graph from Figure 1

The Neo4j developers mention that the nested queries might be especially good if the pattern graph is complicated. In both cases, straightforward and nested, we could further improve the queries by choosing more intelligently the starting node and the order in which we match the edges. Again, we do not apply these modifications since the scope of this work is on how well the database system itself can optimize the query execution. Due to space restrictions and readability issues, we report only the performance of the nested Cypher query since it shows consistently better results on our benchmark than the straightforward implementation.

Clingo (ASP). Answer-set programming (ASP) is a paradigm for declarative problem solving with many applications, especially in the area of artificial intelligence (AI) and knowledge representation & reasoning (KR). In ASP a problem is modeled in the form of logic program in a way such that the so-called stable models of the program correspond to the solutions of the problem. The stable model semantics for logic programs can be computed by ASP solvers like Clingo [16], DLV [17], Smodels [18], or others. We use Clingo v4.2.1 because of its performance at various ASP competitions [19].

In database terminology, Clingo is a *deductive database*, supporting ASP as query language. Data is represented by facts (e.g., vertices and edges) and rules from which new facts (e.g., the embeddings) are derived. For example, the data graph from Figure 1 is given as a text file of facts in the following form.

$$v(0,0). \; v(1,1). \; v(2,0). \; v(3,0). \; e(0,1,0). \; e(0,2,1). \; e(1,2,1). \; e(1,3,0). \; e(2,3,1).$$

The first argument of the vertex predicate v indicates the node ID, the second one the label. For the edge predicate e the first two arguments represent the ID's of the connected nodes and the third argument corresponds to the edge label: Note that we have omitted here half of the edges. To model undirected edges, we have two facts corresponding to each edge. For example, the dual version of the first edge fact above would be $e(1,0,0)$.

The ASP encoding for our running example is shown below:

```
1 {match(0,X) : v(X,0)} 1.
1 {match(1,X): e(Y,X,0)} 1 ← match(0,Y).
← match(1,X), not v(X,1).
1 {match(2,X): e(Y,X,1)} 1 ← match(0,Y).
← match(2,X), not v(X,0).
← node(K,X), node(L,X), K ≠ L.
```

Listing 1.4. ASP query for the pattern graph from Figure 1

In this encoding, we derive a new binary predicate *match* where the first argument indicates the ID of a vertex in the pattern graph and the second argument corresponds to the ID of a vertex in the data graph.

This encoding follows the "guess and check" paradigm. The *match* predicates are guessed as follows. The rule in Line 1 states that from all variables X such that we have a fact $v(X, 0)$, i.e. all vertices with label 0, we choose exactly one at random for our pattern node 0. The rule in Line 2 states that from all variables X such that there exists an edge with label 0 to a node Y which we have chosen as our pattern node 0, we choose exactly one at random for our pattern node 1. Finally, we have constraints in this encoding, which basically throw away results where the guess was wrong. For example, Line 3 is such a constraint which states that a guess for variable X as our pattern node 1 is invalid if the corresponding vertex in the data graph does not have label 1.

4.2   Datasets

We use both synthetic and real data. The synthetic datasets include two types of networks: small-world and erdos renyi networks. *Erdos Renyi Model (ERM)* is a classical random graph model. It defines a random graph as n vertices connected by m edges, chosen randomly from the $n(n-1)/2$ possible edges. The probability for edge creation is given by the parameter p. We use parameter $p = 0.01$. This graph is connected. *Preferential Attachment Model (PAM)*, or small-world model, grows a graph of n nodes by attaching new nodes each with m edges that are preferentially attached to existing nodes with high degree. We use $m = 4$. We choose this graph generation model, since it has been shown that many real-world networks follow this model [20]. In both cases, we generate graphs with 1000 and 10,000 nodes. Vertex labels are assigned randomly. We do not produce edge labels for the synthetic datasets.

The real-world datasets include a citation network and a terrorist organization collaboration network. *The terrorist organization collaboration network (GTON)* is constructed on the basis of Global Terrorism Database[1] which contains 81,800 worldwide terrorist attack events in the last 40 years. In this network, each vertex represents a terrorist organization, and edges correspond to the collaboration of organizations in the common attacks. Vertices are assigned two labels according to the number of recorded events: either 0 if the organization conducted less than 2 attacks, or 1 otherwise. Edges have also two labels depending on the amount of common attacks: either 0 if two organizations collaborated less than twice, or 1 in the other case.

The citation network (HepTh) covers arXiv papers from the years 1992–2003 which are categorized as High Energy Physics Theory. This dataset was part of the KDD Cup 2003 [21]. In this network, a vertex corresponds to a scientific publication. An edge between two vertices indicates that one of the papers cites the second one. We ignore the direction of the citation, and consider the network undirected. As vertex labels, we use the grouped number of authors of the corresponding paper. The edge label corresponds to the absolute difference between publication years of the adjacent vertices.

---

[1] http://www.start.umd.edu/gtd

We summarize the statistics of the constructed data graphs in Table 1.

Table 1. Summary of the datasets

Dataset	Synthetic data				Real data	
	ERM 1000	ERM 10000	PAM 1000	PAM 10000	GTON	HepTh
# vertices	1,000	10,000	1,000	10,000	335	9,162
# edges	4,890	500,065	3,984	39,984	335	52,995
avg degree	9.78	100.01	7.97	8	1.98	11.57
max degree	22	143	104	298	13	430
# vertex labels	2	2	2	2	2	5
# edge labels	–	–	–	–	2	5

4.3 Query Sets

We produce two sets of pattern graphs which are then used in graph pattern matching queries. All the generated pattern graphs are connected. The first set is generated synthetically with a procedure which takes as input the number of vertices and number of edges. The queries in this set have only vertex labels. In the first run of the procedure, we generate queries with five vertices and vary the number of edges from 4 till 10. In the second run, we fix the number of edges to ten and vary the number of vertices from 5 till 11. We generate 20 pattern graphs for each parameter configuration in both cases. We construct the synthetic queries this way in order to verify how the performance of database systems is influenced by the size of the pattern graph: first, we focus on the dependence on the number of edges, and second, on the number of vertices. We call this set of pattern graphs *synthetic patterns*.

The second set of queries is generated specifically for real-world data using graph pattern mining. Thereby we look for patterns which occur at least five times in our data graphs. In this set we specify not only vertex labels but also edge labels. The reason for this set of queries is twofold. First, we can study the performance on pattern graphs with guaranteed embeddings. Second, frequent graph pattern mining is a typical application scenario for graph pattern matching [12]. For example, graph pattern mining together with matching is used to predict citation counts for HepTh dataset in [22].

5 Experimental Results

Experimental Setup. The server and the client application are hosted on the same 64 bit Linux machine. It has four AMD Opteron CPUs at 2.4GHz and 8GB of RAM. The client application is written in Python and uses wrappers to connect to the database systems. The warm-up procedure for PostgreSQL, SPARQL and Neo4j consists of executing several pattern match queries. Since the problem at hand is NP-complete, we limit the execution time for all queries and database systems to three minutes to avoid long-lasting queries and abnormal memory consumption. We use the same order of edges in pattern graphs when encoding them into queries in different database languages. Except for our smallest dataset, GTON, we query only for the first 1000 embeddings. As Lee et al. [4] point out, this limit on embeddings is reasonable in practice. We would like to stress that all systems provide correct answers and agree on the number of discovered embeddings.

Synthetic Data. We report the performance of the database systems for synthetic queries on the four synthetic datasets in Figures 3 and 4. The performance of PostgreSQL is labeled by "SQL". The label "Cypher" stands for the nested Cypher query. Label "SPARQL" and "ASP" show the performance of Jena TDB and Clingo correspondingly. The performance is measured in terms of the execution time per query, and we plot the execution times on a logarithmic scale for better visualization. We also consider how many queries the database system manages to execute within three minutes.

Charts (a) and (c) in the figures correspond to the case where we use pattern graphs with five vertices and change the number of edges from four till ten. We have 20 distinct queries for each category, this means each data point corresponds to the average execution time over 20 queries. Since we consider only connected patterns, all patterns with five vertices and four edges are acyclic. With increasing number of edges, the number of cycles in patterns increases. Patterns with ten edges are complete graphs. Moreover, for these patterns the principle of containment holds: pattern graphs with less edges are subgraphs to some pattern graphs with more edges. Hence, the number of found embeddings can only drop or remain the same as we increase the number of edges.

In charts (b) and (d) from Figures 3 and 4 we start with the same complete pattern graphs as in the last category in charts (a) and (c). Then, by fixing the number of edges to 10, we increase the number of vertices till 11. Pattern graphs with 11 vertices and 10 edges are again acyclic graphs. By construction, the principle of containment does not hold here. Hence, in charts (a) and (c) we investigate how the increasing number of edges influences the performance of the database systems. For charts (b) and (d), the focus is on the dependence between the number of vertices and execution time.

Another aspect studied is the scalability of the database systems with regard to the size of the data graph. Thus, in Figures 3 and 4 charts (a) and (b) correspond to smaller graphs while charts (c) and (d) show results for bigger ones. Since we observe that the performance of the systems also depends on the structure of the pattern, we present the dependence of the run time on the number of cycles in the pattern in Figure 5.

The results indicate that SPARQL is on average better than the others. We observe that the performance of the database systems depends on the following factors: (I) size of the data graph; (II) size of the pattern graph; and (III) structure of the pattern graph.

PostgreSQL shows incoherent results. For example, we can observe a peak in the average run time for the pattern graphs with five vertices and six edges for small datasets in PostgreSQL (see Figure 3a,4a). One reason for this behavior is that in some cases the query optimizer fails in determining the best join strategy (we recall that PostgreSQL offers nested loop-, hash-, and merge join). For example, by disabling the usage of the nested loop join in the configuration of PostgreSQL, we arrive at an average run time of two seconds instead of eight for the dataset PAM1000 for synthetic patterns. However, this trick works only for the smaller graphs. Overall, PostgreSQL does not scale with regard to the size of the pattern graph. We can see it especially in Figure 3b. Furthermore, none of the queries in Figure 3d finished within 3 minutes.

Surprisingly, SPARQL shows the best performance in almost all cases. This complements previous works ([9] and [15]) where RDF-based systems are among the worst

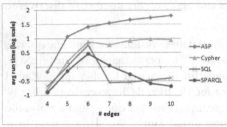
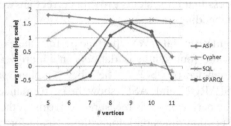
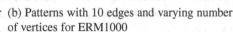

(a) Patterns with 5 vertices and varying number of edges for ERM1000

(b) Patterns with 10 edges and varying number of vertices for ERM1000

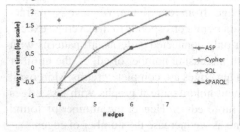
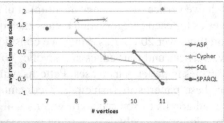

(c) Patterns with 5 vertices and varying number of edges for ERM10000

(d) Patterns with 10 edges and varying number of vertices for ERM10000

Fig. 3. Average run time in seconds on logarithmic scale for ERM

performing systems for the tasks of graph analysis, except our case of graph pattern matching. Our understanding is that, besides solving a different problem, the authors use native access to Neo4j in [15], and access to Neo4j through Cypher via REST API has been shown to be much slower [7]. Also, Renzo et al. [9] chose RDF-3X which seems to perform worse than Jena TDB. Like PostgreSQL, Jena TDB shows incoherence, e.g., there is a peak in average run time for patterns with five vertices and six edges (see Figures 3a and 4a). In Figure 5 we observe that SPARQL can handle pattern graphs with two cycles considerably worse than the others. It seems that the query optimizer of Jena TDB makes worse choices for the matching order of edges, however, it is unclear if such behavior can be configured since the query optimizer is not transparent.

Though SPARQL shows better run times, it cannot handle efficiently complete patterns on ERM10000 like the other database systems (see Figure 3d). SPARQL turned out to be very sensitive towards the changes in the data schema. If we put edges as resources in Jena TDB, it becomes the worst performing database system. Such change in the data schema might be relevant if we introduce more than one property for the edges or we have a string property associated with the edges. At the same time, such drastic changes to the data schema are not required for the other database systems.

Our results show that Clingo is not a scalable system with regard to the size of the data graph. It cannot execute any query for cyclic patterns within three minutes on ERM with 10000 vertices (Figure 3c,3d), but can only handle acyclic patterns on these datasets. Though the size of the pattern affects the run time of Clingo, the decrease in the run time happens mainly due to the growth of cycles in the pattern graph (Figure 5).

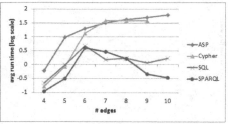 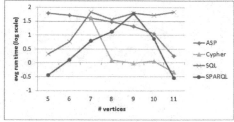

(a) Patterns with 5 vertices and varying number of edges for PAM1000
(b) Patterns with 10 edges and varying number of vertices for PAM1000

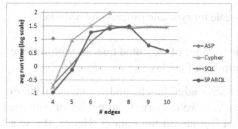 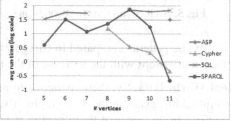

(c) Patterns with 5 vertices and varying number of edges for PAM10000
(d) Patterns with 10 edges and varying number of vertices for PAM10000

Fig. 4. Average run time in seconds on logarithmic scale for PAM

Like for Clingo, cyclic patterns pose the main problem for Neo4j. The average run times grow with the increase of the number of edges in charts (a) and (c), and then drop with the increase of the number of vertices in charts (b) and (d) in Figures 3 and 4. Thus, the worst run time is for complete patterns. This trend is illustrated in Figure 5. Unlike ASP, the dependence between the run time and number of cycles is not linear for Cypher. The issue is that both Cypher and ASP rely on backtracking algorithms. Hence, the starting vertex in the queries is very important for the performance. We could further optimize the queries by choosing the vertex with the least frequent label as the starting point. This change is especially crucial for Neo4j, since its query optimizer is not as mature as the one for PostgreSQL or Clingo. When analyzing the change of average run times between the small and big data graphs in each figure, we see that the performance of ASP drastically drops while Cypher shows the least change in performance among the considered systems. This trend is especially clear for ERM (Figure 3).

As mentioned, database systems not always manage to execute our queries within three minutes. PostgreSQL finishes not all queries with 10 edges on the small PAM and ERM graphs within the time limit. Especially problems arise with patterns which have more than 9 vertices. For example, PostgreSQL finishes only two queries out of 20 on the small PAM graph for pattern graphs with 10 edges and 11 vertices. Furthermore, on both big graphs (PAM10000 and ERM10000) PostgreSQL can execute only a fraction of queries for pattern graphs with more than six edges. Clingo, Neo4j and SPARQL execute all queries for acyclic patterns within the established time limit irrelevant of the size of data and pattern graphs. An interesting observation is that Clingo either executes all queries within the specified category, or none at all. While Neo4j can handle

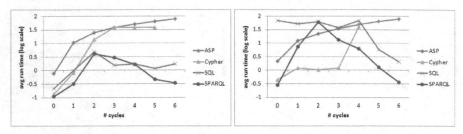

(a) Patterns with 5 vertices and varying num- (b) Patterns with 10 edges and varying number
ber of cycles   of cycles

Fig. 5. Average run time in seconds on log-scale for synthetic patterns on PAM1000

more cyclic patterns compared to Clingo, the number of executed queries within three
minutes still drops with the increase of the number of cycles.

Real World Data. Synthetic patterns do not have edge labels, and for many of them
no embedding exists in the data graphs. Therefore, we construct another set of patterns
by ensuring that each of them occurs at least 5 times in the dataset. Hence, the sets
of generated patterns differ for GTON and HepTh. In both cases graph patterns have
from 5 up to 10 edges, and almost all of them are acyclic. More precisely, the generated
patterns have either the same number of vertices as edges, or they have one vertex more
than edges. The performance for this setting is shown in Figure 6. Again, we use 20
pattern graphs in each category, which corresponds to a data point in the charts, and
report the average run time for each category on the logarithmic scale in the figure.

Clingo and Jena TDB show equally good performance on our smallest dataset GTON
(Figure 6a) with Clingo being better for the bigger pattern graphs. Neo4j and PostgreSQL are considerably worse in this case. However, ASP drops in its performance
on the dataset HepTh (Figure 6b). Clingo does not use any indexes on labels and has
no caching mechanism. This leads to a considerable drop in the performance when the
data graph is especially big. At the same time, SPARQL shows the best run times on
the bigger dataset, though compared to GTON the run times increase.

Surprisingly, PostgreSQL does not provide the best performing results in any case
on the real-world data. We can observe a clear trend that the average run time for PostgreSQL considerably grows with the increase of the number of edges in the pattern
graphs. Moreover, PostgreSQL executes only five queries out of 20 for pattern graph
with 7 edges on HepTh. This observation proves once again that PostgreSQL does not
scale with regard to the size of the pattern graphs. Since the more edges there are in the
pattern graph, the more joins for tables PostgreSQL has to do.

In terms of scalability with the size of the data graph, we conclude that Neo4j shows
the best results. Judging from the results on GTON, we may conclude that there is a
lot of space for improvement for Neo4j. We believe that the query optimizer in Neo4j
could be tuned to narrow the gap between Cypher and SPARQL in this case.

Summary. As a result, we can provide the following insights. In general, Jena TDB
is better for graph pattern matching with regard to the data schema provided in our
benchmark. If we have a very small data graph, Clingo is a good choice. If we have

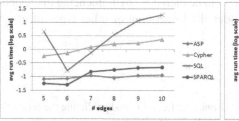 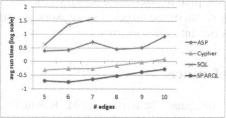

(a) Frequent pattern queries for GTON   (b) Frequent pattern queries for HepTh

Fig. 6. Average run time in seconds on logarithmic scale for GTON and HepTH

a big data graph and pattern graphs are mainly acyclic, Neo4j provides good results. However, in case of big data graphs and cyclic pattern graphs with more than seven edges, none of the studied database systems perform well.

6 Conclusion and Future Work

We have studied how four database systems perform with regard to the problem of graph pattern matching. In our study we use a relational database PostgreSQL, a graph database Neo4j, a deductive database Clingo and RDF-based database Jena TDB. The most mature system among these four is PostgreSQL with Neo4j being the youngest. By conducting extensive experiments on synthetic and real-world datasets we come to the following conclusions.

Clingo does not scale with the size of the data graph. Its performance drastically drops when the data graph becomes bigger. Though the performance of Clingo is not much influenced by the size of the pattern graph, it worsens with the growth of the cycles in the pattern graph. Neo4j cannot efficiently handle cyclic pattern graphs. However, it scales very well with regard to the size of the data graph as well as the size of the pattern graph. The performance of PostgreSQL oscillates generally due to the changes in the join order chosen in the execution plan. Though PostgreSQL shows good perfor mance for cyclic patterns, it does not scale well with regard to the size of the pattern graphs and the size of the data graphs. Jena TDB is in general better than the other database systems on our benchmark. However, it turned out to be the most sensitive system to the changes in the data schema.

In our opinion, the efficiency of databases for solving graph pattern matching tasks is not yet good enough for information systems that deal with real-world big data scenarios. The database systems should consider implementing the state-of-the-art algorithms for this task [4]. For example, the query optimizer of Neo4j cannot ensure the most optimal choice of the starting node, and of the join order for the edges and vertices. The latter holds also for PostgreSQL which can be optimized by tuning up the configuration settings of the server. Furthermore, in all database systems we can configure the servers to achieve better results, but it is unclear what is the best configuration if we need to perform a variety of graph queries and not just graph pattern matching. This calls for further investigation.

We plan to investigate how the efficiency can be increased by influencing the matching order of the edges. Future work includes the integration of other types of graph queries into our benchmark.

References

1. Bringmann, B., Berlingerio, M., Bonchi, F., Gionis, A.: Learning and predicting the evolution of social networks. IEEE Intelligent Systems 25(4), 26–35 (2010)
2. Asnar, Y., Paja, E., Mylopoulos, J.: Modeling design patterns with description logics: A case study. In: Mouratidis, H., Rolland, C. (eds.) CAiSE 2011. LNCS, vol. 6741, pp. 169–183. Springer, Heidelberg (2011)
3. Zou, L., Chen, L., Özsu, M.T., Zhao, D.: Answering pattern match queries in large graph databases via graph embedding. VLDB J. 21(1), 97–120 (2012)
4. Lee, J., Han, W.S., Kasperovics, R., Lee, J.H.: An in-depth comparison of subgraph isomorphism algorithms in graph databases. PVLDB 6(2), 133–144 (2012)
5. Fan, W., Li, J., Ma, S., Tang, N., Wu, Y.: Adding regular expressions to graph reachability and pattern queries. Frontiers of Computer Science 6(3), 313–338 (2012)
6. Xu, J., Chen, H.: Criminal network analysis and visualization. Commun. ACM 48(6), 100–107 (2005)
7. Holzschuher, F., Peinl, R.: Performance of graph query languages: comparison of Cypher, Gremlin and native access in Neo4j. In: Proc. EDBT/ICDT Workshops, pp. 195–204 (2013)
8. Vicknair, C., Macias, M., Zhao, Z., Nan, X., Chen, Y., Wilkins, D.: A comparison of a graph database and a relational database: a data provenance perspective. In: Proc. ACM Southeast Regional Conference, p. 42 (2010)
9. Angles, R., Prat-Pérez, A., Dominguez-Sal, D., Larriba-Pey, J.L.: Benchmarking database systems for social network applications. In: Proc. GRADES, p. 15 (2013)
10. Karp, R.M.: Reducibility among combinatorial problems. In: Proc. Complexity of Computer Computations, pp. 85–103 (1972)
11. Alon, N., Yuster, R., Zwick, U.: Color-coding. J. ACM 42(4), 844–856 (1995)
12. Gallagher, B.: Matching structure and semantics: A survey on graph-based pattern matching. In: Proc. AAAI Fall Symposium on Capturing and Using Patterns for Evidence Detection (2006)
13. Tudorica, B.G., Bucur, C.: A comparison between several NoSQL databases with comments and notes. In: Proc. Roedunet International Conference, pp. 1–5 (2011)
14. Angles, R.: A comparison of current graph database models. In: Proc. ICDE Workshops, pp. 171–177 (2012)
15. Dominguez-Sal, D., Urbón-Bayes, P., Giménez-Vañó, A., Gómez-Villamor, S., Martínez-Bazán, N., Larriba-Pey, J.L.: Survey of graph database performance on the HPC scalable graph analysis benchmark. In: Shen, H.T., Pei, J., Özsu, M.T., Zou, L., Lu, J., Ling, T.-W., Yu, G., Zhuang, Y., Shao, J. (eds.) WAIM 2010. LNCS, vol. 6185, pp. 37–48. Springer, Heidelberg (2010)
16. Gebser, M., Kaufmann, B., Kaminski, R., Ostrowski, M., Schaub, T., Schneider, M.T.: Potassco: The Potsdam answer set solving collection. AI Commun. 24(2), 107–124 (2011)
17. Leone, N., Pfeifer, G., Faber, W., Eiter, T., Gottlob, G., Perri, S., Scarcello, F.: The DLV system for knowledge representation and reasoning. ACM Trans. Comput. Log. 7(3), 499–562 (2006)
18. Syrjänen, T., Niemelä, I.: The Smodels system. In: Eiter, T., Faber, W., Truszczyński, M. (eds.) LPNMR 2001. LNCS (LNAI), vol. 2173, pp. 434–438. Springer, Heidelberg (2001)

19. Alviano, M., et al.: The fourth answer set programming competition: Preliminary report. In: Cabalar, P., Son, T.C. (eds.) LPNMR 2013. LNCS, vol. 8148, pp. 42–53. Springer, Heidelberg (2013)
20. Barabási, A.L., Albert, R.: Emergence of scaling in random networks. Science Magazine 286(5439), 509–512 (1999)
21. Gehrke, J., Ginsparg, P., Kleinberg, J.M.: Overview of the 2003 KDD cup. SIGKDD Explorations 5(2), 149–151 (2003)
22. Pobiedina, N., Ichise, R.: Predicting citation counts for academic literature using graph pattern mining. In: Proc. IEA/AIE, pp. 109–119 (2014)

Ontology Driven Indexing: Application to Personalized Information Retrieval

Vincent Martin[1,2], Emmanuel Bruno[1], and Elisabeth Murisasco[1]

[1] Université de Toulon, CNRS, LSIS, UMR 7296, 83957 La Garde, France
Aix Marseille Université, CNRS, ENSAM, LSIS, UMR 7296, 13397 Marseille, France
{vincent.martin,bruno,murisasco}@univ-tln.fr
[2] Coexel, 131 Avenue Marechal Foch, 83000 Toulon, France
vincent.martin@coexel.com

Abstract. Our work addresses the problem of information retrieval (IR) in an heterogeneous environment by a model driven information retrieval infrastructure that allows each user to explicitly personalize the indexing and retrieval processes. In this paper, we propose an ontological approach to represent a particular user's need, his preferences, his context and his processes within the information system. Then, we define an algorithm to automatically construct personalized IR models. Afterwards, we describe an architecture to handle these models in order to provide personalized answers. Lastly, we show that system responses are entirely personalized and semantically enhanced by the ontological representation behind. These contributions are designed to allow the integration of external knowledge and their fundations enable an adaptation to various applications.

Keywords: Information Retrieval, Ontology, Database, Enterprise context.

1 Context and Motivation

Information retrieval (IR) is a broad domain with a large number of very different applications for very different users with very different knowledge and levels of expertise. Starting from this point, research about IR covers many theories (boolean theory, probability theory, etc.), many domains (knowledge management, natural language processing, etc.) in order to find the best answer(s) for a particular user's need (generally represented as a keyword query). Modern proposals in IR can be separated into two categories: the first one takes into account the context implicitly (for example by using Latent Semantic Analysis / Indexing [4]) and the second uses explicit knowledge to capture the context (for example models which use domain ontologies to contextualize the data [5]).

Our work focuses on the second category. We argue that an IR system can actually be more efficient if the entire system is personalized for each user. Personalization aims at improving the process of retrieval considering the interests

H. Decker et al. (Eds.): DEXA 2014, Part I, LNCS 8644, pp. 242–256, 2014.

of individual users. Ideally, every available information should participate in the construction of the result, including the user expertise (the more expert is a user, the better is the result). However, in many approaches, the personalization only takes place in the query but the whole context is generally not explicitly defined. Furthermore, there is a gap between a real user's need and the way he explains it (either through a keyword query, a natural language query or a structured query). First, because it's very tough to explain precisely and exhaustively an information need and to reiterate the process for every need. Second, because the transcription of an information need in a specific search structure (search form, wildcards, ...) is not obvious. In addition, in many contexts and especially in an enterprise or in a scientific data management context, users have a precise knowledge about their needs (but not enough to be able to perform exact queries). They already know which aspect of information is interesting (for example in a technological vs. an economic context), which interactions between information are useful (for example the relation between a technological event and economic benefits), etc. If this knowledge could be formally represented and soundly integrated in the IR system, query results will be more relevant. In fact, we state that if in a given context, there is a formal representation identical to that of a person, then the system will return results with maximum precision and maximum recall (according to the semantics of the query language).

In this paper, we describe an approach to let the user describe its own indexing process in order to personalize the IR system. Like Model Driven Architecture (MDA) in software design, our intention is to construct a model driven approach for IR. Because our model enables users to represent their knowledge about the IR system, we rely on ontologies [7][1] to represent the knowledge they are interested in.

Our objectives are twofold: first, we shall define a model to represent in an unified way both data and context. Second, we shall propose a global architecture in order to automatically take advantage of the underlying model. In our work, the context represents the knowledge about (i) the information system and its components (indexer, query engine, ...); (ii) the domain data (domain ontologies, knowledge base, ...) and (iii) the users' needs, i.e. the subset of interesting data, resource preferences and resource interactions. Given this definition, a user can have multiple needs and/or he may have several ways to express the same need; consequently, a user may have one or more IR models at a time (because information needs are not static). Of course, some knowledge can be shared between users and contexts like domain ontologies, information system structure, etc.

This paper is organized as follows: section 2 describes the related works; sections 3 and 4 define respectively our model and the architecture for personalized information retrieval. Section 5 gives information about the current implementation and results and section 6 concludes and presents our perspectives.

---

2   Related Works

There are three paradigms for IR models. The first one is the (strict) boolean model in which a term indexes completely or not a document and where a query is completely relevant with a document or not at all. The second paradigm is based on a vector representation. The Vector Space Model (VSM) [13] represents both documents and queries in an $m - dimensional$ vector space (m is the number of unique terms). This model allows partial matchings between queries and documents (or between two documents) and returns results sorted by relevance. Many proposals [4] [14] extend or improve the original vector space model which is still widely used nowadays. The third paradigm is based on the probability theory. Basically, probabilistic IR models [9] try to estimate the probability $P_q(R|d)$, i.e. the probability that the document d be relevant (R is the set of relevant documents) given the query q. Since its first definition in the 60s, many extensions have been proposed: in the 80s, the famous Okapi model has been developed [12] to take into account local and global term frequencies ; in 1990, Turtle and Croft introduce a bayesian network approach for IR [15]; more recently, language models have been applied in IR [11] because they allow a better understanding of the linguistic structure of documents. These three paradigms give a general framework to estimate a similarity between a query and a document without taking into account any external information. Following proposals deal with adding external information and/or knowledge to better address the IR issues.

Since the definition of an ontology in [7], semantic IR models have been proposed to improve search results. A good illustration of this kind of proposal is [5] which presents an approach to enhance information retrieval by using ontologies that describe the data. In this approach, search is done on concepts and concepts are mapped to documents. This proposal is interesting in an IR system that tries to conceptualize objects. However, the overall system can be improved by using knowledge to represent the context jointly with user's profiles (or user's preferences) in order to personalize the information retrieval process. In [10], Mylonas et al. define an approach to automatically capture user's preferences (persistent preferences and live interests) and to provide a context-aware retrieval process. Because the notion of preference is unclear (preferences are rarely true or false), they choose a fuzzy ontology representation in order to handle this uncertainty. Experiments have been done on a medium-scale corpus from the CNN web site and the results give a mean average precision of 0.1353 for the contextual personalization, 0.1061 for the simple personalization (i.e. without context) and 0.0463 with no personalization. These are interesting proposals but the personalization considers only the data and not the information system itself. The Hermeneus Architecture proposed by Fabiano D. Beppler et al. in [2] addresses this issue by using ontologies and knowledge bases to configure the IR system components: the indexing module, the retrieval module, the inference module and the presentation module. In this architecture, the expert configures his IR system by creating an ontology and a knowledge base to describe his knowledge about the domain. Then, the expert can define rules in order to enhance query results by semantic relations (explicit and inferred). In the Hermeneus architecture, each concept has an index and each entry in the index points to an individual. However, there is no personalization of the indexing process itself, i.e.: the index of an ontology instance is always the same regardless its semantic context.

Our objective is to propose a novel approach based on a unique model to handle data and to directly configure the indexing and querying processes. Concerning the configuration of the indexing process, we base our approach on the BlockWeb model [3]. BlockWeb is an IR model for indexing and querying Web pages. In this approach, Web pages are viewed as a block hierarchy where each block has a *visual importance* and can be *permeable* to their neighbour block's content. A language has been proposed to let the user define how importances and permeabilities are assigned. This language allows multiple indexing on the same dataset, which is useful when different applications have to process similar data.

The main goal of our proposal is to generalize the BlockWeb model; (i) to manage other data than Web pages and (ii) to personalize and to improve information retrieval by allowing the user to define multiple indexing strategies according to his needs and his context through ontologies. This approach enables users, who can also be experts, to tune the IR system to fit their levels of expertise.

3   The OnADIQ Model

In this section, we define OnADIQ (Ontological Architecture for Dynamic Indexing and Querying) to address the problem of information retrieval for domain experts. After the model definition, we show its properties and we propose an OWL encoding.

3.1   Model Definition

We propose an ontological representation to model specific information needs. The concept of ontology in computer science has been proposed by Tom Gruber in [7] and it is defined as "*a formal description providing human users a shared understanding of a given domain*" which "*can also be interpreted and processed by machines thanks to a logical semantics that enables reasoning*" [1] (we do not deal with reasoning in this paper). In the OnADIQ model, we represent knowledge about data (*what*) but also knowledge used to directly drive the IR process (*how*): the subset of data to be indexed and the subjective relations between them. As the OnADIQ model is an ontology we talk about *resources* instead of *documents*. We define the OnADIQ model as follows:

Definition 1. *The OnADIQ model is a 7-tuples* $\mathcal{M} = \{\mathcal{R}, \mathcal{P}, \mathcal{H}^C, \mathcal{H}^P, rel, \mathcal{A}, \mathcal{B}\}$ *where*

- $\mathcal{R}$ *is the set of resources (concepts (C), individuals ($\mathcal{I}$) and literals ($\mathcal{L}$)) where each resource has a unique identifier (its URI) ;*
- $\mathcal{P}$ *is the set of properties;*

- $\mathcal{H}^{\mathcal{C}}$ is the concept hierarchy: $\mathcal{H}^{\mathcal{C}} \subseteq \mathcal{C} \times \mathcal{C}$;
- $\mathcal{H}^{\mathcal{P}}$ is the property hierarchy: $\mathcal{H}^{\mathcal{P}} \subseteq \mathcal{P} \times \mathcal{P}$;
- $rel : \mathcal{P} \to (\mathcal{C} \cup \mathcal{I}) \times (\mathcal{C} \cup \mathcal{I} \cup \mathcal{L})$ defines the relation between two resources. A relation has a domain: $\mathcal{P} \to \mathcal{C} \cup \mathcal{I}$ and a co-domain (range): $\mathcal{P} \to \mathcal{C} \cup \mathcal{I} \cup \mathcal{L}$;
- $\mathcal{A}$ is the set of resource importances;
- $\mathcal{B}$ is the set of resource permeabilities.

Our model adds importance ($\mathcal{A}$) and permeability ($\mathcal{B}$) properties to the ontology definition. The α property describes the importance of a resource ($\alpha \in [0, 1]$) and β ($\beta \in [0, 1]$) is the permeability between two resources.

Definition 2. *Importance of a resource in a set of resources in an ontology reflects its contribution in the semantic content of this set.*

We define the importance as a function $\alpha : \mathcal{R} \to \mathbb{R}^+$ where $\mathcal{R}$ is the set of resources in the ontology $\mathcal{M}$. $\alpha(r_i)$ is the importance of the resource $r_i \in \mathcal{R}$. In the context of ontology, α is a property. The higher α of a resource is, the higher the importance of the resource is.

Example: A user which has a full confidence in a information source will affect it a high importance (e.g. : 1). This importance could be computed automatically using, for example, quality criteria.

Definition 3. *Permeability of a resource r to the content of a resource r' reflects the amount of content from r' that enriches the content of r.*

Permeability between a domain resource and a range resource is a function $\beta : \mathcal{R} \times \mathcal{R} \to \mathbb{R}^+$ where $\mathcal{R}$ is the set of resources in the ontology. Permeability is an irreflexive, antisymmetric and transitive relation.

Example: In the membership relation (e.g. : Bob is the father of Mary), a permeability from Bob to Mary exists reflecting the amount of Bob's index which is propagated to Mary's index.

Values for α and β can be subjective and manually set by the user.

3.2 Indexing and Querying

In this sub-section, we present an adaptation of the method proposed for the BlockWeb model [3] to compute resource indexes. Let r be a resource, its index is $\vec{r} = (w(t_1, r), \ldots, w(t_n, r))$ where $w(t_j, r)$ is the weight of the term t_j for the resource r. This index depends on the local index $\vec{rl}$ of a resource r, its importance $\alpha(r)$ and the resources to which it is permeable. The local index can be computed by the well-known $tf.idf$ model adapted to our approach (nothing changes except we replace the term *document* with *resource*). In this sense, we talk about $tf.irf$ (*term frequency, inverse resource frequency*):

$$\vec{rl} = (tf(t_1, r) \times irf(t_1), \ldots, tf(t_n, r) \times irf(t_n)) \tag{1}$$

where

- $tf(t_i, r)$ is the frequency of the term t_i in the resource r. It reflects the importance of the term t_i relative to the resource.
- $irf(t_i)$ is the inverse resource frequency of the term t_i ($rf(t_i)$ is the resource frequency, i.e. the number of resources containing the term t_i). It reflects the importance of the term t_i relative to the corpus.

Several equations can be proposed in order to compute the index of a resource given its importance and its permeable resources. Equations 2 and 3 give two ways to solve this problem.

$$\vec{r} = \alpha(r) \times \left(\vec{rl} + \sum_{k=1}^{m} \beta(r_k, r) \times \vec{r_k} \right) \tag{2}$$

$$\vec{r} = \alpha(r) \times \vec{rl} + \sum_{k=1}^{m} \left(\beta(r_k, r) \times \vec{rl_k} \right) \tag{3}$$

where

- $\alpha(r)$ is the importance of r;
- $\vec{rl}$ is the local index of r;
- $\beta(r_k, r)$ is the permeability of the resource r to the content of r_k;
- m is the total number of resources.

Equation 3 does not take into account the importance of a resource that would have an empty local index ($\vec{rl} = \vec{0}$). This can be an problem when a resource is abstract and used for conceptualization. Equation 2 solves this problem and we prefer it in many cases (we use it in this paper).

Computing all resource indexes is equivalent to solving a system of linear equations whose the form depends on the ontology. Consider the underlying graph of an OnADIQ model and suppose (for simplicity) that the m resources have a unique numeric identifier (from 1 to m). If the graph is acyclic, the system of linear equations has a unique solution and indexes can be computed. Otherwise the solution can only be approximate using iterative methods like the Jacobi method but this aspect is beyond the scope of this paper. For convenience, we note α_i the importance of the resource r_i instead of $\alpha(r_i)$ and $\beta_{i,j}$ the permeability of the resource j to the content of the resource i instead of $\beta(r_i, r_j)$. Equation 2 gives the following systems of linear equations:

$$\begin{cases} \vec{r_1} = \alpha_1 \vec{rl_1} + \alpha_1 \beta_{2,1} \vec{r_2} + \ldots + \alpha_1 \beta_{k,1} \vec{r_k} + \ldots + \alpha_1 \beta_{m,1} \vec{b_n} \\ \vec{r_2} = \qquad\qquad \alpha_2 \vec{rl_2} + \ldots + \alpha_2 \beta_{k,2} \vec{r_k} + \ldots + \alpha_2 \beta_{m,2} \vec{r_m} \\ \ldots \\ \vec{r_k} = \qquad\qquad\qquad\qquad\qquad \alpha_k \vec{rl_k} + \ldots + \alpha_k \beta_{m,k} \vec{r_k} \\ \ldots \\ \vec{r_m} = \qquad\qquad\qquad\qquad\qquad\qquad\qquad\qquad \alpha_m \vec{rl_m} \end{cases} \tag{4}$$

$$\begin{cases} \alpha_1 \vec{rl_1} = \vec{r_1} - \alpha_1\beta_{2,1}\vec{r_2} - \ldots - \alpha_1\beta_{k,1}\vec{r_k} - \ldots - \alpha_1\beta_{m,1}\vec{r_m} \\ \alpha_2 \vec{rl_2} = \qquad\quad \vec{r_2} - \ldots - \alpha_2\beta_{k,2}\vec{r_k} - \ldots - \alpha_2\beta_{m,2}\vec{r_m} \\ \ldots \\ \alpha_k \vec{rl_k} = \qquad\qquad\qquad\quad \vec{r_k} - \ldots - \alpha_k\beta_{m,k}\vec{r_k} \\ \ldots \\ \alpha_m \vec{rl_m} = \qquad\qquad\qquad\qquad\qquad\qquad\quad \vec{r_m} \end{cases} \tag{5}$$

Now we write the system 5 using the matrix notation $A \times W = K$ with

$$A = \begin{pmatrix} 1 & -\alpha_1\beta_{2,1} & \ldots & -\alpha_1\beta_{k,1} & \ldots & -\alpha_1\beta_{m,1} \\ 0 & 1 & \ldots & -\alpha_2\beta_{k,2} & \ldots & -\alpha_2\beta_{m,2} \\ \ldots & \ldots & \ldots & \ldots & \ldots & \ldots \\ 0 & 0 & \ldots & 1 & \ldots & -\alpha_k\beta_{m,k} \\ \ldots & \ldots & \ldots & \ldots & \ldots & \ldots \\ 0 & 0 & \ldots & 0 & \ldots & 1 \end{pmatrix}$$

and

$$K = \begin{pmatrix} \alpha_1 rl_1 \\ \alpha_2 rl_2 \\ \ldots \\ \alpha_k rl_k \\ \ldots \\ \alpha_m rl_m \end{pmatrix}$$

where

- m is the number of resources in the ontology;
- I is the identity matrix $m \times m$;
- $ALPHA$ is the diagonal matrix $m \times m$ where the element at (j,j) is equal to α_j;
- $BETA$ is the $m \times m$ matrix where the element at (k,j) is either $\beta_{k,j}$, if exists, or zero;
- WL is the column matrix $m \times 1$ where the element at $(j,1)$ is the local index of the resource r_j;
- W is the column matrix $m \times 1$ where the element at $(j,1)$ is the final index of the resource r_j.

As we consider only directed acyclic graphs (DAG), $(I - ALPHA \times BETA^t)$ has a non-zero determinant and it is invertible. Then, the system can be solved and indexes could be calculated. We have

$$A = (I - ALPHA \times BETA^t) \text{ and } K = ALPHA \times WL$$

thus

$$(I - ALPHA \times BETA^t) \times W = ALPHA \times WL$$

we conclude

$$W = (I - ALPHA \times BETA^t)^{-1} \times ALPHA \times WL \tag{6}$$

Querying the model and the data is not difficult at all: any method that works on a classic vector space model will work on our model. For example, consider a query q and its vectorized representation $\vec{q} = w(t_1, q), \ldots, w(t_n, q)$ where $w(t_j, q)$ is the weight of the term t_j in the query q; $w(t_j, q)$ can be calculated in different ways (cf. SMART notation [8]). For instance, $\vec{q} = tf(t_1, q), \ldots, tf(t_n, q)$. Similarity between a resource r and a query q can be expressed as follows:

$$sim(r, q) = \cos(\alpha) = \frac{\vec{r} \cdot \vec{q}}{||r|| \times ||q||} = \frac{\sum_{i=1}^{m}(w(t_i, r) \times w(t_i, q))}{\sqrt{\sum_{i=1}^{m} w(t_i, r)^2} \times \sqrt{\sum_{i=1}^{m} w(t_i, q)^2}}$$

3.3 OWL Encoding

OWL (Web Ontology Language) [6][2] is a W3C standard to describe an ontology. Description logics (DL) provide the formal basis of OWL. DL are decidable fragments of first-order logic. We propose an OWL-DL construction in order to encode our model. OWL-DL is based on $\mathcal{SHOIN}(\mathcal{D})$ DL which provides good expressivity and for which reasoning is decidable.

A minimal OnADIQ model is composed of two classes: (i) *DomainResource* is the root class which represents any resources in a data management process. (ii) *Metadata* is the class which represents model metadata: importances and permeabilities. A permeability $\beta(r_i, r_j)$ is an instance of the *Permeability* class with two properties: (i) *PermeabilityDomain* with a domain resource of type *DomainResource* and a co-domain resource of type *Permeability*, (ii) *PermeabilityRange* with a domain resource of type *Permeability* and a co-domain resource of type *DomainResource*. The permeability β value is a property (*permeability*) of class *Permeability*. The value of an importance α is a property of class and sub-classes of *DomainResource*.

3.4 An Illustrative Example

To illustrate the expressiveness of our model, we study the following use case. Suppose that two users, John and Bob, are interested in news about smartphones on Twitter. John is particularly interested in official news on technological breakthroughs in this area. Bob is interested in the manufacturers' e-reputation. Consequently, he wants to know who is the more active on a specific aspect (design, durability, etc.). For example, if Bob wants to retrieve the most satisfied users about a new smartphone, he may submit to the system queries with terms such as *"smartphone + happy"* or *"smartphone + I'm glad"*, etc. in order to find which users write tweets containing these terms. Figure 1 presents an ontology describing the general context of these two applications by representing the fact

---

[2] http://www.w3.org/TR/owl2-overview/

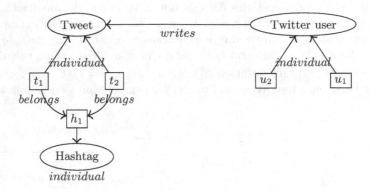

Fig. 1. A model describing an application for indexing tweets and their authors. The ontology describes three concepts (Tweet, Twitter User, Hashtag) and examples of individuals.

that Bob and John are interested by tweets and their authors (*Twitter user*). We also represent the concept of *Hashtag* (i.e. an explicit label in Tweets). To keep the ontology simple, we only represent two users who have published one tweet, each containing the same hashtag.

John and Bob use the same data but they want to index them differently. Consequently, a specific model is needed for their respective information needs. Figure 2 shows two indexing strategies: the 2(a) formally describes John's strategy (tweets are more important than users) and the strategy of 2(b) is the Bob's one (users are more important than tweets). Each resource has an importance α and permeabilities $\beta's$ may exist between resources. For example, the importance of t_1 is set to 1 in John's and Bob's strategies. Permeability β set to 1 from u_1 to t_1 in John's strategy means u_1 index reinforces t_1 index instead of Bob's strategy where it is the opposite (look at the direction of the arrows) with a 0.5 permeability. Both importances and permeabilities can be calculated, for example, depending on the popularity of the author or as in Bob's strategy where the permeability from tweets to users u is $\frac{1}{|\text{user } u \text{ tweets}|}$. Note that *Tweet* and *Twitter user* resources have neither importance nor permeability; this is equivalent to set an importance equal to zero. Therefore, their index are empty and they cannot be retrieved.

The Figure 2(a) describes a model which enhances tweets indexes by their author indexes. The Figure 2(b) presents a model which extends tweets authors indexes by their tweet indexes (notice the direction of the arrow). In both models, tweets are reinforced by the hashtags they contain. In our application, an hashtag is indexed by all tweets in which they appear.

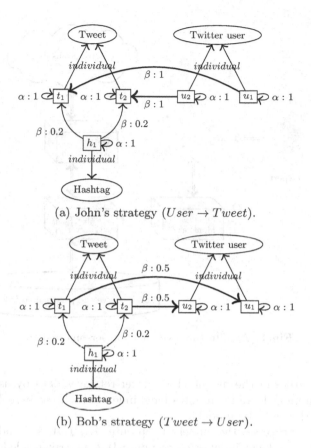

(a) John's strategy ($User \to Tweet$).

(b) Bob's strategy ($Tweet \to User$).

Fig. 2. Two indexing strategies for Twitter

4   Architecture Overview

In this section, we describe an IR architecture that supports the OnADIQ model. The following architecture aims to manage different models (for different needs) and therefore several indexes.

Figure 3 presents the global architecture of our system. This architecture is composed of three modules: the Model Manager module, the Data Acquisition module and the Indexing/Querying module. The model manager is the component that manipulates personalized models: creating new models, updating and reading them in order to extract what information is required and how to process it. Data Acquisition module collects data from heterogeneous sources like web sites, private databases or sensors. Indexing/Querying module is the core of the system. The indexer uses the model manager module to drive its own process and it indexes resources as described in a personalized model. The Indexer achieves its goal in three steps:

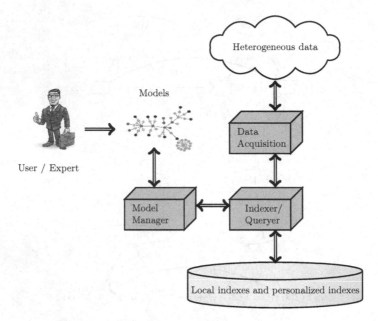

Fig. 3. Architecture overview for personalized IR

1. Given resources in the model, the indexer retrieves data by using the data acquisition module and computes local index of each resource. The data are then seen through the model.
2. The indexer traverses the model to compute $ALPHA, A, I$ and $BETA$ matrices and uses local indexes to compute WL matrix needed to solve the system of linear equations.
3. Finally, the module solves the system of linear equations and computes the W matrix. Then, W is transformed into a usable index.

Indexes computed by the indexer could be represented with an inverted file where each term points to a list of resources. Thereby, the querying module is the same as those used for classical IR.

5   Implementation and Results

We have developed a prototype to evaluate the OnADIQ model in several configurations. This prototype is developed with the Java language and we use Apache Lucene[3] for IR operations and Apache Jena[4] to manipulate ontologies and to reason on them. The prototype exposes an API to manage an instance of the OnADIQ model. SPARQL language[5] is used as the ontology query language.

---

[3] http://lucene.apache.org/
[4] http://jena.apache.org/
[5] http://www.w3.org/TR/rdf-sparql-query/

For this article, we present a direct application of our prototype which follows our illustrative example. Figure 4 presents the stages of our application. In this example, we index data from Twitter. Twitter data was collected over a period of one month using the java library Twitter4J and the dataset was constructed by choosing a set of 14 twitter users who post information about the mobile phone domain in general. We collect tweets, user descriptions and metadata (hashtags) from these 14 twitter users. The dataset is relatively small: 1K tweets, 14 twitter users, 100 hashtags. The data are stored in a database (MySQL) and indexes are computed using our system on top of Apache Lucene. Afterwards, the system automatically constructs a general model according to the OnADIQ definition given in section 3 (without setting importance nor permeability). Then we have simulated three indexing strategies (three personalized models): a first one where tweets are permeable to their authors ($User \rightarrow Tweet$, $\beta = 1$), a second where twitter users are permeable to their tweets ($Tweet \rightarrow User$, $\beta - \frac{1}{\lceil user \ u \ tweets \rceil}$) and a last one where virtual resources $hashtag$ are constructed to reinforce tweets in which they appear ($Hashtag \rightarrow Tweet$, $\beta = 0.2$ and $User \rightarrow Tweet$, $\beta = \frac{1}{\lceil user \ u \ tweets \rceil}$). The last strategy attaches a large permeability between $Hashtag$ and $Tweet$ to reflect the fact that a tweet is mainly indexed by its hashtags. In this context, the resource $hashtag$ refers to a virtual collection which is the aggregation of all tweets that contains the mentioned hashtag. For example, the hashtag "smartphone" refers to a resource which is constructed by all tweets containing the text "#smartphone". This construction leads to large and growing resources in term of textual representation whereas tweets and twitter user are short (140 characters for tweets and 160 for twitter users descriptions). For our experiment, we construct 40k queries consisting in pairs of terms (restricted to the first 300 terms ordered by $tf.idf$, i.e. the 300 most discriminant terms). Then, we evaluate the 40k queries on each personalized index. We average the number of resource types (tweet, twitter user, hashtag) in the top-10 results from the 40k queries for each index. As a reference, we also evaluate the queries on a model without permeability (denoted $Local$). Figure 5 shows these distributions.

First, we see that changing the model has an influence on the distribution of the first results. The local model returns an average of 5.5 results composed of 5.0 tweets and 0.5 users. In the $User \rightarrow Tweet$ model, we see that, when the indexes of the tweets are extended by user indexes, the tweets fit with more general queries; but because user indexes are constructed from a single small user description they remain still specific. On the contrary, when user indexes are extended with tweet indexes ($Tweet \rightarrow User$ model), the users fit with more general queries because a user may have plenty of tweets. In the $Hashtag \rightarrow Tweet$ and $User \rightarrow Tweet$ model, hashtags are added. Their indexes are even more general than user indexes. When a query involves general terms, hashtags (as defined above) are generally better answers than users and as those queries produce more than the limit of 10, users tend to be eliminated.

This experiment shows that our proposal enables a user to express a specific need and that answers match its requirements. This first experiment needs to be extended by IR standard measures (precision and recall); that is why a

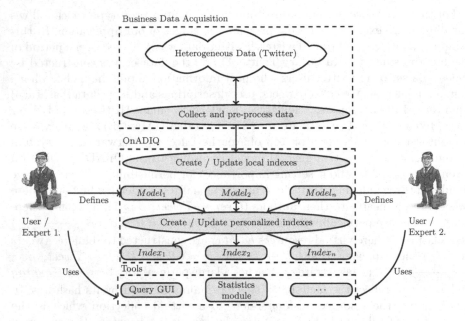

Fig. 4. The OnADIQ model applied on Twitter data

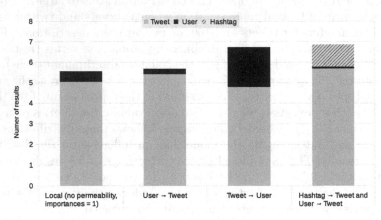

Fig. 5. Average number of tweets, users, hashtags in the top-10 results of 40k two terms queries

preliminary work is to build a dataset (and a groundtruth) like the one used in *TREC Microblog Track*[6] or *TREC Contextual Suggestion*[7], but with context and expressed at a higher level (with ontologies).

---

[6] https://sites.google.com/site/microblogtrack/
[7] https://sites.google.com/site/treccontext/

The OnADIQ model is already applied in an industrial context with our partner *Coexel*[8]. In this context, the same collection of resource is used for different user's needs. For example, some are interested in the technological developments of an area while others are looking for new actors in a specific field. In this sens, OnADIQ helps them to formalize their needs and its properties allow a fine grained approach, required in this domain. We have chosen to present another example (on Twitter data) for intellectual property reasons.

6 Conclusion and Future Works

In this work we present a global approach to address this issue. We applied it on the information retrieval problem in order to personalize for each user the indexing process. To achieve this objective, we propose OnADIQ: an ontology based model to represent resources of interest, their importance and their semantic relationships (permeabilities). The result is a solution that allows each user to construct its own index. These index are then used to query and retrieve existing resources (tweets, twitter users, ...). Because index are also enhanced by the ontology behind, the index allow to query resources that do not actually exist in the original resource collection but which are constructed according to the indexing model (ontology) of the user. The architecture and the current implementation are representative of the expressiveness of our proposal.

To improve this approach, we have to address two issues: the scalability and the amount of effort that the user must produce to construct its system. For the first point, each time a resource changes or each time the user changes its settings, the system have to re-index all resources. To tackle this issue, we are currently working on a probabilistic version of OnADIQ that uses bayesian networks to represents users' needs. One advantage of bayesian networks is that it is not necessary to recompute all the network when a single parameter changes: it's done natively by the network at the query time. For the second point, bayesian networks can be learned automatically from data (structure and probability tables) which significantly reduces the time spent by the user to configure its system. Furthermore, bayesian network are easily understandable and can be modified by a human. Finally, bayesian networks can easily and properly be integrated into a probabilistic information retrieval model to represent, for example, the prior probability of a resource. We will keep the ontology as the high-level model to represent users' needs.

Next step of our work is the development of an expressive language (i) to create and manipulate an OnADIQ model in a declarative way and (ii) to define rules on the model in order to create generic templates (for example, to define how importances and permeabilities are calculated). This kind of language has already been studied in [3] using *XPath*. Later, we will study a query language that combines both exact queries and IR to be able to construct highly expressive semantic queries.

---

[8] http://www.coexel.com/. This work is conducted as a part of a CIFRE Ph.D. thesis.

References

1. Abiteboul, S., Manolescu, I., Rigaux, P., Rousset, M.C., Senellart, P.: Web Data Management. Cambridge University Press (2011)
2. Beppler, F.D., Fonseca, F.T., Pacheco, R.C.S.: Hermeneus: An Architecture for an Ontology-Enabled Information Retrieval (2008)
3. Bruno, E., Faessel, N., Glotin, H., Le Maitre, J., Scholl, M.: Indexing and querying segmented web pages: the blockweb model. World Wide Web 14(5), 623–649 (2011), doi:10.1007/s11280-011-0124-6
4. Deerwester, S., Dumais, S.T., Furnas, G.W., Landauer, T.K., Harshman, R.: Indexing by latent semantic analysis. Journal of the American Society for Information Science 41, 391–407 (1990)
5. Fernández, M., Cantador, I., Lopez, V., Vallet, D., Castells, P., Motta, E.: Semantically enhanced information retrieval: An ontology-based approach. J. Web Sem. 9(4), 434–452 (2011)
6. W3C OWL Working Group. OWL Web Ontology Language: Overview, W3C Recommendation (2009)
7. Gruber, T.R.: Towards principles for the design of ontologies used for knowledge sharing in formal ontology in conceptual analysis and knowledge representation. Kluwer Academic Publishers (1993)
8. Manning, C.D., Raghavan, P., Schtze, H.: Introduction to Information Retrieval. Cambridge University Press, New York (2008)
9. Maron, M.E., Kuhns, J.L.: On relevance, probabilistic indexing and information retrieval. J. ACM 7, 216–244 (1960)
10. Mylonas, P., Vallet, D., Castells, P., Fernández, M., Avrithis, Y.: Personalized information retrieval based on context and ontological knowledge. Knowledge Engineering Review 23(1) (2008)
11. Ponte, J.M., Croft, W.B.: A language modeling approach to information retrieval. In: SIGIR 1998: Proceedings of the 21st Annual International ACM SIGIR Conference on Research and Development in Information Retrieval, pp. 275–281. ACM Press, New York (1998)
12. Robertson, S.E., Walker, S., Jones, S., Hancock-Beaulieu, M.M., Gatford, M.: Okapi at trec-3, pp. 109–126 (1996)
13. Salton, G., Wong, A., Yang, C.S.: A vector space model for automatic indexing. Commun. ACM 18, 613–620 (1975)
14. Singhal, A., Buckley, C., Mitra, M., Mitra, A.: Pivoted document length normalization, pp. 21–29. ACM Press (1996)
15. Turtle, H., Croft, W.B.: Inference networks for document retrieval. In: Proceedings of the 13th Annual International ACM SIGIR Conference on Research and Development in Information Retrieval, SIGIR 1990, pp. 1–24. ACM, New York (1990)

UPnQ: An Architecture for Personal Information Exploration

Sabina Surdu[1], Yann Gripay[1], François Lesueur[1], Jean-Marc Petit[1],
and Romuald Thion[2]

[1] INSA-Lyon, LIRIS, UMR5205
F-69621, France
`firstname.lastname@liris.cnrs.fr`
[2] Université Lyon 1, LIRIS, UMR5205
F-69622, France
`firstname.lastname@liris.cnrs.fr`

Abstract. Today our lives are being mapped to the binary realm provided by computing devices and their interconnections. The constant increase in both amount and diversity of personal information organized in digital files already turned into an information overload. User files contain an ever augmenting quantity of potential information that can be extracted at a non-negligible processing cost. In this paper we pursue the difficult objective of providing easy and efficient personal information management, in a file-oriented context. To this end, we propose the Universal Plug'n'Query (UPnQ) principled approach for Personal Information Management. UPnQ is based on a virtual database that offers query facilities over potential information from files while tuning resource usage. Our goal is to declaratively query the contents of dynamically discovered files at a fine-grained level. We present an architecture that supports our approach and we conduct a simulation study that explores different caching strategies.[1]

Keywords: files information overload, personal information management, potential information, declarative file querying, wrappers.

1 Introduction

Computers have triggered a significant paradigm shift in our lives. Our daily existence is mapped to the binary realm provided by interconnected computing devices. Everyday actions, events, activities are now translated into personal data, most of them being organized in personal files: personal maps of morning joggings in the park, digital playlists of songs from our parents' vinyl record collections, personal health & social care information in Electronic Health Record systems, *etc*. Interactions between citizens and public administration agencies are being progressively dematerialized too [6].

Our research is motivated by the drawbacks of current personal file data management technology. There is a large amount of heterogeneous files storing personal data such as

---

[1] This work is partially funded by the KISS Project (ANR-11-INSE-0005) of the French National Research Agency (ANR).

H. Decker et al. (Eds.): DEXA 2014, Part I, LNCS 8644, pp. 257–264, 2014.
© Springer International Publishing Switzerland 2014

videos, music, semi-structured documents, images, etc. There is also a growing number of technologies that can extract interesting knowledge from these files: image and video processing, data and text mining, machine learning, speech recognition, image pattern recognition, musical analysis, *etc.* Managing user files in this context becomes cumbersome for both developers and end-users. Developers encounter time-consuming difficulties to write applications handling heterogeneous files, while end-users find it difficult to search through all their data to retrieve useful information. Dedicated Personal Information Management (PIM) systems can ease information management on specific domains but usually lack file content querying capabilities. In our previous benchmark [12], we show how data-oriented pervasive application development gets significantly simplified when using declarative queries. Yet, to the best of our knowledge, there are no declarative query languages that can provide application developers with the ability of writing homogeneous queries over heterogeneous user files, that could be the core building block of powerful file-oriented personal data management applications.

We want to be able to pose fine-grained declarative queries at different levels of granularity: structure, metadata and content. We identify the following main challenges: (1) structure heterogeneous files into a homogeneous model; (2) provide a homogeneous interface for different data extraction technologies; (3) use an efficient query execution strategy allowing to query large amounts of files as soon as possible; (4) design a high level declarative query language, abstracting file access issues and enabling optimization techniques.

This paper introduces the UPnQ user files data management approach: a *virtual* database that offers query facilities over *potential data* from files that can be obtained on demand by dedicated wrappers. To this end, we propose a homogeneous representation of heterogeneous files, regardless of their file format, and fine-grained query facilities: a *UPnQ file* is a *physical file* viewed through a *wrapper* that understands its format and semantics. Like SQL over relational DBMSs for most applications, UPnQ is designed to ease application development rather than to serve as a direct end-user interface. This work has been carried out in the framework of the ongoing KISS project[2], devoted to managing personal information on secure, portable embedded devices called Personal Data Servers (PDSs).

The *file model* and the *query language* are introduced in Section 2. The architecture of the UPnQ system is detailed in Section 3. Different query processing strategies are evaluated in Section 4. Section 5 positions our approach with respect to related work.

2   Model and Language

To easily build applications over physical files, we first propose a homogeneous data-centric representation of such files in non-first normal form. Subsequently, we define an SQL-like file-oriented query language, the *UPnQ file query language*. The full expressiveness of SQL can then be applied when querying data from files: join queries can examine related data from different files, aggregate queries can compute statistics over a set of files, *etc.*

---

[2] https://project.inria.fr/kiss/en

2.1 Data Model

Data from UPnQ files are hierarchically organized in *capsules* and *seeds*. The outermost capsule of a file holds all the data from the file. A capsule can contain seeds and / or lists of inner capsules. Seeds provide atomic data. The capsule-and-seeds terminology reflects the (nested) non-first normal form relational representation of files: it enables a coherent description of heterogeneous data in different files.

The schema of a capsule can be described by a tree data structure, whose root is the capsule's name. Figure 1(a) shows the schema of a *Song* capsule. The root node and intermediary nodes (squares on the figure) are capsule names, and leaf nodes (circles) are seed names. Intermediary nodes shown between curly braces represent lists of capsules at the data level, i.e., the instance level of a UPnQ file. Within a capsule, seeds receive atomic values, e.g., *Artist* might have a value of "Beatles", *etc.*

The type of a file is determined by the name of the outermost capsule. A file can also be individually associated with a set of tags, that are usually user-defined and provide further information about the file as seen through the user's eyes. Figure 1(b) shows two UPnQ files (*Yesterday* and *Mr. Tambourine Man*) on user Tom Sawyer's device. Both files have a *Song* file type, due to their outermost capsule. They also contain atomic data at the *Song* capsule level with seeds *Artist, Title, Tempo, etc.* The *Song* capsule also contains a list of *Album* capsules. Each *Album* capsule describes the *Title* and *RecordLabel* of the album, and inner capsules for *Awards* earned by the album.

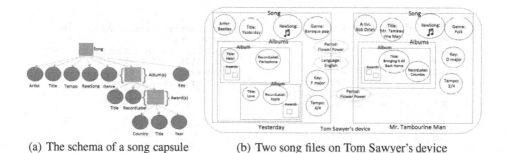

(a) The schema of a song capsule   (b) Two song files on Tom Sawyer's device

Fig. 1. Schema and Instances of Song capsules

2.2 Query Language

We want to express fine-grained queries on data from files sharing a common structure of interest. The query engine is responsible for selecting files providing wanted capsules, and for extracting capsule data from files through corresponding wrapper calls. We extend the SoCQ system [10] to this purpose; the resulting engine allows us to perform fine-grained, lazy data extraction from files, in a relational setting, using SoCQ's *binding patterns* and *virtual attributes* constructs to build XD-Relations (eXtended Dynamic Relations). For UPnQ, SoCQ's concept of SERVICE is replaced with CAPSULE to interact with wrappers through binding patterns in order to provide values for virtual attributes. Our goal is to achieve the UPnQ vision with a UPnQ file engine thanks to the core mechanism of the SoCQ system.

A relational view on file sets. We define a *capsules translation* operator that builds a XD-Relation representing a set of files providing some common features. The purpose of this operator is to translate capsules into a relational, SoCQ-queryable view. We also define a *tags translation* operator, that builds a relation with values for a set of tags associated to a set of files.

Listing 2(a) describes the UPnQ query that creates an XD-Relation based on music files providing *Artist*, *Title*, *Key* and *Album* capsules. Assume the file repository contains three songs. The resulting XD-Relation *FPSongs* is shown in Figure 2(c). There are three capsules that contain capsules *Artist*, *Title*, *Key* and *Album*, all of them being *Song* capsules in this case. We have therefore one tuple for each *Song* capsule. Data are however not materialized yet, as attributes are virtual in this XD-Relation.

Querying files in UPnQ. The next step is to query files' capsules using the binding patterns associated to XD-Relations to retrieve data for virtual attributes. Using previous XD-Relation *FPSongs*, Listing 2(b) describes a UPnQ query that materializes data concerning the Artist, song Title, and the Album Title. Through binding patterns, a query can interact with wrapper invocations and control their behavior. The resulting XD-Relation is depicted in Figure 2(d). Some data, here the albums' record label, do not need to be retrieved from the files for this query.

```
CREATE RELATION FPSongs (
    Song CAPSULE,
    Artist STRING VIRTUAL,
    Title STRING VIRTUAL,
    Album CAPSULE VIRTUAL,
    Album.Title STRING VIRTUAL,
    Album.RecordLabel STRING VIRTUAL )
USING BINDING PATTERNS (
    Artist[Song] ( ) : ( Artist ),
    Title[Song] ( ) : ( Title ),
    Album[Song] ( ) : ( Album,
        Album.Title, Album.RecordLabel ) )
AS TRANSLATE CAPSULES
    PROVIDING Artist, Title, Key, Album
```

(a) a UPnQ Capsules Translation Query

```
SELECT Artist, Title, Album.Title
FROM FPSongs
WHERE Title = "Mr. Tambourine Man"
    or Title = "Yesterday"
USING BINDING PATTERNS
    Artist, Title, Album;
```

(b) a UPnQ Data Extraction Query

Song	Artist	Title	Album	Album.Title	Album.RecordLabel
music: Yesterday.mp3	*	*	*	*	*
music: MrTambourineMan.ogg	*	*	*	*	*
music: DearPrudence.mp3	*	*	*	*	*

(c) XD-Relation result of the UPnQ Capsules Translation Query (* is absence of value)

Artist	Title	Album.Title
Bob Dylan	Mr. Tambourine Man	Bringing It All Back Home
Beatles	Yesterday	Help!
Beatles	Yesterday	Love

(d) XD-Relation result of the UPnQ Data Extraction Query for two songs

Fig. 2. Examples of UPnQ queries – Capsules Translation & Data Extraction

3   Enabling Technologies and Architecture

UPnQ Wrappers. The UPnQ system relies on *parameterizable wrappers* that can be invoked to extract pieces of data from various physical files and expose them, in a homogeneous manner. Dynamic *UPnQ wrappers* present two innovative features: 1) they allow fine-grained data management, and 2) they can be invoked on files on demand, i.e., when a query requires data from a physical file.

There's a multitude of APIs out there that allow manipulating various file formats. The iCal4j API allows the manipulation of iCalendar files, the Apache POI API provides support for Microsoft Office files, *etc.* The development of such APIs supports the fact that our UPnQ wrapper vision is feasible. For these wrappers, we envision an ecosystem of downloadable plugins, like today's application stores.

UPnQ Queries UPnQ files are mapped to relational tuples. Some of the reasons for choosing to project a file environment on the relational canvas can be drawn from [9]: the simplicity, the robustness and the power of the model, which are well-known in the database community for a long time.

The goal of the UPnQ system is to express declarative queries on top of UPnQ files, so queries need to be given control over imperative wrappers. We also need to provide a virtual database over files, where data can be extracted using both lazy and eager approaches. We turn our attention to Pervasive Environment Management Systems (PEMSs) like SoCQ [10] or Active XML [3]. Both systems can homogeneously query distributed heterogeneous entities and allow lazy evaluation of remote service methods. SoCQ is here preferred to Active XML for its SQL-like language.

Architecture. Figure 3 shows the main components of the UPnQ system. The *UPnQ file system* manages the UPnQ wrappers and files, i.e., the available wrappers and files in the system. The *UPnQ relational file engine* contains a query engine that handles relations, smart folders (subset of files defined by dedicated queries) and a cache. During query execution, the engine interacts with the UPnQ file system to get data from physical files via wrappers, potentially using the cache to improve response time.

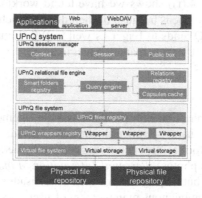

Fig. 3. Architecture of the UPnQ system

4   Experiments on the Query Execution Strategies

We conducted a simulation study to comparatively assess the performance of four different query processing, measuring query execution time, query response time and cumulative CPU usage (unit is an abstract *symbolic instant* (si)). The four assessed strategies are *Eager*, *Lazy*, *UPnQ* and *A-UPnQ* (for *Active UPnQ*). Eager is the typical case of PIM where every file is preprocessed and stored in a database, leading to a high bootstrap cost. Lazy is the opposed case where data is only fetched when needed, leading to a long response time. UPnQ adds a capsule-level caching method to the lazy strategy, improving the response time for queries exhibiting some redundancies. A-UPnQ adds prefetching to UPnQ, allowing to seamlessly query files which have been preprocessed or not and achieving in the end the query response times of the eager strategy.

We simulate a file environment of 50000 files with 100 different file types (10 to 100 capsules). A workload is composed of 1800 queries that examines 1 to 30 capsules from 1 to 1000 files, based on two Zipf popularity curves. Queries arrive at the system following a Poisson process, interleaved with 200 random insertions of 1 to 1000 new files and 100 random updates of 1 to 100 files.

Figure 4 depicts the obtained results for 2 opposite workloads. In Workload W1, queries arrive fast at the system and are likely to overlap to a significant extent. In Workload W2, queries arrive less frequently at the system and have a small overlap. For both workloads, we simulated smart wrappers on expensive files.

As expected, the CPU cost per query (Figures 4.(a) and 4.(b)) is bounded by the Eager and the Lazy extreme strategies. For A-UPnQ, because the system has more idle time in W2, we can see the CPU cost decreases more abruptly than for W1, since the cache fills faster. For both workloads, A-UPnQ eventually catches up with eager, running queries on files that are entirely stored in the cache. The reason UPnQ decreases slower in W2, is that queries don't overlap as much as they do in W1, so cache hits are less likely. The cumulated CPU cost (Figure 4.(c) and Figure 4.(d)) shows the bootstrap disadvantage of Eager, the same for both workloads. A-UPnQ is bounded by Eager. UPnQ has the best performance in this case, since it only processes data retrieval queries and cache invalidation during updates. Query response times for both workloads (Figure 4.(e) and Figure 4.(f)) shows we have found workloads for which A-UPnQ outperforms Eager. This happens because of their different behaviors when processing inserts and updates. UPnQ tends to Eager as well in W2 (Figure 4.(f)). When queries are fast and overlap more, UPnQ surpasses Eager in an obvious manner (Figure 4.(e)).

5   Related Work

File systems are not a viable choice for personal data management, since searching files is either restricted to their name, physical location, or metadata maintained by the file system, or performed on format-dependent tags content by dedicated applications (ID3 for music files, EXIF for photos). *Personal Information Managers* (PIM) are becoming increasingly popular, but display in turn their own disadvantages. Some PIMs like PERSONAL [2] offer a homogenous management of user files but stick to a file granularity for querying. Others like EPIM [1] and MyLifeBits [8] manage more abstract

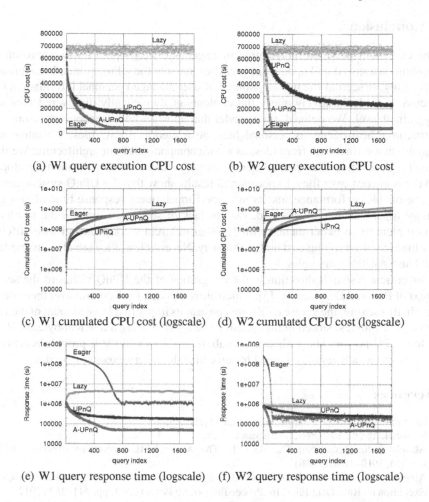

(a) W1 query execution CPU cost   (b) W2 query execution CPU cost

(c) W1 cumulated CPU cost (logscale) (d) W2 cumulated CPU cost (logscale)

(e) W1 query response time (logscale) (f) W2 query response time (logscale)

Fig. 4. Simulation of workloads W1 and W2 with expensive files and smart wrappers (500 runs)

user items and their relations, and then lose the link with concrete user files. Many PIMs can indeed do a great job at easing information management as compared to file systems, but they do so on specific, delimited islands of data, and each PIM manages information in its own, personal way, usually lacking file content querying capabilities. The compartmentalised PIM support issue is raised in [7].

Finally, we mention SQLShare [11] and NoDB [4]. SQLShare is an ad hoc databases management platform which enable SQL querying over data scattered across multiple files. NoDB is big-data-oriented and maintains the main features of a DBMS without requiring data loading: *data-to-query* time is minimized up to the point where query processing directly reads raw files. Both systems however do not address issues related to file-specific model or query language (only classic SQL), and do not propose an architecture to deal with the many file types found in a user repository (only CSV files).

6 Conclusion

In the context of personal information management, we propose a new approach for managing data stored and updated in user files, namely the UPnQ principled approach, as user files are seen as the primary source of user-managed personal information. Our objective is to declaratively query the contents of dynamically discovered files at a fine-grained level. We defined a file model that allows a homogeneous representation of structured heterogeneous personal files, abstracted as UPnQ files. We introduced wrappers that extract data from files, as a core component of our architecture. We then defined a declarative file-oriented query language that allows application developers to express queries over files. Experimental results show that the UPnQ system can attain reasonable performance, and even outperformed query response time of an eager strategy for a reasonable workload. Due to lack of space, we omitted the results of our complete set of experiments, with 16 other different workloads and with different combinations of wrapping cost and granularity. Nevertheless, they are also favorable to UPnQ and A-UPnQ strategies.

Our current research direction is the integration of the UPnQ vision in the secure context of Personal Data Servers [5], which aims at giving back control over their files to users. In this setting, the severe hardware constraints imply that large amount of data (up to several Gigabytes) stored in files needs to be queried using a few Kilobytes of RAM and low CPU power. The embedded database engine should still provide acceptable performance, which entails the need for an embedded query execution model.

References

1. Essential PIM, http://www.essentialpim.com/
2. Personal, https://www.personal.com
3. Abiteboul, S., Benjelloun, O., Milo, T.: The Active XML project: an overview. VLDB J. 17(5), 1019–1040 (2008)
4. Alagiannis, I., Borovica, R., Branco, M., Idreos, S., Ailamaki, A.: NoDB: Efficient Query Execution on Raw Data Files. In: Proceedings of SIGMOD 2012, pp. 241–252 (2012)
5. Allard, T., Anciaux, N., Bouganim, L., Guo, Y., Folgoc, L.L., Nguyen, B., Pucheral, P., Ray, I., Ray, I., Yin, S.: Secure Personal Data Servers: a Vision Paper. PVLDB 3(1), 25–35 (2010)
6. Anciaux, N., Bezza, W., Nguyen, B., Vazirgiannis, M.: MinExp-card: limiting data collection using a smart card. In: Proceedings of EDBT 2013, pp. 753–756 (2013)
7. Boardman, R.: Workspaces that Work: Towards Unified Personal Information Management. In: Proceedings of the 16th British HCI Conference, vol. 2 (2002)
8. Gemmell, J., Bell, G., Lueder, R.: MyLifeBits: a personal database for everything. Commun. ACM 49(1), 88–95 (2006)
9. Gripay, Y.: A Declarative Approach for Pervasive Environments: Model and Implementation. PhD thesis, Institut National des Sciences Appliquées de Lyon (2009)
10. Gripay, Y., Laforest, F., Petit, J.-M.: A Simple (yet Powerful) Algebra for Pervasive Environments. In: Proceedings of EDBT 2010, pp. 359–370 (2010)
11. Howe, B., Cole, G., Khoussainova, N., Battle, L.: Automatic example queries for ad hoc databases. In: Proceedings of SIGMOD 2011, pp. 1319–1322 (2011)
12. Surdu, S., Gripay, Y., Scuturici, V.-M., Petit, J.-M.: P-bench: Benchmarking in data-centric pervasive application development. In: Hameurlain, A., Küng, J., Wagner, R., Amann, B., Lamarre, P. (eds.) TLDKS XI. LNCS, vol. 8290, pp. 51–75. Springer, Heidelberg (2013)

Probabilistic Reuse of Past Search Results

Claudio Gutiérrez-Soto[1,2] and Gilles Hubert[1]

[1] Université de Toulouse, IRIT UMR 5505 CNRS
118 route de Narbonne, F-31062 Toulouse cedex 9
[2] Departamento de Sistemas de Información
Universidad del Bío-Bío, Chile

Abstract. In this paper, a new Monte Carlo algorithm to improve precision of information retrieval by using past search results is presented. Experiments were carried out to compare the proposed algorithm with traditional retrieval on a simulated dataset. In this dataset, documents, queries, and judgments of users were simulated. Exponential and Zipf distributions were used to build document collections. Uniform distribution was applied to build the queries. Zeta distribution was utilized to simulate the Bradford's law representing the judgments of users. Empirical results show a better performance of our algorithm compared with traditional retrieval.

1   Introduction

A wide range of approximations in information retrieval (IR) are devoted to improving the list of documents retrieved to answer particular queries. Among these approaches, we can find solutions that involve efficient assignments of systems to respond to certain types of queries, by applying data mining techniques [1]. Nonetheless, some tasks of data mining can imply not only long periods of time, but also a high cost in money [2]. In addition, solutions that involve an exhaustive analysis of all possible alternatives to find the best answer to a query (i.e., the best precision for each type of query) can be found in IR context. Prior solutions correspond to approaches based on learning techniques (e.g., neural networks, genetic algorithms, and machines support vectors). However, these approaches should imply a high cost in learning time as well as diverse convergence times when the datasets used are heterogeneous [3]. Additionally, characteristics, such as the scopes where these types of algorithms are applied and the performance achieved in different environments, are complex to address [4].

In the IR literature, two types of approaches used in the context of past queries are easily identifiable. The first approaches are based on TREC collections. Most of these approaches use simulation to build similar queries with the aim to provide a suitable framework of evaluation. The second type of approaches rooted in the use of historical queries on the Web, most of which are supported on repetitive queries. As a result, having ad-hoc collections which allow to evaluate the use of past queries in an appropriate way, is a hard task. Therefore, one way to provide an ad-hoc environment for approximations based on past queries is simulation.

H. Decker et al. (Eds.): DEXA 2014, Part I, LNCS 8644, pp. 265–274, 2014.
© Springer International Publishing Switzerland 2014

Our main contribution is a Monte Carlo algorithm, which uses relevant documents from the most similar past query to answer a new query. The algorithm splits the list of retrieved documents from the most similar past query in subsets of documents. Our algorithm is simple to implement and effective. Moreover, it does not require learning time. Documents, query collections, and relevance judgments of users were simulated to built a dataset for evaluating the performance of our algorithm. A wide range of experiments have been carried out. We have applied the Student's paired t-test to support the experimental results. Empirical results show better results of our algorithm (in particular the precision P@10) than traditional retrieval.

The paper is organized as follows. In section 2, related works on past searches, randomized algorithms, and simulation in IR context are presented. In section 3, we present our approach to simulate an IR collection, in the context of past search results. Section 4 details our approach using past search results, with mathematical definitions. In section 5, empirical results are described. Finally, conclusions are presented in section 6.

2   Related Work

Two categories of approaches employed in the context of past queries are easily identifiable. The first category is based on TREC collections. In a recent work [5], a distributed approach is presented in the context of past queries. Similar queries are simulated from a traditional set of queries. Moreover, the judgments of users are omitted. In [6], two strategies aiming at improving precision were implemented. The first strategy corresponds to the combination of results from previous queries, meanwhile the second implies the combination of query models. An extended work is exposed in [7]. The authors address models based on implicit feedback information to increase precision. Implicit feedback information is given by queries and clickthrough history in an active session. It is important to emphasize that TREC collections used here have been modified to evaluate approximations based on past queries.

The second category of approaches focuses on log files in the context of the Web. In [8], an automatic method to produce suggestions based on previously submitted queries is presented. To achieve this goal, an algorithm of association rules was applied on log files. The 95 most popular queries were considered. Nonetheless, the percentage of these 95 queries over 2.3 millions of records is unknown. Hence, it is infeasible to estimate the impact of this approximation. Moreover, [9] claims that there is no easy way to calculate the real effect of approximations founded on association rules. It is mainly due to the complexity to determine the successive queries that belong to the same session (i.e., for the same user). In [10], an approximation based on repeated queries is exposed. The aim is the identification of identical queries executed in the same trace. In [11], two contributions, which take advantages from repeated queries, are presented. The first contribution is aligned on efficiency in execution time and the second is focused on repetitive document access by the search engines.

Simulation to evaluate information retrieval systems (IRSs) is presented as a novel branch of research [12]. Simulation in IR is an automatic method, where documents, query collections, and judgments of users can be built without user intervention [13].

In addition, the IR literature is crammed with contributions based on probabilistic algorithms. The major part of probabilistic algorithms in IR can be categorized in two classes, learning techniques and optimization. Typically, approximations rooted in learning techniques involve the use of Bayesian Networks and their variants. The PrTFIDF algorithm, which is a probabilistic version of TFIDF algorithm is presented in [14]. PrTFIDF provides a new vision of vector space model, where the theorem of total probability, the Bayes' theorem, and a descriptor for every document are used. Final results show a better performance than TFIDF. In [15], the classification of documents in an unsupervised manner is carried out. It uses Poisson distribution according to the query or topic.

Several optimization techniques involve the use of Genetic Algorithms (GA). Inspired by the formula proposed by Salton [16] (where the term weights for documents and queries are the product between the term frequency multiplied by an inverse collection frequency factor), a new fitness function is presented in [17]. Both, vectors of documents and queries are normalized by using the formula. Experimental results show better effectiveness when using this approach than traditional retrieval (i.e., using cosine distance). Eventually, the *Probfuse* algorithm proposed in [18], whose aim is to combine results from several IR algorithms, outperforms the widely used CombMNZ algorithm.

Different to Bayesian Networks and GA, Monte Carlo and Las Vegas algorithms are used usually when the problem is hard to solve like NP problems or when algorithm input is non-deterministic. Las Vegas algorithms provide an answer, which is always correct and where in the worst case the execution time is the same as the deterministic version. In contrast to Las Vegas algorithms, Monte Carlo algorithms give an answer, which can be incorrect (i.e., the algorithm returns *true*, when the answer should be *false*, or vice-versa). When one of these answers is correct, it is called true-biased (the correct answer is *true*) or false-biased (the correct answer is *false*). When both answers can be incorrect, it is called two-sided errors.

Our Monte Carlo algorithm corresponds to the type *two-sided errors*. This is due to the fact that we are not sure about judgments of users with respect to whether a document is either relevant or not relevant regarding the query. Nonetheless, we assume that documents that appear at the top of the result list have more probability to be relevant than documents that appear at the bottom of the list.

3   Simulating IR Collections

Our method consists of two steps, based on prior work [19]. The first step aims at creating terms, documents, and queries. Both Heaps' and Zipf's laws are considered to build document collections. We assume that both processes, elimination of stop words, and stemming were carried out. Due to terms which compose a document can belong to several subjects, Zipf's law is applied to select terms from topics [20]. Exponential distribution can be applied as an alternative to Zipf's law. Then, past queries are created from documents and new queries are built from past queries. In the final step, to simulate judgments provided by users about relevance of documents for a specific query, Bradford's law is applied [21].

The most basic element that composes a document is a term. A term is composed of letters from the English alphabet. Both documents and queries are composed of terms. Each document is unique. Past queries are built from documents and their intersections are empty. A topic (i.e., subject of documents) is defined by terms. Several topics are used to built a document. The intersection among topics is empty. Aiming to build documents, Zipf and Exponential distributions are used to select terms from different topics. Uniform distribution is used to select documents, where terms are selected to built the past queries. Then, past queries are built from the documents. New queries are made up from past queries by either adding a new term or deleting a term. Bradford's law has been applied through Zeta distribution. In order to explain the mechanism to obtain the precision, we can assume two lists of documents. The first list of documents (for the query 1) is composed of the documents $d_1(1), d_3(1), d_5(1), d_6(0), d_8(0), d_9(0)$ where $d_i(1)$ is a relevant document, meanwhile $d_i(0)$ is irrelevant. In the same way, the second list of documents is composed by the documents $d_2(0), d_3(1), d_5(1), d_6(0), d_7(1), d_{10}(0)$. Thus, first, a subset of common documents is found (d_3, d_5 and d_6). Second, from the common subset, relevant documents are determined for both queries by using Bradford's law (d_3 and d_5). Third, Bradford's law is applied for each list by conserving the relevant documents that belong to the common subset (for the first list d_1, d_3 and d_5 are relevant documents). As a consequence, the precision (in our case P@10) is different for both queries.

4   Retrieval Using Past Queries

At the beginning, each submitted query is saved with its documents. Afterwards, each new query is compared with the past queries stored in the system. If there is a past query quite similar, then the relevant documents are retrieved from the most similar past query using our algorithm. Broadly speaking, our algorithm divides the list of documents retrieved from the past query, in groups of power two. For example, if the list of documents comprises 30 documents, the number of documents will be rounded up to the next number in power two, i.e., $n = 32$. Later on, groups of documents are defined as follows. The first group comprises 2^0 documents. The second group involves 2^1 documents, the third group is composed of 2^2 documents, and so on, in such a way that the sum of the documents does not outperform $n = 32$. Thus, the number of groups is 5. The biggest group is composed of documents that appear in the first positions (between the position 1 and 16). The next biggest group comprises documents that appear from the position 17 to 24, and so on. The likelihood of a document to be relevant is determined by two factors: the group it belongs to and its position in the group. The algorithm and a more detailed example are displayed in the next sections.

4.1 Definitions and Notations

Let DB be an IR dataset, composed of a set of documents D, and a set of past queries Q. Besides, let Q' be the set of new queries. $V_N(q)$ is a set of N retrieved documents given q, and $sim(q, d_j)$ is the cosine distance between query q and the document d_j. Besides, $V_N(q) = A(q) \cup A'(q)$, where $A(q)$ corresponds to the set of all relevant documents for the query q. $A'(q)$ is the set all irrelevant documents for the query q. $C = \bigcup c(q, V_N(q))$ is the set of all retrieved documents with their respective queries.

Definition 1. $\partial : R_c(q') \to A(q')$ *is a function, which assigns the most relevant documents to the new query q', such as $q' \in Q'$ and $R_c(q')$ corresponds to a set of retrieved documents, from the most similar past query. (see Definition 2 and Definition 4).*

In addition, let $\|x\|$ be the integer part of a real number x, $\lceil x \rceil$ corresponds to the upper integer of x and $\lfloor x \rfloor$ corresponds to the lower integer of x. $B[N]$ is a binary array such as B has N elements, and $\frac{a}{b}$ is the proportion of values in B (see *Algorithms 1*, lines from 9 to 12), which have the value 1 (true). This array is the base to provide a level of general probability for all documents. Nevertheless, the probability of each document according to the position in $V_N(q)$, is computed by the *Algorithm 1*.

Definition 2. $M(N) = min\{m \mid m \in \mathbb{N} \wedge \sum_{k=0}^{m} 2^k \geq N \wedge N < 2^{m+1}\}$ *be the upper bound set, which involves documents of $V_N(q)$ (in power two) (see Algorithm 1, lines from 2 to 4).*

Definition 3. *Let i be the position of a document in $V_N(q)$, such as the first element ($i = 1$) represents the most similar document, then $f_x(i, N) = min\{x \mid x \in \mathbb{N} \wedge i \leq \sum_{k=1}^{x} \frac{2^{M(N)}}{2^k}\}$, corresponds to the number of set assigned for the document i (see Algorithm 1, line 5).*

Definition 4. *Let*
$$v(i, N) = (2^{M(N)-f_x(i,N)} - 1) - [((\sum_{k=1}^{f_x(i,N)} 2^{M(N)-k}) - i) mod(2^{M(N)-f_x(i,N)})]$$
be the value assigned to i, from 0 to $2^{M(N)-f_x(i,N)}$ (see Algorithm 1, line 15).

Definition 5. $\Phi(i, N) = log_2(2^{M(N)-f_x(i,N)} - v(i, N)) - \|log_2(2^{M(N)-f_x(i,N)} - v(i, N))\|$ *a decimal number, which is $[0, 1[$ (see Algorithm 1, lines from 16 to 19).*

Definition 6. *Let*
$$K(i,N) = \begin{cases} \lceil \Phi(i,N) \rceil * \|M(N) - f_x(i,N)\| & : & if & \Phi(i,N) \geq 0.5 \\ \lfloor \Phi(i,N) \rfloor * \|M(N) - f_x(i,N)\| & : & if & \Phi(i,N) < 0.5 \end{cases} \text{ be the num-}$$
ber of iterations to look for a hit in the array B (see Algorithm 1, lines from 20 to 25).

Thus, $\beta : F(i) \rightarrow \{0,1\}$, is the *hit* and *miss* function.
$$F(i) = \begin{cases} 1 & : & Pr_i(1) = \sum_{l=1}^{K(i,N)} \frac{2^{\|M(N) - f_x(i,N)\|}}{(2^{M(N)})^l}, \\ 0 & : & Pr_i(0) = 1 - Pr_i(1) \end{cases}$$

where $Pr_i(1)$ is the probability of a *hit* (1), and $Pr_i(0)$ corresponds to the probability of a *miss* (0) for the element i.

Our algorithm works as follows. The list of retrieved documents is split in subsets of elements in power two. In our case, $V_N(q)$ has 30 documents, however it can be approximated to 32 documents. Thus, if we apply *Definition 2*, then $M(N) = 5$. Therefore, $V_N(q)$ is split in 5 subsets. In general terms, $Pr_i(1)$ for every subset is different. Specifically on $\frac{2^{\|M(N) - f_x(i,N)\|}}{(2^{M(N)})}$. Thus, the space of possible candidates for the first subset is $\frac{2^{\|5-1)\|}}{(2^5)} = \frac{1}{2}$, for the second subset is $\frac{2^{\|5-2)\|}}{(2^5)} = \frac{1}{4}$ and so on. To show how the probability decreases according to the subsets, two examples are provided, for the first subset and the third subset. The second element of the first subset is $i = 2$, thus applying the *Definition 3*, $f_x(2,30) = 1$, therefore, $v(2,30) = (2^{5-1} - 1) - [\langle (\sum_{k=1}^{1} 2^{5-1}) - 2 \rangle) \bmod 2^{5-1}] = 1$ (see *Definition 4*). Applying *Definition 5*.
  $\Phi(2,30) = log_2(2^{5-1} - 1) - \|log_2(2^{5-1} - 1)\| = 3.906 - 3 = 0.906$. Applying *Definition 11*, $K(2,30) = \lceil \Phi(2,30) \rceil * \|5 - 1\| = 1 * 4 = 4$.
  Thus, $Pr_2(1) = \sum_{l=1}^{4} \frac{2^{\|5-1\|}}{(2^5)^l} = 0.757$. In the same way, $v(26,30) = (2^{5-3} - 1) - [\langle (16 + 8 + 4) - 26 \rangle) \bmod 2^{5-3}] = 1$. Finally, $Pr_{26}(1) = \sum_{l=1}^{2} \frac{2^{\|5-3\|}}{(2^5)^l} = 0.128$

5 Experiments

5.1 Experimental Environment and Empirical Results

The experimental environment was instantiated as follows. The length of a term is between 3 and 7. The length was determined using Uniform distribution. The total number of terms used in each experiment corresponds to 800. A document can contain between 15 and 30 terms. The number of topics used in each experiment is 8. Each topic is defined by 100 terms. Each experiment used 800, 1600, 2400, 3200, and 4000 documents. Terms for a query were between 3 and 8. We built 15 past queries from documents. From the set of past queries, 15 new queries were built. Thereby, we used 30 distinct queries. Simulations were implemented on C language and run on Linux Ubuntu 3.2.9, with Centrino 1350, 1.8 Ghz Intel processor, 1GB RAM, and gcc 4.6.3 compiler.
  Three experimental scenarios were defined. For the first and second experiments, Exponential distributions (with parameters $\theta = 1.0$ and $\theta = 1.5$) were

Algorithm 1. $B[N], A_{Past}(q), V_N(q), q'$

Require: $B[N]$ is a boolean array, $A_{Past}(q)$ is a set of relevant documents for the query q, $V_N(q)$ is the set of retrieved documents for the query q, q' is the most similar query for q
Ensure: $A_{New}(q')$ is a set of relevant documents for the query q'
1:   $A_{New}(q) \leftarrow \emptyset$
2:   **for** $i \leftarrow 0, sum \leftarrow 0, sum < N+1$ **do**
3:        $sum \leftarrow sum + 2^i$
4:   **end for**
5:   $k \leftarrow i - 1$
6:   **for** $i \leftarrow 1, N$ **do**
7:        $B[i] \leftarrow false$
8:   **end for**
9:   **for** $i \leftarrow 1, \frac{N}{2}$ **do**
10:       $j \leftarrow random(1, ..., N)$
11:       $B[j] \leftarrow true$
12: **end for**
13: $l \leftarrow 1$
14: **while do** $k \geq 0\ AND\ l < N$
15:       **for** $i \leftarrow 0, i < 2^k$ **do**
16:            $I \leftarrow 2^k - i$
17:            $u \leftarrow log_2(I)$
18:            $U \leftarrow \|u\|$
19:            $u \leftarrow u - U$
20:            **if** $u - 0, 5 \geq 0$ **then**
21:                 $K \leftarrow \lceil log_2(I) \rceil$
22:            **else**
23:                 $K \leftarrow \lfloor log_2(I) \rfloor$
24:            **end if**
25:            **for** $j \leftarrow 1, j \leq k * K$ **do**
26:                 **if** $2^k * 2 \geq N$ **then**
27:                      $index = N$
28:                 **else**
29:                      $index = 2^k * 2 - 1$
30:                 **end if**
31:                 **if** $B[index] = true$ **then**
32:                      **if** $([idDoc = Position(l\ of\ V_N(q))]\ is\ in\ A_{Past}(q))$ **then**
33:                           $A_{New}(q') \leftarrow A_{New}(q') \cup d_{idDoc}$
34:                           $l \leftarrow l + 1$
35:                      **end if**
36:                 **else**
37:                      $l \leftarrow l + 1$
38:                 **end if**
39:            **end for**
40:       **end for**
41:       $k \leftarrow k - 1$
42: **end while**
43: $return(A_{New}(q'))$

applied to build the collection of documents D. In the third experiment, Zipf distribution (with parameter $\theta = 1.6$) was applied to build D. Simulations of user judgments were carried out under Zeta Distribution. Zeta distribution with parameters $2, 3$, and 4 were applied on the 30 most similar documents with respect to the queries. Besides, the Student's Paired t-Test (Two Samples test) over each average P@10 (our approach with respect to traditional retrieval) were used to support the results. Final results are summarized and displayed in Table 1.

Table 1. Results comparing our approach of reusing past queries with cosine distance

Experiment Distribution to build collection D	Relevance simulation Zeta distribution with parameter S	Percentage of improved queries	Average improvement (Measure: P@10)
Experiment 1 Exponential distribution (with parameter $\theta = 1.0$)	S = 2 S = 3 S = 4	83% 74% 65%	+21%** +17%** +20%**
Experiment 2 Exponential distribution (with parameter $\theta = 1.5$)	S = 2 S = 3 S = 4	77% 73% 65%	+17%** +18%** +16%**
Experiment 3 Zipf distribution (with parameter $\theta = 1.6$)	S = 2 S = 3 S = 4	85% 84% 77%	+21%** +18%** +23%**

** p-value<0.01 (two sample t-test).

5.2 Discussion

Accepted ranges for Zipf's law regarding the distribution of word frequencies in a vocabulary are between 1.4 and 1.8. In our experiments, Zipf distribution was used with value 1.6 to select terms from topics. It is important to emphasize that every time the parameter S of Zeta distribution (to apply Bradford's law) was incremented, both averages of P@10, using *Past Result* (i.e., our approach) and *Cosine* (i.e., traditional IR) declined similarly. Additionally, if the number of queries was increased in the experiments, it should not have different final results. This is because for every past query it exists just one and unique query for which the intersection is not empty. Also, the final results show that increasing the number of documents have no impact on the significance test p-values.

6 Conclusions

In this paper, a new Monte Carlo algorithm for information retrieval using past queries have been presented. It is easy to implement, does not require time to learning, and provides acceptable results improving precision (i.e., P@10). Furthermore, this algorithm can be implemented not only inside an information retrieval system but also as external interface outside search engines. This algorithm relies on reuse of relevant documents retrieved from the most similar past query. In addition, different evaluation scenarios have been simulated. Simulation provides two advantages. First, it provides an ideal environment to evaluate our algorithm. Second it makes possible to build not only document and query collections but also relevance judgments for documents given a query. Empirical results showed better precision (P@10) of our algorithm compared with traditional retrieval.

References

1. Bigot, A., Chrisment, C., Dkaki, T., Hubert, G., Mothe, J.: Fusing different information retrieval systems according to query-topics: a study based on correlation in information retrieval systems and tree topics. Inf. Retr. 14(6), 617–648 (2011)
2. Gray, P., Watson, H.J.: Present and future directions in data warehousing. SIGMIS Database 29(3), 83–90 (1998)
3. Nopiah, Z.M., Khairir, M.I., Abdullah, S., Baharin, M.N., Arifin, A.: Time complexity analysis of the genetic algorithm clustering method. In: Proceedings of the 9th WSEAS International Conference on Signal Processing, Robotics and Automation, ISPRA 2010, Stevens Point, Wisconsin, USA, pp. 171–176. World Scientific and Engineering Academy and Society, WSEAS (2010)
4. Kearns, M.J.: The Computational Complexity of Machine Learning. PhD thesis, Harvard University, USA, Cambridge, MA, USA (1989)
5. Cetintas, S., Si, L., Yuan, H.: Using past queries for resource selection in distributed information retrieval. Technical Report 1743, Department of Computer Science, Purdue University (2011)
6. Shen, X., Zhai, C.X.: Exploiting query history for document ranking in interactive information retrieval. In: Proceedings of the 26th Annual International ACM SIGIR Conference on Research and Development in Informaion Retrieval, SIGIR 2003, pp. 377–378. ACM, New York (2003)
7. Shen, X., Tan, B., Zhai, C.: Context-sensitive information retrieval using implicit feedback. In: Proceedings of the 28th Annual International ACM SIGIR Conference on Research and Development in Information Retrieval, SIGIR 2005, pp. 43–50. ACM, New York (2005)
8. Fonseca, B.M., Golgher, P.B., de Moura, E.S., Ziviani, N.: Using association rules to discover search engines related queries. In: Proceedings of the First Conference on Latin American Web Congress, LA-WEB 2003, pp. 66–71. IEEE Computer Society, Washington, DC (2003)
9. Baeza-Yates, R., Hurtado, C., Mendoza, M.: Query recommendation using query logs in search engines. In: Lindner, W., Fischer, F., Türker, C., Tzitzikas, Y., Vakali, A.I. (eds.) EDBT 2004. LNCS, vol. 3268, pp. 588–596. Springer, Heidelberg (2004)
10. Teevan, J., Adar, E., Jones, R., Potts, M.A.S.: Information re-retrieval: repeat queries in yahoo's logs. In: Proceedings of the 30th Annual International ACM SIGIR Conference on Research and Development in Information Retrieval, SIGIR 2007, pp. 151–158. ACM, New York (2007)
11. Garcia, S.: Search Engine Optimisation Using Past Queries. PhD thesis, RMIT University, Australia (2007)
12. Clough, P., Sanderson, M.: Evaluating the performance of information retrieval systems using test collections. Information Research 18(2) (2013)

13. Huurnink, B., Hofmann, K., de Rijke, M., Bron, M.: Validating query simulators: An experiment using commercial searches and purchases. In: Agosti, M., Ferro, N., Peters, C., de Rijke, M., Smeaton, A. (eds.) CLEF 2010. LNCS, vol. 6360, pp. 40–51. Springer, Heidelberg (2010)
14. Joachims, T.: A probabilistic analysis of the rocchio algorithm with tfidf for text categorization. In: Proceedings of the Fourteenth International Conference on Machine Learning, ICML 1997, pp. 143–151. Morgan Kaufmann Publishers Inc., San Francisco (1997)
15. Chan, E.P., Garcia, S., Roukos, S.: Probabilistic modeling for information retrieval with unsupervised training data. In: Proceedings of the Fourth International Conference on Knowledge Discovery and Data Mining (KDD), pp. 159–163. AAAI Press (1998)
16. Salton, G., Buckley, C.: Readings in information retrieval. In: Sparck Jones, K., Willett, P. (eds.) Readings in Information Retrieval, pp. 355–364. Morgan Kaufmann Publishers Inc., San Francisco (1997)
17. Radwan, A.A.A., Latef, B.A.A., Ali, A.M.A., Sadek, O.A.: Using genetic algorithm to improve information retrieval systems. World Academy of Science, Engineering and Technology 17, 1021–1027 (2008)
18. Lillis, D., Toolan, F., Mur, A., Peng, L., Collier, R., Dunnion, J.: Probability-based fusion of information retrieval result sets. Artif. Intell. Rev. 25(1-2), 179–191 (2006)
19. Gutiérrez-Soto, C., Hubert, G.: Evaluating the interest of revamping past search results. In: Decker, H., Lhotská, L., Link, S., Basl, J., Tjoa, A.M. (eds.) DEXA 2013, Part II. LNCS, vol. 8056, pp. 73–80. Springer, Heidelberg (2013)
20. Poosala, V.: Zipf's law. Technical Report 900 839 0750, Bell Laboratories (1997)
21. Garfield, E.: Bradford's Law and Related Statistical Patterns. Essays of an Information Scientist 4(19), 476–483 (1980)

A Fine-Grained Approach for Extracting Events on Microblogs

Lizhou Zheng[1,3], Peiquan Jin[1,3], Jie Zhao[2], and Lihua Yue[1,3]

[1] School of Computer Science and Technology,
University of Science and Technology of China, Hefei, China
[2] School of Business, Anhui University, 230601, Hefei, China
[3] Key Laboratory of Electromagnetic Space Information,
Chinese Academy of Sciences, China
jpq@ustc.edu.cn

Abstract. Microblog platforms like Twitter have been the important sources for news events extraction. Existing works on event extraction on microblogs usually used keywords, entities, or selected microblog posts to represent events, which cannot give a detailed description for the extracted events, e.g., "when and where did the event happen? ", "who were involved in the event?", etc. In this paper, we aim at providing a fine-grained event extraction on microblogs. In particular, we focus on extracting the 5W1H features (i.e., *when, where, who, what, whom,* and *how*) for events on microblogs. We first perform a clustering step to partition the microblog posts into several event clusters. After that, we extract the 5W1H features for those clusters using different algorithms. Our approach is evaluated on two microblog datasets crawled from Sina Weibo, which is the most popular microblog platform in China. The experiment results demonstrate the effectiveness of our approach.

Keywords: Microblog, Event extraction, 5W1H elements.

1   Introduction

Microblog platforms have been one of the major sources for new events detection and spreading. A lot of works on event detection and analysis over microblogs [1-3] have been done in recent years. However, existing studies always use some selected keywords or entity names, or some selected posts to describe an event [4], which can only provide a much coarse description for events. For example, the words "Li Na" and "Champion" are not sufficient to describe the recent event on Li Na's winning the Champion on the Australia Tennis Open on January 2014.

In this paper, we consider the concept of 5W1H which refer to *when, where, who, what, whom* and *how* in traditional news report [5, 6] to provide fine-granular representation for events on microblogs. Extracting the 5W1H features on news articles has been studied for years, and there are approaches like *SRL* (Semantic Role Labeling), *NER* (Name Entity Recognition), and *rule-based* approaches [7, 8]. However, there are special difficulties to extract the 5W1H features on microblog posts (e.g. short words and abbreviations, informal structure and multiple topics within one post).

H. Decker et al. (Eds.): DEXA 2014, Part I, LNCS 8644, pp. 275–283, 2014.

Following these observations and analysis, in this paper we propose novel approaches to extract the 5W1H features for events. The basic idea is to first construct event clusters and then conduct a rule-based method on each cluster to determine the 5W1H details. The main contributions of the paper can be summarized as follows:

(1) We propose a new framework that aggregate all posts in an event cluster to perform 5W1H extraction since the whole cluster posts corpus contain rich information about the event rather than one or very few selected posts. (Section 3.1 and 3.2)

(2) We propose a multi-granular similarity-based algorithm to extract *when* and *where* from event clusters. And for *who*, *what* and *whom*, we present a new term clustering and linking algorithm. (Section 3.3)

(3) We conduct experiments on real data sets. The results show that the proposed framework has a high precision in event clustering and details extraction. (Section 4).

2   Related Work

Event detection on microblog has been a research focus in recent years [1, 4]. For example, a system called TEDAS [1] is proposed to detect crime and disaster related events. The description of event is a critical issue in event extraction from microblogs. Previous works mainly use selected words or phrases, named entitiesor a set of microblogs to represent events [4]. But people are willing to know the details of an event, such as "*who/whom* were involved?", "*when* and *where* did it happen?", "*what* was the essential information in the event?". Those information can be denoted as 5W1H features, the primary metric for traditional news reporting [5].

Traditional event extraction was mainly towards news articles. The major task of event extraction has been formulated by MUC [8] and ACE [10]. In [10], Heng Ji et al. proposed a scheme to improve the performance of tackling the ACE's task. In [11], the authors utilized the *Maximum Entropy* method in argument recognition. In recent years, [5] introduced valency grammar and extracted structural semantic information from online news corpus. In [6], the authors extracted the 5W1H elements of Chinese news by employing a key event identification algorithm.

In a brief conclusion, existing works on extracting the 5W1H features for events were mainly for news articles. Event extraction on microblogs is a more challenging task because of the special characteristics of microblogs and the lack of event-related knowledge. Therefore, new methods must be proposed for news event extraction.

3   Fine-Grained Event Extraction from Microblogs

3.1   Overview

Fig.1. shows the framework of our methods. We first crawl microblogs and then perform preprocessing and clustering on microblogs. The preprocessing includes word segmentation, POS tagging using NLPIR (http://nlpir.org), and removing posts without entities. We finally extract the 5W1H elements of events for each event cluster.

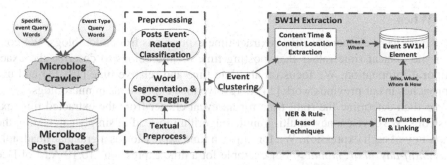

Fig. 1. Framework of fine-grained event extraction from microblogs

3.2 Event Clustering

In the event-clustering step, we first divide the microblogs into small daily sets and perform clustering for each set. Since it's not the focus of our work, we use simple algorithm to perform clustering. We mainly consider three types of similarities.

Temporal and Spatial Similarity. We extract time expressions from microblogs content, and represent them in different time granularities (e.g. *day, half-day, hour,* and *minute*). Content locations are also represented using granularities like *province, city, country,* and *local.* The similarity is then measured in each granularity.

Entity Similarity. Since name entities place an important role in detecting and describing an event, we calculate the similarity between entities in two microblog posts. We use the *Minimum Edit Distance (MED)* to measure the similarity between two entities *ent1* and *ent2*, which is defined in (1).

Other Similarity. We *also* compute the similarity of hashtags as well as other nouns and verbs. The definition of this kind of similarity is the same as shown in (1).

$$SIM\_ENT = \sum_{p1,p2} \max_{\substack{ent1\in p1 \\ ent2\in p2}} (1 - \frac{MED(ent1,ent2)}{\max(ent1.length, ent2.length)}) \quad (1)$$

Finally we sum up the three similarities with weighted parameters (tuned as 0.1, 0.7 and 0.2 respectively) to obtain the final similarity score. The algorithm starts with merging the posts with the highest similarity score and for two post clusters, we compute the average similarity between posts in the two clusters. The algorithm stops when similarity between each two clusters are below the threshold. We do not use flat clustering algorithms like *K-means* since it is difficult to predict the value of cluster number *k* beforehand.

3.3 Extracting the 5W1H Features

The 5W1H features for events are extracted from the event clusters we built in Section 3.2. In this section, we will discuss our methods on extracting 5W1H elements.

(1) **When**

To extract *when* element, we first extract time expressions from microblogs. We consider the content time rather than posting time of microblogs to describe the exact temporal information. We focus on absolute time and relative time [12, 13] and use approaches in our previous work [14] to resolve time expressions in microblogs.

We then determine the right time for an event cluster from the extracted time expressions. We devise a new multi-granular algorithm instead of simply calculating the frequency of each expression. We formulate a time expression as a granular quadruple <*day, half-day, hour, minute*>. For example for a time expression "10:10 A.M of Dec 29, 2013", the granular quadruple is <2013-12-29, Morning, 10 A.M, 10:10 A.M >.

Given the quadruples of the time expressions for each event cluster, we calculate the frequencies of time elements in each granularity, as well as their co-occur matrix. We finally present the algorithm shown in Fig.2 to determine the right event time.

Algorithm. *EventTime*

Input: (1) Day-level time point set *Dset*, Half-Day-level time point set *HDset*, Hour-level time point set *Hset*, Minute-level time point set *Mset*, along with their frequency.
(2) The co-occurrence matrix of each two adjacent levels.
(3) The minimum ratio threshold α and the post count of a cluster N.

Output: the result time expression *resstr*.

1:     Get most frequent day-level time point *mdtp* from *Dset*;//day-level
2:     **if** freq(*mdtp*)< $\alpha*N$ **then return** *resstr*;
3:     **else** *resstr=mdtp*;
4:     Get the most frequent half-day-level time point *mhdtp* from *HDset*;//half-day-level
5:     Get the most frequent co-occur day-level time point of *mhdtp*, named *mcodtp*;
6:     **if** freq(*mhdtp*)>= $\alpha*N$ **and** *mdtp==mcodtp* **then** *resstr* +=*mhdtp*;
7:     **else return** *resstr*;
8:     Get the most frequent hour-level time point *mhtp* from *Hset*;//hour-level
9:     Get the most frequent co-occur half-day-level time point of *mhtp*, named *mcohdtp*;
10:    **if** freq(*mhtp*)>=α*N **and** *mhdtp==mcohdtp* **then** *resstr* +=*mhtp*;
11:    **else return** *resstr*;
12:    Get the most frequent minute-level time point *mmtp* from *Mset*;//minute-level
13:    Get the most frequent co-occur hour-level time point of *mmtp*, named *mcohtp*;
14:    **if** freq(*mmtp*)>= $\alpha*N$ **and** *mhtp==mcohtp* **then** *resstr* +=*mmtp*;
15:    **else return** *resstr*;
16:    **return** *resstr*;

Fig. 2. The algorithm of extracting event time

(2) **Where**

The extraction of *where* is similar to that of *when*. We first extract location expressions using NLPIR. We then define four levels of location granularities, i.e., *province, city, county,* and *local*. We manually create a Chinese location gazzetteer with hierarchical relation (*province-city-county/district*) [15]. We also combine consecutive nouns following a location entity to construct the *local* locations. For example, "安徽/*ns* 医科/*n* 大学/*n*"("Anhui/*ns* Medical/*n* University/*n*") will be transformed into "安徽/*ns* 医科大学/*n*"("Anhui/*ns* Medical University/*n*").

(3) Who, What, and Whom

For the extraction of *who*, *what*, and *whom*, some approaches have been for English sentences [9]. Here we present a *term clustering* method to put different expressions about the same meaning into one term cluster and a novel method to link different clusters. An example is shown in Fig.3.

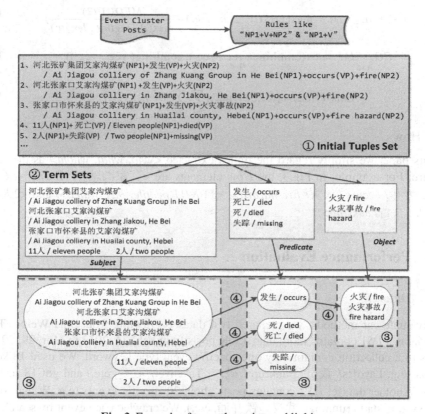

Fig. 3. Example of term clustering and linking

We first get the initial <*subject, predicate, object*> and < *subject, predicate* > tuples by employing some tailor-made rules like "*NP1+VP+NP2*" and "*NP1+VP*".

We then extract term sets for each component in the tuples (i.e., *subject, predicate,* and *object*) as well as the co-occur frequency between terms in different sets (i.e. *subject-predicate, predicate-object* and *subject-object* co-occur matrix).

For the next step we conduct clustering for each component of term set by considering textual and co-occurrence similarity. The textual similarity between two term sets is the average similarity of terms (computed by (2)) in two term sets, here *LCS* is the *longest common subsequence*. For co-occurrence, we compute the cosine similarity between two co-occur vectors of terms, as shown in (3). We sum up the two similarities with a parameter α tuned as 0.5 in our experiment. We finally utilize a hierarchical method which is the same as in Section 3.2 to perform clustering.

Finally, we link clusters to obtain candidate *who-what-whom* tuples. We utilize the term co-occur frequency shown in Equation (5) to link each subject cluster with a predicate cluster, and link each predicate cluster with an object cluster. We rank the links based on the average frequency of terms in the link. We finally retain at most K ($K=5$) links as output and for each event.

$$TEXTUAL\_SIM_{t1,t2} = \max(LCS(t1,t2).length, 1 - \frac{MED(t1,t2)}{\max(t1.length, t2.length)}) \tag{2}$$

$$CO\_SIM_{t1,t2} = \cos(combine(\underset{t1,M1}{vec}, \underset{t1,M2}{vec}), combine(\underset{t2,M1}{vec}, \underset{t2,M2}{vec})) \tag{3}$$

$$Link(c1,c2) = \{(c1,c2) \mid \underset{c2}{\max}(\sum_{t1 \in c1}\sum_{t2 \in c2} freq(t1,t2))\} \tag{4}$$

(4) How

In this paper we simply regard the tuples extracted *<who, what, whom>* as the *how* feature. For example, in Fig.3, the *How* elements are *<艾家沟煤矿/ Ai Jiagou Colliery, 发生/ occurs, 火灾/ fire hazard>*, *<11人/11 people, 死亡/ died>*, and *<2人/ 2 people, 失踪/ missing>*.

4   Performance Evaluation

4.1   Dataset

To evaluate our methods, we prepared two datasets crawling from Sina Weibo. The first dataset *DS*1 contains more than 450K posts from Feb. 24, 2013 to Mar. 29, 2013. Posts in a collection contain only one specific event type keyword. We used 18 keywords (e.g. Layoffs/ 裁员, Bankrupt/ 破产, Takeover/ 收购, etc.) and got 18 sets of posts. Another dataset *DS*2 contains posts about 20 specific events (e.g. Wuhan subway system starts running/ 武汉地铁开通, etc.). We crawled each event by searching for posts containing several keywords about the specific event.

4.2   Result of Event Clustering

We first present the result of event clustering. For *DS*1, there are clusters which do not represent an event, e.g., advertisement, movie reviews, etc. Therefore, we manually remove the non-event clusters from *DS*1 and finally get 278 event clusters. For *DS*2, since almost all of the posts in a collection is about one specific event, we regard each post collection as an event cluster and there are 20 event clusters in *DS*2.

4.3   5W1H Extraction

(1)   When and Where

The experimental results for *when* and *where* are shown in Table 1 and Table 2. The baseline method is regarding the expression with highest frequency in an event cluster as the result. We also compared the results in different granularities. Given a true event time point, e.g., "9:30 AM Mar. 15, 2014", if the extracted expression is like "8:00 AM Mar. 15, 2014", we regard the *day* and *half-day* features are correct but the *hour* and *minute* features are not extracted. We define *recall* as the right cluster time points divided by the number of event clusters. We just compute the precision for the baseline method since we do not perform granular process on it. We can see that our methods perform better than the baseline method in most cases.

Table 1. Results of *when* extraction

Dataset		Our Algorithm				Baseline
		Day	Half-Day	Hour	Minute	
DS1	Precision	86.64%	76.92%	69.33%	71.03%	Presicion=
	Recall	76.98%	50.36%	37.41%	27.34%	70.90%
DS2	Precision	88.89%	73.68%	56.25%	70%	Presicion=
	Recall	80%	70%	45%%	35%	60%

Table 2. Results of *where* extraction

Dataset		Our Algorithm				Baseline
		Province	City	Country	Local	
DS1	Presicion	80.54%	83.74%	78.18%	53.6%	69.47%
	Recall	53.60%	37.05%	15.48%	24.10%	
DS2	Precision	85%	91.67%	61.54%	71.43%	65%
	Recall	85%	55%	40%	50%	

(2)   Who, What, and Whom

We use two metrics to measure our methods of extracting the *who, what, and whom* elements. The metrics are based on the accuracy of <*who, what, whom*> tuples. The first metric is the accuracy of main tuple which we defined as a tuple that describes the main component of an event, e.g., in Fig.3, the main tuple is <艾家沟煤矿/ *Ai Jiagou Colliery*, 发生/ *occurs*, 火灾/ *fire hazard*> since the other two tuples are sub-details of the event. We then check whether the output tuples of an event cluster contains a main tuple. The next metric is the average accuracy of the tuples we extracted from all the event clusters. Here we define *recall* as the average count of true tuples we extracted from each event cluster. Table 3 shows the performance of extracting who, what and whom elements. Here the baseline method only considers the frequency of <*subject, predicate, object*> and < *subject, predicate* > tuples. The experiment results show that our proposed methods perform better.

Table 3. Results for *<who, what, whom>* Tuples

Dataset	Our Algorithm			Baseline		
	Main Tuple Accuracy	Average Tuple Accuracy	Recall	Main Tuple Accuracy	Average Tuple Accuracy	Recall
DS1	68.71%	61.43%	2.43	56.83%	46.11%	1.85
DS2	70%	59.26%	2.4	45%	45.16%	1.4

5   Conclusion

In this paper, we aim at extracting details about events on microblogs. We introduced some new algorihtms to extract the 5W1H features for events. Particularly, we emplied a clustering approach to construct event clusters from the original microblogs, and presented a multi-granular method to extract *when* and *where* information for events. We also proposed a term clustering and linking method to extract the *who*, *what*, and *whom* elements. The experimental results on two real microblog datasets demonstrated the superiority of our methods when compared to the baseline methods.

Our future work will be focused on utilizing out-linking information in microblogs to extract the *how* information for events. We will also improve the performance of 5W1H extraction and try to analyze the evolution of an event using 5W1H features.

Acknowledgement. This paper is supported by the National Science Foundation of China (No. 61379037 and No. 71273010), the National Science Foundation of Anhui Province (No. 1208085MG117), and the OATF project in USTC.

References

1. Li, R., Lei, K., Khadiwala, R., Chang, K.C.: TEDAS: A Twitter-based event detection and analysis system. In: Proc. of ICDE, pp. 1273–1276 (2012)
2. Zhang, L., Zhang, Z., Jin, P.: Classification-based prediction on the retweet actions over microblog dataset. In: Wang, X.S., Cruz, I., Delis, A., Huang, G. (eds.) WISE 2012. LNCS, vol. 7651, pp. 771–776. Springer, Heidelberg (2012)
3. Zheng, L., Yang, K., Yu, Y., Jin, P.: Predicting Age Range of Users over Microblog Dataset. International Journal of Database Theory and Application 6(6), 85–94 (2013)
4. Chakrabarti, D., Punera, K.: Event summarization using tweets. In: Proc. of ICWSM (2011)
5. Wang, W., Zhao, D., Zou, L., Wang, D., Zheng, W.: Extracting 5W1H event semantic elements from Chinese online news. In: Chen, L., Tang, C., Yang, J., Gao, Y. (eds.) WAIM 2010. LNCS, vol. 6184, pp. 644–655. Springer, Heidelberg (2010)
6. Wang, W.: Chinese news event 5W1H semantic elements extraction for event ontology population. In: Proc. of WWW, pp. 197–202 (2012)
7. Choudhury, S., Breslin, J.: Extracting semantic entities and events from sports tweets. In: Proc. of #MSM, pp. 22–32 (2011)
8. Chinchor, N., Marsh, E.: Muc-7 information extraction task definition. In: MUC-7 (1998)
9. ACE (Automatic Content Extraction). Chinese Annotation Guidelines for Events. National Institute of Standards and Technology (2005)

10. Ji, H., Grishman, R.: Refining Event Extraction through Cross-Document Inference. In: Proc. of ACL, pp. 254–262 (2008)
11. Tan, H., Zhao, T., Zheng, J.: Identification of Chinese event and their argument roles. In: Proc. of CIT Workshops, pp. 14–19 (2008)
12. Lin, S., Jin, P., Zhao, X., Zhao, J., Yue, L.: Extracting Focused Time for Web Pages. In: Gao, H., Lim, L., Wang, W., Li, C., Chen, L. (eds.) WAIM 2012. LNCS, vol. 7418, pp. 266–271. Springer, Heidelberg (2012)
13. Lin, S., Jin, P., Zhao, X., Yue, L.: Exploiting Temporal Information in Web Search. Expert Systems with Applications 41(2), 331–341 (2014)
14. Zhao, X., Jin, P., Yue, L.: Automatic Temporal Expression Normalization with Reference Time Dynamic-Choosing. In: Proc. of COLING, pp. 1498–1506 (2010)
15. Zhang, Q., Jin, P., Lin, S., Yue, L.: Extracting Focused Locations for Web Pages. In: Wang, L., Jiang, J., Lu, J., Hong, L., Liu, B. (eds.) WAIM 2011. LNCS, vol. 7142, pp. 76–89. Springer, Heidelberg (2012)

Generating Data Converters to Help Compose Services in Bioinformatics Workflows

Mouhamadou Ba[1], Sébastien Ferré[2], and Mireille Ducassé[1]

[1] IRISA/INSA Rennes
20 Avenue des Buttes de Coesmes, 35708 Rennes cedex, France
[2] IRISA/Université de Rennes 1
263 Avenue Général Leclerc, 35042 Rennes cedex, France
mouhamadou.ba@irisa.fr

Abstract. Heterogeneity of data and data formats in bioinformatics often entail a mismatch between inputs and outputs of different services, making it difficult to compose them into workflows. To reduce those mismatches bioinformatics platforms propose ad'hoc converters written by hand. This article proposes to systematically detect convertibility from output types to input types. Convertibility detection relies on abstract types, close to XML Schema, allowing to abstract data while precisely accounting for its composite structure. Detection is accompanied by an automatic generation of converters between input and output XML data. Our experiment on bioinformatics services and datatypes, performed with an implementation of our approach, shows that the detected convertibilities and produced converters are relevant from a biological point of view. Furthermore they automatically produce a graph of potentially compatible services with a connectivity higher than with the ad'hoc approaches.

1 Introduction

Heterogeneity of data and data formats in bioinformatics often entail a mismatch between inputs and outputs of different services, making it difficult to compose them into workflows [1]. Formats to represent input and output data can be textual or based on XML technologies. Textual formats, often specific to a few services have the advantage to be human readable [2] but they do not facilitate automatic processing. Several platforms for bioinformatics data analysis use textual formats. They provide *converters*, namely special services for format conversion, which have to be manually defined (see, for example, Emboss [3], Galaxy [4], Mobyle [5]). According to the analysis of Wassink et al [6] on the Taverna Workflows, over 30% of services in life science workflows are used for the conversion of data. When composing services, users can get lost in specific converters required to transfer data between services. It is difficult to find appropriate converters because they are often mixed with other services. Furthermore, the lack of descriptions for many converters complicates the choice. Users are often forced to create new converters, it is time-consuming and error prone.

Formats based on XML technologies describe data types independently of tools [7,8]. BioXSD, for example, proposes a standard representation of basic bioinformatics data [9]. It also allows meta-information to be added from ontologies, increasing the accuracy of representations. Yet, XML-based formats alone are not sufficient to

H. Decker et al. (Eds.): DEXA 2014, Part I, LNCS 8644, pp. 284–298, 2014.

solve the problem of data matching. On the one hand, most formats are textual, thus it is important to be able to match services using XML *and* textual formats. On the other hand, even if all formats were XML standardized, it remains to solve the n:m matching problem [10], corresponding to the matching and conversion between two *composite structures*, i.e. XML trees.

Work related to data mismatches are numerous. For example, Li et al. [11] classify service composition mismatches. They distinguish between functional and nonfunctional mismatches on the syntactic and semantic levels. Elizondo et al. [12], with similar categorizations of data mismatches, automate the detection of data mismatches in end-user compositions. They identify existing converters able to solve mismatches. Stroulia et al. [13] provide a method to assess the similarity between two web services, based on the structure of their data types and operations as well as on their textual descriptions. Those work show the benefits of taking into account the structure of data to detect and solve mismatches. However, they do not fully exploit the composite structure of data. They mostly access internal nodes of XML trees, and do not traverse and synthesize XML trees in a systematic and recursive way, thus limiting their applicability when the data types are complex. Furthermore, they are not able to generate new converters. They focus on the discovery of existing converters and services. Some approaches [14,15] , in bioinformatics, propose to use ontologies such as EDAM [16] and myGrid ontology [17]. They facilitate matching by defining descriptions and hierarchies on formats and data types. A first limit of those solutions is that convertibility of data must be manually and explicitly declared in ontologies. A second limit is that composite types are not much taken into account. For example, the work of Dibernado et al. [14] is the only one, we are aware of in bioinformatics, that uses the structure of data. However, it only uses the *hasA* relation to decompose types.

This article proposes to systematically detect convertibility from output types to input types. Convertibility detection relies on abstract types, close to XML Schema, allowing to abstract data while precisely accounting for its composite structure. The main contribution is the definition of convertibility rules that exploit composition and decomposition as well as specialization and generalization of types. Furthermore, the rules automatically generate a complete specification of the matching between input and output types. That specification allows to generate converters between input and output XML data. We report an experiment on bioinformatics services with an implementation of our approach. We manually specified the inputs and outputs of services using abstract types, where each service is here understood as a function from input(s) to output(s) as proposed by Missier et al [18]. The detected convertibilities were analyzed with a team of the GenOuest[1] bioinformatics platform. They have been reckoned relevant from a biological point of view. When adding a new service, with our approach, it is sufficient to define or reuse abstract types for input and output data, and the new service will be automatically integrated in the global system. At present, in order to achieve the same goal, many converters have to be manually developed for services which are (not immediatly) compatible. That is significantly heavier than specifying a few abstract types. Furthermore, identifying the compatible services by hand is already a challenge while our approach does it automatically. As a consequence, our approach automatically

---

[1] http://www.genouest.org/

produces a graph of potentially compatible services with a connectivity higher than with the ad'hoc approaches.

In the following, Section 2 introduces the abstract representation of types. Section 3 defines the convertibility rules. Section 4 shows how to instantiate our method to bioinformatics. Section 5 presents the experiment in bioinformatics. Section 6 compares our approach with related work.

2   Representation of Types

In this section we present the language used to describe the types of data. It is defined from an open set of primitives and a fixed set of type constructors. From a semantic point of view, a type denotes a set of XML values. An XML value is a sequence of XML nodes. An XML sequence may be empty or contain a single node. An XML node is either an XML element or a textual element (CDATA). An XML element is made of a tag and a content, which is a sequence of XML nodes. Our language of types follows the main XML Schema constructs, but to simplify the presentation we use regular expressions inspired by the work of Hosoya et al. [19]. In the following, types are denoted by uppercase letters (e.g., T, T_1) and function $XML(T)$ defines the semantics of type T by a set of XML values.

- **Primitive types**: $T = p$, where p is a primitive type. Primitive types are the basic ingredients to build other types. Their structure is atomic, they are not decomposable. XML instances of a primitive type are CDATA (text). For example, *int* is a primitive type representing integers.
- **Constructor *tag***: $T = t[T_1]$ where t is a tag. This expression denotes XML elements whose tag is t and whose content is of type T_1 : $XML(t[T_1]) = \{<t> x_1 </t> |\ x_1 \in XML(T_1)\}$. Tags provide semantics for data and can be bound to concepts of ontologies. In XML Schema, constructor *tag* can be expressed by a tag or by the attribute *name* of tag *xs:element*.
- **Constructor *empty***: $T = \varepsilon$. The empty XML sequence: $XML(\varepsilon) = \{\varepsilon\}$.
- **Constructor *tuple***: $T = T_1 T_2$. This type expression denotes XML sequences that are the concatenation of instances of T_1 and instances of T_2 : $XML(T_1 T_2) = \{x_1 x_2 \mid x_1 \in XML(T_1), x_2 \in XML(T_2)\}$. That constructor is used to define composite types and sequences. In XML Schema, constructor *tuple* corresponds to tags *xs:complexType* and *xs:sequence*.
- **Constructor *union***: $T = T_1 | T_2$. This type expression denotes the union of instances of T_1 and instances of T_2 : $XML(T_1 | T_2) = XML(T_1) \cup XML(T_2)$. Constructor *union* can be used, for example, to consider several different types and make some treatments without distinction. In XML Schema, that constructor corresponds to tag *xs:choice*.
- **Constructor *list***: $T = T_1 +$. This type expression denotes the non-empty sequences of instances of type T_1 (homogeneous lists): $XML(T_1+) = \{x_1 \ldots x_n \mid n \geq 1 \wedge x_1, \ldots, x_n \in T_1\}$. In XML Schema, that constructor corresponds to the *minOccurs* and *maxOccurs* attributes associated with tag *xs:sequence*.
- **Constructor *optional***: $T = T_1?$. This type expression is equivalent to $T = T_1|\varepsilon$.

$$\frac{f_p : p_1 \rightarrow_p p_2}{f : p_1 \rightarrow p_2 \quad f(x) = f_p(x)} \quad \text{(PRIMITIVE)}$$

$$\frac{t_a \rightarrow_t t_b \quad f_1 : A \rightarrow B}{f : t_a[A] \rightarrow t_b[B] \quad f(x) = element(t_b, f_1(content(x)))} \quad \text{(TAGCHANGE)}$$

$$\frac{f_1 : A \rightarrow B}{f : t[A] \rightarrow B \quad f(x) = f_1(content(x))} \quad \text{(TAGREMOVAL)}$$

$$\frac{}{f : A \rightarrow \varepsilon \quad f(x) = \varepsilon} \quad \text{(EMPTY)}$$

$$\frac{f_1 : A \rightarrow B_1 \quad f_2 : A \rightarrow B_2}{f : A \rightarrow B_1 B_2 \quad f(x) = concat(f_1(x), f_2(x))} \quad \text{(CONCAT)}$$

$$\frac{f_1 : A_1 \rightarrow B}{f : A_1 A_2 \rightarrow B \quad f(x) = (let\ x_1, x_2 - select(x, A_1, A_2)\ in\ f_1(x_1))} \quad \text{(LEFTSELECTION)}$$

$$\frac{f_2 : A_2 \rightarrow B}{f : A_1 A_2 \rightarrow B \quad f(x) = (let\ x_1, x_2 = select(x, A_1, A_2)\ in\ f_2(x_2))} \quad \text{(RIGHTSELECTION)}$$

$$\frac{f_1 : A_1 \rightarrow B \quad f_2 : A_2 \rightarrow B}{f : A_1 | A_2 \rightarrow B \quad f(x) = (case\ (x : A_1)\ then\ f_1(x)\ |\ (x : A_2)\ then\ f_2(x))} \quad \text{(PRECHOICE)}$$

$$\frac{f_1 : A \rightarrow B_1}{f : A \rightarrow B_1 | B_2 \quad f(x) - f_1(x)} \quad \text{(LEFTPOSTCHOICE)}$$

$$\frac{f_2 : A \rightarrow B_2}{f : A \rightarrow B_1 | B_2 \quad f(x) = f_2(x)} \quad \text{(RIGHTPOSTCHOICE)}$$

$$\frac{f_1 : A \rightarrow B}{f : A+ \rightarrow B+ \quad f(x) = map(f_1, x)} \quad \text{(MAP)}$$

$$\frac{f_1 : A \rightarrow B}{f : A+ \rightarrow B \quad f(x) = (let\ x_1 = choose(x, A)\ in\ f_1(x_1))} \quad \text{(CHOICE)}$$

$$\frac{f_1 : A \rightarrow B}{f : A \rightarrow B+ \quad f(x) = f_1(x)} \quad \text{(SINGLETON)}$$

Fig. 1. Convertibility rules and definitions of generated converters

3 Convertibility Rules

Figure 1 lists all the rules that specify when a type A is convertible to a type B. They also define associated converters as functions from A to B. Those rules form a natural deduction system whose judgements are in the form $f : A \rightarrow B$, i.e. f is a converter from an XML value of type A to an XML value of type B, and hence A is convertible to B. A judgement $f : A \rightarrow B$ holds true if and only if it is possible to build a proof tree with that judgement at the root, and where each node instantiates a rule. The deduction system works by structural induction on couples of types (A, B), covering

Table 1. Utility functions on XML values

Function	Input	Output	Description
content	XML	XML	returns the content of an XML element
element	tag, XML	XML	builds an XML element given a tag and an XML content
concat	XML, XML	XML	returns the concatenation of two XML sequences
select	XML, type, type	XML, XML	splits an XML sequence in two parts matching given types
map	converter, XML	XML	applies a converter to each node of an XML sequence and returns the concatenation of the results
choose	XML, type	XML	returns any element matching a given type from an XML sequence

all combinations of type constructors for which convertibility is possible. The rules depend on conversion axioms for tags ($t_a \to_t t_b$), and on converters between primitive types ($f_p : p_1 \to_p p_2$). Those base conversions depend on the application domain, and correspond, for instance, to well-known conversion functions (e.g., from floats to integers). By default, we assume that $t \to_t t$ for every tag t, and $f_p : p \to_p p$ with $f_p(x) = x$ for every primitive type p. The definitions of converters in rules make use of utilitary functions on XML values, which are described in Table 1. Figure 2 shows a proof convertibility.

Primitive rule. Rule (PRIMITIVE) allows the use of a primitive converter f_p when the two types are primitive types. That rule handles the conversion of the leaves of XML trees (CDATA nodes).

Tag rules. These rules handle the conversion from and to XML elements. Rule (TAGCHANGE) defines converters from an XML element x to another XML element $element(t_b, f_1(content(x)))$ by applying a domain-dependent tag conversion (here, from t_a to t_b), and by recursively applying a converter f_1 to the content of x. Function *content* gives access to the content of an XML element, and function *element* builds the new element from the converted tag and converted content. Rule (TAGREMOVAL) define converters from an XML element to an XML sequence by ignoring the tag, and recursively converting the content.

Empty and tuple rules. These rules handle conversions of XML sequences, i.e. constructors *empty* and *tuple*. Rule (EMPTY) says that any XML value x can be converted to the empty XML sequence ϵ. Rule (CONCAT) defines converters that first apply the converters f_1 and f_2 to the source value x, and then concatenates the two results $f_1(x)$ and $f_2(x)$ with function *concat*, hence producing an XML sequence. Rules (LEFTSELECTION) and (RIGHTSELECTION) define converters that select respectively the left and right part of the source data, an XML sequence, and convert it to the target data. This is useful when only a part of the source data is necessary to produce

seq[ns[acgt] species[string] version[int]]+ → seq[organism[string] ns[ACGT]]+
(MAP) $f(x) = map(f_1, x)$
   1. seq[ns[acgt] species[string] version[int]] → seq[organism[string] ns[ACGT]]
   (TAGCHANGE) $f_1(x) = element(seq, f_{1.2}(content(x)))$
      1.1. seq $\rightarrow_t$ seq
      1.2. ns[acgt] species[string] version[int] → organism[string] ns[ACGT]
      (CONCAT) $f_{1.2}(x) = concat(f_{1.2.1}(x), f_{1.2.2}(x))$
         1.2.1. ns[acgt] species[string] version[int] → organism[string]
         (RIGHTSELECTION) $f_{1.2.1}(x) = (let\ x_1, x_2 = select(x)\ in\ f_{1.2.1.1}(x_2))$
            1.2.1.1. species[string] version[int] → organism[string]
            (LEFTSELECTION) $f_{1.2.1.1}(x) = (let\ x_1, x_2 = select(x)\ in\ f_{1.2.1.1.1}(x))$
               1.2.1.1.1. species[string] → organism[string]
               (TAGCHANGE) $f_{1.2.1.1.1}(x) = element(organism, f_{1.2.1.1.1.2}(content(x)))$
                  1.2.1.1.1.1. species $\rightarrow_t$ organism
                  1.2.1.1.1.2. string $\rightarrow_p$ string
                  (PRIMITIVE) $f_{1.2.1.1.1.2}(x) = x$
         1.2.2. ns[acgt] species[string] version[int] → ns[ACGT]
         (LEFTSELECTION) $f_{1.2.2}(x) = (let\ x_1, x_2 = select(x)\ in\ f_{1.2.2.1}(x_1))$
            1.2.2.1. ns[acgt] → ns[ACGT]
            (TAGCHANGE) $f_{1.2.2.1}(x) = element(ns, f_{1.2.2.1.2}(content(x)))$
               1.2.2.1.2. acgt $\rightarrow_p$ ACGT
               (PRIMITIVE) $f_{1.2.2.1.2}(x) = uppercase(x)$

Fig. 2. An example proof tree of convertibility between two kinds of sequence lists

the target data. The selection of the parts (function *select*) is guided by the sub-types of the source sequence.

Union rules. These rules handle conversions from and to unions of types. Rule (PRECHOICE) defines converters that produce a target data using a different converter depending on the type of source data (A_1 or A_2). This is useful when the source data can have different structures (union type). Rules (LEFTPOSTCHOICE) and (RIGHTPOSTCHOICE) choose a converter to a target sub-type, when the target type is an union. This is useful when the target data has several acceptable structures.

List rules. The remaining rules handle conversions from and to lists. Rule (MAP) define converters from a source list to a target list where a same converter is applied to each element of the list. Function *map* is used to perform iteration over list elements, and concatenation of converted elements. Rule (CHOICE) defines converters that first choose an element of a list, and then recursively apply a converter to it. This is useful when a single element is expected while a list is provided. Rule (SINGLETON) defines converters that produce singleton lists from a source element, after recursively applying a converter to it. This is useful when a list is expected while a single element is provided.

For a given couple (A, B) of type expressions, several rules may be applicable. In that case, it is sufficient that one of them leads to a success to prove the convertibility

$f(x) = map(f_1, x)$
where $f_1(x) = element(\text{seq}, concat($
      $let\ x_1, x_2 = select(content(x), \text{ns[acgt]}, (\text{species[string] version[int]}))$
       $in\ let\ x_{21}, x_{22} = select(x_2, \text{species[string]}, \text{version[int]})$
        $in\ element(\text{organism}, content(x_{21})),$
      $let\ x_1, x_2 = select(content(x), \text{ns[acgt]}, (\text{species[string] version[int]}))$
       $in\ element(\text{ns}, uppercase(content(x_1)))))$

Fig. 3. The generated converter for example of Figure 2

from A to B. Figure 2 details the proof of convertibility between two kinds of biological sequence lists. In the source list, sequences are made of a nucleotide sequence, a species name, and a version number, while in the target list, sequences are made of an organism, and a nucleotide sequence. Another difference is that nucleotide sequences are lowercase in the source list (primitive type *acgt*), and uppercase in the target list (primitive type *ACGT*). In the proof (Figure 2), we assume that a primitive converter is available to convert from lowercase to uppercase (see step 1.2.2.1.2.), and domain knowledge tells us that tag *species* can be replaced by *organism* (see step 1.2.1.1.1.). Each item in Figure 2 is the conclusion of a rule, and the sub-items are the hypotheses of the rule. At each item, the converter function is defined with calls to the converter function of sub-items. After inlining the definition of intermediate functions in the main function f, we obtain the full definition of f in Figure 3. For concision, we omit the second and third arguments of function *select*.

Our rule system exhibits two kinds of non-determinism: (1) in the generation of converters, and (2) in the definition of converters. Firstly, given two types A and B, the system may generate several converters from A to B, i.e. several solutions to the conversion problem. This is a common feature of rule systems. For example, a converter $f : AA \rightarrow A$ can be produced by either Rule (LEFTSELECTION) or Rule (RIGHTSELECTION): in the former, the left part of the source value is selected, while in the latter, the right part is selected. In practice, one converter must be chosen, which could be done through user interaction. Secondly, a generated converter may produce several target value for a same source value. This non-determinism comes from some utility functions, and the *case* construct. Function *select* may find different ways to split the source value in two parts. Function *choose* has as many results as elements in the list. The *case* construct has two results when the two conditions are satisfied, when x matches both types A_1 and A_2. This second form of non-determinism could be used to express iteration in a workflow. For example, assuming that service S_1 produces lists of sequences, and service S_2 consumes one sequence at a time, the converter generated by Rule (CHOICE) could be a way to express that S_2 must be iterated over the results of S_1, and the output of S_2 could be considered to be the list of individual results.

We implemented our rule system in a program that decides the convertibility between any two type expressions, and generates converters from data matching the first type expression to data matching the second type expression. The algorithm is directly derived from the above rules and combines *pattern matching* on type expressions to identify constructors, and recursive calls on type sub-expressions. The examination of rules shows that recursive calls always involve smaller couples of expressions, which ensures termination of the program in all cases. Our generated converters are represented in XQuery, which makes them executable. We have chosen XQuery because it is a suitable language to process XML documents. To account for non-determinism, the result of our program will be a collection of converters, and each converter will be a function from a XML value to a collection of XML values.

4 Instantiation to Bioinformatics

Depending on requirements and on the nature of the data in bioinformatics platforms, many formats are available. To represent genomics data, various textual formats (e.g. FastQ, BED)[2] and XML formats (e.g. BioXSD[3], phyloXML[4]) are provided. Textual formats are the most commonly used. In addition to data formats, ontologies, such as EDAM [5] have been proposed to organize and classify resources including data types and formats. Our work starts from these resources to define input and output types of services. We abstract data types focusing on the information contents and composite structure of data. Figure 4 shows simple examples of types defined manually

```
> Accession = accession[string]
> SimpleSequence = simpleSequence[string]
> NucleotideSequence = ns[SimpleSequence]
> AminoAcidSequence = as[SimpleSequence]
> Biosequence = NucleotideSequence | AminoAcidSequence
> ComplexBiosequence = complexBiosequence[
        sequence[Biosequence]
        species[string] source[string] name[string]
        version[string] note[string]?]
> ComplexProteinSequence = complexProteinSequence[
        sequence[AminoAcidSequence]
        species[string] source[string] name[string]
        version[string] note[string]?]
> ListOfComplexBiosequences = ComplexBiosequence+
```

Fig. 4. Examples of bioinformatics types

from existing bioinformatic formats. *Accession* and *SimpleSequence* are simple types representing, respectively, an accession number and a raw sequence defined with a constructor tag and a primitive. In the same way, *NucleotideSequence* (representing nucleotide sequences) and *AminoAcidSequence* (representing amino acid sequences) are defined using *SimpleSequence*, they specialize the sequences.

---

[2] http://genome.ucsc.edu/FAQ/FAQformat.html/
[3] http://bioxsd.org/
[4] http://www.phyloxml.org/
[5] http://edamontology.org

Their union forms *Biosequence*, a biological sequence generalizing the sequences. *ComplexBiosequence* and *ComplexProteinSequence* are composite types holding several types through constructor *tuple*. They are biological sequences containing required (e.g., *sequence[Biosequence]*) and optional (e.g., *note[string]?*) contents, *ComplexProteinSequence* being more specific than *ComplexBiosequence*. *ListOfComplexBiosequences* defines a list of *ComplexBiosequence* using constructor *list*. The other types we use are defined in the same way as the above types. Labels are inspired from the EDAM ontology.

Compared to data types and textual formats used on platform EMBOSS[6], our types can define accession numbers allowing to represent, for example, sequence and database references. They can represent raw sequences as in plain text format, single sequences as in gcc format, one or several sequences (e.g., alignment of sequences) as in FASTA format, as well as a simple sequence associated to its annotations and features as in EMBL format. Our types can also represent lists of files and differentiate the nature of information contained in files, for example, nucleotide sequence versus amino acid sequence. We take into account data types and formats commonly used for inputs and ouputs of services on platform EMBOSS. Most platforms we visited use the same categories of data types and formats. Compared to XML formats such as BioXSD, our abtraction represents contents at a higher abtraction level. We only consider information relevant for our matching between input and output data of services. We ignore, for example some type attributes and type restrictions irrelevent for current input and ouput data used in services. If necessary, they can easily be added.

Abstraction of types is straightforward for XML formats thanks to XML schemas. For textual formats, informal specifications must be studied to derive a structural representation. Our experiment with genomics types shows that type expressions recur frequently, they easily be reused after being defined once. Since the most common data types are defined, there is increasingly less types to define. The BioXSD initiative defines several data types for common bioinformatics web services. Specialized XML formats, such as phyloXML [8] for phylogenetic data and PDBML [20] for systems biology, exist for sub-domains of bioinformatics. Moreover, XML alternatives are provided for some textual formats (e.g., GFF [21]) and some platforms define their own XML format (e.g., Uniprot XML [22]). For our experiment, the abstraction of types is done manually but the spreading of the above mentioned solutions will facilitate the task. We can even, expect automatic or semi-automatic abstraction processes. The defined types represent inputs and outputs of current genomics services. Tables 2 and 3 show exemples of services and types. In our approach, adding a new service requires two steps. Firstly, identify the abstract types used as inputs and ouputs of the service. Secondly, implement, if they do not already exist, the converters between XML abstract format and concrete formats, since our types define an XML format. Unlike other approaches, it is not neccessary to define the converters for all pairs of existing formats.

Table 2. Examples of services

Service	Source	Inputs	Outputs	task
Blast	BioXSD	biosequence database URI	biosequence	search
Blastp	EMBL-EBI	Fasta sequence (typed) database URI	BlastResult	search
ClustalWFastaCollection	BioMoby	Fasta files	MSF	alignment
ClustalW	BioXSD	biosequence ($>= 2$)	alignment	alignment
maskfeat	EMBOSS	EMBL sequence	Fasta sequence	handling

Table 3. Examples of formats and types

Type	Type in our representation	Represents
Fasta sequence	ComplexProteinSequence | ComplexBiosequence	(typed) sequences
BlastResult, Fasta files	ListOfComplexBiosequences | ListOfBiosequences	a list of sequences
EMBL sequence	AnnotatedSequence	an annotated sequence
BioXSD biosequence	Biosequence | ComplexBiosequence	one sequence
BioXSD Alignment, MSF	SequenceAlignment	an alignement of sequences
Database URI	DatabaseReference	a reference to database

5 Experiment

We used our algorithm to build a graph of connections between services in bioinformatics. We selected 30 services from platforms EMBOSS [3], EBI (European Bioinformatics Institute) [23], BioMoby [24] and services adopting the BioXSD format. Other variations and similar categories are provided in platforms but, as mentionned above, their input and output types do not change in general, and they would add little to our experiment.

From our service selection, using our matching algorithm, we automatically generated a graph of the connections between the services. Figure 5 shows an excerpt from the obtained graph. The complete graph and service list are available online[7]. The graph shows services, input/output types and conversions between types. Services are represented by rectangles, input/output types by ellipses and conversions by diamonds. Services are associated with their inputs and outputs respectively by incoming and outgoing arrows. Similarly, each conversion is associated with a source type and target type, it materializes an automatically detected conversion between two types. Connectivity of services in the graph materializes processing chains where output data are transformed according to the need of the service inputs. Conversions from external

---

[6] http://emboss.sourceforge.net/docs/#Themes, last visited June 6, 2014.
[7] http://www.irisa.fr/LIS/Members/moba/graph/view

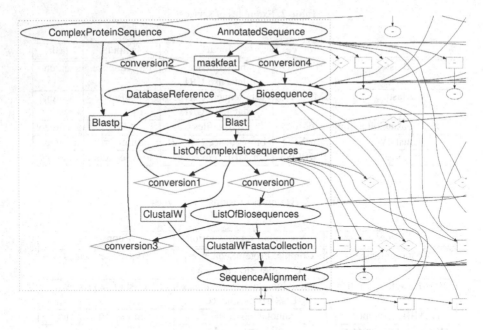

Fig. 5. Excerpt of the graph of links between services

formats to our representation are not shown in the graph. Our algorithm finds direct links when the services use the same representation (the same type) to define the same data. This is the case, for example, with the link between services *Blast* and *ClustalW*. Indirect links correspond to a *conversion$_i$*. In the following, each *conversion$_i$*(A, B) specifies a function to transform data of type A to data of type B. With *conversion0(ListOfComplexBiosequences, ListOfBiosequence)*, a list of simple sequences is derived from a list of complex sequences. The function converts each element of the list and produces a new list of the converted elements. The elements of the list are converted using Rule (LEFTSELECTION) (or Rule (RIGHTSELECTION)), and the new list is produced by Rule (MAP). *Conversion1(ListOfComplexBiosequences, Biosequence)* combines rules (CHOICE) and (LEFTSELECTION) (or (RIGHTSELECTION)) to go from a list of complex protein sequences to each simple biological sequence of the list. A simple biological sequence being a component of a complex biological sequence. *Conversion2(ComplexProteinSequence, Biosequence)* shows the generalization and specialization of types. Our algorithm detects that a sequence of proteins is also a biological sequence. The conversion uses (LEFTPOSTCHOICE) (or (RIGHTPOSTCHOICE)) to obtain a complex biological sequence from the complex protein sequence and uses Rule (LEFTSELECTION) (or (RIGHTSELECTION)) to obtain a biological sequence from the complex biological sequence. Our algorithm also individually considers the elements of a list. With *conversion3(ListOfBiosequences, Biosequence)* each element of a list of sequences can be selected, it is done by rule (CHOICE). With *conversion4(AnnotedSequence, Biosequence)*, a simple sequence is derived from an annotated sequence. A composite type is decomposed and some components are selected to feed services. This conversion corresponds to rules (LEFTSELECTION) and (RIGHTSELECTION). In above conversions, rules (TAGCHANGE), (TAGREMOVAL) and (PRIMITIVE) are used to convert between primitives and tags.

The complete graph contains 264 links between services out of the 900 possible links between 30 services (30x30). Our program therefore finds numerous links, but remains specific enough to be useful. Among those links, 88 are direct links, i.e., do not imply any conversion. Our convertibility relation therefore enables a three-fold increase of the number of links between services. The 30 services use 26 different types, among which 10 types are in fact ad-hoc and not decomposable (e.g., pictures, reports). Our program has identified 27 possible conversions between the 16 composite types, out of the 256 possible ones (16x16). This again shows that our approach is both productive and specific.

We presented the graph of the experiment to developers and users of the GenOuest bioinformatics platform. They pointed out that "it is remarkable that the central role of sequence alignment is so visible in the graph". They stated that "the produced graph has a pedagogical interest". Indeed, in a typical course handout[8], a graph produced by hand related to a library of bioinformatics services contains similar services and connections as our graph. However, being more complete in the modeling of input and output data, our approach offers more flexibility on input and output types. Thus our algorithm provides more explicit connections and differentiation between categories of services. It also reveals possible conversions between input and output data, that creates new connections. Moreover, our graph is machine processable, it shows a proof of the connections created between services and can quickly take into account new changes on data types and services. With a growing set of currently over 1500 available tools, it is unlikely that people can produce the graph by hand. Our main objective is to guide biologist when composing workflows. One perspective is to take into account other aspects of services. When constructing a workflow, input and output types play a central role in selection of services by setting constraints on applicable services, but are not the only criteria for selection biologist. It is also necessary to represent the services by their functional and non-functional properties (e.g., bioinformatics task performed, quality of results, provenance, efficiency, popularity). The GenOuest developpers nevertheless mentioned that a user with domain knowledge would already find useful support in the produced graph to select services for a workflow among the possibilities given by the graph. Thus, they validated that the graph generated by our approach detects relevant information and produces, in a systematic way, knowledge usually acquired by experience. The costly step of the approach is the production of abstract types currently done by hand. They highlighted some interesting perspectives. Our data abstractions could be enriched from ontologies, especially EDAM, which will significantly facilitate the type abstraction step and allow integrating business facets. In addition, data in Genomics (e.g., phylogenetic trees) and in others domains (e.g., metabolism) have to be added to the experiment.

---

[8] Presentation of services of Wisconsin Package- Olivier Collin - CNRS Roscoff - Formation Génopole Ouest - november 2002.

6   Related Work

The development of automatic solutions for service composition is a response to the time-consuming and error prone methods currently used to manage service selection and service mediation during composition of services. Mediating incompatible services requires an identification of the categories of mismatches. The work of Li et al. [11] provides a multi-dimension classification of mismatches. It identifies syntactic and semantic mismatches of functional and non-functional properties. Mismatches of functional properties include signature mismatches that occur on the structure and semantics of parameters such as data types. That systematic classification of service composition mismatches helps understand the problem. It also helps find appropriate solutions for each case. Based on a similar categorization of mismatches, Velasco-Elizondo et al. [12] propose to resolve data mismatches. They automate the identification of existing converters and the determination of the relevant ones for users QoS expectations. It is oriented by formats and structures but it does not allow to consider parts of a data to determine a converter. Stroulia et al. [13] propose to assess similarity between WSDL (Web Service Description Language) specifications based on the structure of their data types, messages, operations and textual descriptions. That approach exploits the structure of data types but is designed for service discovery, it does not focus on data conversion and service mediation. The previous contributions take into account the syntax (structure) of data but they are not designed to generate converters, and they do not allow to recursively reason on the composite structure of data.

Many contributions use semantic description of resources and propose methods, more or less automatic, to manage inputs and outputs when composing services. Our approach is similar to Dibernado et al. [14] to manage input and output types of services. They take into account composite types based on the relations of generalization and composition of ontologies. However, their approach only uses the *hasA* relation to (de)compose types. It does not explicitly take into account constructors tag, union and optional. It allows to extract elements of the left hand type but does not decompose the right hand type during matching. However, being based on ontological resources, it can benefit from new relations. One of our perspectives is to use ontologies to complete the description of the types beyond our structural description. Lebreton et al. [15] propose to check the semantic compatibility of service parameters, using the semantic description of resources to match parameters. They, thus, confirmed the benefits of using the semantic technologies already demonstrated in earlier work [25,26]. Their work handles generalization and specialization matching but does not (de)compose types. Galaxy [4] uses a library of converters to manage links between services. Matching of service parameters is based on data formats. Services, in their implementation, take into account several formats. When a format is not provided, a format converter defined by hand is used. In that approach, there is a strong dependency between services and formats, which limits the help that tools may provide to users. In addition, as discussed above, textual formats do not promote automation. Separating data types and formats is the subject of much work, for example Kalas et al. [9]. To facilitate interoperability of tools, common XML-based formats are proposed to represent bioinformatics data. At present, few implementations use these technologies. A generalization of their use would strengthen our approach, as it would facilitate the specification of abstract data types. Possibilities offered by XML technologies to represent complex data and relationships associated to domain ontologies may be used to provide a pivot language for conversion between heteregeneous data formats.

7 Conclusion

Existing solutions to face data mismatches when constructing workflows take into account composite types in a limited way. Furthermore, conversions from output data to input data of services are often done by converters manually defined. This makes it difficult to compose services into workflows. In this paper we presented an approach that systematically detects convertibility from output types to input types. We have defined convertibility rules that exploit (de)composition as well as specialization and generalization of types. The rules automatically generate converters between input and output XML data. An experiment on bioinformatics services showed that the detected convertibilities and produced converters are relevant from a biological point of view. Furthermore, they automatically produce a graph of potentially compatible services with a connectivity higher than with the ad'hoc approaches.

Acknowledgment. We thank Olivier Collin, Olivier Dameron, Francois Moreews and Olivier Sallou for their expertise in bioinformatics services and workflows and for enriching discussions.

References

1. Oinn, T., Greenwood, M., Addis, M., Ferris, J., Glover, K., Goble, C., Hull, D., Marvin, D., Li, P., Lord, P.: Taverna: Lessons in creating a workflow environment for the life sciences. Concurrency and Computation: Practice and Experience 18(10), 1067–1100 (2006)
2. Gundersen, S., Kalas, M., Abul, O., Frigessi, A., Hovig, E., Sandve, G.K.: Identifying elemental genomic track types and representing them uniformly. BMC Bioinformatics 12, 494 (2011)
3. Rice, P., Longden, I., Bleasby, A.: Emboss: the european molecular biology open software suite. Trends in Genetics 16(6), 276–277 (2000)
4. Goecks, J., Nekrutenko, A., Taylor, J., Team, T.G.: Galaxy: a comprehensive approach for supporting accessible, reproducible, and transparent computational research in the life sciences. Genome Biology 11(8), R86 (2010)
5. Ménager, H., Gopalan, V., Néron, B., Larroudé, S., Maupetit, J., Saladin, A., Tufféry, P., Huyen, Y., Caudron, B.: Bioinformatics applications discovery and composition with the mobyle suite and mobyleNet. In: Lacroix, Z., Vidal, M.E. (eds.) RED 2010. LNCS, vol. 6799, pp. 11–22. Springer, Heidelberg (2012)
6. Wassink, I.H.C., van der Vet, P.E., Wolstencroft, K., Neerincx, P.B.T., Roos, M., Rauwerda, H., Breit, T.M.: Analysing scientific workflows: Why workflows not only connect web services. In: SERVICES I, pp. 314–321 (2009)
7. Seibel, P.N., Krüger, J., Hartmeier, S., Schwarzer, K., Löwenthal, K., Mersch, H., Dandekar, T., Giegerich, R.: Xml schemas for common bioinformatic data types and their application in workflow systems. BMC Bioinformatics 7, 490 (2006)

8. Han, M.V., Zmasek, C.M.: phyloxml: Xml for evolutionary biology and comparative genomics. BMC Bioinformatics 10, 356 (2009)
9. Kalas, M., Puntervoll, P., Joseph, A., Bartaseviciute, E., Töpfer, A., Venkataraman, P., Pettifer, S., Bryne, J.C., Ison, J.C., Blanchet, C., Rapacki, K., Jonassen, I.: Bioxsd: the common data-exchange format for everyday bioinformatics web services. Bioinformatics 26(18)
(2010)
10. Embley, D.W., Xu, L., Ding, Y.: Automatic direct and indirect schema mapping: Experiences
and lessons learned. SIGMOD Record 33(4), 14–19 (2004)
11. Li, X., Fan, Y., Jiang, F.: A classification of service composition mismatches to support service mediation. In: GCC, pp. 315–321 (2007)
12. Velasco-Elizondo, P., Dwivedi, V., Garlan, D., Schmerl, B., Fernandes, J.M.: Resolving data
mismatches in end-user compositions. In: Dittrich, Y., Burnett, M., Mørch, A., Redmiles, D.
(eds.) IS-EUD 2013. LNCS, vol. 7897, pp. 120–136. Springer, Heidelberg (2013)
13. Stroulia, E., Wang, Y.: Structural and semantic matching for assessing web-service similarity.
Int. J. Cooperative Inf. Syst. 14(4), 407–438 (2005)
14. DiBernardo, M., Pottinger, R., Wilkinson, M.: Semi-automatic web service composition for
the life sciences using the biomoby semantic web framework. Journal of Biomedical Informatics 41(5), 837–847 (2008)
15. Lebreton, N., Blanchet, C., Claro, D.B., Chabalier, J., Burgun, A., Dameron, O.: Verification
of parameters semantic compatibility for semi-automatic web service composition: a generic
case study. In: Taniar, D., Pardede, E., Nguyen, H.-Q., Rahayu, J.W., Khalil, I. (eds.) Int.
Conf. on Information Integration and Web Based Applications and Services, pp. 845–848.
ACM (2010)
16. Ison, J.C., Kalas, M., Jonassen, I., Bolser, D.M., Uludag, M., McWilliam, H., Malone, J.,
Lopez, R., Pettifer, S., Rice, P.M.: Edam: an ontology of bioinformatics operations, types of
data and identifiers, topics and formats. Bioinformatics 29(10), 1325–1332 (2013)
17. Wolstencroft, K., Alper, P., Hull, D., Wroe, C., Lord, P.W., Stevens, R.D., Goble, C.A.: The
myGrid ontology: bioinformatics service discovery. Int. Journal of Bioinformatics Research
and Applications 3(3), 303–325 (2007)
18. Missier, P., Wolstencroft, K., Tanoh, F., Li, P., Bechhofer, S., Belhajjame, K., Pettifer, S.,
Goble, C.A.: Functional units: Abstractions for web service annotations. In: SERVICES, pp.
306–313. IEEE Computer Society (2010)
19. Hosoya, H., Vouillon, J., Pierce, B.C.: Regular expression types for xml. In: ICFP, pp. 11–22
(2000)
20. Westbrook, J.D., Ito, N., Nakamura, H., Henrick, K., Berman, H.M.: Pdbml: the representation of archival macromolecular structure data in xml. Bioinformatics 21(7), 988–992 (2005)
21. Dowell, R.D., Jokerst, R.M., Day, A., Eddy, S.R., Stein, L.: The distributed annotation system. BMC Bioinformatics 2, 7 (2001)
22. The universal protein resource (uniprot) in 2010. Nucleic Acids Research 38(Database-
Issue), 142–148 (2010)
23. McWilliam, H., Valentin, F., Goujon, M., Li, W., Narayanasamy, M., Martin, J., Miyar, T.,
Lopez, R.: Web services at the european bioinformatics institute-2009. Nucleic Acids Research 37(Web-Server-Issue), 6–10 (2009)
24. Wilkinson, M.D., Links, M.: Biomoby: An open source biological web services proposal.
Briefings in Bioinformatics 3(4), 331–341 (2002)
25. Sirin, E., Hendler, J., Parsia, B.: Semi-automatic composition of web services using semantic descriptions. In: Web Services: Modeling, Architecture and Infrastructure Workshop in
ICEIS, vol. 2003. Citeseer (2003)
26. Ríos, J., Karlsson, T.J.M., Trelles, O.: Magallanes: a web services discovery and automatic
workflow composition tool. BMC Bioinformatics 10, 334 (2009)

A ToolBox for Conservative XML Schema Evolution and Document Adaptation

Joshua Amavi, Jacques Chabin, Mirian Halfeld Ferrari, and Pierre Réty

Univ. Orléans, INSA Centre Val de Loire, LIFO EA 4022, FR-45067 Orléans, France
{joshua.amavi,jacques.chabin,mirian,pierre.rety}@univ-orleans.fr

Abstract. We propose an algorithm that computes a mapping to obtain a conservative extension of original local schemas. This mapping ensures schema evolution and guides the construction of a document translator.

1 Introduction

Our goal is to establish a multi-system environment composed by a global central system which is a *conservative* evolution of local ones, capable of processing changes that can then be transmitted to local systems. The communication should be possible in both directions: local-to-global and global-to-local. We allow independent local services to continue working on their own data, with their own tools while permitting diagnosis and changes based on a general and complete view of all services. This scenario requires tools for dealing with type evolution and document adaptation. It can be useful as a temporary configuration, deferring complete integration until local systems are ready, or as a flexible architecture adopted by the enterprise. In this context, we suppose that $S_1, \ldots, S_n$ are local systems which deal with sets of XML documents $X_1, \ldots, X_n$, respectively, and that inter-operate with a global, integrated system S. Each set X_i conforms to schema or type constraints $\mathcal{D}_i$, while $\mathcal{D}$ is an extended type (of S) that accepts any local document from $\mathcal{D}_i$. We assume that the global system S may evolve to S', accepting more documents or rejecting some original ones. Our goal is to propose tools allowing automatic type transformation accompanied by automatic document translation. We implement a platform[1] where all our proposed tools will be available (refer to [3] for details):

- *ExtSchemaGenerator*([5]) extends a given schema G, seen as a regular tree grammar, into a new grammar G' respecting the following property: the language generated by G' is the smallest set of unranked trees that contains the language generated by G and the grammar G' is a Local Tree Grammar (LTG) or a Single-Type Tree Grammar (STTG).

---

[1] From previous work: *ExtSchemaGenerator*, available on
http://www.univ-orleans.fr/
lifo/Members/rety/logiciels/RTGalgorithms.html *XMLCorrector*, available on
http://www.info.univ-tours.fr/~savary/English/xmlcorrector.html

H. Decker et al. (Eds.): DEXA 2014, Part I, LNCS 8644, pp. 299–307, 2014.

- *XMLCorrector*([2]) corrects an XML document *w.r.t.* schema constraints expressed as a DTD (or an LTG). The corrector reads the entire XML tree t to propose solutions. *XMLCorrector* finds all solutions within a given threshold *th*.
- *MappingGen*: We propose an algorithm that applies the ideas of [5] to generate a mapping from one schema G, seen as a regular tree grammar, to an extended schema G' which will be an LTG. The resulting schema mapping m is a sequence of operations on grammar rules that indicates, step by step, how to transform G into G' following the approach in [5]. Given a mapping m we can easily compute its inverse m^{-1} or compose it to other mappings; allowing schemas to evolve.
- *XTraM*: Based on a given mapping m (from schema S to T), we propose a method to translate an XML document (or tree) t, valid *w.r.t.* S into a document t' valid *w.r.t.* T. The edit distance between t and t' is no higher than a given positive threshold *th*. Moreover, t' is the closest tree to t, obtained by changing t according to the schema modifications imposed by m. For each edit operation on S, to obtain T, we analyse what should be the corresponding update on document t. When this update violates validity, we use *XMLCorrector* to propose corrections to the subtree involved in the update.

2   Schema Evolution

An XML document is an *unranked tree*, defined in the usual way as a mapping t from a set of positions $Pos(t)$ to an alphabet Σ. We deal with XML schema represented as RTG (regular tree grammar) $G = (N, \Sigma, S, P)$, where: N is a finite set of *non-terminals*; Σ is a finite set of *terminals*; S is a set of *start symbols*, where $S \subseteq N$ and P is a finite set of *production rules* of the form $X \to a\,[R]$, where $X \in N$, $a \in \Sigma$, and R is a regular expression over N. As usual, in this paper, our algorithms start from grammars in reduced and normal form. Among RTG we are interested in local tree grammars (LTG) which have the same expressive power as DTD. We refer to [5] for details.

Conservative XML Type Extension (ExtSchemaGenerator). In order to compute an LTG that extends minimally a given RTG, we follow the idea of *ExtSchemaGenerator*, presented in [5]. This method is very simple when dealing with the generation of an LTG from an RTG: replace each pair of competing non-terminals by a new non-terminal, until there are no more competing non-terminals. The regular expression of a new non-terminal rule is the disjunction of the regular expressions associated to competing non-terminals.

Consider three hospital services (patient and treatment service, insurance service and bill service), each one having its own LTG (or DTD) as schema. Figure 1(lines 1-5) shows the RTG obtained by the union of the production rules of all these three grammars while Figure 1(lines 3-6) shows the resulting LTG. The obtained LTG is an extension of the original RTG since it generates *all* trees generated by the original RTG and possibly others as well (refer to example of Figure 3). Clearly, the obtained grammar is also an extension of each hospital

1    $H_1 \to hospital[I_1^*]$ $H_2 \to hospital[I_2^*]$ $H_3 \to hospital[I_3^*]$
2    $I_1 \to info[P \mid T]$ $I_2 \to info[C \mid Pol]$ $I_3 \to info[B]$
3    $P \to patient[S \cdot N \cdot V^*]$ $C \to cover[S \cdot PN]$ $B \to bill[S \cdot It^* \cdot D]$
4    $V \to visitInfo[Id \cdot D]$ $Pol \to policy[PN \cdot Id^*]$ $It \to item[Id \cdot PZ]$
5    $T \to treatment[Id \cdot TN \cdot PR]$ $PR \to procedure[T^*]$
6    $H_1 \to hospital[I_1^* \mid I_1^* \mid I_1^*]$ $I_1 \to info[(P \mid T) \mid (C \mid Pol) \mid B]$

Fig. 1. Lines 1-5 : RTG obtained from the union of production rules of grammars. Lines 3-6 : LTG obtained by algorithm in [5] from the RTG (lines 1-5).

service grammar. In Figure 3(a) we find a document valid $w.r.t.$ the billing local schema. It is also valid $w.r.t.$ the global schema of Figure 1(lines 3-6).

Schema Mappings. We say that a *source* schema (or grammar) evolves to a *target* schema. A *schema mapping* is specified by an operation list, denoted as an *edit script*, that should be performed on source schema in order to obtain the target schema. We propose an algorithm that generates a mapping to translate an RTG G into an LTG G', following the lines of [5] . Our mapping is composed by a sequence of *edit operations* that should be applied on the rules of grammar G in order to obtain G'.

In the following definition, let ed be an *edit operation* defined on RTG G. We denote by $ed(G)$ the RTG obtained by applying ed on G. Each edit operation is associated with a cost that can be fixed according to the user's priority. The cost of an edit script is the sum of the costs of the edit operations composing it.

Definition 1 (Edit Script). An *edit script* $m = \langle ed_1, ed_2, \ldots ed_n \rangle$ is a sequence of *edit operations* ed_k where $1 \leq k \leq n$. Let G be an RTG, an *edit script* $m = \langle ed_1, ed_2, \ldots ed_n \rangle$ is defined on G iff there exists a sequence of RTG $G_0, G_1, \ldots, G_n$ such that: (i) $G_0 = G$ and (ii) $\forall 1 \leq k \leq n$, ed_k is defined on G_{k-1} and $ed_k(G_{k-1}) = G_k$. Hence, $m(G) = G_n$. The empty *edit script* is denoted $\langle \rangle$. The cost of an edit script m is defined as $cost(m) = \Sigma_{i=1}^n (cost(ed_i))$.   □

Definition 2 (Schema Mapping). A schema mapping is a triple $\mathcal{M} = (S, T, m)$, where S is the source schema, T is the target schema, and m is an *edit script* that transforms S into T ($i.e.$, $m(S) = T$). We say that $\mathcal{M}$ is syntactically *specified by*, or, *expressed* by m.   □

Edit Operations. In this section, the problem of changing one RTG into another is treated as a tree editing problem.

Tree Representation for Production Rules. We represent the right-hand side of a production rule $X \to a[R]$ as a tree denoted t_X^r such that $t_X^r = a(t_R)$. The root of t_X^r is the terminal a which has only one subtree t_R. For example, in Figure 2, the tree on the top left corner is t_I^r with $I \to info[T.(Y.Co)]$.

Definition 3 (Well Formed Tree). A tree t representing the right-hand side of a production rule is well formed iff the following conditions are verified: (i) the root is a terminal symbol with only one child; (ii) the leaves nodes are in

$N \cup \{\epsilon\}$ and (iii) the internal nodes are in the set $\{|, ., *\}$ having exactly one child (if it is in $\{*\}$) or at least one child, in the other cases.   □

Elementary Edit Operations. In [3], we define elementary edit operations by using rewriting. Here we give an intuition of some of these operations. Given an RTG $G = (N, \Sigma, S, P)$ in normal form, an *elementary edit operation ed* is a partial function that transforms G into a new RTG G'. The *elementary edit operation ed* can be applied on G only if *ed* is defined on G. Below we distinguish four types of *elementary edit operations* on RTG, where $A \in N$:

1. Edit operations to modify the set of start symbols S: $\mathtt{set\_startelm}(A)$ and $\mathtt{unset\_startelm}(A)$ to add/delete the non-terminal A to/from S.
2. Edit operations to modify non-terminal or terminal symbols in a content model. For example, $\mathtt{ins\_elm}(X, A, u.i)$: (cf. Figure 2($ed_1$)) applies the rewrite rule $\begin{smallmatrix} op \\ x \quad y \end{smallmatrix} \longrightarrow \begin{smallmatrix} op \\ x \quad A \quad y \end{smallmatrix}$ to insert A in the position $u.i$ of tree representing the production rule associate to X;
$\mathtt{del\_elm}(X, A, u.i)$: (cf. Figure 2($ed_2$)) is the inverse operation. The other operations are: $\mathtt{rel\_root}(X, a, b)$: (cf. Figure 2($ed_3$)) to rename the root from a to b; $\mathtt{rel\_elm}(X, A, B, u)$: (cf. Figure 2($ed_4$)) to rename non-terminal A at position u into B.
3. Edit operations to modify operator symbols in a content model are similar to those described in the previous item, but apply on operator nodes. See for instance Figure 2(ed_5), (ed_6) and (ed_7).
4. Edit operations to modify the set of production rules P are: $\mathtt{ins\_rule}(A, a)$ that adds the new production rule $A \rightarrow a\,[\epsilon]$ to P and the non-terminal A to S, where $A \notin N$ and $\mathtt{del\_rule}(A, a)$ its inverse.

After each edit operation, the sets Σ and N are automatically updated to contain all and only the terminal (resp. non-terminal) symbols appearing in P.

Proposition 1. *Let G and G' be two RTG. There exist an edit script, composed only by elementary edit operation, that transforms G into G'.*   □

Non-elementary Edit Operations. For readability and cost estimation, we define short-cut operations, *i.e.*, operations seen as a one-block operation but equivalent to a sequence of *elementary edit operations*, such as $\mathtt{ins\_tree}(X, R, u.i)$, $\mathtt{ins\_treerule}(A, a, R)$ and inversely $\mathtt{del\_tree}(X, R, u.i)$, $\mathtt{del\_treerule}(A, a, R)$.

Operation Cost. For each *edit operation ed*, we define a non-negative and application-dependent cost. On the one hand, we assume that operations that do not change the language generated by the RTG G on which they were applied, are 0-cost. Their goal is just to simplify a given regular expression. For instance, $\mathtt{del\_opr}(X, opr, u.i)$ where $t_X^r(u) = t_X^r(u.i) = opr$ and $\mathtt{del\_opr}(X, opr, u.i)$ where $t_X^r(u.i) \in \{|, .\}$ and $t_X^r(u.i)$ has exactly one child, are 0-cost operations. On the other hand, we suppose that an *elementary edit operation* costs 1, while a *non-elementary edit operation* costs 5.

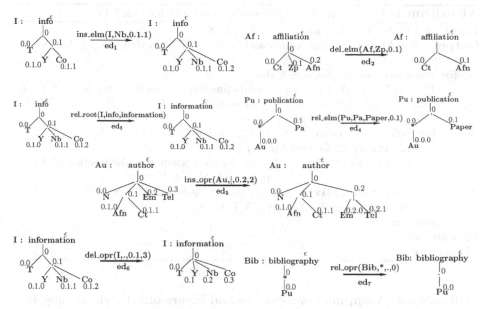

Fig. 2. Example of *elementary edit operations*

Generating a Schema Mapping (MappingGen). Algorithm 1 generates a mapping that converts an RTG in an LTG by following the ideas in [5], explained in Section 2. This algorithm starts by determining a set of competing non-terminals EC_a (lines 2-3). Then we can take arbitrarily in EC_a, one of these non-terminals (say X_0) to represent all others, *i.e.*, when merging rules of competing terminals, one non-terminal name is chosen to represent the result of the merge (line 4). Recall that edit operations always deal with a production rule in its tree-like format. The new production rule of X_0 is built in two steps. We add an OR operation as the parent of its original regular expression $reg(X_0)$ (line 5) and then we insert all regular expressions associated with its competing non-terminals as siblings of $reg(X_0)$ (line 7). In line 8 we just replace, in all production rules, non-terminals in EC_a by X_0. Original rules of non-terminals in EC_a are deleted (line 10) after, possibly, adjusting start symbols (line 9).

Proposition 2. *Let m be the mapping obtained by Algorithm 1 from an RTG G. The language $L(m(G))$ is the least LTL that contains $L(G)$. Moreover, the grammar $m(G)$ equals the one obtained by ExtSchemaGenerator.*   □

Going Further with Mappings to Support Schema Evolution. In [6], it was shown how two fundamental operators on schema mappings, namely composition and inversion, can be used to address the mapping adaptation problem in the context of schema evolution. Let $\mathcal{M}_1$ be a mapping between XML schemas S and T. When S or T evolve, $\mathcal{M}_1$ shall be adapted. By using composition and inversion operators, one can avoid mapping re-computation. We now precise the notions of composition and inversion in our context.

Algorithm 1. A mapping for transforming an RTG into an LTG

Input: A Regular Tree Grammar $G = (NT, \Sigma, S, P)$
Output: An *edit script* m between G and the LTG G' such that $L(G) \subseteq L(G')$
1: $m := \langle \rangle$
2: **for each** terminal symbol $a \in \Sigma$ **do**
3:      $EC_a = \{X_0, \ldots, X_k\}$ is a set of competing non-terminals where $term(X_i) = a$
4:      Non-terminal X_0 is chosen to represent $X_0, \ldots, X_k$
5:      Add $\texttt{ins\_opr}(X_0, |, 0, 1)$ to m
6:      **for each** non-terminal $X_i \in \{X_1, \ldots, X_k\}$ **do**
7:          Add $\texttt{ins\_tree}(X_0, reg(X_i), 0.i)$ to m
8:          Add $\texttt{rel\_elm}(Y, X_i, X_0, u)$ to m, for all u where u is the position of X_i
                 in the rule $Y \to b\,[R] \in P$
9:          Add $\texttt{set\_startelm}(X_0)$ to m where $X_0 \notin S$ and $X_i \in S$
10:         Add $\texttt{del\_treerule}(X_i, a, reg(X_i))$ to m
11:     **end for**
12: **end for**
13: **return** m

Definition 4 (Mapping composition and inversion). Given two mappings $\mathcal{M}_1 = (S, T, m_1)$ and $\mathcal{M}_2 = (T, V, m_2)$, the composition of $\mathcal{M}_1$ and $\mathcal{M}_2$ is the mapping $\mathcal{M}_1 \circ \mathcal{M}_2 = (S, V, m_1 . m_2)$. If $m_1 = \langle ed_1, \cdots, ed_n \rangle$, then the inverse of mapping $\mathcal{M}_1$ is the mapping $\mathcal{M}_1^{-1} = (T, S, m_1^{-1})$ where $m_1^{-1} = \langle ed_n^{-1}, \cdots, ed_1^{-1} \rangle$ and $ed_k^{-1} (1 \leq k \leq n)$ is defined in [3][2].

3   Adapting XML Documents to a New Type

Correcting XML Documents (XMLCorrector) In [2], given a well-formed XML tree t, a schema G and a non negative threshold th, *XMLCorrector* finds every tree t' valid *w.r.t.* G such that the edit distance between t and t' is no higher than th. Contrary to most other approaches, [2] considers the correction as an enumeration problem rather than a decision problem and computes all the possible corrections on t. The algorithm, proved to be correct and complete in [2], consists in fulfilling an edit distance matrix which stores the relevant edit operation sequences allowing to obtain the corrected trees. The theoretical exponential complexity of *XMLCorrector* is related to the fact that edit sequences and the corresponding corrections are generated and that the correction set is complete.

In this paper, contrary to [2], we do not consider all the possible corrections on t. The correction of XML documents is guided by a given mapping. For each edit operation on S, to obtain T, we analyse what should be the corresponding update on document t. When this update violates validity, we use *XMLCorrector* to propose corrections to the subtree involved in the update.

---

[2] Notice that inverses are defined in the intuitive way as, for example,
$\texttt{ins\_elm}(X, A, u) \rightleftharpoons \texttt{del\_elm}(X, A, u)$ or $\texttt{rel\_opr}(X, p, q, u) \rightleftharpoons \texttt{rel\_opr}(X, q, p, u)$.

Document Translation Guided by Mapping (XTraM). Our method consists in performing a list of changes on XML documents, in accordance with the edit operations found in the mapping. For example, adding or deleting a regular expression in a rule under the operator '.' is a mapping operation that provokes, respectively, the insertion or the deletion of a subtree in an originally valid XML tree (to maintain its validity). Similarly, renaming a non-terminal A by B, provokes the substitution of the subtree generated by A into the subtree generated by B. When local correction on XML subtrees are needed, *XMLCorrector* is used to ensure document validity.

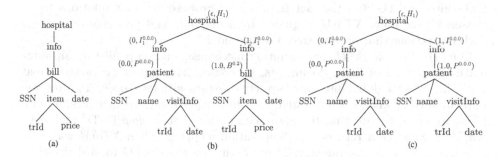

Fig. 3. (a) XML tree valid *w.r.t.* the billing local schema. (b) XML tree valid *w.r.t.* the global schema of Figure 1(lines 3-6). (c) Tree resulting from the translation of (b) into the patient local schema (cf. Figure 1(lines 1-5)). Trees (b) and (c) are annotated.

Consider an XML tree t valid *w.r.t.* schema S and a mapping m from S to T. Our method can be summarized in two steps:

1. Since t belongs to the language $L(S)$, it is possible to associate a non-terminal A with each tree node position p generated by this non-terminal. We analyse t, detect each non-terminal and annotate it with its corresponding position u in the used production rule. This annotation respects the format (p, A^u). For example, in Figure 3(b), we notice that the tree node *bill* is generated by the non-terminal B whose position in $t_{I_1}^r$ is 0.2, noted as $(1.0, B^{0.2})$.
2. Each edit operation ed in m activates a set of modifications on t. When ed transforms a grammar into a new grammar containing the previous one, the set of modifications is empty. Otherwise, our method consists in traversing t (marked as in step 1) in order to find the tree positions which may be affected due to ed. Modifications on t are defined according to each edit operation and are not detailed here due to the lack of space. Obviously, if no position is affected, t does not change.

Figure 3(b) shows an XML document concerning patients and bills. This document is valid *w.r.t.* the global schema but not valid *w.r.t.* to any local schemas. Translating the document of Figure 3(b) into a document respecting the patient schema we obtain the document of Figure 3(c).

4   Related Work and Concluding Remarks

Much other work deals with schema evolution. In [6] second order logic is needed to express some mapping compositions. This approach is the basis of [1,8,12]. We believe that the use of edit operations makes our approach simpler than theirs and gets on well with our previous work concerning XML document correction. Proposals, such as those in [7,9,10,4,11], use edit operations. ELaX [10] and Exup [4] are a domain-specific language that proposes to handle modifications on XSD and to express such modifications formally. An important originality of our approach is the *automatic* generation of a *conservative extension* of an RTG into an LTG and the fact that it may propose different solutions to be chosen by the user. *XTraM* is guided by a mapping and produces documents with corrections that do not exceed a threshold.

Our *ToolBox* offers schema evolution mechanisms accompanied by an automatic adaptation of XML documents. Its conservative aspect guarantees great flexibility when a global integrated system co-exists with local ones. A prototype, implemented in Java, is been tested. As a first experiment, we have produced an LTG, in $24ms$, by merging the grammars obtained from *dblp* DTD[3] and *HAL* XSD[4]. *MappingGen* returned a 19-operation mapping. Then *XTraM* was used to adapt a 52-node document valid *w.r.t.* the computed LTG toward the *HAL* grammar, giving, in this case, 36 solutions in 22.6 *s*. As in this test, all possible translations can be considered, but the user may also interfere in an intermediate step, making choices before the end of the complete computation - guiding and, thus, restricting the number of solutions.

References

1. Amano, S., Libkin, L., Murlak, F.: XML schema mappings. In: Proceedings of the 28th ACM Symposium on Principles of Database Systems, PODS 2009, pp. 33–42. ACM, New York (2009)
2. Amavi, J., Bouchou, B., Savary, A.: On correcting XML documents with respect to a schema. The Computer Journal 56(4) (2013)
3. Amavi, J., Chabin, J., Halfeld Ferrari, M., Réty, P.: A toolbox for conservative XML schema evolution and document adaptation. CoRR abs/1406.1423 (2014)
4. Cavalieri, F., Guerrini, G., Mesiti, M.: Updating XML schemas and associated documents through Exup. In: Proceedings of the IEEE 27th International Conference on Data Engineering, ICDE 2011, pp. 1320–1323. IEEE Computer Society, Washington, DC (2011)
5. Chabin, J., Halfeld Ferrari, M., Musicante, M.A., Réty, P.: Conservative Type Extensions for XML Data. In: Hameurlain, A., Küng, J., Wagner, R. (eds.) TLDKS IX. LNCS, vol. 7980, pp. 65–94. Springer, Heidelberg (2013)
6. Fagin, R., Kolaitis, P.G., Popa, L., Tan, W.C.: Schema mapping evolution through composition and inversion. In: Bellahsene, Z., Bonifati, A., Rahm, E. (eds.) Schema Matching and Mapping, pp. 191–222. Springer (2011)

---

[3] http://dblp.uni-trier.de/xml/dblp.dtd
[4] http://import.ccsd.cnrs.fr/xsd/generationAuto.php?instance=hal

7. Horie, K., Suzuki, N.: Extracting differences between regular tree grammars. In: Proceedings of the 28th Annual ACM Symposium on Applied Computing, SAC 2013, pp. 859–864. ACM, New York (2013)

8. Jiang, H., Ho, H., Popa, L., Han, W.S.: Mapping-driven XML transformation. In: Proceedings of the 16th International Conference on World Wide Web, WWW 2007, pp. 1063–1072. ACM, New York (2007)

9. Leonardi, E., Hoai, T.T., Bhowmick, S.S., Madria, S.K.: Dtd-diff: A change detection algorithm for dtds. Data Knowledge Engineering 61(2), 384–402 (2007)

10. Nösinger, T., Klettke, M., Heuer, A.: XML schema transformations. In: Decker, H., Lhotská, L., Link, S., Basl, J., Tjoa, A.M. (eds.) DEXA 2013, Part I. LNCS, vol. 8055, pp. 293–302. Springer, Heidelberg (2013)

11. Suzuki, N., Fukushima, Y.: An XML document transformation algorithm inferred from an edit script between DTDs. In: Proceedings of the 19th Conference on Australasian Database, ADC 2008, vol. 75, pp. 175–184. Australian Computer Society, Inc., Darlinghurst (2007)

12. Yu, C., Popa, L.: Semantic adaptation of schema mappings when schemas evolve. In: Proceedings of the 31st International Conference on Very Large Data Bases, VLDB 2005, pp. 1006–1017. VLDB Endowment (2005)

Refinement Correction Strategy for Invalid XML Documents and Regular Tree Grammars

Martin Svoboda* and Irena Holubová (Mlýnková)

XML and Web Engineering Research Group,
Faculty of Mathematics and Physics, Charles University in Prague
Malostranske namesti 25, 118 00 Prague 1, Czech Republic
svoboda@ksi.mff.cuni.cz

Abstract. It was shown that real-world XML documents often contain errors that make their processing complicated or impossible. Having a potentially invalid XML document, our goal is to find its structural corrections with respect to a provided schema represented in DTD or XSD languages. In this paper we propose several improvements of algorithms we introduced earlier, extend our model to support the full expressive power of regular tree grammars and we still guarantee that all the minimal corrections are found regardless the extent of document invalidity. According to experiments, our new correction algorithm outperforms the existing ones by more than two times, achieves near linear execution times in practice and works successfully on documents of sizes up to two orders of magnitude higher than those assumed so far.

1 Introduction

High number of XML documents [3] together with distributed and dynamic nature of the Web are probably only some of the reasons why these documents often contain various types of errors, as it was detected not only by Mlýnková et al. [6]. Processing of such documents may then become more difficult or not possible at all. Therefore, correcting documents themselves instead of adjusting well-functioning processing algorithms seems to be a good strategy.

In this paper we limit ourselves only to the structural correction of invalid XML documents. Assume hence that we have one well-formed document and a schema in DTD [3] or XML Schema (XSD) [4] languages to which this document should conform. When it does not abide by all its structural requirements on elements and their intended nesting, our goal is to find its suitable corrections.

Our work is based on Bouchou et al. [2]. Their correction model uses edit operations via which new subtrees can be recursively inserted and existing ones repaired or deleted. Sequences of sibling nodes are generated and inspected dynamically, one by one, step by step, always following allowed finite automaton transitions. Next, we were also inspired by uncertain querying with possible and certain answers by Staworko and Chomicky [8]. Beside that, other inconsistencies are discussed as well. For example, functional dependencies by Flesca et al. [5] or keys and multivalued dependencies by Tan et al. [11].

---

* Corresponding author.

H. Decker et al. (Eds.): DEXA 2014, Part I, LNCS 8644, pp. 308–316, 2014.
© Springer International Publishing Switzerland 2014

In this work we return back to the model and algorithms we introduced earlier [9,10] and significantly revise and improve them. On the contrary to [2] and its revision by Amavi et al. [1], we inspect allowed node sequences statically within recursively nested correction multigraphs that allow us to transform the problem of finding corrections to the problem of finding shortest paths.

Contributions of this paper can be summarized as follows:

- We separated correction strategies from execution approaches and obtained and mutually compared a whole set of correction algorithms.
- We assumed the expressive power of the whole class of regular tree grammars and proposed important improvements of the refinement strategy.
- We implemented all the newly proposed correction algorithms and made this implementation and source files publicly available.
- We conducted experiments over documents of sizes of even 100,000 nodes, i.e. sizes up to 2 orders of magnitude higher than considered so far.
- We detected that our new algorithm is 2 times faster and has near linear execution times in practice despite the polynomial worst case complexity.

In Section 2 of this paper we describe basics of our correction model. Then we outline the proposed correction algorithms in Section 3 and also provide results of conducted experiments in Section 4. Finally, Section 5 concludes.

2   Corrections

We first describe how XML documents and schemata are modeled and then we move forward to basics of our correction model.

2.1   Data Trees and Regular Tree Grammars

Assuming that $\mathbb{N}_0^*$ is a set of all finite words over the set of non-negative integers $\mathbb{N}_0$, we model a particular XML document as a *data tree* $\mathcal{T} = (D, lab, val)$, where $D \subset \mathbb{N}_0^*$ is an *underlying tree* capturing the tree structure, *lab* is a *labeling function* assigning element names $\mathbb{E}$ to nodes D and *val* is a partial *value function* assigning data values $\mathbb{V}$ to data leaf nodes from D.

Next, we model structural restrictions of a particular DTD or XSD schema using a *regular tree grammar* [7] defined as a tuple $\mathcal{G} = (N, T, S, P)$, where N is a set of *nonterminal symbols*, T a set of *terminal symbols*, $S \subseteq N$ is a set of *starting symbols*, and, finally, P is a set of *production rules*, each of the form $[t, r \rightarrow n]$, where $t \in T$, $n \in N$ and r is a deterministic (1-unambiguous) regular expression over N with its language $L(r)$. Without loss of generality, for each $t \in T$ and $n \in N$ there exists at most one $[t, r \rightarrow n] \in P$.

Data tree $\mathcal{T}$ is *valid* against $\mathcal{G}$, if there exists at least one *interpretation tree* $\mathcal{N} = (D, int)$, where *int* is a function $D \rightarrow N$ such that $\forall p \in D$ there exists a production rule $[t, r \rightarrow n] \in P$ satisfying $int(p) = n$, $lab(p) = t$ and $int(p.0).int(p.1)\ldots int(p.k) \in L(r)$, $k = fanOut(p)-1$. If $p = \epsilon$, then $int(p) \in S$.

Example 1. Suppose we have the following XML document:

```
<a> <x><c/></x> <d><c/></d> <d><c/><a/></d> </a>
```

It corresponds to a data tree $\mathcal{T} = (D, lab, val)$ in Figure 1(a) with an underlying tree $D = \{\epsilon, 0, 0.0, 1, 1.0, 2, 2.0, 2.1\}$, element labels $lab = \{(\epsilon, a), (0, x), (0.0, c),$ $(1, d), (1.0, c), (2, d), (2.0, c), (2.1, a)\}$ and data values $val = \emptyset$.

Next, let $\mathcal{G} = (N, T, S, P)$ be a regular tree grammar with nonterminals $N = \{A, B, C, D_A, D_B\}$, terminals $T = \{a, b, c, d\}$, starting symbols $S = \{A, B\}$ and rules $\mathcal{F}_1 = [a, C.D_A{}^* \to A]$, $\mathcal{F}_2 = [b, D_B{}^* \to B]$, $\mathcal{F}_3 = [c, \epsilon \to C]$, $\mathcal{F}_4 = [d, C^* \to D_A]$ and $\mathcal{F}_5 = [d, A|B|C \to D_B]$. $\mathcal{T}$ is not valid against $\mathcal{G}$.

2.2   Edit Operations and Tree Distances

An *edit operation* is a partial function that transforms a data tree $\mathcal{T}_0 = (D_0, lab_0, val_0)$ into a data tree $\mathcal{T}_1 = (D_1, lab_1, val_1)$, denoted as $\mathcal{T}_0 \xrightarrow{e} \mathcal{T}_1$. For suitable $p \in \mathbb{N}_0^*$ we introduce operations $addLeaf(p, a)$ to insert a new leaf node $a \in \mathbb{E}$ to a position p, $removeLeaf(p)$ to remove a leaf node p, and $renameNode(p, a)$ to change a label of a node p to $a \in \mathbb{E}$. By composing edit operations into sequences, entire new subtrees can be inserted, existing ones deleted or recursively repaired. Let $\mathcal{T}_0, \ldots, \mathcal{T}_n$ be data trees and $e_1, \ldots, e_n$ edit operations such that $\forall i \in \mathbb{N}_0$, $0 \leq i < n$: $\mathcal{T}_i \xrightarrow{e_{i+1}} \mathcal{T}_{i+1}$ are correctly defined. Then $S = \langle e_1, \ldots, e_n \rangle$ is an *edit sequence* transforming $\mathcal{T}_0$ into $\mathcal{T}_n$. Given an edit operation e, we define $cost(e)$ to be a positive *cost* of e. The lower the cost is, the smaller the changes are. Intuitively, $cost(S) = \sum_{i=1}^{k} cost(e_i)$.

Suppose now that $\mathcal{T}_1$ and $\mathcal{T}_2$ are two data trees and $\mathcal{S}$ is a set of all edit sequences capable to transform $\mathcal{T}_1$ into $\mathcal{T}_2$. We define a data tree *distance* of $\mathcal{T}_1$ and $\mathcal{T}_2$ to be $dist(\mathcal{T}_1, \mathcal{T}_2) = \min_{E \in \mathcal{S}} cost(E)$. Given a grammar $\mathcal{G}$ and its language $L(\mathcal{G})$ of all data trees valid against $\mathcal{G}$, we define a distance between $\mathcal{T}_1$ and $L(\mathcal{G})$ as $dist(\mathcal{T}_1, L(\mathcal{G})) = \min_{\mathcal{T}_2 \in L(\mathcal{G})} dist(\mathcal{T}_1, \mathcal{T}_2)$.

2.3   Grammar Contexts and Correction Intents

Having a data tree $\mathcal{T}$ and a grammar $\mathcal{G}$, our goal now is to efficiently find all its corrections with the minimal cost, i.e. $dist(\mathcal{T}, L(\mathcal{G}))$. We initiate processing of $\mathcal{T}$ at its root node ϵ and continue towards its leaves. Assume now that we are at a particular node p and our goal is to correct it, i.e. to correct a sequence of its child nodes $u = \langle u_1, \ldots, u_k \rangle$. This means we have already decided which label t this node p should have and to which nonterminal n it should be mapped, i.e. we have selected a rule $\mathcal{F} = [t, r \to n]$ with a regular expression r to which u should conform, i.e. we have defined a *grammar context* $\mathcal{C}_{t,n}$ to be used.

The idea of correcting u against r is directly motivated by the traditional Levensthein metric and correction of ordinary words, i.e. is based on a traversal of the state space of the automaton $\mathcal{A}_r = (Q, \Sigma, \delta, q_0, F)$ recognizing $L(r)$, where Q is a *set of states*, Σ is an *input alphabet*, δ is a partial *transition function* $Q \times \Sigma \to Q$, $q_0 \in Q$ is an *initial state*, and $F \subseteq Q$ is a set of *accepting states*.

At the beginning we have the entire sequence u unprocessed and we are at the initial state q_0. The goal is to process u completely and terminate at any of the accepting states F. Now assume that we have already processed the first $0 \leq s \leq k$ nodes of u, we are at some automaton state q_s and there is a node u_{s+1} ahead in the input (if any). All the actions we can consider now are defined by values of the transition function δ defined from the current state q_s.

Without further details, let us now denote the outlined goal of correcting u as a *correction intent* $\mathcal{I}$. According to the introduced edit operations, we distinguish between **insert** (to insert a new subtree), **delete** (to delete an existing one), **repair** and **rename** (to recursively process an existing one, optionally renaming its root node) and **correct** (to initiate the entire $\mathcal{T}$ correction) intent types.

2.4 Correction Multigraphs and Intent Repairs

Now the question is, how to evaluate a particular correction intent $\mathcal{I}$, i.e. to statically represent and inspect all its nested intents. For this purpose we build a *correction multigraph* $\mathcal{M}_\mathcal{I} = (V, E, v_S, V_T)$, where (V, E) forms a directed multigraph, V as a set of *vertices* corresponds to the set of all the sequence u positions combined with automaton $\mathcal{A}_r$ states Q, E as a set of *edges* corresponds to all $NestedIntents(\mathcal{I})$, $v_S = (0, q_0)$ is a *source vertex*, $v_S \in V$, and, finally, $V_T = \{v_T \mid v_T = (k, q_T), q_T \subset F\}$ is a set of *target vertices*.

Having $\mathcal{M}_\mathcal{I}$ for $\mathcal{I}$, we can now transform the problem of correcting the node sequence u to the problem of finding paths in $\mathcal{M}_\mathcal{I}$; in particular the shortest paths P_{v_S,V_T}^{min} from the source vertex v_S to any of the target vertices in V_T. Having found them, we encapsulate them into an *intent repair* structure $\mathcal{R}_\mathcal{I}$ with *cost* defined as an overall *correction cost* of $\mathcal{I}$ and with $\mathcal{N}_\mathcal{I}$ as a subgraph of (V, E) induced by P_{v_S,V_T}^{min}. When a repair for the starting intent is computed, the entire data tree $\mathcal{T}$ correction is finished. We could formally translate its repair to get all the edit sequences, but we propose to let the user choose the best correction directly without the need of their unnecessary unfolding.

Example 2. When assuming unit costs for all edit operations, Figure 1 depicts all the 5 minimal corrections of the data tree $\mathcal{T}$ against grammar $\mathcal{G}$ from Example 1, all of them with the cost equal to 3. The translated edit sequences are:

$S_1 = \langle addLeaf(0, c), renameNode(1, d), renameNode(3.1, c)\rangle$,
$S_2 = \langle addLeaf(0, c), renameNode(1, d), removeLeaf(3.1)\rangle$,
$S_3 = \langle renameNode(0, c), removeLeaf(0.0), renameNode(2.1, c)\rangle$,
$S_4 = \langle renameNode(0, c), removeLeaf(0.0), removeLeaf(2.1)\rangle$ and
$S_5 = \langle renameNode(\epsilon, b), renameNode(0, d), removeLeaf(2.1)\rangle$.

3 Algorithms

Now we describe how correction multigraphs are actually constructed using 3 different correction strategies (DEF for *default*, EXP *exploring*, RFN *refinement*), 5 execution approaches (N1 for *nesting*, I1 and IN for *invoking* and F1 and FN for *forwarding*), and 2 signature handling modes (E for *enabled* and D *disabled*).

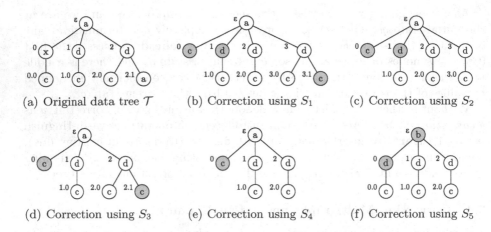

(a) Original data tree $\mathcal{T}$ (b) Correction using S_1 (c) Correction using S_2

(d) Correction using S_3 (e) Correction using S_4 (f) Correction using S_5

Fig. 1. Original data tree $\mathcal{T}$ and all its minimal corrections against grammar $\mathcal{G}$

3.1 Correction Strategies

Default Given an intent $\mathcal{I}$ to be evaluated, we construct the entire multigraph $\mathcal{M}_\mathcal{I}$ with all its vertices and edges and request recursive evaluation of all its nested intents. Acquiring their repairs and costs, we simply start searching for all the shortest paths P_{v_S,V_T}^{min} using the traditional Dijkstra's algorithm.

Exploring We start with an empty multigraph $\mathcal{M}_\mathcal{I}$ and integrate the Dijkstra's algorithm directly into the multigraph exploration. Therefore we explore and evaluate only those nested intents that are really required to find P_{v_S,V_T}^{min}.

Refinement So far the Dijkstra's algorithm was always provided with fully resolved costs on all the involved edges. Now we describe a strategy that relies only on cost approximations and repairs evaluated only partially. In other words, our goal is to scatter the evaluation into smaller refinement steps. Thus we can reveal unpromising ways of correction even to the depth of the recursive intent nesting.

We say that an edge e is *closed* if the associated repair $\mathcal{R}_\mathcal{I}$ is already fully evaluated. Otherwise e is *open*. Next, assume we have a reached vertex v and E_v^{in} are all explored edges ingoing to v. We say that v is *complete*, if there exists at least one closed edge in E_v^{in} and there are no open edges in E_v^{in} at all or the distance to v using them would be greater than the minimal distance to v using only closed edges from E_v^{in}. Otherwise v is *incomplete*.

When searching for the shortest paths P_{v_S,V_T}^{min}, the loop over the reached vertices from the Dijkstra's algorithm changes only a bit. When a selected vertex v is complete, we explore all its outgoing edges and initialize their (now empty) repairs without requesting their evaluation, since their first cost estimations are available immediately (because of the intended edit operations). In case vertex v is incomplete, we request evaluation of the most promising open edges (those providing minimal distance to v). We also limit their evaluation by assigning appropriate quotas via which we control and interrupt the evaluation right at the moment when the required progress is attained.

3.2 Intent Signatures

The main problem so far is that we are often forced to evaluate different or even the same intents despite the resulting repairs can be exactly the same. Without further details, let us introduce the notion of an *intent signature*. When two intents $\mathcal{I}_1$ and $\mathcal{I}_2$ have the same signatures $sig(\mathcal{I}_1) = sig(\mathcal{I}_2)$, their repairs will always be identical $\mathcal{R}_{\mathcal{I}_1} = \mathcal{R}_{\mathcal{I}_2}$. Hence, whenever a new repair is computed, we put it into a store using its intent signature as a key; and before a newly explored intent is about to be evaluated, we first try to fetch the repair from the store.

3.3 Execution Approaches

Whereas correction strategies define which nested intents are explored and requested for the (full or partial) evaluation, and so they differ in their (horizontal or vertical) pruning capabilities, execution approaches, on the other hand, technically define how evaluation requests are created and how their execution is initiated and terminated – in just one or multiple threads in parallel (1 or N).

Nesting We directly call the recursive correction procedure. After it attains its goal, it reaches its end, the context is returned back and the processing continues.

Invoking We acquire a worker thread, initiate its execution and go to sleep. After the nested thread attains its goal, it notifies our own thread. Then the nested thread terminates, we are woken up and the processing continues.

Forwarding We acquire a worker thread, initiate its execution and terminate our own thread. After the nested thread attains its goal, it acquires a new thread to resume our processing. Then the nested thread terminates and we continue.

4   Experiments

All the proposed correction strategies, execution approaches and signature handling modes are free to be combined. We managed to integrate all of them into just one universal implementation in Java and made it publicly available[1]. XML documents used in the experiments were generated by our proprietary generator and are based on a single type tree grammar from Example 1.

Our objective is to experimentally compare all the described configurations and show that the revised refinement strategy can outperform our older algorithms (DEF-N1-D and EXP-N1-E [9]; RFN-F1-E without improvements [10]).

---

[1] http://www.ksi.mff.cuni.cz/~svoboda/projects/corrector/

Intent Signatures We first confirm the effect of enabled signatures on reducing the number of intents that need to be evaluated (denoted as tasks). We computed averages over 10 data trees, each with 100 nodes, average maximal fan-out 4 and average minimal and maximal depths 6 and 11 respectively. For example, in case of the DEF strategy, the difference can be higher than 2 orders of magnitude. Thus, caching of repairs is absolutely necessary, even for very small documents.

Horizontal Pruning Now, we would like to verify the impact of EXP and RFN strategies on reducing the number of created tasks due to the limited need of correction multigraph exploration. According to averages over documents of 1,000 nodes (10 instances with average maximal fan-out 5 and average depths 8 and 14 resp.), EXP saved 7% and RFN even 31% of tasks comparing to DEF.

Multigraph Sizes The number of created tasks is in a direct relation with characteristics of constructed correction multigraphs. For instance and over the same documents as above, RFN was able to reduce the number of explored vertices by 49% and edges by 68%, both compared to DEF which always constructs the entire multigraphs. In general, multigraph sizes are reasonably small.

Vertical Pruning Let us now focus on the dual pruning effect of the RFN strategy. Using documents of 1,000 to 10,000 nodes (always 10 instances with average maximal fan-out 5 and depths starting at 8 and ending at 20 resp.), we showed not only that just 56% of explored edges led to the creation of new tasks (due to signatures), but at least 28% of these tasks had not been called even once, since their first and immediate cost estimation was sufficient enough.

Execution Times When focusing on execution times, DEF cannot be better than EXP. However, different execution approaches now start to play their important role. All the presented experiments were run on a casual laptop (Intel Core i5 M560, 2.67 GHz, 2 cores, HT, 64-bit, 4 GB 1333 MHz DDR3 memory) with Windows 7 Professional SP1 and Java SE 7 Update 45.

Figure 2 shows average execution times (without the repair translation phase) for documents of 1,000 to 10,000 nodes (10 instances, maximal fan-out 7, depths

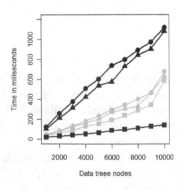

Fig. 2. Average execution times

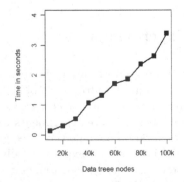

Fig. 3. Times for the RFN strategy

from 7 to 16, 5 ignored runs and then 10 measured ones). Gray lines are used for EXP, black for RFN strategies; squares for N1, circles for IN and triangles for FN approaches. Unlike our expectations, both IN and FN approaches did not perform well due to the high parallelism overhead and required synchronization.

Finally, despite the polynomial worst case complexity, nearly linear average execution times in Figure 3 over documents of 10,000 to 100,000 nodes (20 instances, 5 ignored and 5 measured runs) confirm that the best correction configuration seems to be the newly proposed RFN-N1-E.

5   Conclusion

Having a potentially invalid XML document and its schema in DTD or XSD, our correction model and algorithms guarantee that we are always able to find all its minimal structural corrections regardless the extent of its invalidity, and that we are able to find them without any parameters from the user. In this paper we confirmed our original suggestion that the refinement strategy is perspective enough to become a basis of our newly proposed correction algorithm that significantly outperforms all the existing ones.

Acknowledgments. This work was supported by Charles University Grant Agency grant 4105/2011 and project SVV-2014-260100.

References

1. Amavi, J., Bouchou, B., Savary, A.: On Correcting XML Documents with Respect to a Schema. The Computer Journal (2013); 10.1093/comjnl/bxt006
2. Bouchou, B., Cheriat, A., Ferrari, M.H., Savary, A.: Integrating Correction into Incremental Validation. In: BDA (2006)
3. Bray, T., Paoli, J., Sperberg-McQueen, C.M., Maler, E., Yergeau, F.: Extensible Markup Language (XML) 1.0, 5th edn. (2008), http://www.w3.org/TR/xml/
4. Fallside, D.C., Walmsley, P.: XML Schema Part 0: Primer, 2nd edn. (2004), http://www.w3.org/TR/xmlschema-0/
5. Flesca, S., Furfaro, F., Greco, S., Zumpano, E.: Querying and Repairing Inconsistent XML Data. In: Ngu, A.H.H., Kitsuregawa, M., Neuhold, E.J., Chung, J.-Y., Sheng, Q.Z. (eds.) WISE 2005. LNCS, vol. 3806, pp. 175–188. Springer, Heidelberg (2005)
6. Mlynkova, I., Toman, K., Pokorny, J.: Statistical Analysis of Real XML Data Collections. In: Proceedings of the 13th International Conference on Management of Data, vol. 6, pp. 20–31 (2006)
7. Murata, M., Lee, D., Mani, M., Kawaguchi, K.: Taxonomy of XML Schema Languages using Formal Language Theory. ACM Transactions on Internet Technology 5(4), 660–704 (2005)
8. Staworko, S., Chomicki, J.: Validity-Sensitive Querying of XML Databases. In: Grust, T., Höpfner, H., Illarramendi, A., Jablonski, S., Fischer, F., Müller, S., Patranjan, P.-L., Sattler, K.-U., Spiliopoulou, M., Wijsen, J. (eds.) EDBT 2006. LNCS, vol. 4254, pp. 164–177. Springer, Heidelberg (2006)

9. Svoboda, M., Mlýnková, I.: Correction of Invalid XML Documents with Respect to Single Type Tree Grammars. In: Fong, S. (ed.) NDT 2011. CCIS, vol. 136, pp. 179–194. Springer, Heidelberg (2011)
10. Svoboda, M., Mlýnková, I.: An Incremental Correction Algorithm for XML Documents and Single Type Tree Grammars. In: Benlamri, R. (ed.) NDT 2012, Part I. CCIS, vol. 293, pp. 235–249. Springer, Heidelberg (2012)
11. Tan, Z., Zhang, Z., Wang, W., Shi, B.: Computing Repairs for Inconsistent XML Document Using Chase. In: Dong, G., Lin, X., Wang, W., Yang, Y., Yu, J.X. (eds.) APWeb/WAIM 2007. LNCS, vol. 4505, pp. 293–304. Springer, Heidelberg (2007)

The Absolute Consistency Problem of XML Schema Mappings with Data Values between Restricted DTDs

Yasunori Ishihara, Hayato Kuwada, and Toru Fujiwara

Osaka University, Japan
{ishihara,h-kuwada,fujiwara}@ist.osaka-u.ac.jp

Abstract. This paper proposes a restricted class of DTDs, called $MDC^{?+\#}$-DTDs, such that the absolute consistency problem of XML schema mappings is solvable in polynomial time. To the best of the authors' knowledge, tractability results on absolute consistency have been obtained only for XML schema mappings between nested-relational DTDs, which are a proper subclass of non-recursive, disjunction-free DTDs. $MDC^{?+\#}$-DTDs are a proper superclass of nested-relational DTDs, where recursion and disjunction are allowed but every label can appear at most once or unboundedly many times as children of a node. We show that the absolute consistency problem of XML schema mappings between $MDC^{?+\#}$-DTDs is solvable in polynomial time if the dependencies are specified by (1) self, child, parent, following-sibling, preceding-sibling, and data value equality operators; or (2) self, child, following-sibling, preceding-sibling, qualifier, and data value equality operators.

1 Introduction

XML is a markup language for describing structured document and used for a variety of applications and database systems. Many XML schemas and DTDs have been defined so far. Some of them are similar, and some of them are still evolving. An *XML schema mappings* is the basis of data transformation from one schema to another caused by XML data exchange or XML schema evolution. More formally, an XML schema mapping $\mathcal{M}$ is a triple (D_S, D_T, Σ) of a *source schema* D_S, a *target schema* D_T, and a set Σ of *dependencies*, each of which is a pair of *document patterns*. $\mathcal{M}$ represents a set of pairs of source and target documents that satisfy all the dependencies in Σ.

There are several research issues on XML schema mappings. The absolute consistency problem is one of the most important ones. An XML schema mapping is said to be *absolutely consistent* if every source document has a corresponding target document. That is, absolute consistency guarantees that every source document can be transformed into a target document according to the given dependencies.

H. Decker et al. (Eds.): DEXA 2014, Part I, LNCS 8644, pp. 317–327, 2014.

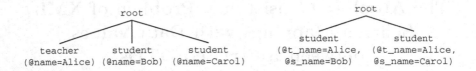

Fig. 1. An example of T_1 **Fig. 2.** An example of T_1'

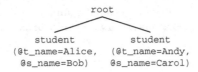

Fig. 3. An example of T_2

Example 1. Consider the following source schema D_S and target schema D_T:

```
Dₛ: root -> class              Dₜ: root -> class
    class -> teacher student*      class -> student*
    teacher : @name                student : @t_name, @s_name
    student : @name
```

Here, `teacher : @name` represents that element `teacher` has an attribute `@name`. Consider the following document patterns:

$$p = \texttt{class[teacher(@name} = x)]\texttt{[student(@name} = y)],$$
$$p' = \texttt{class/student(@t\_name} = x, \ \texttt{@s\_name} = y).$$

p represents that the root node has a child element `class`, which has a child element `teacher` with value x for attribute `@name` and a child element `student` with value y for attribute `@name`. p' represents that the root node has a child element `class`, which has a child element `student` with value x for attribute `@t_name` (the name of the student's teacher) and value y for attribute `@s_name` (the name of the student). The "/" in p' represents concatenation of document patterns.

$(D_S, D_T, \{p \longrightarrow p'\})$ is absolutely consistent because for each document T_1 conforming to D_S, there is a document T_1' conforming to D_T such that for every part of T_1 that matches p, there is a part of T_1' that matches p' (see Figs. 1 and 2 for example). On the other hand, consider the case where the roles of source and target are swapped. Then, $(D_T, D_S, \{p' \longrightarrow p\})$ is not absolutely consistent because there is a tree T_2 that has `student` elements with different `@t_name` values (see Fig. 3).

The absolute consistency problem was originally addressed by Amano et al. [1]. They also investigated the computational complexity of the absolute consistency problem under several restrictions of document patterns and DTDs.

To the best of the authors' knowledge, tractability results have been obtained only for XML schema mappings between *nested-relational* DTDs, which are a proper subclass of non-recursive, disjunction-free DTDs. More precisely, nested-relational DTDs are non-recursive DTDs such that each content model is in the form of $\hat{l}_1 \cdots \hat{l}_m$, where all l_i's are distinct labels and each $\hat{l}_i$ is in the form of l_i, $l_i{}^*$, $l_i{}^+$, or $l_i{}^?$. Here, document patterns must be written without sibling axes.

Our purpose is to find practically wide classes of DTDs and document patterns for which the absolute consistency problem is solvable in polynomial time. We propose MDC$^{?+\#}$-DTDs, which are a superclass of nested-relational DTDs. In each content model of MDC$^{?+\#}$-DTDs, disjunction must be in the scope of $*$ or $+$, and each symbol outside the scope of $*$ or $+$ must not appear more than once. For example, in a content model $a^?(b|a)^+$, the first a is outside the scope of $*$ or $+$. But a appears twice in the content model. So, this content model does not meet the condition of MDC$^{?+\#}$ DTDs. In another content model $a^?(b|c)^+$, the label appearing outside the scope of $*$ or $+$ is only a, and it appears only once in the content model. So, this content model meets the condition of MDC$^{?+\#}$-DTDs. Then we show that the absolute consistency problem of XML schema mappings between MDC$^{?+\#}$-DTDs is solvable in polynomial time if the dependencies are specified by (1) self, child, parent, following-sibling, preceding-sibling, and data value equality operators; or (2) self, child, following-sibling, preceding-sibling, qualifier, and data value equality operators.

Because of the space limitation, most of the technical details are omitted in this paper. Refer to [2] for a longer version of this paper.

2   Preliminaries

2.1   XML Documents and DTDs

An XML document is represented by an unranked labeled ordered tree with data values. The label of a node v, denoted $\lambda(v)$, corresponds to a tag name. We extend λ to a function on sequences, i.e., for a sequence $v_1 \cdots v_n$ of nodes, let $\lambda(v_1 \cdots v_n) = \lambda(v_1) \cdots \lambda(v_n)$. Let Dom be a countable domain of data values, and let $\rho_{@a}(v)$ denote the data value of *attribute* @a of node v.

A regular expression over an alphabet Γ consists of constants ϵ (empty sequence) and the symbols in Γ, and operators $\cdot$ (concatenation), $*$ (repetition), $|$ (disjunction), $?$ (zero or one occurrence), $+$ (one or more occurrences), and $\#$ (either or both). Here, $\#$ is an $(m+l)$-ary operator and $(a_1, \ldots, a_m)\#(b_1, \ldots, b_l)$ is equivalent to $a_1 \cdots a_m b_1^? \cdots b_l^? | a_1^? \cdots a_m^? b_1 \cdots b_l$. We exclude $\emptyset$ (empty set) because we are interested in only nonempty regular languages. The concatenation operator is often omitted as usual. The string language represented by a regular expression e is denoted by $L(e)$.

A regular expression e is $MDC^{?+\#}$ if

– every disjunction $|$ appearing in e is in the scope of $*$ or $+$, and
– for each symbol a appearing in e, if the occurrence of a is not in the scope of $*$ or $+$, then a appears only once in e.

The *length* of an MDC$^{?+\#}$ regular expression $e = e_1 e_2 \cdots e_n$ is defined as the number n of subexpressions of the top-level concatenation operator, and denoted by $len(e)$. Moreover, i $(1 \le i \le len(e))$ is called a *position* and each e_i is called the i-th subexpression of e.

A *DTD* D is a quintuple (Γ, A, r, P, R), where

- Γ is a finite set of labels,
- A is a finite set of attribute names,
- $r \in \Gamma$ is the root label,
- P is a mapping from Γ to the set of regular expressions over Γ, and
- R is a mapping from Γ to the power set 2^A of A.

$P(l)$ is called the *content model* of label $l \in \Gamma$ and represents the possible label sequence of the children of a node with label l. $R(l)$ represents the attributes that a node with label l has.

A tree T *conforms* to a DTD $D = (\Gamma, A, r, P, R)$ if

- the root label of T is r,
- for each node v of T and its children sequence $v_1 v_2 \cdots v_n$, it holds that $\lambda(v_1 v_2 \cdots v_n) \in L(P(\lambda(v)))$, and
- for each node v of T, $\rho_{@a}(v)$ is defined if and only if $@a \in R(\lambda(v))$.

Let $TL(D)$ denote the set of all the trees conforming to D. In this paper, we assume that every DTD $D = (\Gamma, A, r, P, R)$ contains no useless symbols. That is, for each $l \in \Gamma$, there is a tree T conforming to D such that the label of some node of T is l.

A DTD $D = (\Gamma, A, r, P, R)$ is an *MDC$^{?+\#}$-DTD* if $P(l)$ is MDC$^{?+\#}$ for each $l \in \Gamma$.

The size of a regular expression is the number of constants and operators appearing in the regular expression. The size of a DTD is the sum of the sizes of all content models.

2.2   XPath Expressions

Let us fix a countable set X of *variables*. The syntax of an XPath expression p is defined as follows:

$$p ::= \chi :: l \mid \chi :: l(\alpha) \mid p/p \mid p[p],$$
$$\chi ::= \cdot \mid \downarrow \mid \uparrow \mid \rightarrow^+ \mid \leftarrow^+,$$
$$\alpha ::= @a = x \mid \alpha, \alpha ,$$

where $l \in \Gamma$, $@a \in A$, and $x \in X$. Each $\chi \in \{\cdot, \downarrow, \uparrow, \rightarrow^+, \leftarrow^+\}$ is called an *axis*. Also, a subexpression in the form of $[p]$ is called a *qualifier*. An expression in the form of $\chi :: l$ or $\chi :: l(\alpha)$ is said to be *atomic*. If no variable in p appears more than once in p, then p is said to be *linear*. The *size* of an XPath expression p is defined as the number of atomic subexpressions in p.

The semantics of an XPath expression over a tree T is defined as follows, where p is regarded as a binary predicate on paths from the root of T. In what follows,

v_0 denotes the root of T, and v and v' denote nodes of T. Also, parameters of p (e.g., w, wv, etc.) are nonempty sequences of nodes of T starting by v_0, unless otherwise stated. Let σ be a mapping called *valuation* from X to *Dom*.

- $(T, \sigma) \models (\cdot :: l)(wv, wv)$ if path wv exists in T and $\lambda(v) = l$.
- $(T, \sigma) \models (\downarrow:: l)(w, wv')$ if path wv' exists in T and $\lambda(v') = l$.
- $(T, \sigma) \models (\uparrow:: l)(wvv', wv)$ if path wvv' exists in T and $\lambda(v) = l$.
- $(T, \sigma) \models (\rightarrow^+:: l)(wv, wv')$ if paths wv and wv' exist in T, v' is a following sibling of v, and $\lambda(v') = l$.
- $(T, \sigma) \models (\leftarrow^+:: l)(wv, wv')$ if paths wv and wv' exist in T, v' is a preceding sibling of v, and $\lambda(v') = l$.
- $(T, \sigma) \models (\chi :: l(@a = x))(w, w'v')$ if $(T, \sigma) \models (\chi :: l)(w, w'v')$ and $\rho_{@a}(v') = \sigma(x)$.
- $(T, \sigma) \models (\chi :: l(\alpha, \alpha'))(w, w')$ if $(T, \sigma) \models (\chi :: l(\alpha))(w, w')$ and $(T, \sigma) \models (\chi :: l(\alpha'))(w, w')$.
- $(T, \sigma) \models (p/p')(w, w')$ if $(T, \sigma) \models p(w, w'')$ and $(T, \sigma) \models p'(w'', w')$ for some w''.
- $(T, \sigma) \models (p[p'])(w, w')$ if $(T, \sigma) \models p(w, w')$ and $(T, \sigma) \models p'(w', w'')$ for some w''.

A tree T *satisfies* an XPath expression p under a valuation σ if there is a path w such that $(T, \sigma) \models p(v_0, w)$, where v_0 is the root node of T. An XPath expression p is *satisfiable* under a DTD D if some $T \in TL(D)$ satisfies p under some valuation σ.

Let p° denote the XPath expression obtained by replacing every $\chi :: l(\alpha)$ in p with $\chi :: l$. It is not difficult to see that p is satisfiable under a DTD D if and only if p° is satisfiable under D.

Following the notation of [3], a subclass of XPath is indicated by $\mathcal{X}(\_)$. For example, the subclass with child axes and qualifiers is denoted by $\mathcal{X}(\downarrow, [\ ]_\wedge)$. Here, we use $[\ ]_\wedge$ rather than $[\ ]$ to explicitly indicate that qualifiers in this paper cannot specify disjunctive meanings.

2.3 XML Schema Mappings

A *dependency* is an expression of the form $p \longrightarrow p'$, where p is a linear XPath expression and p' is an arbitrary XPath expression. p and p' are called *document patterns*. An *XML schema mapping* $\mathcal{M}$ is a triple (D_S, D_T, Σ), where D_S is a DTD representing a source schema, D_T is a DTD representing a target schema, and Σ is a finite set of dependencies.

Hereafter, XML schema mappings are referred to as just *mappings*. Originally, in [1,4,5], document patterns are defined by means of tree patterns. However, we use XPath expressions in order to exploit many results on XPath satisfiability.

A pair of trees $T_S \in TL(D_S)$ and $T_T \in TL(D_T)$ *satisfies* a dependency $p \longrightarrow p'$ if the following condition is satisfied: If T_S satisfies p under a valuation σ, then T_T satisfies p' under σ. The set of all pairs of trees which satisfy all dependencies of $\mathcal{M}$ is denoted by $[\![\mathcal{M}]\!]$.

Definition 1. $\mathcal{M} = (D_S, D_T, \Sigma)$ *is absolutely consistent if, for every tree* $T_S \in TL(D_S)$, *there is a tree* $T_T \in TL(D_T)$ *such that* $(T_S, T_T) \in [\![\mathcal{M}]\!]$.

3   Deciding Absolute Consistency

In this section, we first focus on the absolute consistency problem for $\mathcal{M} = (D_S, D_T, \Sigma)$ such that Σ is a singleton (i.e., Σ has only one element).

If $\mathcal{M}$ is a mapping without data values, absolute consistency of $\mathcal{M}$ is reducible to XPath satisfiability [6]. However, as shown in the following example, we have to take account of *diversity* and *equality* of variables as well as satisfiability if $\mathcal{M}$ is a mapping with data values.

Example 2. Let $\Gamma = \{r, b, c, d\}$, $A = \{@a\}$, and $R(r) = R(b) = R(c) = R(d) = \{@a\}$. Consider the following two DTDs $D_S = (\Gamma, A, r, P_S, R)$ and $D_T = (\Gamma, A, r, P_T, R)$:

- $P_S(r) = bc^*d$, $P_S(b) = P_S(c) = P_S(d) = \epsilon$, and
- $P_T(r) = bcd^*$, $P_T(b) = P_T(c) = P_T(d) = \epsilon$.

Also, let

- $p_1 = \downarrow :: c(@a = x)$,
- $p_2 = \downarrow :: b(@a = x)/\uparrow :: r/\downarrow :: d(@a = y)$, and
- $p_3 = \downarrow :: b(@a = x)/\uparrow :: r/\downarrow :: b(@a = y)$.

First, let us see the diversity of values assigned to variables. Consider $\mathcal{M}_1 = (D_S, D_T, \{p_1 \longrightarrow p_1\})$. As pointed out in [5], $\mathcal{M}_1$ is not absolutely consistent. This is because there is a tree $T \in TL(D_S)$ with two nodes v and v' whose labels are c but the values of $@a$ are different while every tree in $TL(D_T)$ has exactly one node whose label is c. Hence, there is no tree T' such that $(T, T') \in [\![\mathcal{M}_1]\!]$. In other words, x has diversity under D_S but not under D_T.

Next, let us see the equality of variables. Consider $\mathcal{M}_2 = (D_S, D_T, \{p_2 \longrightarrow p_3\})$. Again, $\mathcal{M}_2$ is not absolutely consistent. This is because there is σ with $\sigma(x) \neq \sigma(y)$ such that a tree in $TL(D_S)$ satisfies p_2 under σ while every σ such that a tree in $TL(D_T)$ satisfies p_3 under σ must be with $\sigma(x) = \sigma(y)$. In other words, p_3 under D_T requires more equality than that p_2 under D_S does.

In what follows, we introduce *schema graphs*, and then, characterize the diversity and equality of variables. Next, we show a necessary and sufficient condition for absolute consistency using the characterization. Finally, we show that the necessary and sufficient condition is decidable in polynomial time.

3.1   Schema Graphs

The schema graphs [7] are originally defined for DC-DTDs, i.e., DTDs without ?, +, and # such that each disjunction | is in the scope of *. To define schema graphs for MDC$^{?+\#}$-DTDs, we use the following mapping δ:

- $\delta(\epsilon) = \epsilon$,
- $\delta(a) = a$ for each $a \in \Sigma$,
- $\delta(e_1 \cdot e_2) = \delta(e_1) \cdot \delta(e_2)$,
- $\delta(e^*) = (\delta(e))^*$,
- $\delta(e_1 | e_2) = \delta(e_1) | \delta(e_2)$,

- $\delta(e^?) = \delta(e)$,
- $\delta(e^+) = (\delta(e))^*$, and
- $\delta((e_{11}, \dots, e_{1m}) \# (e_{21}, \dots, e_{2l}))$
  $= \delta(e_{11}) \cdots \delta(e_{1m}) \cdot \delta(e_{21}) \cdots \delta(e_{2l})$.

Intuitively, δ removes all the ? operators, and replaces all the + and # operators with $*$ and $\cdot$ (concatenation) operators, respectively. For example, $\delta(a^+((b\#(c\#d))a^*)^?) = a^*bcda^*$. For a DTD $D = (\Gamma, A, r, P, R)$, let $\delta(D)$ denote the DTD $(\Gamma, A, r, \delta \circ P, R)$, where $(\delta \circ P)(l) = \delta(P(l))$ for each $l \in \Gamma$. Note that if D is MDC$^{?+\#}$, then each content model of $\delta(D)$ is in the form of $e_1 \cdots e_n$, where each e_i is either a single symbol in Γ or $(e_i')^*$ for some e_i'.

Now, we introduce schema graphs for MDC$^{?+\#}$-DTDs.

Definition 2. *The* schema graph *[7]* $G = (U, E)$ *of MDC$^{?+\#}$-DTD* $D = (\Gamma, A, r, P, R)$ *is a directed graph defined as follows:*

- *A node* $u \in U$ *is either*
  - $(\bot, 1, -, r)$, *where* $\bot$ *is a new symbol not in* Γ, *or*
  - (a, i, ω, b), *where* $a, b \in \Gamma$, $1 \leq i \leq len((\delta \circ P)(a))$ *such that* b *appears in the* i-th *subexpression* e_i *of* $(\delta \circ P)(a)$, *and* $\omega = $ "$-$" *if* e_i *is a single symbol in* Γ *and* $\omega = $ "$*$" *otherwise.*

  The first, second, third and fourth components of u *are denoted by* $\lambda_{par}(u)$, $pos(u)$, $\omega(u)$, *and* $\lambda(u)$, *respectively. Especially,* $\lambda(u)$ *is called the* label *of* u. λ_{par}, pos, *and* λ *are extended to functions on sequences.*
- *An edge from* u *to* u' *exists in* E *if and only if* $\lambda(u) = \lambda_{par}(u')$.

Let e be an MDC$^{?+\#}$ regular expression. We say that a symbol b is *singular* in e if b is not in the scope of $*$ or $+$ in e. By the definition of MDC$^{?+\#}$, each singular symbol appears only once in e. Let $D = (\Gamma, A, r, P, R)$ be an MDC$^{?+\#}$-DTD and suppose that a symbol b is singular in $P(a)$. Then, we can uniquely determine the node u of the schema graph of D such that $\lambda_{par}(u) = a$ and $\lambda(u) = b$. Moreover, for such u, it holds that $\omega(u) = $ "$-$". Hence, a node u of a schema graph is *singular* if $\omega(u) = $ "$-$". Also, a node v of a tree $T \in TL(D)$ is *singular* if v corresponds to a singular node of the schema graph of D.

3.2 Analyzing Variable Constraints

The purpose of here is to characterize diversity and equality of variables. We first introduce an equivalence relation called variable constraint relation.

Definition 3. *A* variable constraint relation *(VC relation for short)* η *for an XPath expression* p *and an MDC$^{?+\#}$-DTD* D *is an equivalence relation over variables appearing in* p *and a finite number of non-empty paths from* $(\bot, 1, -, r)$ *on the schema graph of* D. *A valuation* σ *is consistent with a VC relation* η *if* $\sigma(x) = \sigma(y)$ *for each pair* (x, y) *of variables in* η.

First, let us focus on equality of variables. We would like to define a VC relation $\hat{\eta}$ characterizing the equality of variables. Precisely, σ is consistent with $\hat{\eta}$ if and only if some tree satisfies p under σ. To do this, we define a satisfaction relation $\models_{\text{MDC}}$ between schema graphs and XPath expressions. Here, we augment the parameters of p by VC relations.

Actually, it is not necessary to keep all variable-constraint information. Only the information on nodes which is *certainly* revisited must be handled by η. Let s be the "current path" on G of the analysis. By using upward axes, each node on the "current path" is certainly revisited. Moreover, from such nodes, singular nodes are certainly revisited. So, those information must be maintained by η. On the other hand, there is no way to *certainly revisit the other nodes* in our XPath class. So, such information does not have to be maintained.

Now, we provide the formal definition of $\models_{\text{MDC}}$. Let u, u', etc. be nodes of G, and let s, s', etc. be nonempty sequences of nodes of G starting by $(\perp, 1, -, r)$, unless otherwise stated. Let $\eta|_{\text{sgl},s}$ denote the same equivalence relation as η except that the path domain of $\eta|_{\text{sgl},s}$ is restricted to s' such that s' is a prefix of s followed by a (possibly empty) sequence of singular nodes of G.

Definition 4. *A satisfaction relation $\models_{\text{MDC}}$ between a schema graph G and an XPath expression $p \in \mathcal{X}(\cdot, \downarrow, \uparrow, \rightarrow^+, \leftarrow^+, [\ ]_\wedge, =)$ is defined as follows (some rules are omitted due to the space limitation):*

- $G \models_{\text{MDC}} (\downarrow::\ l)((s, \eta), (su', \eta|_{\text{sgl},su'}))$ *if path su' exists in G, $\lambda(u') = l$, and $\eta = \eta|_{\text{sgl},s}$.*
- $G \models_{\text{MDC}} (\uparrow::\ l)((suu', \eta), (su, \eta|_{\text{sgl},su}))$ *if path suu' exists in G, $\lambda(u) = l$, and $\eta = \eta|_{\text{sgl},suu'}$.*
- $G \models_{\text{MDC}} (\rightarrow^+::\ l)((su, \eta), (su', \eta|_{\text{sgl},su'}))$ *if*
  - $\lambda_{par}(u) = \lambda_{par}(u')$,
  - $\lambda(u') = l$,
  - $pos(u) < pos(u')$ *if $\omega(u) = $ "$-$" and $pos(u) \leq pos(u')$ if $\omega(u) = $ "$*$",*
  - $\eta = \eta|_{\text{sgl},su}$.
- $G \models_{\text{MDC}} (\chi ::\ l(@a = x))((s, \eta), (s', \eta'))$ *if $G \models_{\text{MDC}} (\chi ::\ l)((s, \eta), (s', \eta''))$ and η' is the finest equivalence relation induced by $(\eta'' \cup \{(s', x)\})|_{\text{sgl},s'}$.*
- $G \models_{\text{MDC}} (p/p')((s, \eta), (s', \eta'))$ *if there is a pair (s'', η'') such that $G \models_{\text{MDC}} p((s, \eta), (s'', \eta''))$ and $G \models_{\text{MDC}} p'((s'', \eta''), (s', \eta'))$.*
- $G \models_{\text{MDC}} (p[p'])((s, \eta), (s', \eta'''))$ *if there are s'', η', and η'' such that $G \models_{\text{MDC}} p((s, \eta), (s', \eta'))$ and $G \models_{\text{MDC}} p'((s', \eta'), (s'', \eta''))$, and η''' is the finest equivalence relation induced by $(\eta' \cup \eta'')|_{\text{sgl},s'}$.*

It is easy to see that, from the definition of $\models_{\text{MDC}}$, if $G \models_{\text{MDC}} p((s, \eta), (s', \eta'))$, then $\eta|_{\text{sgl},s} = \eta$ and $\eta'|_{\text{sgl},s'} = \eta'$.

Suppose that an MDC$^{?+\#}$-DTD D and an XPath expression p are given. Let G be the schema graph of D. There may exist several η' such that $G \models_{\text{MDC}} p(((\perp, 1, -, r), \emptyset), (s', \eta'))$, but $\eta'|_{\text{sgl},\epsilon}$ is uniquely determined if exists. We say that $\eta'|_{\text{sgl},\epsilon}$ is *the VC relation of p under D.*

The following theorem formally characterizes the equality of variables:

Theorem 1. *Let $\hat{\eta}$ be the VC relation of $p \in \mathcal{X}(\cdot, \downarrow, \uparrow, \rightarrow^+, \leftarrow^+, [\,]_\wedge, =)$ under an $MDC^{?+\#}$-DTD D. Valuation σ is consistent with $\hat{\eta}$ if and only if some tree in $TL(D)$ satisfies p under σ.*

Next, let us focus on diversity of variables. Let $\hat{\eta}$ be the VC relation of $p \in \mathcal{X}(\cdot, \downarrow, \uparrow, \rightarrow^+, \leftarrow^+, [\,]_\wedge, =)$ under an $MDC^{?+\#}$-DTD D. We say that variable x appearing in p has *diversity* under D if there is no path s on the schema graph of D such that $(x, s) \in \hat{\eta}$. The following theorem states the correctness of this characterization of diversity:

Theorem 2. *Let $p \in \mathcal{X}(\cdot, \downarrow, \uparrow, \rightarrow^+, \leftarrow^+, [\,]_\wedge, =)$ and D be an $MDC^{?+\#}$-DTD. Variable x appearing in p has diversity under D if and only if there is some tree in $TL(D)$ satisfying p under both σ and σ' such that $\sigma(x) \neq \sigma'(x)$.*

3.3   Deciding Absolute Consistency Efficiently

Now, we have a necessary and sufficient condition for absolute consistency:

Theorem 3. $\mathcal{M} = (D_S, D_T, \{p \longrightarrow p'\})$ *is absolutely consistent if and only if the following three conditions hold:*

- *If p° is satisfiable under D_S, then p'° is satisfiable under D_T.*
- *For each pair (x, y) of variables appearing in p, if $(x, y) \notin \hat{\eta}_S$ then $(x, y) \notin \hat{\eta}_T$, where $\hat{\eta}_S$ and $\hat{\eta}_T$ are the VC relations of p under D_S and of p' under D_T, respectively.*
- *For each variable x appearing in p, if x has diversity under D_S, then x has diversity under D_T.*

In what follows, we show that the condition stated in Theorem 3 is decidable in polynomial time if $p \in \mathcal{X}(\cdot, \downarrow, \uparrow, \rightarrow^+, \leftarrow^+, =)$ or $p \in \mathcal{X}(\cdot, \downarrow, \rightarrow^+, \leftarrow^+, [\,]_\wedge, =)$. It suffices to show that $\hat{\eta}_S$ and $\hat{\eta}_T$ can be computed in polynomial time. For both cases, the same technique as [8] are used. We only describe the brief idea here.

Let p be in $\mathcal{X}(\cdot, \downarrow, \uparrow, \rightarrow^+, \leftarrow^+, =)$. Our algorithm eval_1 runs in a top-down manner with respect to the parse tree of p, and computes the set of the second parameters (s', η') of p for a given set of first parameters (s, η). Let B denote a set of pairs of a path on G and a VC relation. Formally:

$$\mathrm{eval}_1(p, B) = \begin{cases} \{(s', \eta') \mid G \models_{\mathrm{MDC}} p((s, \eta), (s', \eta')) \\ \qquad \text{for each } (s, \eta) \in B\} \text{ if } p \text{ is atomic,} \\ \mathrm{eval}_1(p_2, \mathrm{eval}_1(p_1, B)) \qquad \text{if } p = p_1/p_2. \end{cases}$$

Such B and (s', η') are shown to be unique up to the labeling function λ. eval_1 runs in polynomial time using this property.

Let p be in $\mathcal{X}(\cdot, \downarrow, \rightarrow^+, \leftarrow^+, [\,]_\wedge, =)$. We can define an algorithm eval_2 so that it runs in a bottom-up manner with respect to the parse tree of p, and computes the set of all the pairs $((s, \eta), (s', \eta'))$ such that $G \models_{\mathrm{MDC}} p((s, \eta), (s', \eta'))$. Again, we use the uniqueness up to the labeling function. Moreover, p has no upward axis, we have to maintain just the last node of s and s'. eval_2 runs in polynomial time using these properties.

Theorem 4. *Absolute consistency of* $\mathcal{M} = (D_S, D_T, \{p \longrightarrow p'\})$ *is decidable in polynomial time if p and p' are in $\mathcal{X}(\cdot, \downarrow, \uparrow, \rightarrow^+, \leftarrow^+, =)$ or $\mathcal{X}(\cdot, \downarrow, \rightarrow^+, \leftarrow^+, [\ ]_\wedge, =)$.*

3.4 Multiple Dependencies Case

Consider $\mathcal{M} = (D_S, D_T, \Sigma)$, where $\Sigma = \{p_i \longrightarrow p_i' \mid 1 \le i \le k\}$. Without loss of generality, we can assume that distinct dependencies do not share variables and all p_i's are satisfiable under D_S.

Lemma 1. $\mathcal{M} = (D_S, D_T, \Sigma)$ *is absolutely consistent if and only if $\mathcal{M}' = (D_S, D_T, \{p \longrightarrow p'\})$ is absolutely consistent, where $\Sigma = \{p_i \longrightarrow p_i' \mid 1 \le i \le k\}$, $p = \cdot :: r[p_1] \cdots [p_k]$, and $p' = \cdot :: r[p_1'] \cdots [p_k']$.*

p and p' can be defined using parent axes instead of qualifiers because we have only child axis as downward axes. Thus, we have the next theorem:

Theorem 5. *Let $\mathcal{M}$ be a mapping (D_S, D_T, Σ), where D_S and D_T are $MDC^{?+\#}$-DTDs. Absolute consistency of $\mathcal{M}$ is decidable in polynomial time if*

- *all the source document patterns of Σ are in $\mathcal{X}(\cdot, \downarrow, \uparrow, \rightarrow^+, \leftarrow^+, =)$ or in $\mathcal{X}(\cdot, \downarrow, \rightarrow^+, \leftarrow^+, [\ ]_\wedge, =)$; and*
- *all the target document patterns of Σ are in $\mathcal{X}(\cdot, \downarrow, \uparrow, \rightarrow^+, \leftarrow^+, =)$ or in $\mathcal{X}(\cdot, \downarrow, \rightarrow^+, \leftarrow^+, [\ ]_\wedge, =)$.*

4   Conclusion

This paper has proposed a restricted class of DTDs, called MDC$^{?+\#}$-DTDs, and the absolute consistency problem of XML schema mappings between MDC$^{?+\#}$-DTDs is solvable in polynomial time if the dependencies are specified by (1) self, child, parent, following-sibling, preceding-sibling, and data value equality operators; or (2) self, child, following-sibling, preceding-sibling, qualifier, and data value equality operators. As future work, we are planning to find wider classes of DTDs and XPath expressions such that absolute consistency is still tractable.

References

1. Amano, S., Libkin, L., Murlak, F.: XML schema mappings. In: Proc. 28th PODS, pp. 33–42 (2009)
2. Ishihara, Y., Kuwada, H., Fujiwara, T.: The absolute consistency problem of XML schema mappings with data values between restricted DTDs (2014), http://www-infosec.ist.osaka-u.ac.jp/~ishihara/papers/dexa14-full.pdf
3. Benedikt, M., Fan, W., Geerts, F.: XPath satisfiability in the presence of DTDs. Journal of the ACM 55(2) (2008)
4. Arenas, M., Libkin, L.: XML data exchange: Consistency and query answering. Journal of the ACM 55(2) (2008)

5. Arenas, M., Barcelo, P., Libkin, L., Murlak, F.: Relational and XML Data Exchange. Morgan & Claypool (2010)
6. Kuwada, H., Hashimoto, K., Ishihara, Y., Fujiwara, T.: The consistency and absolute consistency problems of XML schema mappings between restricted DTDs. In: Ishikawa, Y., Li, J., Wang, W., Zhang, R., Zhang, W. (eds.) APWeb 2013. LNCS, vol. 7808, pp. 228–239. Springer, Heidelberg (2013); (An extended version is to appear in World Wide Web Journal)
7. Ishihara, Y., Morimoto, T., Shimizu, S., Hashimoto, K., Fujiwara, T.: A tractable subclass of DTDs for XPath satisfiability with sibling axes. In: Gardner, P., Geerts, F. (eds.) DBPL 2009. LNCS, vol. 5708, pp. 68–83. Springer, Heidelberg (2009)
8. Ishihara, Y., Suzuki, N., Hashimoto, K., Shimizu, S., Fujiwara, T.: XPath satisfiability with parent axes or qualifiers is tractable under many of real-world DTDs. In: Proc. 14th DBPL (2013), http://arxiv.org/abs/1308.0769

k-Anonymity of Microdata with NULL Values*

Margareta Ciglic, Johann Eder, and Christian Koncilia

Alpen-Adria-Universität Klagenfurt,
Department of Informatics Systems,
Klagenfurt, Austria
{firstname.lastname}@aau.at

Abstract. Releasing, publishing or transferring microdata is restricted by the necessity to protect the privacy of data owners. K-anonymity is one of the most widespread concepts for anonymizing microdata but it does not explicitly cover NULL values frequently found in microdata. We study the problem of NULL values (missing values, non-applicable attributes, etc.) for anonymization in detail, present a set of new definitions for k-anonymity explicitly considering NULL and analyze which definition protects from which attacks. We show that an adequate treatment of missing values in microdata can be easily achieved by an extension of generalization algorithms and show that NULL aware generalization algorithms have less information loss than standard algorithms.

Keywords: privacy, microdata, k-anonymity, NULL-values, missing values.

1 Introduction

Detailed data collections are an important resource for research, for fact based governance, or for knowledge based decision making. In the field of statistical databases any collection of data with detailed information on entities, in particular persons and organizations, is called microdata.

A crucial requirement for the release of microdata is the preservation of the privacy of the data owners which is protected by laws and regulations. Furthermore, for data collections requiring the willingness of data owners to share (donate)their data, studies [8] clearly indicate that the protection of privacy is one of the major concerns of data owners and decisive for a consent to donate data [11]. For protecting privacy from linkage attacks the concept of k-anonymity [23] received probably the widest attention. It's core idea is to preserve privacy by hiding each individual in a crowd of at least k members. Many anonymization algorithms implementing these concepts were developed.

Surprisingly, neither the original definition of k-anonymity nor any of the many anonymization algorithms deals with unknown, or missing values (NULL

---

* The work reported here was supported by the *Austrian Ministry of Science and Research* within the project BBMRI.AT and the *Technologie- und Methodenplattform für die vernetzte medizinische Forschung e.V. (TMF)* within the project ANON.

H. Decker et al. (Eds.): DEXA 2014, Part I, LNCS 8644, pp. 328–342, 2014.

values in database terms) in microdata. We could not find a single source discussing the problem of NULL values in microdata for anonymization. Recent surveys [17] or textbooks [10] do not mention NULL values or missing values. However, all techniques and algorithms we found, explicitly or implicitly require that all records with at least one NULL value have to be removed from a table before it can be anonymized ([16,15,14,25,21,27,13,3], and many more). There is only some treatment of NULL values in form of suppressed values, i.e. NULL values resulting from removing ("suppressing") data in the course of anonymization procedures. Attack possibilities on suppressed rows can be found in [26] and [20]. A discussion about suppression of values in single cells can be found in [6,19,1]. However, neither of these approaches discusses the problem of missing values in the original data or of non-existing values due to non-applicable attributes.

NULL values, nevertheless, are not exceptional in microdata, e.g. they appear frequently in datasets for medical research [2,7,28]: Some attributes might not be applicable for each patient. A patient might have refused to answer some questions in a questionnaire or could not be asked due to physical or mental conditions. In an emergency situation some test might not have been performed, etc., etc.

Anonymization by generalization and suppression of data cause loss of information. The aim of reducing this information loss triggered many research efforts. The ignorance of NULL values in anonymization algorithms results in dropping rows from a table, causing a considerable loss of information. Furthermore, dropping rows with NULL values also could introduce some bias in the dataset which is not contained in the original table. This is of course unfortunate for further analysis of the data (for example in evidence based medicine) and might compromise the statistical validity of the results. For example, dropping rows with a NULL value in the field *occupation* would skip all children from the data set and introduce an age bias which was not present in the original dataset.

In this paper we provide a thorough grounding for the treatment of NULL values in anonymization algorithms. We show that we can reduce the problem of NULL values in k-anonymity to different definitions of matching between values and NULL values. We show that generalization algorithms, widely used for anonymization can be easily extended to cover NULL values and we show that this extension reduces information loss during anonymization.

2   *k*-Anonymity Revisited

A detailed collection and representation of data on information subjects is called microdata - as opposed to data in less detail like statistical data. For this paper a microdata table is a multiset of rows [12]. We can classify the attributes in the schema of such table in four categories: (1) identifiers: all attributes which uniquely identify a row in the table, (2) quasi identifier: all attributes which an adversary might know, (3) sensitive attributes: attributes with values that should not be inferable by an adversary and (4) all other data. For this paper we assume that the identifiers have already been removed from a table and that the schema of a table includes a set of quasi-identifiers Q which we denote by $Q_1, \ldots, Q_n$.

The aim of anonymization is to assure that a table can be published without opening an adversary the possibilities to gain additional knowledge about the data subjects.

Table 1 shows our running example for such a table with the quasi-identifiers Gender, Height, Job, and ZIP and the sensitive attribute Condition.

Table 1. Original table

	Gender	Height	Job	ZIP	Condition
A	f	165	NULL	9020	Cancer
B	m	187	mayor	9020	Hepatitis
C	f	163	clerk	9020	Flu
D	m	NULL	techn.	9020	Pneumonia
E	m	183	NULL	9020	Malaria
F	m	189	pilot	9020	Gastritis

Samarati and Sweeney[23] proposed an approach to preserve the privacy of a data owner by hiding each data owner in a crowd of at least k individuals such that an adversary might not get detailed information about an individual but only information about a group of k individuals. The larger the k, the smaller the possible information gain of an adversary.

In [24] k-anonymity is defined as follows: 'Each release of data must be such that every combination of values of quasi-identifiers can be indistinctly matched to at least k individuals'. The term *indistinct match* is not defined explicitly, nevertheless, it is clear from the context that two rows match, if they have identical values in the quasi-identifiers. However, missing values are not mentioned. We basically follow this definition here, and analyze, how rows of a table match in case some values are NULL. Hence we formalize the notion of k-anonymity, dependent on some match operator.

Definition 1. *(k-Anonymity) Let T be a table and Q the set of quasi identifier attributes and let $\sim$ be a match predicate on T. T is k-anonymous with respect to $\sim$, iff $\forall t \in T : |\{t'|t \sim t'\}| \geq k$.*

3 k-Anonymity with NULL Values

3.1   NULL Values

NULL values [18] are the standard way of representing missing information in database tables. We can distinguish three kinds of NULL values: (1) attribute not applicable: in this case there is no value for this attribute for this row in the world represented in the database. (2) missing value: there exist a value in the world, but it is not contained in the database. (3) no information: it is not known whether the value exist in the world or not. In SQL the semantics of NULL is "no information".

For the following considerations we follow the treatment of NULL values in SQL [12]. This means in particular, that a comparison of a NULL value with any other value never results in TRUE and there is a special unary predicate *is*NULL to test for NULL values.

3.2   Matching NULL Values

For matching of NULL values we have the following options:

- basic match: NULL values do not match with NULL values nor with any other value.
- extended match: NULL values match with NULL values.
- maybe match: NULL values match with any value including NULL values.

In the original definition of *k*-anonymity and in the current anonymization algorithms basic match is used. It is in accordance with the definition in SQL, where 'A=B' is not true, if A or B are NULL values. Extended match treats a NULL value like any other value. Maybe match sees NULL values as wildcards for matching. It corresponds to Codd's maybe selection [5], where rows are returned, if the selection predicate is true for a substitution of the NULL values.

3.3   Basic Match

We call the match used in [24], where rows with NULL values are discarded, *basic match*, and formally define it as follows:

Definition 2. (Basic match) *Let T be a table and Q the set of quasi identifier attributes.*

$$t_1 \sim_b t_2 :\iff \forall q \in Q : t_1[q] = t_2[q]$$

Table 2 shows the result of the anonymization of table 1 to a 2-anonymous table. The table has only 3 rows, as all rows of the original table which contain NULL values had to be removed before the generalization. For the rest of this paper we always follow the full-domain generalization scheme [24,22,15] in our examples, however, the considerations are applicable to all algorithms for *k*-anonymization.

3.4   Extended Match

In extended match NULL values are treated like any other value, in particular, a NULL value only matches with another NULL value but not with any values from the domains of the attributes.

Definition 3. (Extended match) *Let T be a table and Q the set of quasi identifier attributes of T. For two rows $t_1, t_2 \in T$ we define the extended match as*

$$t_1 \sim_e t_2 :\iff \forall q \in Q : t_1[q] = t_2[q] \lor (t_1[q]is\text{NULL} \land t_2[q]is\text{NULL})$$

Table 2. 2-anonymity with basic match

	Gender	Height	Job	ZIP	Condition
A					
B	all	all	all	9020	Hepatitis
C	all	all	all	9020	Flu
D					
E					
F	all	all	all	9020	Gastritis

The extended matching definition can be used to extend existing anonymization algorithms. First we have to extend all generalization hierarchies with a branch with the value NULL on each level of the hierarchy below the root 'ALL'. Using these extended hierarchies we can apply the generalization method again and receive Table 3 which is 2-anonymous with respect to the extended match. Note that in contrast to the basic match no row has been lost.

Table 3. 2-anonymity with extended match

	Gender	Height	Job	ZIP	Condition
A	all	all	NULL	9020	Cancer
B	all	all	admin	9020	Hepatitis
C	all	all	admin	9020	Flu
D	all	all	technical	9020	Pneumonia
E	all	all	NULL	9020	Malaria
F	all	all	technical	9020	Gastritis

The aim of k-anonymity is to prevent attacks on released data, in particular, record linking attacks [10], i.e. joining a table with some known information to associate values of sensitive attributes with some data owner. In particular, matching any record containing quasi-identifiers with the released table should result in no or at least k hits. It is easy to see that this requirement is fulfilled, if any query posed on the released table in the form of "Select * From T where search_condition" yields 0 or at least k result rows, if the search condition only contains predicates on quasi-identifiers.

Theorem 1. *(link-safe) Let T be a table and Q a set of quasi identifier attributes and let $\pi_Q T$ be the projection of T on Q. If T is k-anonymous with respect to $\sim_e$ then for all search conditions p the query "Select * From $\pi_Q T$ where p" returns 0 or at least k rows.*

Proof. The theorem follows from the observation that if a row $t \in \pi_Q T$ satisfies the search condition of the query then all rows matching t according to the extended match also satisfy the search condition. Because T is k-anonymous with respect to $\sim_e$ there are at least k such rows.

3.5   Maybe Match

When we extend the domain of the match predicate to also consider NULL values we can build on the treatment of NULL values in Codd's *maybe* operations for the relational algebra [5]. The maybe selection operator does not only return those rows for which the selection predicate is satisfied but also all those rows which satisfy the selection predicate if NULL values are replaced by suitable values.

Applying the concept of 'maybe' selects to the matching of rows, we define the maybe match as follows: NULL matches both with NULL and other values. NULL values in the rows can be used as wildcards in both directions.

Definition 4. *(maybe match) Let T be a table and Q the set of quasi identifier attributes of T. For two rows $t_1, t_2 \in T$ we define the maybe match as*

$$t_1 \sim_m t_2 :\Longleftrightarrow \forall q \in Q : t_1[q] = t_2[q] \vee (t_1[q] isNULL \vee t_2[q] isNULL).$$

The maybe match is not transitive and does not lead to an equivalence partitioning of the table (in contrast to basic and extended match). We call a row t and it's matching rows the *k*-group of t. *k*-groups of different rows might overlap without being equal.

Applying maybe matching in the generalization method we compute the table shown in Table 4.

Table 4. 2-anonymity with maybe match

	Gender	Height	Job	ZIP	Condition
A	f	161-180	NULL	9020	Cancer
B	m	181-200	mayor	9020	Hepatitis
C	f	161-180	clerk	9020	Flu
D	m	NULL	techn.	9020	Pneumonia
E	m	181-200	NULL	9020	Malaria
F	m	181-200	pilot	9020	Gastritis

Let us now analyze whether this table is safe. First we try an attack on missing values.

Hampering Reconstruction[20] is an attack that shows how a value that was suppressed in the anonymization process can be reconstructed. We extend it here to cover also missing values in the original table. For example, if an adversary knows that Daniel's data are in the table and Daniel is 205 cm tall, then he can associate row D with Daniel.

Hampering reconstruction requires that a value for some attribute exists in the real world, but is not recorded in the database. It shows that tables where NULL values have the semantics of missing values may be compromisable.

Next we show that there are also attacks possible on NULL values which have the semantics 'not applicable', i.e. for which no value exists in the real world.

We introduce the novel NULL-*identifier attack* which uses knowledge whether an attribute is applicable for some row. For example, let us assume that an adversary knows that Alice, a female patient, is not employed, and therefore the value in the Job attribute has to be NULL. He can thus query Table 4 with the search condition 'Gender = "f" and Job is NULL' and retrieves row A.

The NULL-identifier attack leverages on the knowledge that a certain value does not exist. Therefore, the row in the table has to have the value NULL in the corresponding attribute. Anonymization based on maybe match is thus not safe against NULL identifier attacks.

With hampering reconstruction and NULL-identifier attack we show that tables which are k-anonymous with respect to maybe match are not safe from linking attacks. Both attacks exploit situations where there are less than k NULL values in some attribute within a k-group.

3.6   Right Maybe Match

We restrict the definition of the maybe match, such that within a k-group a NULL value in some attribute has to appear at least k times, but use the wildcard character of NULL for matching other values with NULL.

Definition 5. *(right maybe match) Let T be a table and Q the set of quasi identifier attributes of T. For two rows $t_1, t_2 \in T$ we define the right maybe match as*

$$t_1 \sim_r t_2 :\Longleftrightarrow \forall q \in Q : t_1[q] = t_2[q] \vee t_2[q] \, is\text{NULL}.$$

The right maybe match relation is no longer symmetrical. It is also not transitive and does not define an equivalence partitioning on a table.

The result of anonymization with right maybe match is shown in the example of Table 5.

Table 5. 2-anonymity with right maybe match

	Gender	Height	Job	Condition
A	all	all	NULL	Cancer
B	all	all	mayor	Hepatitis
C	all	all	clerk	Flu
D	all	all	technician	Pneumonia
E	all	all	NULL	Malaria
F	all	all	pilot	Gastritis

Now let's analyze, whether the right maybe match admits tables which are safe. We first observe that there are queries on the quasi-identifier projection of a table which yield less than k results. For example, the search condition 'Job = "technician"' would only return row D. However, this query might lead to a wrong result for an adversary, because it is possible, that the 'true' value in the Job attribute is also "technician" for the rows A and E but just missing. Therefore, a rational adversary would use a maybe query instead ('Job = "technician" or Job is NULL').

It is possible to show that right maybe match is safe for all maybe queries.

We introduce *singularity attack* to show that straight (not maybe) queries make sense and possibly compromise the data. A singularity attack uses knowledge that some value is unique, at least for some combinations of other attribute values. For example, ZIP code and Job may be compromised, when the job = "Mayor" and there is never more than one mayor in a town (i.e. per ZIP code). In such cases an adversary will use straight queries rather than maybe queries and can compromise tables which are *k*-anonymous with respect to the right maybe match.

The singularity attack shows that tables which are *k*-anonymous with respect to right maybe match are not safe against attacks.

4 An Extended Generalization Algorithm

Anonymization by generalization [24] is based on a generalization hierarchy for each quasi-identifier. The procedure in general is as follows: When a row does not match at least $k-1$ other rows, then some attribute values are generalized, i.e. replaced with the parent of this value in the generalization hierarchy defined for each domain (resp. each attribute). In the case of local recoding this is done for individual rows, for global recoding or full-domain generalization scheme [10] the generalization is performed equally for all the rows. This is repeated until the table is *k*-anonymous or the highest level of generalization is reached in all attributes. A shortcoming of this method is that outliers in the microdata can lead to a very course grain generalization. To avoid information loss caused by such outliers full domain generalization is mostly accompanied with row suppression where up to a given number or percentage rows are suppressed, i.e. rows are removed from the table or all values in these rows are replaced by ALL (resp. NULL). To avoid attacks: as pointed out by [26] at least *k* rows have to be suppressed. It is easy to see that in general several tables qualify as result. The aim is now to compute the table with the lowest information loss. The problem is known to be NP complete [19]. For global recoding the complexity of the method is exponential in the sum of all levels in the generalization hierarchies of the quasi identifier attributes. Many algorithms have been proposed which apply heuristics to reduce complexity and which apply different measures for information loss to efficiently compute "good" anonymized tables.

The detailed analysis of NULL values and their matching operators allows to extend these generalization algorithms with minimal effort to cover also NULL values in the input table using the k-anonymity definition with extended match. There are only two extensions necessary: (1) the generalization hierarchies of each quasi identifier is extended with an additional branch below the root that contains a NULL value in each level of the hierarchy. (2) k-anonymity ist tested with extended match. With these extensions, the anonymization algorithms can accept microdata with NULL values without any preprocessing. It is noteworthy that the complexity of the algorithms are not changed by this extension.

Figure1 shows, how the generalization hierarchies of our running example were extended, in order to apply extended matching.

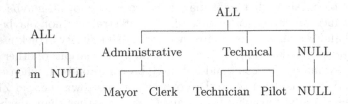

Fig. 1. Generalization hierarchies for NULL value handling in the running example

We implemented a flexible and customizable tool, ANON, for computing k-anonymous and ℓ-diverse tables based on full domain generalization row suppression based on the algorithms [24,26]. The contribution of ANON is on one hand that it computes the k-anonymous and ℓ-diverse generalization of a given table with minimal information loss, where information loss can be defined application specific in a fine grained way when defining the generalization hierarchies [4]. We implemented both anonymization with basic match and with extended match such that ANON offers two ways of handling NULL values: removing rows with NULL values before anonymization (basic match), or treating NULL values as any other value (extended match).

5   Experiments

In Table 1 we showed in a tiny example the differences resulting from applying different match definitions. Here we analyze the differences between anonymization with basic match and with extended in more detail and more exhaustive. Since in case of basic match whole rows have to be removed (suppressed) from the table if they contain a NULL value we expect an improvement in the information content of the results through the application of extended match. Our hypothesis, therefore, was that anonymization with extended match has less information loss than anonymization with basic match.

To test the hypothesis we conducted a series of 3168 experiments using the anonymization tool ANON applied to datasets derived from the Adult Database from the UCI Machine Learning Repository [9], with varying parameter settings and varying ratios of NULL values. We included 8 quasi identifiers (first 7 are also used in [16]) in the following order: *age, sex, race, marital_status, workclass, education, native_country, occupation*. According to this order, tables with $n <$ 8 quasi identifiers contain the first n quasi identifiers. Generalized values were provided with help of taxonomy trees for all quasi identifiers except the *age*, where we used ranges with steps {5, 10, 20, 100}.

The Adult Database itself contains several NULL values in the attributes *workclass, occupation* and *native_country*. To assure that every table has the right target amount of randomly placed NULL values used in the experiments (and not more than that), we first eliminated the rows with NULL values from the original table to obtain a common base for all test tables. From this base table with 45222 records we created 88 test tables with NULL values as a result of combination of the number of quasi identifiers (1 to 8) and the percentage of randomly inserted NULL values (0.1%, 0.5%, 1%, 2.5%, 5%, 7.5%, 10%, 15%, 20%, 25%, 30%). We used random number generation in Java for determination of cells in the table where NULL values were inserted.

For the anonymization runs of the 88 tables we used following parameters:

- *k*-parameter: 2, 3, 4, 5, 10, 15, 20, 50, 100
- max. allowed suppression: 0%, 1%
- matching: basic match, extended match

The max. allowed suppression specifies the quota for suppressing rows (to avoid adverse effects of outliers). For basic match this is in addition to the rows with NULL values which are removed in a preprocessing step.

The information loss of a generalized table is calculated with the following formula:

$$IL = \frac{s}{n} + \sum_{i=1}^{m} \frac{gl_{\alpha_i}}{ht_{\alpha_i}} \times \frac{1}{m} \times \frac{n-s}{n}$$

n number of rows
s number of removed rows
m number of quasi identifier attributes
gl_{α_i} generalization level of the attribute α_i in the generalized table
ht_{α_i} height of the generalization hierarchy of the attribute α_i
         (number of total generalization levels of α_i)

With this formula we consider two major factors that cause information loss: (1) generalization and (2) row suppression. Since we use global recoding (all values of an attribute α_i are generalized to the same hierarchy level), the information loss caused by generalization is simply the sum of information losses over the attributes α_i ($\sum_{i=1}^{m} \frac{gl_{\alpha_i}}{ht_{\alpha_i}} \times \frac{1}{m}$), multiplied by the percentage of remaining rows to which the generalization is applied ($\frac{n-s}{n}$). Information loss caused by row suppression is given by the percentage of eliminated rows ($\frac{s}{n}$).

We show first representative comparisons of the information loss between basic match and extended match without and with row suppression. In the figures results of extended match is shown in light gray bar and those of basic match in black bars. Each bar represents the information loss of one anonymization run.

Fig. 2 and Fig. 3 show anonymizations without row suppression (max. supp. = 0%). Anonymizations with extended match (light gray bars) tend to have constant information loss whereas the information loss of anonymizations with basic match (black bars) has a growing trend with increasing percentage of NULL values. This behavior can be explained with 2 influence factors: (1) number of removed rows (in case of basic match) and (2) outlier rows and the corresponding high generalization. If the percentage of NULL values is low, the rows with NULL

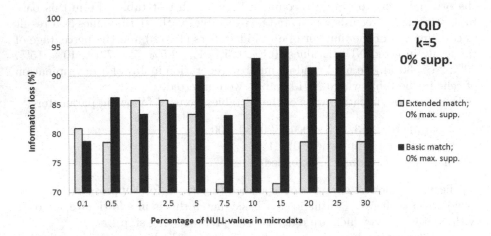

Fig. 2. Impact of the percentage of NULL values on the information loss (7 quasi identifiers, k-parameter = 5, 0% max. row suppression)

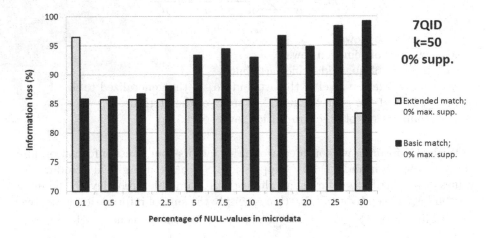

Fig. 3. Impact of the percentage of NULL values on the information loss (7 quasi identifiers, k-parameter = 50, 0% max. row suppression)

values are outlier rows, causing information loss to grow if extended match is used. If basic match is used instead, those "NULL-outliers" are simply removed and thus do not cause massive generalizations. On the other end, where the ratio of NULL values is high, rows with NULL values are not outliers anymore. Therefore, they do not increase the information loss if extended match is used. For basic match, however,information loss is proportional to the ratio of rows with NULL values leading to an increase in information loss with increasing ratios of NULL values.

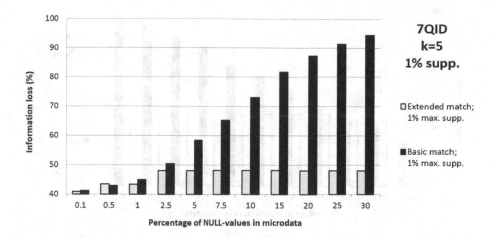

Fig. 4. Impact of the percentage of NULL values on the information loss (7 quasi identifiers, *k*-parameter = 5, 1% max. tuple suppression)

Fig. 4 and Fig. 5 show anonymizations with the same setup as those in Fig. 2 and Fig. 3, but with row suppression of up to 1%. Here extended match is no longer so sensible on NULL outliers and results in an almost constant information loss over increasing ratio of NULL values (light gray bars) while information loss grows drastically for basic match (black bars). That for low ratios of NULL values (below 1%) basic match is slightly better than extended match might be due that for basic match more rows are removed (number of rows with NULL plus 1% of the rows without NULL).

Figure 6 shows the aggregated results of all 3168 anonymizations in our experiment. Each bar represents the average difference in information loss of anonymizations with basic match and anonymizations with extended match, calculated over all 8 quasi identifiers and all 9 different *k*-parameters. The light gray bars represent the setups without row suppression (max. supp. = 0%) and the dark gray bars the setups with 1% max. suppression.

To summarize the results: The experiments showed that the best method in general was extended match with 1% row suppression. For very low ratios of NULL values basic match was slightly better. Extended match without row suppression performs worse for low ratios of NULL values because it suffers from the generalizations caused by NULL outliers. Basic match was only favorable for very low numbers of NULL values and the quality of the results deteriorates with increasing ratios of NULL values caused by the removal of all rows with NULL values. Furthermore, the information loss for extended match with row suppression did not seem to be influenced by the number of NULL values in the data set, as shown by the almost constant information loss over varying ratios of NULL values.

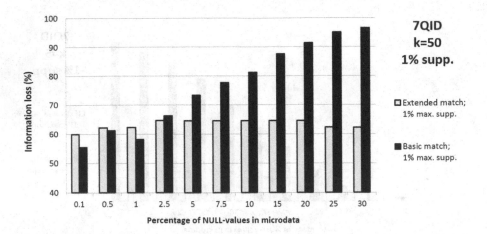

Fig. 5. Impact of the percentage of NULL values on the information loss (7 quasi identifiers, k-parameter $= 50$, 1% max. tuple suppression)

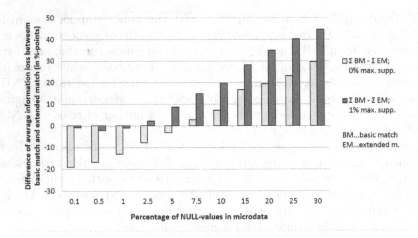

Fig. 6. Average advantage in %-points of information loss of the extended match over the basic match, depending on the percentage of NULL values in a table. In the positive y-area the extended match outperforms the basic match.

6   Conclusions

NULL values (missing values, not applicable attributes) appear frequently in microdata. Surprisingly, current anonymization algorithms require that all rows containing NULL values are removed from a table before it can be anonymized. We analyzed the effects of including NULL values in the definition of k-anonymity in detail and showed that the extended match where NULL values match (only) with other NULL values is a viable approach for extending *k*-anonymity to cover missing values. We introduced two new attacks that show that a further relaxation of the match operator which interprets NULL values as wildcards in the sense of Codd's maybe select leads to tables which can be attacked successfully. The extension of *k*-anonymity to tables with NULL values reduces the information loss induced by the removal of rows with NULL values by current anonymization algorithms and avoids the introduction of biases. Experiments showed that extended match reduces information loss for an generalization algorithm with row suppression considerably. The improvement is significant for large. The definition of *k*-anonymity we propose here can be used easily as basis for extending anonymization algorithms to also cover tables with NULL values in an adequate and save way.

References

1. Aggarwal, G., Feder, T., et al.: Approximation algorithms for k-anonymity. In: Proc. Int. Conf. on Database Theory, ICDT 2005 (2005)
2. Asslaber, M., Abuja, P.J., et al.: The genome austria tissue bank. Pathobiology: Journal of Immunopathology, Molecular, and Cellular Biology 74(4) (2007)
3. Bayardo, R.J., Agrawal, R.: Data privacy through optimal k-anonymization. In: Proc. of the 21st Int. Conf. on Data Engineering, ICDE 2005. IEEE Computer Society (2005)
4. Ciglic, M., Eder, J., Koncilia, C.: ANON - a flexible tool for achieving optimal k-anonymous and ℓ-diverse tables. Technical report (2014), http://isys.uni-klu.ac.at/PDF/2014-ANON-Techreport.pdf
5. Codd, E.F.: Extending the database relational model to capture more meaning. ACM Trans. Database Syst. 4(4) (1979)
6. Cox, L.H.: Suppression methodology and statistical disclosure control. Journal of the American Statistical Association 75(370) (1980)
7. Eder, J., Dabringer, C., Schicho, M., Stark, K.: Information systems for federated biobanks. Transactions on Large-Scale Data-and Knowledge-Centered Systems 1(1) (2009)
8. Eder, J., Gottweis, H., Zatloukal, K.: It solutions for privacy protection in biobanking. Public Health Genomics 15(5) (2012)
9. Frank, A., Asuncion, A.: UCI machine learning repository (2010), http://archive.ics.uci.edu/ml
10. Fung, B.C.M., Wang, K., Fu, A.W.-C., et al.: Introduction to Privacy-Preserving Data Publishing: Concepts and Techniques, 1st edn. Chapman & Hall/CRC (2010)
11. Gaskell, G., Gottweis, H., Starkbaum, J., et al.: Publics and biobanks: Paneuropean diversity and the challenge of responsible innovation. European Journal of Human Genetics 21(1) (2013)
12. ISO. ISO/IEC 9075-2:2011 Information technology — Database languages — SQL — Part 2: Foundation, SQL/Foundation (2011)
13. Iyengar, V.S.: Transforming data to satisfy privacy constraints. In: Proc. of the 8th Int. Conf. on Knowledge Discovery and Data Mining, KDD 2002. ACM (2002)
14. Kifer, D., Gehrke, J.: Injecting utility into anonymized datasets. In: Proc. of the Int. Conf. on Management of Data, SIGMOD 2006. ACM (2006)

15. LeFevre, K., DeWitt, D.J., Ramakrishnan, R.: Incognito: efficient full-domain k-anonymity. In: Proc. of the Int. Conf. on Management of Data, SIGMOD 2005. ACM (2005)
16. Machanavajjhala, A., Kifer, D., Gehrke, J., et al.: L-diversity: Privacy beyond k-anonymity. ACM Trans. Knowl. Discov. Data 1(1) (2007)
17. Matthews, G.J., Harel, O.: Data confidentiality: A review of methods for statistical disclosure limitation and methods for assessing privacy. Statistics Surveys 5 (2011)
18. Meyden, R.: Logical Approaches to Incomplete Information: A Survey. In: Chomicki, J., Saake, G. (eds.) Logics for Databases and Information Systems. The Kluwer Int. Series in Eng. and Comp. Science, vol. 436 (1998)
19. Meyerson, A., Williams, R.: On the complexity of optimal k-anonymity. In: Proc. 23rd ACM Symp. on Princ of Database Systems, PODS 2004. ACM (2004)
20. Ohrn, A., Ohno-Machado, L.: Using boolean reasoning to anonymize databases. Artificial Intelligence in Medicine 15(3) (1999)
21. Park, H., Shim, K.: Approximate algorithms for k-anonymity. In: Proc. of the ACM Int. Conf. on Management of Data, SIGMOD 2007. ACM (2007)
22. Samarati, P.: Protecting respondents' identities in microdata release. IEEE Trans. on Knowl. and Data Eng. 13(6) (2001)
23. Samarati, P., Sweeney, L.: Generalizing data to provide anonymity when disclosing information. In: Proc. of the 17th ACM Symp. on Principles of Database Systems, PODS 1998. ACM (1998)
24. Samarati, P., Sweeney, L.: Protecting privacy when disclosing information: k-anonymity and its enforcement through generalization and suppression. Technical report (1998)
25. Sun, X., Wang, H., Li, J., et al.: Enhanced p-sensitive k-anonymity models for privacy preserving data publishing. Trans. Data Privacy 1(2) (2008)
26. Sweeney, L.: Achieving k-anonymity privacy protection using generalization and suppression. Int. J. Uncertain. Fuzziness Knowl.-Based Syst. 10(5) (2002)
27. Tian, H., Zhang, W.: Extending l-diversity to generalize sensitive data. Data & Knowledge Engineering 70(1) (2011)
28. Wichmann, H.-E.E., Kuhn, K.A., Waldenberger, M., et al.: Comprehensive catalog of european biobanks. Nature Biotechnology 29(9) (2011)

Adaptive Security for Risk Management
Using Spatial Data

Mariagrazia Fugini[1], George Hadjichristofi[2], and Mahsa Teimourikia[1]

[1] Department of Electronics, Information and Bioengineering,
Politecnico di Milano, Milan, Italy
{mariagrazia.fugini,mahsa.teimourikia}@polimi.it
[2] Department of Computer Science and Engineering, Frederick University,
Nicosia/Limassol, Cyprus
com.hg@frederick.ac.cy

Abstract. This paper presents the design principles for adaptive security for areas where changing conditions trigger events signaling risks that might require modifying authorizations of risk management teams. Spatial resources and information of the areas to be protected are considered in sample scenarios, and principles of security design are introduced building on ABAC (Attribute Based Access Control). Adaptivity of security rules applying to subjects who intervene in the risk area is the core of our security model so as to make it responsive to risks by dynamically granting privileges to subjects to access resources.

Keywords: environment risk, adaptive security, context, ABAC, xacml, spatial data.

1 Introduction

Resource and people management for safety in risky environments is increasingly relevant [1]. In particular, during risky events, data and individual confidentiality and privacy should be preserved, while allowing dynamic adaptation of security rules (e.g., augmenting permissions of risk management teams) to face risks. Furthermore, spatial data are currently widely used to monitor and manage various aspects of people's life, and various countries have been setting up their own Spatial Data Infrastructures and Geographical Information Systems [2]. On the other hand, adaptivity of security models are topics currently popular in various areas of research, such as data management and web applications [3].

In this paper, we address adaptation of security rules to environmental risks: subjects can receive enhanced access privileges temporarily on resources to handle the risk, and then return to the "normal" situation having these privileges revoked. The proposed security model takes into account events occurring in a monitored area, which may lead to a risky situation and may modify the security needs. For example, if a risk of fire arises, monitoring cameras should be enabled to provide detailed images at a higher level of precision than usual; namely, security rules should be *adaptive* to the area sensed. We model adaptivity by introducing *Contexts* that

H. Decker et al. (Eds.): DEXA 2014, Part I, LNCS 8644, pp. 343–351, 2014.

indicate which security rules apply in a given situation without violating the need-to-know principles. Context allow operations on objects by subjects to be expanded while remaining within a security domain. Contexts are *activated* for risk and *deactivated* upon the conclusion of risk. Security modeling is based on Attribute-Based Access Control (ABAC) where security attributes include spatial information and are considered at various levels of detail so that they can be inspected at various zoom levels according to the severity of the risk and according to the defined access privileges. Based on ABAC, we address the definition of an XML Schema for subject/object entities to be used with XACML policy language [4]. For risk modeling, we rely on our proposed solutions in [5]. In [6], we have considered aspects related to security of spatial information, referring to GIS. In this paper we complement that approach by focusing on adaptive security derived from knowledge about an area where geo referenced spatial objects are included.

2   Related Work

The issue of providing security to people and locations according to what happens in an area is an open issue [7]. Security of physical objects and data, which have a location in an area, is treated in works such as [8], where cloud computing services and subjects' authorizations to geographic data are studied. The location datasets are transformed before being uploaded to the service provider. Authors in [9] propose to enhance security of spatial data in information sharing systems based on workflow services using XML key management, XML digital signatures, and geospatial extensible access control markup languages.

Research on spatial data security focuses on security of data management, sharing and transmission, and on Subject access control. However, few papers tackle *adaptive authorizations* considering what happens in the environment. Secure data management in GIS repositories is a relevant issue [9,10]. Role-Based Access Control (RBAC) [11] has been extended for spatial data management in the GEO-RBAC model [12] where spatial entities model Objects, Subject's positions, and geographically bounded roles. Roles are activated based on the position of the Subject only, and other aspects such as time, risks, emergencies, etc. do not affect the authorization decisions like in our proposal.

Coming to security models, recently, there has been considerable interest in Attribute Based Access Control (ABAC) [13] due to the limitations of the dominant and mostly used models such as Mandatory Access Control (MAC), Discretionary Access Control (DAC), and Role-Based Access Control (RBAC) [10]. ABAC takes into account the attributes of entities (subjects and objects), environmental conditions, and operations to authorize a certain request. ABAC can successfully encompass the benefits of MAC, DAC, and RBAC while surpassing their issues [11]. This research adopts the ABAC model [13], where fine-grained authorization is possible with no need to explicitly define the relationship between each object and subject, as in RBAC. XACML policy language [4] is adopted since it avoids conflicts between policies and rules. Additionally, the proposed model allows us to activate/deactivate security rules through the use of Contexts, as security domains, which include security rules of interest for given event(s) that can include a risky or emergency situation. Authors in [11] elaborate on risk-based adaptive access control (RAdAC) to semi-automatically adjust security risk to provide access to resources accounting for operational needs, risk factors, and situational factors. The risks treated here are security risks relating to the dynamic balance between the need to access information in view of mission priorities, risk, and cost of information compromise, and the overall operational and threat status of the system.

3   Security and Risks

We define the *Subject* as an entity taking actions in the system, namely requiring the execution of an operation upon an *Object*. We define the *Object* (also referred to as a *resource*) as a geo-referenced (spatial) entity to be protected from unauthorized use, such as data, devices, services, physical objects, areas, etc. Access control or authorization is the decision to permit or deny a *Subject* access to *Objects*[1]. *Privileges* represent the "authorized behavior of a subject"; they are defined by an authority. We have *elementary actions* (e.g., "read" privilege on environment objects, which will map into "view", "read", "zoom in/out" privileges, depending on the technology used for monitoring the environment, and the set of privileges defined therein) and *complex activities* (e.g., re-position a camera or rescue a person). For example, in an airport, there are parking areas, main buildings and so on, which can suffer from various types of risks. The airport itself is a spatial object with blueprints, security exits, surveillance sensors, and localization devices. The *Security Manager* is a *Subject* in charge of monitoring the airport and of planning/executing risk interventions. The *Security Staff* is a *Subject* to be cleared to access services that locate a risky event, or people/objects exposed to risk. The *Security Manager* has the highest clearance and can access Objects with virtually no limitations in a risk context; the *Security Staff* has lower clearance and can execute some security actions (e.g., launch an alarm) on a limited set of objects. These *Subjects* can receive an upgrade in their security level if a risk occurs: for instance, the *Security Staff* can gain the zoom-in privilege on additional areas upon risks and have this privilege revoked when the risk ceases.

Contexts establish the *security policies*, namely for which *Subject* attributes which operations can be executed on which *Objects* attributes. Contexts in the airport can be: *Risk; Flight* (some flights can be blocked); *Cargo Context* (ground personnel can operate in reserved areas). By monitoring the environment, some events are triggered, which activate and/or deactivate Contexts allow dynamic changing the security rules and/or the subject/object attributes (e.g., to increase the security clearance attribute of a subject).

4   Security Model

When an environment at risk is considered, complex problems regarding planning and management of security need be handled since the authorizations of subjects can vary

---

[1] The terms *access control, authorization* and *security* will be used synonymously throughout this paper. In particular, security is equaled to confidentiality and we place less emphasis on integrity aspects and other security properties.

dynamically e.g., to locate and analyze situations for decision making (*context analysis*) about risks. Moreover, security problems related to spatial data are becoming more and more imperative, especially in public security, or in applications for smart environments [9]. In what follows, we describe our security model for risks, including spatial objects attributes in the security model.

4.1   Security Model Components

We give the security model components and outline the access control mechanisms, for which we refer to the architecture of security controls in Figure 1 (details on the architecture are in [4]).

Subject s: a user, application or process wanting to perform an action on a resource/object. A subject holds *three groups* of attributes: 1) *General Attributes*: define the general characteristics of a subject, such as its identity, name, job title, and etc. 2) *Geo Attributes*: define the location and spatial properties of a subject, such as location, reachable positions, etc.   Geo attributes can be given at various levels of granularity, i.e., for privacy reasons, the exact location of the subject might be hidden while the subject's logical position, usual location, and the places that the subject is allowed to have access to are visible; 3) *Security Attributes*: define the security-related properties of the subject, such as security clearances, highest-possible security clearance, privileges that the subject can grant/receive in different circumstances, roles which can be active at a given instant, and so on. Security attributes include conditions such as time restriction for the role to be active: under a time restriction, the role should not remain active after the time has elapsed the defined boundaries. As an example, a Security Manager in the airport can be described in XML as in Figure 2.

Although the geo location of the subject is available, we can know only in which building the Security Manager is, rather than the detailed coordinates of his/her position in an area, since the granularity is defined to be at the building level. He/she can be usually found in office T21, and he/she is allowed also to be in building no. 2 while he/she cannot access (*DeniedPosition*) the cargo area. The *secClearance* attribute is defined in terms of levels Ln, in which $n \in \mathbb{N}$, where the smaller the index the lower the security clearance of the subject. The Security Manager in this example has security clearance L4 and can have a maximum security clearance equal to L7 so that during the dynamic assignment of security clearances this subject has a threshold. His role of *security manager* is deactivated at a given time instant (and can be activated in case of a risky event), and the role of manager is active only during his *office hours*.

Object o: in our definition, objects are any resource to be protected. However, in this paper we are particularly concerned with spatial objects, namely entities which are geo-referenced. Spatial Objects (*SOb*s) are entities (maps, streets, buildings, areas) analyzed in special locations. Objects hold three groups of attributes (*OA*): 1) *General Attributes* can be object specific and differ depending on the type of the object; 2) *Geo Attributes,* support the identification of places where resources are located and record relevant quantities and densities. Geographic Attributes are for instance geo referenced coordinates (latitude, longitude), and levels of granularity (they exist at various zooming levels) available in the repository, and the objects in the vicinity; 3) *Security Attributes* define restrictions on information privacy, owner, level of sensitivity, and so on.

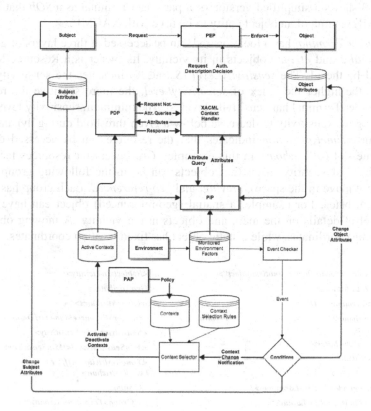

Fig. 1. Access Control Mechanism with Contexts

```
<Subject            xmlns="urn:schemas:entities:subject">
<GeneralAttributes><ID>SM1ID</ID></GeneralAttributes>
    <GeoAttributes>
      <Coordinates>
        <Longitude>23.55</Longitude>
<Latitude>32.22</Latitude>
        <Granularity>Building</Granularity>
      </Coordinates>
      <CommonPos><Position>OfficeT21</Position></CommonP
os>
        <AllowedPos><Position>BuildingN2</Position>...
</AllowedPos>
        <DeniedPos><Position>Cargo</Position></DeniedPos>
      </GeoAttributes>
      <SecurityAttributes>
        <SecurityClearance>L4</SecurityClearance>
```
```
        <MaxSecClearance>L7</MaxSecClearance>
        <Roles>
          <Role>
            <RoleName>Security
Manager</RoleName>
            <RoleStatus>Deactive</RoleStatus>
          </Role>
          <Role>
            <RoleName>Manager</RoleName>
            <RoleStatus>Active</RoleStatus>
<TimeRestriction>OfficeHours</TimeRestriction>
          </Role>
        </Roles>
      </SecurityAttributes>
    </Subject>
```

Fig. 2. Subjects Security Attributes XML Schema

Figure 3 shows a simplified version of a passenger terminal as a *SOb* that is based on our XML schema of an object entity to be used with XACML.

Passenger Terminal has a location and can be accessed at three layers of granularity. *Terminal 2* and *etc.* are objects in its vicinity. Its owner is a Resource Manager referenced by the id. The *sensitivityLevel (S_n, $n \in \mathbb{N}$)* indicates the sensitivity of the resources: the higher the index of *sensitivityLevel*, the more sensitive the resource. The Passenger Terminal has sensitivity level S_4. The minimum sensitivity level can be S_3, avoiding the sensitivity to decrease below a given threshold during dynamic level assignments. *TimeRestriction* indicates that the resource can be accessed during a certain time slot (*office hours* in this example). *Groups* cluster resources facilitating the definition of security rules. E.g. objects can be in the following groups: *static* (they do not move in the space), *moving*, and *geo-referenced*. Each group has its own peculiar attributes. For example, a spatial (geo-referenced) object can have coordinates, level of details on the map, and objects in the vicinity. A moving object has time-varying coordinates, while a static object has fixed location coordinates.

```
<Object xmlns="urn:schemas:entities:object">          </ObjectsInVicinity>
   <GeneralAttributes>                                 </GeoAttributes>
      <ID>Terminal1</ID>                               <SecurityAttributes>
         <Description>Passenger          Termin-          <owner>ResourceManagerID</owner>
al</Description>                                          <Sensitivity>S4</Sensitivity>
   </GeneralAttributes>                                  <MinSensitivity>S3</MinSensitivity>
   <GeoAttributes>                                       <TimeRestriction>Office
      <Coordinates>                           Hours</TimeRestriction>
         <Longitude>23.55</Longitude>                    <Groups>
         <Latitude>32.22</Latitude>                         <Group>Terminal</Group>
         <DetailLayer>3</DetailLayer>                       <Group>StaticObject</Group>
      </Coordinates>                                        <Group>SpatialObject</Group>
      <ObjectsInVicinity>                                </Groups>
         <Object>                                       </SecurityAttributes>
            <ID>Terminal2</ID>                          </Object>
         </Object>
```

Fig. 3. Objects Security Attributes XML Schema

Actions and Activities a: these are operations[2] that can be executed by Subjects on Objects in a given Context. We consider: *simple operations* (read, write, execute, zoom-in/out); *complex operations*, called *activities*, which model a task, a process, an application, or a physical action. Examples of activities in an airport are "Redirect the airplane to another runway", or "Turn the fire alarm protocol on". In the schema below we define the zoom-in action; for actions also we specify our own XML schema to be used with XACML policy language.

---

[2] "Operation" here denotes a security privilege.

```
<Action xmlns="urn:schemas:entities:action">
    <ID>zoom-in</ID>
    <Name>Zoom-in</Name>
    <Description>Zooming in, to view more detailed leveled of a spatial object</Description>
</Action>
```

In order to receive a permission to execute an action/activity, a request is submitted to a Policy Enforcement Point (PEP), as in Figure 1. This request is specified by three elements: the requesting subject (access subject), the action/activity to be permitted, and the object to be accessed. A sample request is the *Security Manager* (subject) with id of *SM1ID*, wants to *zoom-in* (action) on the *Terminal* area (object) with id = *Terminal1*.

Context c and Security Rule r: The Context delimits which security rules apply when risks occur. To adopt XACML, security rules in each Context are defined by the *DefineRule(a, c, authDecision)* function, where 'a', 'c' and 'authDecision' are the action/activity, context, and authorization decision, respectively. These rules exist in isolation in the Policy Administration Point (PAP) (Figure 1). As an example, a rule can specifythat the *Risk Manager* subjects, endowed with security clearance $>= L_3$, can *turn on* (activity) the *alarms* (object o) whose sensitivity level is $<=S_4$. Thus, for $s \in S$ *and* $o \in O$ we state:

$$r_1: DefineRule(turnon, RiskContext, permit) \leftarrow$$
$$(s.Role = "RiskManager") \wedge$$
$$(s.SecurityClearance > "L_3") \wedge$$
$$(o.Group = "alarm") \wedge$$
$$(o.SensitivityLevel < "S_4").$$

Risk rk: Some factors that change dynamically can signal the occurrence of a risk situation, which can be recognized by monitoring the environment [5] based on parameters such as: *type, level,* and *location* of the risk factors. It is then possible to decide how to adapt the security rules to handle the risk *rk*. Risk detection is performed in the environment by monitoring factors which possible trigger events.

Event ev: Changes in the environment captured by monitoring devices can trigger events (see Figure 1). Events activate/deactivate *Contexts*. Events may also cause the modification of the attributes of *subjects* and *objects* within the Context, according to what the Subjects/Objects relationships in the Context were *before* the event and those to be established *after* the event to manage the risk. Events play a key role in dynamic adaptation of security in response to changes in the environment.

4.2   Adaptive Security

Events are recognized by a generic "event checker" module (Figure 1) which requires risk management for an *a-priori* unpredictable situation. Re-assignment of security rules is carried out dynamically as soon as the event is identified. Permissions related to security are associated to an *authorization policy*, which determines which security rules apply during risk management. In particular, specific *events* due to some *conditions* may activate/deactivate *Contexts* as well as change the attributes of Subject/Object, as in Figure 1. Activation/deactivation of *Contexts* adaptively determines which security rules apply according to the detected *event(s)*. There can be multiple rules per Context and different Contexts can share the same rules. Since we adopt XACML as the policy language, such conflicts can be avoided using the policy combining rules defined in XACML. As we mapped our contexts into the definition of <Policy> entity in XACML, the set of activated contexts are relevant to the <Policy-Set> entity, and therefore, the policy combination algorithms can be applied to avoid conflicts between the activated contexts. For events, we adopt *ECA (Event-Condition-Action)* [14] rules which indicate that in case of an *event*, if the *condition* holds, then a certain action(s) should take place[3]. The event is triggered by the change in the environment conditions and is detected by the event checker of Figure 1. The Action is the activation/deactivation of a Context(s) or/and modifications in the attributes of subjects/objects by the function *ChangeAttr(attribute, subject/object, value)*. To activate/deactivate contexts dynamically, we have *context selection rules*, as in Figure 1, that are pre-defined (and that apply at the occurrence of events (dynamically at run time). A function *ContextSelection(ev,Context-activate,Context-deactivate)* is the template for Context activation/deactivation.

As an example, suppose we have the following ECA statement:

> *Event :* '*fire*'
>
> *Conditions :* (rk.Type: 'explosion')∧(rk.Level: 'high')∧(e.TimeOfDay: 'AfterOfficeHours')∧
> (e.locateSubjects(em.Position) != 0)
>
> *Actions:* Activate RiskContext, Deactivate FlightContext, ChangeAttr(s.SecurityClearance,
> s.Role: 'RiskManager', 'L_0'), ChangeAttr(e.TimeRestriction, e.TimeRestriction :'OfficeHour',
> 'none')

The conditions indicate the case of *ev fire*, and risk *rk* of type *explosion* with *high* level of danger, if subjects (people) are in the environment *e*, and the office hour is elapsed, then, the *risk* context should be activated, and the *flight* context deactivated[4] . And subject *s* with role *Risk Manager* should get a higher *clearance L_0* while the time restrictions should be removed from object *o,* which had such an attribute. Considering *S, O, A,* and*C* as the set of all subjects, objects, actions/activities (operations), and contexts, respectively, we define a policy model based on ABAC. Regarding the single subject $s \in S$, object $o \in O$, action/activity $a \in A,$ and context $c \in C$ a security rule is defined at the security level as follows:

> *SecurityRule: CheckAccess (s:S, a:A, o:O, c: C)*

Considering the *operation*, attributes related to the *subject*, *object* and the rules in the *context*, *CheckAccess* returns a tuple <*s, a, o, c*> meaning that the operation $a \in A$ is allowed for subject $s \in S$ on object $o \in O$ in context $c \in C$. If such a tuple is not found, the action is denied.

---

[3] Note that here the "action" as in the ECA paradigm, is different from "action" defined in our model (denoting the operation).

[4] We assume that activation and deactivation procedures exist to check if the context is already activated or is not

5   Conclusions and Future Work

This paper presented adaptive security modeling motivated by the need for smart environments to dynamically authorize actors in facing risks. Based on the ABAC and on XACML policy language, security rules can change dynamically according to Contexts which delimit how subjects can enlarge their access privileges on (spatial) objects on the basis of predefined security policies. We intend to focus on the topics of binding environmental and spatial information, on the dynamics of assigning authoritative roles to administrators, and on ways to handle conflicting Context switching. Future work focuses on implementation of the XML schemas to be included in our developed web application described in [5]).

References

1. Chourabi, H., Nam, T., Walker, S., Gil-Garcia, J., Mellouli, S., Nahon, K., Scholl, H.: Understanding smart cities: An integrative framework. In: The 45th Hawaii International Conference on System Science (HICSS), pp. 2289–2297 (2012)
2. Murti, K., Tadimeti, V.: A simplified GeoDRM model for SDI services. In: The International Conference on Communication, Computing & Security (2011)
3. René, M., Schmidtke, H., Sigg, S.: Security and trust in context-aware applications. In: Personal and Ubiquitous Computing, pp. 1–2 (2014)
4. Rissanen, E.: eXtensible access control markup language (XACML) version 3.0., OASIS standard (2012)
5. Fugini, M., Raibulet, C., Ubezio, L.: Risk assessment in work environments: modeling and simulation. Concurrency and Computation: Practice and Experience 24(18), 2381–2403 (2012)
6. Dessì, N., Fugini, M., Garau, G., Pes, B.: Architectural and security aspects in innovative decisional supports. In: ITAIS 2013 Conf. (2013)
7. Li, G.: Research on security mechanism of sharing system based on geographic information service. In: The International Conference on Information Engineering and Applications (IEA), pp. 345–351 (2013)
8. Smith, K.: Environmental hazards: assessing risk and reducing disaster. Routledge (2013)
9. Tompson, J., Kennedy, S.: Where exactly is the target market? Using geographic information systems for locating potential customers of a small business. Entrepreneurial Practice Review 2(4) (2013)
10. Jin, X., Krishnan, R., Sandhu, R.: A unified attribute-based access control model covering DAC, MAC and RBAC. In: Cuppens-Boulahia, N., Cuppens, F., Garcia-Alfaro, J. (eds.) DBSec 2012. LNCS, vol. 7371, pp. 41–55. Springer, Heidelberg (2012)
11. Kandala, S., Sandhu, R., Bhamidipati, V.: An attribute based framework for risk-adaptive access control models. In: Sixth International Conference on Availability, Reliability and Security, ARES (2011)
12. Xiong, Z., Xu, J., Wang, G., Li, J., Cai, W.H.: UCON application model based on role and Security. In: ARES (2011)

Secure and Efficient Data Placement in Mobile Healthcare Services

Anne V.D.M. Kayem[1,*], Khalid Elgazzar[2], and Patrick Martin[2]

[1] Dept. of Computer Science, University of Cape Town,
Rondebosch 7701, Cape Town, South Africa
akayem@cs.uct.ac.za
[2] School of Computing, Queen's University
Kingston, ON, K7L, 3N6, Canada
{elgazzar,martin}@cs.queensu.ca

Abstract. Mobile health services offer access to patient data, reported by body sensors, that is stored locally on a mobile phone. Storage capacity limitations however, make storing complete copies of the mobile data challenging and using backup cloud storage is impractical as well as insecure when Internet access is intermittent. We propose a novel approach, based on fragmentation and caching, to secure and efficient data management in mobile health services. Fragmentation classifies the data by confidentiality and attribute affinity. While caching prioritizes data fragments by frequency of access in order to compute a minimum dataset of information to store on the mobile phone. Our simulation results demonstrate that data fragmentation improves query response time by almost 25% and, caching, by an additional 35% on various query workloads on the fragmented data when compared to cloud centric un-fragmented data.

Keywords: Healthcare-as-a-Mobile-Service, Fragmentation, Prioritization, Data Security, Mobile Data, Cloud Computing.

1   Introduction

Electronic health data management is a cost effective health care management strategy. Existing solutions use a combination of smart devices and the Internet to manage health care information. However, these solutions do not work well in areas where Internet access is intermittent. In these areas mobile phones offer a low cost and portable alternative to regular computers [1] and so, health care providers are exploring mobile Health (m-Health) services as a cost-effective approach to increased health data accessibility [2].

M-health services offer patient data accessibility anytime and anywhere, which is advantageous for rural and developing world scenarios where access to the Internet can sometimes be non-existent or unreliable. By maintaining a local (mobile) copy of a patient's medical history cases of misdiagnosis, particularly in

* Funding for this research was provided by the University of Cape Town and the National Research Foundation (NRF) of South Africa and the National Science and Engineering Research Council (NSERC) of Canada.

H. Decker et al. (Eds.): DEXA 2014, Part I, LNCS 8644, pp. 352–361, 2014.

emergency situations, can be avoided. While this offers the advantage of portability one caveat is that mobile phones are limited storage-wise so, storing complete medical records on the phone is challenging. Additionally, guaranteeing the
enforcement of the HP's security policies on the mobile data is difficult, making
HP's hesitant to encourage the use of m-health systems.

In this paper, we address the storage and security issues with a mechanism
based on fragmentation and caching. Fragmentation is used to segment the data
based on confidentiality, and affinity constraints. Confidentiality enforces the security requirements of the HP on the data fragments while affinity aids efficient
query processing on the mobile-end. Caching is used to maintain a priority queue
of data fragments ordered by relevance and frequency of access. Storage management is optimized by placing a copy of only the most frequently accessed data
fragments on the mobile phone while a complete copy of the data is archived on
the cloud. Data security is enforced by imposing confidentiality constraints to
control data sharing. Our simulation results demonstrate that data fragmentation improves the query response time by almost 25%, while caching supported
by attribute affinity offers an additional 35% improvement in query response
time which is efficient.

The rest of the paper is structured as follows, Section 2 presents related work
on data outsourcing. Section 3 describes our proposed solution and Section 4,
presents our simulation results. We offer concluding remarks and directions for
future work in Section 5.

2   Related Work

Considerable work has gone in to managing outsourced data securely and efficiently. Damiani et al. [3,4] triggered work in this area with the idea of metadata
management of encrypted outsourced databases. They showed that metadata
is integral to retrieving information to respond accurately to outsourced data
queries. Other works on querying encrypted outsourced data in untrustworthy
environments include Hacigumus et al.'s [5,6] and De Capitani Di Vimercati et
al.'s [7] on encryption policies for outsourced data, supported by hierarchical
cryptographic key management schemes [8]. Other works focus on efficient security policy updates on read-intensive data at the service provider's end [9,10,11]
and on minimizing query failure rates on read intensive data [12,13,14]. While
previous work is mainly on server-side read-intensive data, the m-health care scenario involves both client and server-end write and read-intensive data. As well,
storage limitations make relying on manual mobile to cloud data outsourcing
mechanisms impractical. Therefore, a secure and efficient approach to m-health
data management is needed to minimize the cost of updates and queries.

3   Secure Fragmentation and Caching

As illustrated in Figure 1 in our healthcare-as-a-mobile-service framework, data
is collected via body sensors and stored locally on the patient's mobile device.

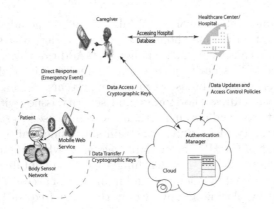

Fig. 1. Healthcare-as-a-Mobile-Service Architecture

This data is accessible, via a mobile web service (MWS), to a caregiver[1]. MWSs offer an interoperable interface that facilitates data sharing via ubiquitous protocols and data formats such as HTTP and XML. This is advantageous because it allows HPs and patients to use a uniform communication platform irrespective of the device or operating system. Initially, all of the patient's data is stored on the HP's end where it is categorized into files based on pre-specified confidentiality and attribute affinity requirements. For consistency with similar solutions in the literature [7,15,16,10], we consider that a cryptographic key management scheme is implemented to enforce the access control policies on these files. As De Capitani Di Vimercati et al. [16] proposed, the data owner (healthcare provider (HP)) protects the data that is transferred for storage and management to a service provider (or cloud) by encrypting the data. This ensures that the data remains secret even to the service provider. To enforce non-repudiation, the service provider imposes a second layer of encryption on the data and shares the key used with the HP. All users wanting to access the data receive two keys from the HP, one that is used to decrypt the encryption layer imposed by the cloud service provider and the other key, to decrypt the encryption layer imposed by the HP. An authentication manager (AM) controls access to the cloud data by verifying all access requests. Depending on the storage capacity of the patient's mobile phone, the data stored at the HP is further fragmented to form a "summarized" copy that is transferred to the mobile phone. To ensure that the data is kept secure both on the cloud and on the mobile phone it is important to model the security requirements of the HP by applying confidentiality constraints on the sets of attributes in the patient's data. A well-defined confidentiality constraint can be defined formally as follows:

Definition 1. *(Well-Defined Confidentiality Constraints)*
Given a set of attributes $A = \{a_0, ..., a_i, a_j, ..., a_n\}$, and a set of confidentiality constraints $C = \{c_0, ..., c_i, c_j, ..., c_m\}$ where m is the maximum number of confidentiality constraints that have been specified, and n the maximum number of

---

[1] A caregiver: person who administers health related operations on patients.

attributes that have been specified, C is said to be well-defined if and only if $\forall c_i$, $c_j \in C$, $i \neq j$ $c_i \nsubseteq c_j$ and $c_i \subseteq A$.

The confidentiality constraints indicate which attributes need to be kept together and protected from unauthorized access. For instance, {Name, DOB} are considered to be sensitive attributes that are correlated and so must be kept in the same fragment of the data.

Based on the confidentiality constraints, the HP fragments the data into a set of disjoint fragments F_i represented by a tree $(T, \prec)$, where $T = \{F_0, F_1, ..., F_{n-1}\}$. By definition, $F_i \prec F_j$ implies that F_j contains attributes that hold information that is more important or relevant to the patient's current condition than the attributes contained in F_i. A formal definition of fragmentation is the following.

Definition 2. *(Fragmentation.) Given a relation schema R, we say that a set of data fragments F is formed by subdividing R into subsets of tuples that obey a set of well-defined confidentiality constraints. The set F is correct and complete if each tuple of R is mapped unto at least one tuple of F. The resulting fragments have the same schema structure as R, but differ in the data they contain.*

Importance is derived from the specification of the confidentiality constraints. For instance, if $c_0 \equiv F_0$ and $c_1 \equiv F_1$, then, $F_0 \prec F_1$ implies that c_0 contains more sensitive information and it is assumed that F_0 is more important than F_1. To protect against inference attacks and linking attacks no attribute must appear in the clear form in more than one fragment [17].

Since multiple security policies might exist, creating fragments solely according to the confidentiality constraints can result in several small but disjoint fragments that are expensive to query. Aggregating fragments with similar confidentiality constraints to form bigger data fragments optimizes query execution. However, finding a fragment that contains all the relevant information required to make query execution on the mobile device efficient and maximize the utility of mobile device resources such as battery power is provably NP-hard [18] so, we use a heuristic.

3.1 Fragmentation Heuristics

Our heuristic for optimal fragmentation relies on the notions of marginal gain and fragment distance. Marginal gain (MG) is used to decide whether or not to merge data fragments based on the hit rate per fragment considered. The MG is computed by evaluating the number of hits (queries) on a fragment say F_i, in comparison to another fragment say, F_j. We compute MG as follows:

$$\text{MG}(F_i, F_j) = \text{Hits}(F_i) - \text{Hits}(F_j) \quad ......(1)$$

When $\text{MG}(F_i, F_j) > 0$ the implication is that F_i has a higher number of hits than F_j, and when $\text{MG}(F_i, F_j) < 0$ the implication is that F_j has a higher number of hits than F_i. A value of 0 for $\text{MG}(F_i, F_j)$ indicates that both fragments receive the same hit rate.

As a further step, based on the computed MG, we need to determine whether or not merging the fragments will reduce the query failure rate. We do this by evaluating the distance, Dist (F_i, F_j), between two fragments F_i and F_j from which the MG was computed. This distance is computed as follows

$$\text{Dist}\,(F_i, F_j) = \sqrt{\sum_{h}^{k} (\text{Hits}\,(a_{i,h}) - \text{Hits}\,(a_{j,h}))^2} \ldots\ldots(2)$$

where Hits $(a_{i,h})$ is the number of hits on attribute h in fragment F_i, $h \geq 1$, and $k \geq 1$ such that k represents the highest number of attributes in the largest fragment. A small distance value, that is lower than a predefined threshold value, indicates that merging both fragments of data is advantageous whereas a high distance value indicates the reverse. The assumption here is that the corresponding attributes in F_i and F_j are similar.

Frequently accessed attributes are placed on the mobile device to ensure that the information can be made available to the caregiver in cases of emergency when access to the Internet is temporarily unavailable due to bandwidth limitations. Regularity of access provides added information to decide which attributes of the data to store on the mobile phone. Data consistency, both on the mobile phone and the cloud, is maintained via updates to the medical data that are affected during periods when access to the network occurs. For instance, when a patient moves into an area with good Internet coverage, we might want to launch bandwidth intensive operations such as down/up loads of data.

We must now consider the problem of optimizing mobile storage management which is provably NP-Hard [18], and so we use a heuristic approach to address this issue.

3.2 Caching for Mobile Storage Management

We maximize storage utilization by categorizing the data fragments according to frequency of access and caching the data to maintain historical information on the fragments of data that are necessary for successful queries. The cache is modeled to store the data fragments hierarchically based on an importance metric that is guided by the confidentiality constraints. In the hierarchy, there are n cache levels where $n \geq 1$, and the hierarchy is organized as a rooted tree which is attached to the storage partition on the mobile device. In the cache hierarchy each node contains a data fragment. The most recent and frequently accessed data is stored at the higher levels of the hierarchy while the least frequently accessed and, by comparison, older information is stored at the lower levels of the hierarchy.

A fragment gets moved to a lower priority cache level if it is not accessed frequently and to a higher level, if the reverse is true or if it contains new material. To avoid data loss, a buffering operation is used to temporarily store the data during the change. Each data fragment is assigned a priority according to frequency of access. Initially a new fragment is assigned the highest priority. Depending on access frequency, the fragment will get assigned either a higher or lower priority. In the case of a lower priority assignment, the affected fragment gets moved to a lower cache level. While a higher priority assignment will result in the fragment being moved to highest possible position in the hierarchy by comparison to the other cache priority assignments. For simplicity, we assume that caches are stationary and that the data fragments are moved between them. In addition, each cache holds at most one data fragment at a time.

We use a priority queue that is modeled as a binary heap to implement our caching hierarchy. In the caching hierarchy, the data fragments are stored in the form of a complete binary tree and are ordered to ensure that the most important (high priority) data fragment is always easily accessible from the root node. Initially, the caching hierarchy is populated on a first come first served basis so that the first data fragment to be processed is stored at the root node and the last, at a leaf node. The caching hierarchy is evaluated at regular intervals and re-ordered to place the most frequently accessed fragments at the upper levels of the hierarchy. We compute the importance, I, of a data fragment, as follows:
$I = \frac{\text{Hits}(F_i)}{U_{F_i}}$, where U_{F_i} denotes the freshness of F_i, calculated in terms of the number of time units (seconds, minutes, days...) for which the data has been on the system, and $\text{Hits}(F_i)$ is the number of accesses/queries involving the data fragment F_i. As illustrated in Figure 2, based on fragment importance, we use

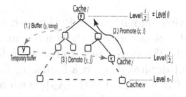

$i,j = 1...n$, where n = maximum number of nodes (caches) in hierarchy

Fig. 2. Caching Hierarchy: Operation

the following three operations to order the hierarchy:

1. BUFFER (y, temp): Move fragment y from cache$_i$ at level $\lfloor \frac{i}{2} \rfloor$ to a temporary buffer
2. PROMOTE (x, i): Move fragment x from cache$_j$ to cache$_i$ at level $\lfloor \frac{i}{2} \rfloor$
3. DEMOTE (y, j): Move fragment y from the buffer to cache$_j$ at level $\lfloor \frac{j}{2} \rfloor$

We sort the cache hierarchy by iteratively evaluating each level of the hierarchy and moving the data fragments upwards or downwards based on importance.

Example 1. As an example of how our cache hierarchy sorting algorithm works, in Figure 3 integer values are assigned randomly to each node to denote the current importance rating of the data fragment at that position in the hierarchy. We iterate through all the levels of the hierarchy rearranging the data fragments to ensure that the fragment with the highest importance rating is located at the top of the binary tree that represents the priority queue. Notice that in Step 3, the fragment with a rating of 9 has been progressively moved up to the root position.

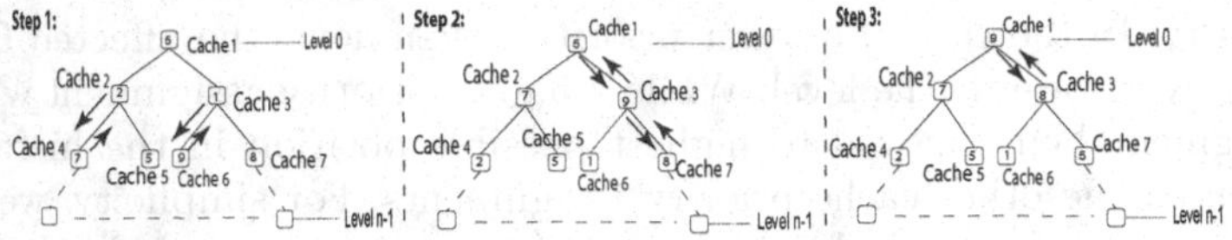

Fig. 3. Caching Hierarchy: Storage and Sorting

When a new data fragment needs to be inserted in the hierarchy, in accordance with the requirement that the data be stored in the form of a complete binary tree, we place the new data fragment in the last position of the caching hierarchy. For instance, using a binary heap of size $H[max]$ would imply that the next empty slot would be the current size of the heap augmented by one (i.e. $H[max + 1]$).

Finally, to decide which fragments of data to offload to the cloud, we begin by sorting the data fragments in order of maximum importance. A threshold value that is defined by the mobile device user is used to determine how many of the fragments need to be transferred. Using this value the caching hierarchy management algorithm will use a delete operation to delete data fragments from the caching hierarchy and have these fragments transferred to the mobile device.

4   Results

We implemented our proposed algorithms, in Python, on a hybrid platform comprising a mobile device and a cloud server. Python has an embedded lightweight database engine SQLite [19]. We developed a prototype to evaluate the performance of our approach and deployed it on a Samsung Galaxy II I9100 smartphone (Dual-core 1.2 GHz Cortex-A9, 1 GB RAM) with a rooted Android 4.0.4 platform, connected to a Wi-Fi network. This device consumes approximately 1.3 Watt per second to send data over the wireless link. The cloud server is represented by an Amazon EC2 virtual machine of the type 't1.large' with an EC2 pre-configured image (AMI) of Ubuntu Server 12.04 LTS, 64 bits.

In our first experiment, we used a set of 32 attributes, such as Electrocardiography (ECG), Oxygen Saturation (SPO2), Temperature, and Blood Sugar (Glucose). We generated 30 - 90 fragments based on 30 confidentiality constraints, attribute affinity, and data freshness. We split the data vertically based on the confidentiality constraints and then calculated the marginal gain between the various data fragments based on attribute access frequency. As mentioned in Section 3.1, the marginal gain and distance were computed to decide which data fragments to merge to minimize the query failure rate and increase response time. Each data fragment was also segmented horizontally based on data freshness (in terms of date and time). The fragments are generated using database views, and each view represents a fragment of size between 100KB and 100MB, containing 1000 - 1000000 records. We queried the database with 1000 randomly generated queries. Figure 4 shows the average query response time for un-fragmented and fragmented data. We note that data fragmentation improves the query response time by almost 25%. Furthermore, due to the impact of attribute affinity, caching offers an additional performance gain of about 35%.

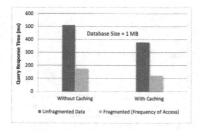

Fig. 4. Query Response time: Fragmented versus Un-fragmented Data

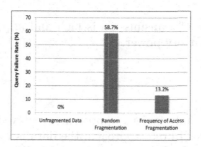

Fig. 5. Query Failure Rate: Fragmented versus Un-fragmented Data

Figure 5 shows the query failure rate on un-fragmented data and fragmented data. In the latter case we compare our data fragmentation approach to a random data fragmentation approach. The query failure rate is the number of failed queries in comparison to the total number of queries. The results reveal that our approach outperforms the random data fragmentation approach, where both the attributes and fragments that get to remain on the mobile device are chosen randomly. Our approach results in 13.2% query failure rate in contrast to 58.7% for the random fragmentation approach.

The un-fragmented data yields a 100% query success rate by keeping the entire dataset on the mobile device. We attribute this difference in success rates to the fact that queries on the fragmented data experience delays in going out to the cloud copy of the data and hence are not able to provide a response in a time window that is complaint to the quality of service agreements. Other factors to which this might be attributed include low bandwidth or inaccessibility due to network failures. Figure 6 depicts the energy cost of transferring data to the cloud. The energy consumption is directly proportional to the amount of transferred data and available bandwidth on wireless link between the mobile device and the cloud data storage provider. The energy consumption is calculated using $E = \frac{D}{B} \times p_t$ where E represents the total energy consumption in Watts, D indicates the size of transferred data, B represents the available bandwidth, and p_t is the energy consumption unit that the device consumes to transfer data over its network interface per second. The rationale behind this equation is that a high bandwidth reduces the energy consumed for data transmission since more data is transmitted in a short period and so lowers the energy consumption per second. Whereas under low bandwidth conditions we avoid data transfers

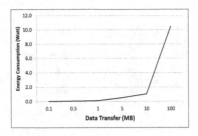

Fig. 6. Data Transfer: Mobile Device to Cloud

because the energy cost of transmission is high and impacts negatively on the
limited mobile device battery power.

5   Conclusion

We presented a secure and efficient approach, supported by cloud storage, to
managing m-health data. In our approach, the data is fragmented with an importance rating metric and security is cryptographically enforced. Mobile storage
limitations, are handled by using an "aging" metric to decide which portions of
the data to offload to the cloud. Our results indicate that fragmentation improves
query response time by almost 25% while caching offers a 35% performance improvement in query response time on various query workloads.

   As future work we plan to extend our approach to multimedia data from multiple sources (text, xrays, etc). We will consider different media types in relation
to optimizing storage management and battery usage. Further experimentation
to evaluate usability is also needed.

References

1. Kujawski: Latest mobile phone statistics from africa and what this means,
   http://www.mikekujawski.ca/2009/03/16/
   latest-mobile-statistics-fromafrica-and-what-this-means/
2. MHealth: Mobilizing innovation for global health,
   http://www.mhealthalliance.org/
3. Damiani, E., De Captani Di Vimercati, S., Jajodia, S., Paraboschi, S., Samarati,
   P.: Balancing confidentiality and efficiency in untrusted relational dbmss. In: Proceedings of the 10th ACM Conference on Computer and Communications Security,
   CCS 2003, pp. 93–102. ACM, New York (2003)
4. Damiani, E., De Captani Di Vimercati, S., Finetti, M., Paraboschi, S., Samarati, P., Jajodia, S.: Implementation of a storage mechanism for untrusted dbmss.
   In: Proceedings of the Second International IEEE Security in Storage Workshop,
   Washington DC, USA (May 2003)
5. Hacigümüş, H., Iyer, B., Mehrotra, S.: Providing database as a service. In: Proceedings of the 18th International Conference on Data Engineering, San Jose, California, USA (February 2002)

6. Hacigümüş, H., Iyer, B., Chen, L., Mehrotra, S.: Executing sql over encrypted data in the database-service-provider model. In: Proceedings of the 2002 ACM SIGMOD International Conference on Management of Data, SIGMOD 2002, pp. 216–227. ACM, New York (2002)

7. De Capitani Di Vimercati, S., Foresti, S., Jajodia, S., Paraboschi, S., Samarati, P.: Over-encryption: Management of access control evolution on outsourced data. In: Proceedings of the 33rd International Conference on Very Large Databases. VLDB 2007, pp. 123–134. VLDB Endowment (2007)

8. Atallah, M.J., Blanton, M., Fazio, N., Frikken, K.B.: Dynamic and efficient key management for access hierarchies. ACM Trans. Inf. Syst. Secur. 12(3), 18:1–18:43 (2009)

9. De Capitani Di Vimercati, S., Foresti, S., Jajodia, S., Paraboschi, S., Samarati, P.: Encryption policies for regulating access to outsourced data. ACM Trans. Database Syst. 35(2), 12:1–12:46 (2010)

10. Kayem, A., Martin, P., Akl, S.G.: Effective cryptographic key management for outsourced dynamic data sharing environments. In: Proceedings of the 10th Annual Information Security Conference (ISSA 2011), Johannesburg, South Africa, August 15 -17, pp. 1–8. IEEE (2011)

11. Samanthula, B., Howser, G., Elmehdwi, Y., Madria, S.: An efficient and secure data sharing framework using homomorphic encryption in the cloud. In: Proceedings of the 1st International Workshop on Cloud Intelligence, vol. 8, pp. 1–8 (2012)

12. Ciriani, V., De Capitani Di Vimercati, S., Foresti, S., Jajodia, S., Paraboschi, S., Samarati, P.: Fragmentation design for efficient query execution over sensitive distributed databases. In: Proceedings of the 29th IEEE International Conference on Distributed Computing Systems (ICDCS), pp. 32–39 (2009)

13. Ciriani, V., De Capitani Di Vimercati, S., Foresti, S., Jajodia, S., Paraboschi, S., Samarati, P.: Combining fragmentation and encryption to protect privacy in data storage. ACM Trans. Inf. Syst. Secur. 13(3), 22:1–22:33 (2010)

14. Foresti, S.: Preserving Privacy in Data Outsourcing. Advances in Information Security, vol. 51. Springer, New York (2011)

15. Samarati, P., De Capitani Di Vimercati, S.: Data protection in outsourcing scenarios: Issues and directions. In: Proceedings of the 5th ACM Symposium on Information, Computer and Communications Security, ASIACCS 2010, pp. 1–14. ACM, New York (2010)

16. De Capitani Di Vimercati, S., Foresti, S., Jajodia, S., Paraboschi, S., Samarati, P.: Encryption policies for regulating access to outsourced data. ACM Trans. Database Syst. 35(2), 12:1–12:46 (2010)

17. Pashalidis, A., Meyer, B.: Linking anonymous transactions: the consistent view attack. In: Danezis, G., Golle, P. (eds.) PET 2006. LNCS, vol. 4258, pp. 384–392. Springer, Heidelberg (2006)

18. Cormen, T.H., Stein, C., Rivest, R.L., Leiserson, C.E.: Introduction to Algorithms, 2nd edn. McGraw-Hill Higher Education (2001)

19. SQLite, http://www.sqlite.org/

Link-Based Viewing of Multiple Web API Repositories

Devis Bianchini, Valeria De Antonellis, and Michele Melchiori

Dept. of Information Engineering University of Brescia
Via Branze, 38 - 25123 Brescia (Italy)
{devis.bianchini|valeria.deantonellis|michele.melchiori}@unibs.it

Abstract. Web API sharing, fueled by large repositories available online, is becoming of paramount relevance for agile Web application development. Approaches on Web API sharing usually rely on single Web API repositories, which provide complementary Web API descriptions, on which different search facilities can be implemented. In this paper, we propose a framework for defining a linked view over multiple repositories and for searching their content. In particular, we apply Linked Data principles to publish repository contents and identify semantic links across them in order to exploit complementary Web API descriptions. We discuss how Web API sharing across multiple repositories, based on such a link-based view, may benefit from selection criteria that combine several aspects in Web API characterization. A preliminary evaluation based on two popular public Web API repositories is presented as well.

1 Introduction

Selection and aggregation of reusable Web APIs is gaining momentum as a new development style for building new, value-added applications. Web APIs are specific kinds of digital resources that can be shared over the Web. They may be general purpose, provided by third parties, and they can be used as a productivity tool for on-the-spot problems, that is, situational applications that require few resources in terms of investment costs and development time. Existing approaches for Web API selection rely on APIs registered within the ProgrammableWeb[1] repository, because of its popularity. However, although ProgrammableWeb constitutes a well-known meeting point for the community of mashup developers, it is mainly focused on a feature-based description of Web APIs, classifying them through categories and tags and storing features like the adopted data formats, protocols and the list of mashups that have been developed using the Web APIs. There exist different repositories that emphasize distinct aspects to be considered for Web API sharing. For instance, Mashape, a cloud API hub[2] leveraging a twitter-like organization, associates each Web API with the list of developers who adopted or declared their interests for it and technical details

---

[1] http://www.programmableweb.com/
[2] https://www.mashape.com/

H. Decker et al. (Eds.): DEXA 2014, Part I, LNCS 8644, pp. 362–376, 2014.

for Web API invocation are provided as well. Similarly, other repositories (e.g., `theRightAPI`) provide a community-oriented perspective on Web APIs. In this scenario, distinct repositories act as information silos where: (i) complementary or partially overlapping Web API characterizations are stored; (ii) the same Web APIs or mashups are registered multiple times within different repositories; (iii) similarity between Web APIs and mashups across different repositories cannot be exploited to enrich search results. Various approaches have been proposed to integrate Web APIs with the Linked Data cloud, but they require models that are difficult and time-consuming to build [1] and generally are based on a single repository. On the other hand, publishing Web API repository contents as Linked Data may be useful to overtake the issues highlighted above. Moreover, publishing Web API descriptions as (open) Linked Data offers the opportunity for semantically enriching them by making explicit their social, technical and terminological aspects with links from the Web API description to relevant parts of the Web of Data. A further completion could be, when available, linking them to descriptions specifically conceived for consuming and producing linked data [2–4].

In this paper, we present a novel framework built according to Linked Data principles aimed at enhancing effective cross-repository Web API browsing and search for mashup development. The framework is based on an unified model for Web APIs providing a linked view on the contents of each repository. Such a model makes the content of repositories machine processable and accessible through non-proprietary tools (e.g., SPARQL endpoints). Therefore, it enables Web API search in a transparent way with respect to the localization of each repository. This unified model is based on three modeling perspectives that we introduced in [5] for Web APIs in a single repository. Within the framework, a *Linker* is in charge of identifying and storing semantic links across repositories according to identity and similarity criteria. Semantic links are published as Linked Data and can be exploited to perform Web API search over multiple repositories. thus taking benefit from existing approaches on query processing over Linked Data [6] also in the Web API selection scenario.

The paper is organized as follows: in Section 2 we provide the motivations of our approach, through the presentation of the running example and the discussion about related work; in Sections 3 and 4 we detail the design of the unified vocabulary and linking criteria on which the Linker is based; in Section 5 we describe how to exploit semantic links for Web API search through the Web API Search Engine; we discuss implementation issues and preliminary evaluation of the approach in Section 6; finally, Section 7 closes the paper.

2 Motivations

2.1 Running Example

As a running example, we refer to the `ProgrammableWeb` (PW) and `Mashape` (MP) repositories. We chose these two repositories since they are the most popular ones and provide complementary and partially overlapping features in Web

API descriptions so better illustrating the advantages and the problems arising in using both of them. In particular, let us consider a web designer who is charge of including a face recognition functionality to access the private area of his/her web site. Since developing this kind of applications from scratch would require very specific competencies, let us suppose that the designer looks for a Web API in the PW repository. Using the `face` and `recognition` keywords, the designer finds 15 matches. Among them, the `LambdaLabs Face` API is not assigned to any mashup. Since the goal of the designer's search on PW is to make easier the integration of a new Web API in an existing web site, the `LambdaLabs Face` API may not be considered. Nevertheless, on the MP repository, the same Web API has 1385 consumers and 1304 followers (that is, developers who declared their interest in the Web API). It is the most popular Web API out of the 106 APIs retrieved on MP using the same keywords. This additional information could influence the designer's choice. It is thus very important for the web designer to be able to identify and exploit correspondences between Web APIs in different repositories. However, this task may be not trivial. In fact, the same Web API may be classified into different categories or tagged with different tags across distinct repositories (for instance, the `LambdaLabs Face` API is classified in the `media` category on MP, while it is classified in the `Photos` category on PW). Moreover, the categorizations adopted in the two repositories are very different (68 categories on PW, 18 categories on MP, none of them organized in a hierarchy, 10 overlapping categories, but no explicit semantics). Finally, in an ideal situation, there should be specific properties to be used for identifying the same Web API description across different repositories, such as the URL. Unfortunately, data in Web API repositories is not often as complete as it could be: for example, the `LambdaLabs Face` API is registered with slightly different names on the two repositories and it is associated with the `http://api.lambdal.com/` URL on PW and to the `http://www.lambdal.com/developers/` URL on MP. Therefore, to merge together the search results coming from the two Web API repositories, the web designer must carefully inspect and compare each API description. We propose an approach that defines semantic links based on identity and similarity criteria between elements across different repositories in order to overcome these issues.

2.2   Related Work

Related efforts in the literature about Linked Web services or Linked Web APIs for discovery purposes focused on the semantic annotation of service features [7] through domain ontologies to publish service descriptions on the Web of Data [2, 8, 9], on the enrichment of the characterization of Web APIs to enable them to consume and produce Linked Data [4, 10], and on publishing as linked resources each single API invocation [11, 3].

The iServe platform described in [9] proposes adopting the Minimal Service Model (MSM), that is, a simple RDF(S) integration ontology that captures the maximum common denominator between existing conceptual models for services. In this model, Linked Data techniques are used for semantic annotation of the

I/O parameters of the interfaces of Linked Web services or Linked Web APIs with ontologies taken from the Web of Data. Comparison between I/Os based on semantic annotations is performed to infer semantic relationships between Web APIs. In [1], an interactive Web-based interface enables domain experts to rapidly create semantic models of services and Web APIs, using an expressive vocabulary that can be seen as an evolution of the MSM and includes also lowering and lifting rules, that is mappings, between the semantic model and the concrete API or service, that in this way is also able to consume and produce RDF data. All these approaches are characterized by a complex semantic model, where the semantic annotation of I/Os usually inhibits their wide adoption. This is especially critical in Web API repositories that have highly variable content. Moreover, the description of Web APIs at this level of details is not always available and, although semi-automatic solutions have been proposed in [1], the contents of repositories vary too fastly to enable this kind of approaches.

With respect to these approaches, our aim is to publish as Linked Data sources the Web API repositories as a whole, enabling discovery of Web APIs as resources. Therefore, we rely on information stored within multiple and public available repositories without requiring a semantic annotation activity of Web APIs. Similarity and identity links are automatically set to enable cross-repository resource sharing. Moreover, our unified view is not focused on Web APIs only, but also on mashups where APIs have been included according to the multiple perspectives described in [5], thus enabling more in-depth searching facilities. Approaches, which discuss how to process or generate Linked Data, such as Linked Open Services [11, 3] and Linked Data Services [4, 10], have not been designed for Web API search on multiple repositories. Linked Open Services (LOS) and Linked Data Services (LDS) are still focused on I/O modeling.

Among the most popular approaches and tools for link identification, we mention the Silk framework [12], a toolkit to link entities across different data sources. Nevertheless, the user-defined metrics presented in [12] can be built combining generic similarity measures only, and cannot be specifically applied to our domain of interest.

3   Modeling Web Mashup Resources

In order to obtain a link-based view of repository content we look for same/similar resources and define links among them. For comparing resources we assume a multi-perspective model introduced in [5] where features characterizing resources in existing repositories are considered. According to the multi-perspective model a resource can be described through:

– a functional characterization of the Web API obtained through a top-down classification, according to fixed categories, a bottom-up tagging through semantic tags associated by designers, and a set of technical features (e.g., protocols or data formats) used to further characterize the API (*component perspective*);

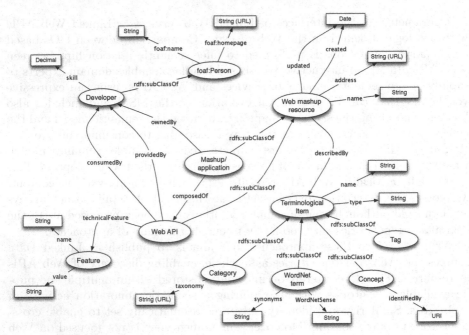

Fig. 1. RDF representation of the Web mashup resource unified model

- any existing mashups that include the Web API, described through the other APIs that compose them and tags associated by the designers with mashups as well (*application perspective*);
- the ratings assigned to the Web API by designers who used it in their own mashups (*experience perspective*).

The basic concepts and relationships of the multi-perspective model are represented in Figure 1. In the model, both a Web API and a mashup (composed of APIs) are defined as subclasses of *Web mashup resource*, denoted by its URL, a human-readable name, the date in which resource has been created, and the time of the last upgrade. Each Web mashup resource is associated with a set of *terminological items*, which correspond to: (a) categories extracted from top-down classifications imposed within a given repository where the resource is registered; (b) a term with an explicit semantics, either a term extracted from a terminological thesaurus like WordNet or a concept extracted from an ontology in the Semantic Web context; and (c) a simple keyword or tag without an explicit representation of semantics. Each terminological item is described by a name, a type (namely, C for categories, WD for WordNet terms, O for ontological concepts, K for simple keywords or tags) and a set of properties:

- if the item is a category, the item name corresponds to the category name and it has a property that identifies the taxonomy or the classification to which the category belongs;

- if the item is extracted from WordNet, it has two properties, namely the list of synonyms of the term and the meaning associated with the WordNet sense to which the term belongs;
- if the item is a concept extracted from an ontology, its name corresponds to the concept name; it is further specified through its URI, which identifies the definition of the concept within the ontology where it is provided;
- if the item is a keyword or a tag without an explicit semantics, the item has no properties and its name corresponds to the keyword or the tag name.

We distinguish keywords and tags as follows: tags are assigned by designers, aimed at classifying resources in a folksonomy-like style, but keywords are recurrent terms extracted from resource textual descriptions using common IR techniques. Each Web mashup resource is also associated with ratings assigned by designers, who may be characterized by their development skill, either self-declared, as shown in [5], or estimated, using techniques such as those proposed in [13]. Designers may be either Web API providers, Web mashup owners, or Web API consumers who rate resources according to their personal opinion. Web APIs can be further characterized through technical features, such as protocols or data formats.[3]

To enrich the model, we rely on external ontologies, when available, following the methodological guidelines suggested in [14]. For instance, we modeled designers using the foaf:Person class from the FOAF (Friend of a Friend) ontology,[4] where contact information of each foaf:Person is composed of the name (foaf:name) and the homepage address (foaf:homepage).

4 Defining Links Among Web Mashup Resources

Formally, we represent a link $\mathcal{L}$ between Web mashup resources, either within the same repository or across different repositories, as follows:

$$\mathcal{L} = \langle \text{type}, s\_URI, t\_URI, \text{conf}, [\text{when}] \rangle \qquad (1)$$

where s_URI and t_URI are the URIs of the resources that are source and target of the link, respectively, $\text{conf} \in [0,1]$ is the confidence to set the link (obtained through identity and similarity metrics evaluation depending on the link type, as we describe below) and the optional element when denotes when the link has been established and therefore published as Linked Data. The when clause is exploited to filter out or check links older than a predefined number of days. During Web API search, the designer can choose a threshold $\text{conf}' \in [0,1]$ such that only links with $\text{conf} \geq \text{conf}'$ are considered. We have identified two link types, sameAs and simAs, which we describe as follows:

- sameAs link between Web mashup resources, to denote that two resources registered in the repositories refer to the same software artifact; this kind of link

---

[3] See, for example, the technical description of the LambdaLabs Face API at:
   http://www.programmableweb.com/api/lambdalabs-face
[4] http://xmlns.com/foaf/spec/

makes sense only across different repositories, since we assume that the publication of the same Web mashup resource in the same repository is not allowed;
- simAs link between two Web mashup resources, established on the basis of a comparison between their terminological items; a further specification of this kind of link is provided to distinguish among similar APIs and similar mashups; in the former case, also technical features (such as protocols and data formats) are taken into account to establish the link; in the latter one, we must also consider that the more two mashups are similar, the more similar the Web APIs that compose them are; the simAs link can be set either within the same repository or across different repositories.

Each kind of link is identified through proper metrics evaluation, detailed in the following. A threshold $\tau \in [0,1]$ is set in order to publish each link as shown in Section 6. In our approach, we fixed $\tau = 0.5$ to avoid storing loosely related resources. If the result of metrics evaluation is equal or greater than $\tau = 0.5$, the conf parameter introduced in Equation (1) is set to this value and the link is stored in the *Link Repository* as we discuss in Section 6.

The sameAs link between Web mashup resources. We say that two Web mashup resources across different vocabularies *reference* the same software component if they present the same URL. In this case, the link is set and conf $= 1.0$. Unfortunately, URL mismatches like the ones described in the motivating example make the use of additional similarity computations necessary. In particular, to manage these mismatches, the criteria used to set the sameAs link between Web mashup resources is based upon a *host name similarity* metric (*HostSim*). This metric is computed between the host names of two URLs (e.g., http://api.lambdal.com and http://www.lambdal.com/developers/, except for the scheme (e.g., http://) and the fragment (e.g., /developers/) [15]. Each host name is composed of substrings separated by dots, e.g., www.lambdal.com. Host name similarity is computed using the Dice formula in Equation (2):

$$HostSim(URI_1, URI_2) = 2 \cdot \frac{|URI_1 \cap URI_2|}{|URI_1| + |URI_2|} \in [0,1] \qquad (2)$$

where $|URI_1 \cap URI_2|$ denotes the maximum number of common substrings within URI_1 and URI_2 (in the exact order) starting from the top-level domain (e.g., com), $|URI_i|$ denotes the number of substrings within URI_i. The Dice formula is used to normalize the metric in the [0,1] range. For instance, $HostSim$(www.lambdal.com, api.lambdal.com) $= 2 \cdot \frac{2}{3+3} = 0.67$, because the two host names share the two lambdal.com substrings (underlined) on a total of 3+3 substrings. Since the same domain may be assigned to several Web mashup resources, the *HostSim* computation is also applied to the URLs of the designers who provided or own the resource (for Web APIs or mashups, respectively) and is combined with the string similarity between resource names, using one of the classical string distance metrics. It is out of the scope of our framework to propose a particular metric for string comparison. The framework includes the most common metrics [16]. In the future we may explore tuning strategies further. The overall metric applied on two Web mashup resources res_1 and res_2, provided or owned by designers d_1 and d_2, respectively, is defined as:

$$
\begin{aligned}
& 0.4 \cdot HostSim(res_1.\textbf{address}, res_2.\textbf{address}) + \\
& 0.4 \cdot HostSim(d_1.\textbf{homepage}, d_2.\textbf{homepage}) + \\
& 0.2 \cdot StringSim(res_1.\textbf{name}, res_2.\textbf{name}) \in [0, 1]
\end{aligned}
\tag{3}
$$

where, for instance, $res_1.\textbf{address}$ is the value of property **address** for the resource res_1; in our preliminary evaluation, for $StringSim()$ computation, we chose the Levenshtein distance metric. On the basis of the results of this experimentation, in the equation (3) $HostSim()$ is weighted more than string similarity between resource names. In fact, the latter is relevant only if we are within the same or very close domains. The value of the overall metric in this case will be assigned to the `conf` parameter.

The simAs link between Web mashup resources. The similarity between two Web mashup resources is deeply rooted in the comparison of their terminological items, that is:

$$
TermSim(res_1, res_2) = \frac{2 \cdot \sum_{t_1 \in \mathcal{T}_1, t_2 \in \mathcal{T}_2} itemSim(t_1, t_2)}{|\mathcal{T}_1| + |\mathcal{T}_2|} \in [0, 1]
\tag{4}
$$

where we denote with $\mathcal{T}_i$ the set of terminological items used to characterize res_i, t_1 and t_2 are terminological items, $|\mathcal{T}_i|$ denotes the number of items in the set $\mathcal{T}_i$ and $itemSim(\cdot)$ values are aggregated through the Dice formula. The point here is how to compute $itemSim(t_1, t_2) \in [0, 1]$ given the different types of involved terminological items.

The algorithm for the $itemSim(\cdot)$ calculation is shown in Algorithm (1). When the types of t_1 and t_2 coincide, proper metrics from the literature are used for the comparison (see rows 1-8). In all the other cases, a comparison between the names of terminological items using the $StringSim(\cdot)$ metric is performed (row 18), except for the case of WordNet terms, that are expanded considering all the synonyms (rows 10-17)in order to look for a better matching term in the synset. In this version of the framework, we considered only the concept name in the case of ontological concepts. Future work will be devoted for refining this part, by expanding the set of terms extracted from the ontological concept with the name of other concepts connected through semantic relationships in the ontology.

For establishing `simAs` links between Web APIs, the $TermSim(\cdot)$ metric is equally balanced with the technical feature similarity $TechSim(\cdot) \in [0, 1]$, which evaluates how much the two Web APIs have common technical features, that is, $ApiSim(res_1, res_2) = 0.5 \cdot TermSim() + 0.5 \cdot TechSim()$ ($\in [0, 1]$). The value of the $ApiSim()$ metric in this case will be assigned to the `conf` parameter. Feature values are compared only within the context of the same feature. For example, if res_1 presents {XML, JSON, JSONP} as data formats and {REST} as protocol,

---

Algorithm 1. The $itemSim(\cdot)$ calculation algorithm

Input : Two terminological items t_1 and t_2.
Output: The calculated $itemSim(t_1, t_2)$ value.

1 **if** (t_1.type == C) **and** (t_2.type == C) **then**
2     $itemSim(t_1, t_2) = Sim_{cat}(\mathbf{t_1}, \mathbf{t_2})$ (using the Sim_{cat} defined in [5]);

3 **else if** (t_1.type == WD) **and** (t_2.type == WD) **then**
4     $itemSim(t_1, t_2) = Sim_{tag}(t_1, t_2)$ (using the Sim_{tag} defined in [5]);

5 **else if** (t_1.type == O) **and** (t_2.type == O) **then**
6     $itemSim(t_1, t_2) =$ H-MATCH(t_1, t_2) (using the H-MATCH function
     given in [17]);

7 **else if** (t_1.type == K) **and** (t_2.type == K) **then**
8     $itemSim(t_1, t_2) = \alpha \cdot StringSim(t_1, t_2)$ (using the Levenshtein measure);

9 **else**
10   **if** t_1.type == WD **then**
11   t_1.bagOfWords = expandWithSynonyms(t_1);

12   **else**
13   t_1.bagOfWords = t_1.name;

14   **if** t_2.type == WD **then**
15   t_2.bagOfWords = expandWithSynonyms(t_2);

16   **else**
17   t_2.bagOfWords = t_2.name;

18    $itemSim(t_1, t_2) = max_{i,j}\{StringSim(t_1^i, t_2^j)\}$, where $t_1^i \in t_1$.bagOfWords
     and $t_2^j \in t_2$.bagOfWords;

19 **return** $itemSim(t_1, t_2)$;

---

while res_2 presents {XML, JSON} as data formats and {REST, Javascript, XML} as protocols, the $TechSim()$ value is computed as

$$\frac{2 \cdot [|\{\text{XML, JSON, JSONP}\} \cap \{\text{XML, JSON}\}| + |\{\text{REST}\} \cap \{\text{REST, Javascript, XML}\}|]}{|\{\text{XML, JSON, JSONP}\}| + |\{\text{XML, JSON}\}| + |\{\text{REST}\}| + |\{\text{REST, Javascript, XML}\}|} = 0.67$$
(5)

In this example, XML is used both as data format and as XML-RPC protocol and it is considered separately in the two cases.

For establishing simAs links between mashups, the $TermSim(\cdot)$ metric is equally weighted with the mashup composition similarity $MashupCompSim(\cdot) \in [0, 1]$, which measures the degree of overlapping between two mashups as the number of common or similar APIs between them, that is

$$MashupCompSim(res_1, res_2) = \frac{2 \cdot \sum_{i,j} ApiSim(res_1^i, res_2^j)}{|res_1| + |res_2|}$$
(6)

where res_1^i and res_2^j are two Web APIs, used in res_1 and res_2 mashups, respectively, $|res_1|$ (resp., $|res_2|$) denotes the number of Web APIs in res_1 (resp.,

res_2) mashup and $ApiSim() = 1.0$ by construction when the two Web APIs are the same. Therefore, $MashupSim(res_1, res_2) = 0.5 \cdot TermSim() + 0.5 \cdot MashupCompSim()$ ($\in [0, 1]$). The value of the $MashupSim()$ metric in this case will be assigned to the `conf` parameter.

5 Exploiting Links among Web mashup Resources

In this section, we present an applicative scenario for link-based view on repositories by defining and implementing a Web API search process. The process exploits the representation of repositories given according to the model of Section 3. The search process is formalized in Algorithm (2). A developer submits a query by specifying a set of keywords, a set of desired technical features, and an optional set of Web APIs that have been selected by the developer to be aggregated with the Web API to search for. Once results are obtained by merging the contents coming from considered repositories matching the query, they are filtered with respect to a set of required technical features, their appropriateness with respect to a given mashup (see below) or their popularity based on the number of developers and followers who are interested in a given Web API of the result. A query is formally defined as $\mathcal{Q} = \langle \mathcal{K}_\mathcal{Q}, \mathcal{F}_\mathcal{Q}, \mathcal{M}_\mathcal{Q} \rangle$, where $\mathcal{K}_\mathcal{Q}$ is the set of keywords, $\mathcal{F}_\mathcal{Q}$ is a set (possibly empty) of pairs $\langle$`tech_feature=value`$\rangle$, and $\mathcal{M}_\mathcal{Q}$ (possibly empty) is a mashup (that is, a set of Web APIs).

In the prototype implementation, in order to build an answer $\mathcal{R}(\mathcal{Q})$, a set of SPARQL queries are issued on the Virtuoso Universal Server[5]. Additionally, since we assume that the web designer is not confident with SPARQL, the Web interface lets the designer insert the components of the query $\mathcal{Q}$ and generates corresponding SPARQL queries.

When a query $\mathcal{Q}$ is evaluated, the content of each repository of Web APIs is inspected searching for those Web APIs that include in their terminological equipments at least one of the keywords specified in $\mathcal{Q}$ (rows 1-3). The inspection of terminological equipments is performed through the InTERM predicate that checks the presence of a keyword in the name attribute of a terminological item (see the model in Fig. 1) and, in case of a WordNet term, in the set of synonyms. In a second phase, for each retrieved Web API the `simAs` links are used to include in the result also those Web APIs that are similar to the retrieved ones (rows 4-5). In a third phase, descriptions of pairs of Web APIs related by `sameAs` links are merged by building a unified representation. Moreover, the original descriptions are removed from the result (rows 6-12).

An API r^i belonging to the query result $\mathcal{R}(\mathcal{Q})$ is modeled as 4-tuple:

$$\langle api\_URIs, r^i_\mathcal{M}, r^i_\mathcal{D}, r^i_\mathcal{F} \rangle \tag{7}$$

where api_URIs are the URIs assigned to the Web API in the repositories, $r^i_\mathcal{M}$ is the set of mashups where the Web API has been used, identified by their URIs, $r^i_\mathcal{D}$ is the set of developers, consumers and followers who used the Web API (e.g., developers of mashups which contain the Web API are used for PW), $r^i_\mathcal{F}$ is the set of technical features of the Web API.

---

[5] http://virtuoso.openlinksw.com/

Algorithm 2. Linked Web API search

Input : the query $\mathcal{Q} = \langle \mathcal{K}_\mathcal{Q}, \mathcal{F}_\mathcal{Q}, \mathcal{M}_\mathcal{Q} \rangle$.
Output: $\mathcal{R}(\mathcal{Q})$, where $r^i \in \mathcal{R}(\mathcal{Q})$ is a 5-tuple $\langle \text{api\_URIs}, r^i_\mathcal{M}, r^i_\mathcal{D}, r^i_\mathcal{F} \rangle$.

1 **foreach** *Web API Repository* $\mathcal{S}$ **do**
2   **foreach** $k \in \mathcal{K}_\mathcal{Q}$ **do**
3        $\mathcal{R}(\mathcal{Q}) \leftarrow$ Web APIs $\mathcal{W}$ from $\mathcal{S}$ such that $\text{InTERM}(k, \mathcal{W})$;

4 **foreach** *Web API* $\mathcal{W} \in \mathcal{R}(\mathcal{Q})$ **do**
5   add $\mathcal{W}'$ to $\mathcal{R}(\mathcal{Q})$ such that W `simAs` W' with `conf` $\geq$ `conf`';

6 **foreach** *Web API* $\mathcal{W} \in \mathcal{R}(\mathcal{Q})$ **do**
7   **foreach** *Web API* $\mathcal{W}' \in \mathcal{R}(\mathcal{Q})$ **do**
8   **if** *(W `sameAs` W' with* `conf` $\geq$ `conf`'*)* **then**
9   build $\mathcal{W}''$ by merging $\mathcal{W}$ and $\mathcal{W}'$;
10   add W" to R(Q);
11   remove $\mathcal{W}$ from $\mathcal{R}(\mathcal{Q})$;
12   remove $\mathcal{W}'$ from $\mathcal{R}(\mathcal{Q})$;

13 **if** $\mathcal{F}_\mathcal{Q} \neq \emptyset$ **then**
14   filter $\mathcal{R}(\mathcal{Q})$ with respect to the set $\mathcal{F}_\mathcal{Q}$;

15 **if** $\mathcal{M}_\mathcal{Q} \neq \emptyset$ **then**
16   rank Web APIs in $\mathcal{R}(\mathcal{Q})$ according to their appropriateness wrt $\mathcal{M}_\mathcal{Q}$;
17   rank equally appropriate Web APIs in $\mathcal{R}(\mathcal{Q})$ according to their popularity;

18 **else**
19   rank Web APIs in $\mathcal{R}(\mathcal{Q})$ according to their popularity;

20 **return** $\mathcal{R}(\mathcal{Q})$;

The last search steps of Algorithm (2) concern filtering and ranking of search results. Retrieved Web APIs are filtered out according to the set of required features $\mathcal{F}_\mathcal{Q}$ if specified in $\mathcal{Q}$ (rows 13-14). Finally, search results are ranked according to their appropriateness with respect to the target mashup $\mathcal{M}_\mathcal{Q}$ if specified in $\mathcal{Q}$ (rows 12-14) and according to their popularity (row 16). Specifically, we define the similarity between two mashups $\mathcal{M}_1$ and $\mathcal{M}_2$ (as sets of Web APIs) using a formula according to the same rationale of Equation (4):

$$MashupSim(\mathcal{M}_1, \mathcal{M}_2) = \frac{2 \cdot |\mathcal{M}_1 \cap \mathcal{M}_2|}{|\mathcal{M}_1| + |\mathcal{M}_2|} \tag{8}$$

where $|\mathcal{M}_1 \cap \mathcal{M}_2|$ denotes the number of common Web APIs in the two mashups and $|\mathcal{M}_i|$ the number of Web APIs in the mashup $\mathcal{M}_i$. Given the set $r^i_\mathcal{M}$ of mashups of a search result r^i, if $r^i_\mathcal{M} \neq \emptyset$, the appropriateness of r^i with respect to the mashup $\mathcal{M}_\mathcal{Q}$ is given by $max_j\{MashupSim(\mathcal{M}_\mathcal{Q}, \mathcal{M}_j)\}$, where $\mathcal{M}_j \in r^i_\mathcal{M}$. If $r^i_\mathcal{M} = \emptyset$, then ranking based on appropriateness is not performed. Popularity of a result r^i is measured as the number of developers in $r^i_\mathcal{D}$. For ranking purposes, the designer may choose, through the search interface, to give priority to the appropriateness or to the popularity of results.

Example. Let us consider again the `LambdaLabs Face` face recognition API. In particular, suppose that after retrieving Web APIs from repositories (rows 1-3), this API is present in both PW and MP. In particular, the descriptions of the API in these sets is given in Table 1. Note that total number of followers and consumers of this API in `Mashape` is reported and that the number of developers that used the API in `ProgrammableWeb` is 0.

Table 1. Descriptions of `LambdaLabs Face` face recognition API

	PW	MP
URI	{http://api.lambdal.com}	{http://www.lambdal.com/developers}
$r_{\mathcal{M}}^i$	{}	{}
$\lvert r_{\mathcal{D}}^i \rvert$	0	2689
$r_{\mathcal{F}}^i$	{REST, JSON}	{}

A `sameAs` link is already set in the Link Repository based on the linking criteria presented in the previous section. Hence, these Web API descriptions are merged and added to $\mathcal{R}(\mathcal{Q})$ (rows 9-10). Specifically, the merged description presented in the query result will be, according to (7):

$$\langle \{\texttt{http} : //\texttt{api}..., \texttt{http} : //\texttt{www}...\}, \{\}, r_{\mathcal{D}}^i, \{REST, JSON\}\rangle$$

Because $r_{\mathcal{M}}^i$ is empty, the ranking based on mashup appropriateness is not performed, as explained above. On the contrary, if $\mathcal{M}_{\mathcal{Q}}$ is empty, the ranking based on the popularity, given by $\lvert r_{\mathcal{D}}^i \rvert$, of this API will be rather high.

Preliminary evaluation. We performed a preliminary evaluation of the search process based on classical IR measures of precision and recall. The aim is to check the capability of our approach to provide improved search results with respect to the separated use of the available repositories. A more extensive experimentation on the system implementing the process will be performed as future work. As a proof of concept, we started from two popular repositories, namely `ProgrammableWeb` (PW) and `Mashape` (MP), considered for the running example in Section 2.1. Observing the two repositories and their differences, we note that their search results strongly depend on the tags and categories used for search. Experiments have been run on an Intel laptop, with 2.53 GHz Core 2 CPU, 2GB RAM and Linux OS. We manually selected all the relevant Web APIs on face recognition stored within the PW and MP repositories and we collected the sets of tags and categories for the two Web APIs. We then issued several queries using different subsets of tags and categories: on the PW repository only, on the MP repository only, and on both the repositories through our approach. In Table 2 we report an excerpt of results of this preliminary experimentation, using the following subsets of tags: ⟨face,recognition⟩ and ⟨facial,detection⟩. We note that even if we consider the union of the results from the PW and MP repositories, queried separately, our approach presents better precision and recall results. The other aspect that can take advantage of

Table 2. Preliminary evaluation results

	⟨face,recognition⟩		⟨facial,detection⟩	
	Precision	Recall	Precision	Recall
ProgrammableWeb	0.72	0.64	(no results)	(no results)
Mashape	0.68	0.69	0.47	0.43
Union of PW and MP results (invoked separately)	0.73	0.70	0.40	0.41
Results from the joint use of PW and MP through our	0.93	0.91	0.77	0.70

our approach is the identification of corresponding Web APIs across different repositories: for instance, 75% of the face detection Web APIs that have been registered in both the repositories present different URLs. The action to reconcile Web APIs, if manually performed, would be time-consuming and error-prone, due to the dynamic nature of the two sources.

6   The Framework Architecture

The framework architecture is shown in Figure 2. The *Web API meta-repository* aims at enabling uniform access to the contents of individual repositories. It is based on the RDF Quad Store of the Virtuoso Universal Server, on which our approach is implemented and it includes the unified vocabulary and the *Link Repository*, to store similarity and identity links across distinct Web API repositories.

Once an RDF vocabulary for our unified model has been defined, the Virtuoso Sponger tool[6] has been properly configured in order to retrieve resources from the Web API repositories, according to their conceptualization in the vocabulary. In this way, resources are directly retrieved from the repositories by relying on the Virtuoso update procedures. These procedures are executed off-line and are combined with link maintenance strategies mentioned in Section 3. To perform its tasks, Sponger relies on the *Virtuoso Content Crawler (VCC)*, which executes a periodic update of cached contents of PW and MP repositories, and a set of *cartridges* which are used to extract RDF tuples from the retrieved sources. A cartridge is composed of an interface for invocation (Cartridge Hook), an extractor to obtain (non-RDF) data from the source and a mapper that looks for correspondences between data extracted from the source and the RDF vocabulary that has been built for the source. The mapper is based on an XSLT document. Cartridges are stored within the *Cartridge Registry*. Cartridges have been designed to invoke specific methods made available by public repositories to query their contents[7]. It is clear that the effort of configuring a cartridge has to be done only one time for each repository that we want to include.

Currently, ProgrammableWeb and Mashape cartridges are available. Other proprietary Web API repositories may be provided by enterprises, which can inte-

---

[6] http://www.openlinksw.com/dataspace/doc/dav/wiki/Main/VirtSponger

[7] See, for instance, http://api.programmableweb.com for the ProgrammableWeb repository or http://www.mashape.com/mashaper/mashape#!documentation for the Mashape repository.

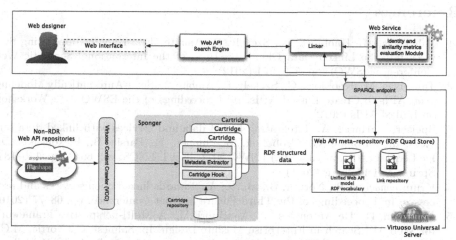

Fig. 2. The framework architecture

grate their own components with Web APIs made available within public repositories, thus adopting an "app store" development approach to reduce investment
costs and development time for those applications that do not shape strategic
decisions. In this case, new cartridges must be added to the system according
to a modularized architecture. The contents stored within the Web API meta-
repository are only accessible through the *SPARQL Endpoint*. The *Linker* is in
charge of populating the Link Repository by performing evaluation of the metrics designed to identify semantic links across the repositories. To this aim, the
Linker relies on the *Identity and Similarity Evaluator*, which implements the
metrics as a Web Service.

Finally, the *Web API Search Engine* implements the search process by issuing
SPARQL queries on the Virtuoso Universal Server in order to access the contents
of different repositories. Due to the dynamic nature of Web API repositories, a
links maintenance mechanism has been combined with the update procedures
implemented by the Virtuoso Content Crawler.

7   Concluding Remarks

In this paper, we discussed an approach to link, according to Linked Data principles, contents of multiple Web API repositories, and how to exploit these links
for Web API search purposes. The approach is based on a unified model for Web
mashup resources. As future work, we plan to quantify how the productivity of
Web designers is increased through the use of multiple repositories for Web API
selection, where different repositories focus on complementary Web mashup resource descriptions. We have run preliminary experiments to test effectiveness
of Web API search in terms of precision and recall, using two popular public
repositories as a proof of concept.

References

1. Taheriyan, M., Knoblock, C., Szekely, P., Ambite, J.L.: Rapidly Integrating Services into the Linked Data Cloud. In: Proc. of the International Semantic Web Conference (ISWC), pp. 559–574 (2012)
2. Taheriyan, M., Knoblock, C., Szekely, P., Ambite, J.: Semi-Automatically Modeling Web APIs to Create Linked APIs. In: Proceedings of the ESWC 2012 Workshop on Linked APIs (2012)
3. Speiser, S., Harth, A.: Integrating linked data and services with linked data services. In: Antoniou, G., Grobelnik, M., Simperl, E., Parsia, B., Plexousakis, D., De Leenheer, P., Pan, J. (eds.) ESWC 2011, Part I. LNCS, vol. 6643, pp. 170–184. Springer, Heidelberg (2011)
4. Krummenacher, R., Norton, B., Marte, A.: Towards linked open services and processes. In: Proceedings of the Third Future Internet Conference, pp. 68–77 (2010)
5. Bianchini, D., De Antonellis, V., Melchiori, M.: A Multi-perspective Framework for Web API Search in Enterprise Mashup Design. In: Salinesi, C., Norrie, M.C., Pastor, Ó. (eds.) CAiSE 2013. LNCS, vol. 7908, pp. 353–368. Springer, Heidelberg (2013)
6. Hartig, O., Langegger, A.: A database perspective on consuming linked data on the web. In: Datenbank-Spektrum, pp. 57–66 (2010)
7. Bianchini, D., De Antonellis, V., Melchiori, M., Salvi, D.: Semantic-enriched service discovery. In: Proc. of the 22nd International Conference on Data Engineering (ICDE), pp. 38–47 (2006)
8. Bianchini, D., De Antonellis, V.: Linked Data Services and Semantics-enabled Mashup. In: On Semantic Search on the Web, pp. 281–305. Springer (2012)
9. Pedrinaci, C., Liu, D., Maleshkova, M., Lambert, D., Kopecky, J., Domingue, J.: iServe: a Linked Services Publishing Platform. In: Proceedings of ESWC Ontology Repositories and Editors for the Semantic Web (2010)
10. Norton, B., Krummenacher, R.: Consuming Dynamic Linked Data. In: Proc. of First International Workshop on Consuming Linked Data (2010)
11. Speiser, S., Harth, A.: Towards Linked Data Services, in: Proc. of the 9th International Semantic Web Conference, ISWC (2010)
12. Volz, J., Bizer, C., Gaedke, M., Kobilarov, G.: Discovering and Maintaining Links on the Web of Data. In: Bernstein, A., Karger, D.R., Heath, T., Feigenbaum, L., Maynard, D., Motta, E., Thirunarayan, K. (eds.) ISWC 2009. LNCS, vol. 5823, pp. 650–665. Springer, Heidelberg (2009)
13. Malik, Z., Bouguettaya, A.: RATEWeb: Reputation Assessment for Trust Establishment among Web Services. VLBD Journal 18, 885–911 (2009)
14. Villazón-Terrazas, B., Vilches, L., Corcho, O., Gómez-Pérez, A.: Methodological Guidelines for Publishing Government Linked Data. Springer, Heidelberg (2011)
15. Qi, X., Nie, L., Davison, B.: Measuring Similarity to Detect Qualified Links. In: Proc. of the 3rd Int. Workshop on Adversarial Information Retrieval on the Web, pp. 49–56 (2007)
16. Cohen, W., Ravikumar, P., Fienberg, S.: A comparison of string distance metrics for name-matching tasks. In: IJCAI-2003 Workshop on Information Integration on the Web, pp. 73–78 (2003)
17. Castano, S., Ferrara, A., Montanelli, S.: Matching Ontologies in Open Networked Systems: Techniques and Applications. Journal on Data Semantics 2, 25–63 (2006)

On Materialized sameAs Linksets

Marco A. Casanova[1], Vânia M. P. Vidal[2], Giseli Rabello Lopes[1],
Luiz André P. Paes Leme[3], and Livia Ruback[1]

[1] Department of Informatics, PUC-Rio, Rio de Janeiro, RJ – Brazil
{casanova,grlopes,lrodrigues}@inf.puc-rio.br
[2] Computer Science Department, UFC, Fortaleza, CE – Brazil
vvidal@lia.ufc.br
[3] Computer Science Institute, UFF, Niterói, RJ – Brazil
lapaesleme@ic.uff.br

Abstract. The Linked Data initiative promotes the publication of previously isolated databases as interlinked RDF datasets, thereby creating a global scale data space. Datasets are frequently interlinked using some automated matching process that results in a *materialized sameAs linkset*, that is, a set of links of the form *(s, owl:sameAs, o)*, which asserts that *s* denotes the same resource as *o*. This paper proposes strategies to reduce the cognitive overhead of creating materialized sameAs linksets and to correctly maintain them. The paper also outlines an architecture to improve the support for materialized sameAs linksets.

Keywords: RDF dataset interlinking, Linked Data, view update.

1 Introduction

The Linked Data initiative [2] promotes the publication of previously isolated databases as interlinked RDF datasets, or simply *datasets*, thereby creating a global scale data space, known as the Web of Data. Therefore, at the heart of the Linked Data initiative stands the problem of interlinking datasets.

Datasets are frequently interlinked using some automated matching process, whose result is a *materialized linkset*, that is, a set of links that is automatically created and explicitly stored. Of special interest is the case of *materialized sameAs linksets* consisting of links of the form *(s, owl:sameAs, o)*, which asserts that *s* denotes the same resource as *o*. In this paper, we propose strategies to reduce the cognitive overhead of creating materialized sameAs linksets and to correctly maintain them. The paper also outlines how triple stores could improve the support for materialized sameAs linksets.

Consider a materialized sameAs linkset *L* interlinking datasets *T* and *U*. To simplify the creation of *L*, we advocate that the administrator of each dataset, who understands how the dataset is organized, should pre-define views that act as resource catalogues – sets of resources with useful properties. The task of the user starts by selecting a view *v* over *T* and a view *w* over *U*, that define catalogues of comparable resources, and using *v* and *w* to create *L*. We therefore suggest replacing the difficulties of understanding how *T* and *U* are modeled by selecting pre-defined views.

H. Decker et al. (Eds.): DEXA 2014, Part I, LNCS 8644, pp. 377–384, 2014.

To address the maintenance of L, we propose an incremental strategy, justified for two reasons: (1) L is computed by a (complex) instance matching process between property values obtained from T and U using views; and (2) L does not contain the property values that generated the sameAs links. The architecture to support materialized sameAs linksets we suggest separates the problem of detecting updates on the datasets in a *view controller* component from the problem of maintaining the materialized sameAs linkset, which is the responsibility of a *linkset controller*.

Briefly, as for related work, several tools have been specifically designed to create linksets [3][8][13]. Tools have also been developed to recommend datasets with a high probability of interlinking [4][5][7], thereby reducing the cost of the interlinking process. The introduction of views, as suggested in Section 3, would simplify the configuration of the tools designed to create links. Furthermore, the link recommendations tools would benefit from the publication of view metadata. In another direction, tools, such as DSNotify [10], have been designed to inform database administrators about dataset changes and to allow them to preserve link integrity. DSNotify is closely related to the *view controller* component discussed in Section 4. Finally, it has been shown that incremental maintenance generally outperforms full view re-computation [11][12]. In fact, we argued that the linkset maintenance problem might be addressed in much the same way as the materialized view maintenance problem.

The paper is organized as follows. Section 2 introduces basic concepts and a simple example used in the paper. Section 3 covers how to create materialized sameAs linksets. Section 4 discusses how to maintain sameAs linksets and suggests how triple stores could better support them. Finally, Section 5 contains the conclusions.

2   Basic Concepts

An *RDF dataset*, or simply a *dataset*, is a set of RDF triples. A resource identified by an IRI (*Internationalized Resource Identifier*) s *is defined in* a dataset T iff s occurs as the subject of a triple in T. A triple in T *defines a property of* s iff the triple is of the form (s, p, v) and p is not *rdf:type*.

Let T and U be two datasets. A *link* from T to U is an RDF triple (s, p, o) such that s is defined in T, o is defined in U and p is not *rdf:type*. We say that T *is linked to* U, or that U *is linked from* T, iff there is at least a link from T to U. A *linkset* from T to U is a set of links from T to U. A *sameAs link* is a link of the form $(s, owl:sameAs, o)$, which asserts that s denotes the same resource as o.

We will adopt a simple example based on two datasets. The first one, the *Lattes dataset*, represents CVs of Brazilian researchers and was extracted from the Lattes platform. We refer to this dataset as *BrCV* and assume that it is available at http://lattes.br/ (a fictitious IRI). The second dataset, the *Semantic Web Conference Corpus*, contains triples about the main conferences and workshops in the area of Semantic Web research. We refer to this dataset as *SWCC* and assume that it is available at http://data.semanticweb.org/. *BrCV* uses a specific ontology, which we call the *Lattes* ontology, and *SWCC* uses the *Semantic Web Conference* (*SWC*) ontology [6].

A SPARQL query F is a *simple property path query*, or a *simple* query, iff

- The CONSTRUCT clause of F has exactly one template of the form " $?x$ rdf:type C " and a list of templates of the form "$?x\, P_k\, ?p_k$", where C is a class and P_k is a property, for $k=1,...,n$; we say that $V_F=\{C,P_1,...,P_n\}$ is the *vocabulary* of F.
- F contains a single FROM clause, specifying the dataset used to evaluate F.
- The WHERE clause of F is a list "$c, p_1,..., p_m, f_1,..., f_n$", where
  - c is of the form " $?x$ rdf:type D ", where C and D are not necessarily equal
  - for each $k=1,...,m$, the expression p_k is a sequence property path of the form "$?x\, R_k^1/ .../R_k^{n_k}\, ?p_k$", where R_k^i is an IRI or the inverse of a path consisting of a single IRI, for $i=1,...,n_k$
  - for each $l=1,...,n$, the expression f_l is a SPARQL filter restricting the variables used in the path expressions.

Let T be the dataset specified in the FROM clause of F. When evaluated against T, the simple query F returns a set of triples, which we denote $F[T]$.

A *simple view definition* is a pair $v = (V_F, F,$ where

- F is a simple SPARQL query, called the *view mapping*, whose FROM clause specifies the dataset over which v is evaluated.
- V_F is the vocabulary of F, also called the *view vocabulary* (hence, V_F consists of a single class and an ordered list of properties).

A *materialization* of v is the process of computing the set $F[T]$ and explicitly storing it as part of a dataset.

A *linkset view definition* is a quintuple $l = (p, F, G, \pi, \mu,$ where

- p is an object property
- F and G are simple queries whose vocabularies have the same cardinality n and whose FROM clauses specify the datasets over which l is evaluated
- π is a permutation of $(1,...,n)$, called the *alignment* of l
- μ is a $2n$-relation, called the *match predicate* of l

Let $V_F=\{C,P_1,...,P_n\}$ be the vocabulary of F and $V_G=\{D,Q_1,...,Q_n\}$ be the vocabulary of G. Intuitively, π indicates that, for each $k=1,...,n$, the match predicate will compare values of P_k with values of Q_m, where m = $\pi(k)$.

We also admit a linkset view definition of the form $l = (p, v, w, \pi, \mu,$ where $v=(V_F,F)$ and $w=(V_G,G)$ are simple view definitions. This alternative form translates to the previous one, if we replace v and w by their queries, F and G, respectively.

Let T and U be the datasets specified in the FROM clauses of F and G, respectively. We say that l is *evaluated over T and U* and that l is *from T to U*. The linkset view definition l induces a set of triples, denoted $l[T,U]$, as follows:

$(s, p, o) \in l[T,U]$ iff there are triples $(s,rdf:type,C), (s,P_1,s_1),...,(s,P_n,s_n) \in F[T]$ and $(o,rdf:type,D), (o,Q_1,o_1),...,(o,Q_n,o_n) \in G[U]$ such that $(s_1,..., s_n, o_{m1},..., o_{mn}) \in \mu$ where $m_k = \pi(k)$, for each $k=1,...,n$

A *materialization* of l is the process of computing the set $l[T,U]$ and explicitly storing it as part of a dataset. Again, we could expand the abstract notation to indicate the dataset and provide a name for the materialization of a linkset view definition.

3   Creating Materialized sameAs Linksets

To create a materialized sameAs linkset L between T and U, the user should:

(SA1) Specify a sameAs linkset view definition over T and U.

(SA2) Materialize l.

Step (SA1) requires that the user understand how datasets T and U are modeled, which properties may act as identifiers for the classes, etc. Based on this analysis, the user will create queries F and G (an example is given below). Step (SA2) is usually costly and may return only approximate values.

Using our running example, assume that the user wants to create a sameAs linkset, *sameRe*, between researchers in the *SWCC* dataset and those represented in the *BrCV* dataset by their CVs. Then, he may match the person's name from both datasets and, to disambiguate, use the homepage of the organization the person works for. The user starts by specifying a sameAs linkset view definition $l = (owl:sameAs, F_{SWCC}, G_{BrCV}, \pi, \mu,$ where F_{SWCC} is the SPARQL query

```
1.   PREFIX  rdf: <http://www.w3.org/1999/02/22-rdf-syntax-ns#>
2.   PREFIX foaf: <http://xmlns.com/foaf/0.1/>
3.   CONSTRUCT { ?x rdf:type foaf:Person           .
4.                ?x foaf:firstName          ?fn .
5.                ?x foaf:lastName           ?ln .
6.                ?x foaf:workplaceHomepage ?op  }
7.   FROM <http://data.semanticweb.org/>
8.   WHERE
9.   { ?x rdf:type foaf:Person          .
10.    ?x foaf:firstName          ?fn .
11.    ?x foaf:lastName           ?ln .
12.    ?x foaf:member/foaf:page   ?op  }
```

G_{BrCV} is the SPARQL query

```
13. PREFIX  rdf: <http://www.w3.org/1999/02/22-rdf-syntax-ns#>
14. PREFIX foaf: <http://xmlns.com/foaf/0.1/>
15. PREFIX   la: <http://onto.lattes.br/>
16. CONSTRUCT { ?x rdf:type foaf:Person           .
17.               ?x foaf:firstName          ?fn .
18.               ?x foaf:lastName           ?ln .
19.               ?x foaf:workplaceHomepage ?op  }
20. FROM <http://lattes.br/>
21. WHERE
22. { ?x rdf:type la:Curriculum    .
23.    ?x foaf:firstName          ?fn .
24.    ?x foaf:lastName           ?ln .
25.    ?x la:refersToWorkedFor/la:refersToOrg/foaf:homepage ?op }
```

the alignment π is the identity permutation and the match predicate μ is defined as

$$(s_1,\ldots, s_n, o_1,\ldots, o_n) \in \mu \text{ iff } \sigma(s_k, o_k) \geq \alpha, \text{ for each } k=1,\ldots,n$$

where σ is the 3-gram distance and $\alpha = 0.5$.

The strategy we propose to reduce the cognitive overhead of creating materialized sameAs linksets rests on the idea that the responsibility for Step (SA1) should not be placed on the user that is trying to create the linkset, but it should be shared with the administrator of each dataset. That is, the administrator should publish one or more view definitions over the dataset such that

(V1) Each view definition is simple.

(V2) Each view definition includes a mapping to the underlying dataset.

(V3)   Each view is accompanied by metadata that describe the set of instances represented in the view (using the VoID vocabulary [1], for example) and indicate how its vocabulary is pre-aligned with standard vocabularies.

Thus, a user who wishes to create links between T and U may browse the published metadata to find simple view definitions $v=(V_F,F)$ and $w=(V_G,G)$, thereby simplifying Step (SA1). As an example, consider again the problem of creating a sameAs linkset, *sameRe*, between researchers represented in the *SWCC* dataset and researchers represented in the *BrCV* dataset by their CVs. For disambiguation purposes, one identifies a researcher by his name and the homepage of the organization he works for. The creation of the *sameRe* linkset will be considerably simplified if the administrators of the *SWCC* and *BrCV* datasets define views that capture the properties that qualify researchers in their datasets.

For example, the administrator of *SWCC* may create a simple view definition $ReSWCC = (V_F,F_{SWCC},$ where $V_F = \{foaf{:}Person,\ foaf{:}firstName,\ foaf{:}lastName,\ foaf{:}workplaceHomepage\}$ and F_{SWCC} is the query already used at the beginning of this section. Likewise, the administrator of *BrCV* may create a simple view definition $ReBrCV = (V_G,G_{BrCV},$ where $V_G = \{foaf{:}Person,\ foaf{:}firstName,\ foaf{:}lastName,\ foaf{:}workplaceHomepage\}$ and G_{BrCV} is the SPARQL query previously defined. Lastly, the user creates the sameAs linkset view definition $l = (owl{:}sameAs,\ ReSWCC,\ ReBrCV,\ \pi,\ \mu,$ as before, except that he will use the view definitions and will trivially specify the alignment π as the identity permutation. This simple example illustrates how view definitions help define materialized linksets.

4   Maintaining Materialized sameAs Linksets

We may address the maintenance of materialized sameAs linksets using strategies similar to those devised to maintain materialized views: *rematerialize L*, when updates are applied to T or U; *incrementally maintain L*, based on the updates on T or U; *invalidate* links in L that are affected by updates on T or U. We do not consider versioning L since we would have to assume that T and U are also versioned, which leads a different set of problems. To decide which is best among these alternatives, observe that: (1) L is computed by a (potentially complex) matching process between property values obtained from T and U using queries; and (2) L does not contain the triples capturing the property values that generated the sameAs links.

Hence, by (1), rematerialization is potentially costlier than incremental maintenance or link invalidation. But, in either case, by (2), it is impossible to maintain or invalidate links in L in the presence of updates on T and U. Indeed, by (2), L does not contain enough information that would permit: (i) detecting when an update on T or U invalidates a sameAs link in L; (ii) recomputing a sameAs link after an update on T or U occurs.

Suppose that L is obtained by the materialization of $l = (owl{:}sameAs,\ v,\ w,\ \pi,\ \mu,$ where $v=(V_F,F)$ and $w=(V_G,G)$ are view definitions over T and U, respectively. To incrementally maintain L, we have to analyze how updates on T and U affect L and to compute the changes that must be applied to L.

Let u be an update on T (the discussion for updates on U is entirely similar). Let T_{old} and T_{new} be the states of T before and after the update. The strategy we adopt to incrementally maintain L briefly goes as follows:

(1) (Before u is executed) Compute the set R of resources defined in $F[T_{old}]$ whose properties in $F[T_{old}]$ are affected by u.

(2) (After u is executed) For each $s \in R$,

    a. Delete from L all sameAs links of the form $(s, owl:sameAs, o)$, for some o.

    b. If s is still defined in $F[T_{new}]$, then

        i. Retrieve the (new) properties of s defined in $F[T_{new}]$.

        ii. Try to match the (new) properties of s with those of a resource o in $G[U]$; if a match is found, add $(s, owl:sameAs, o)$ to L.

We illustrate this strategy with the example of Section 3, which creates a sameAs linkset, *sameRe*, as a materialization of $l = (owl:sameAs, ReSWCC, ReBrCV, \pi, \mu$.

Step 1. As an example, suppose that u is the following update on *SWCC*:

```
1.   WITH <http://data.semanticweb.org>
2.   DELETE { ?x foaf:page "www.pucrio.br"  }
3.   INSERT { ?x foaf:page "www.puc-rio.br" }
4.   WHERE
5.   { ?x foaf:page "www.pucrio.br" }
```

The set R is the result of the following query F_R, synthesized from F_{SWCC} and u:

```
6.   SELECT ?x
7.   FROM <http://data.semanticweb.org>
8.   WHERE
9.   { ?x rdf:type foaf:Person   .
10.    ?x foaf:member/foaf:page "www.pucrio.br" }
```

Note that F_R must be executed before u is. Assume that F_R returns just two IRIs:

$R = \{$ `<http://example/re1>`, `<http://example/re2>` $\}$

Step 2. We proceed by deleting from L all sameAs links of the form $(s, owl:sameAs, o)$, where $s \in R$. This is expressed as the following SPARQL deletion (assume that L is stored in `<http://sameRe.org>`):

```
11. WITH <http://sameRe.org>
12. DELETE WHERE{ { <http://example/re1> owl:sameAs ?x } UNION
13.                { <http://example/re2> owl:sameAs ?x } }
```

To retrieve the properties of `<http://example/re1>`, we use a query F_{re1}, synthesized from F_{SWCC}:

```
14. CONSTRUCT {<http://example/re1> rdf:type foaf:Person          .
15.            <http://example/re1> foaf:firstName          ?fn .
16.            <http://example/re1> foaf:lastName           ?ln .
17.            <http://example/re1> foaf:workplaceHomepage ?op  }
18. WHERE
19.  { GRAPH <http://data.semanticweb.org>
20.       { <http://example/re1> rdf:type foaf:Person          .
21.         <http://example/re1> foaf:firstName          ?fn .
22.         <http://example/re1> foaf:lastName           ?ln .
23.         <http://example/re1> foaf:member/foaf:page ?op  } }
```

To retrieve the properties of `<http://example/re2>`, we use a query F_{re2}, likewise synthesized from F_{SWCC}. Note that, since the matching process has to access triples stored in *BrCV* through view *ReBrCV*, if insertions into *SWCC* are frequent, then it would be useful to explicitly materialize *ReBrCV*. Otherwise, it would be advantageous to maintain *ReBrCV* as a virtual set of triples.

We stress that the possibility of appropriating solutions proposed for materialized RDF view maintenance was indeed one of the major motivations for introducing views to address materialized sameAs linkset maintenance. Furthermore, we note that the use of views defined by simple property path queries was a pragmatic decision since the matching tools adopt property paths to specify the values to be matched.

Finally, we suggest two types of components, called *view controllers* and *linkset controllers*, to include in triple stores to improve the support for dataset interlinking (see Figure 1). The *view controller* for a view *v* over a dataset *T*, has the following functionality:

- Accept registrations from *linkset controllers* that will consume data through *v*.
- Offer a SPARQL endpoint to access *v* (useful to create a linkset using *v*).
- Monitor each update on *T* that affects *v* and create a set *R* of IRIs (as in Step 1 discussed in this section).
- Send *R* to the *linkset controllers* registered with itself (useful to maintain a linkset created using *v*); the *view controller* may send the sets of IRIs one-by-one, or in batch, starting from a given timestamp.

The *linkset controller* for a linkset *L*, defined over views *v* and *w*, has the following functionality:

- Register itself with the *view controllers* for *v and w*.
- Initialize *L* by accessing *v* and *w*.
- Receive (or request) sets of IRIs from the *view controllers* for *v* and *w* and update *L* accordingly (as in Step 2 discussed in this section); the *linkset controller* may receive the sets one-by-one, or request them in batch.

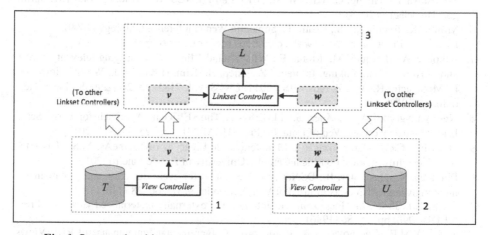

Fig. 1. Suggested architecture to create and maintain materialized sameAs linksets

5   Conclusions

In this paper, we analyzed the major issues that dataset interlinking, using materialized sameAs linksets, raises. We introduced views to simplify the creation of sameAs linksets and showed that materialized sameAs linkset maintenance can appropriate the solutions designed for materialized view maintenance. Then, we proposed two types of components that triple stores should include to improve the support for materialized sameAs linksets, both at the creation and the maintenance stages.

The discussion in this paper benefited from early implementations of a sameAs linkset maintenance tool [9]. As for future plans, we will work with more general classes of views to create and maintain materialized (generic) linksets.

Acknowledgments. This work was partly supported by CNPq, under grants 160326/2012-5, 303332/2013-1 and 57128/2009-9, and by FAPERJ, under grants E-26/170028/2008 and E-26/103.070/2011.

References

1. Alexander, K., Cyganiak, R., Hausenblas, M., Zhao, J. (2011) Describing Linked Triplesets with the VoID Vocabulary. W3C Interest Group Note (March 3, 2011)
2. Berners-Lee, T.: Linked Data (2006),
   `http://www.w3.org/DesignIssues/LinkedData.html`
3. Isele, R., Jentzsch, A., Bizer, C.: Efficient Multidimensional Blocking for Link Discovery without losing Recall. In: Proc. WebDB (2011)
4. Leme, L.A.P.P., Lopes, G.R., Nunes, B.P., Casanova, M.A., Dietze, S.: Identifying candidate datasets for data interlinking. In: Daniel, F., Dolog, P., Li, Q. (eds.) ICWE 2013. LNCS, vol. 7977, pp. 354–366. Springer, Heidelberg (2013)
5. Lopes, G.R., Leme, L.A.P.P., Nunes, B.P., Casanova, M.A., Dietze, S.: Recommending Tripleset Interlinking through a Social Network Approach. In: Lin, X., Manolopoulos, Y., Srivastava, D., Huang, G. (eds.) WISE 2013, Part I. LNCS, vol. 8180, pp. 149–161. Springer, Heidelberg (2013)
6. Möller, K., Bechhofer, S., Heath, T.: Semantic Web Conference Ontology (2009),
   `http://data.semanticweb.org/ns/swc/ontology`
7. Nikolov, A., d'Aquin, M., Motta, E.: What should I link to? Identifying relevant sources and classes for data linking. In: Pan, J.Z., Chen, H., Kim, H.-G., Li, J., Wu, Z., Horrocks, I., Mizoguchi, R., Wu, Z. (eds.) JIST 2011. LNCS, vol. 7185, pp. 284–299. Springer, Heidelberg (2012)
8. Ngonga Ngomo, A.-C., Auer, S.: LIMES - A Time-Efficient Approach for Large-Scale Link Discovery on the Web of Data. In: Proc. IJCAI 2011, pp. 2312–2317 (2011)
9. Ourofino, C.G.: Materialização e Manutenção de Ligações owl:sameAs. M.Sc. Dissertation. Dept. Informatics, Pontifical Catholic University of Rio de Janeiro (2013)
10. Popitsch, N., Haslhofer, B.: DSNotify – A Solution for event detection and link maintenance in dynamic triplesets. Journal of Web Semantics 9(3), 266–283 (2011)
11. Staudt, M., Jarke, M.: Incremental maintenance of externally materialized views. In: Proc. VLDB 1996, pp. 75–86 (1996)
12. Vidal, V.M.P., Casanova, M.A., Cardoso, D.S.: Incremental Maintenance of RDF Views of Relational Data. In: Meersman, R., Panetto, H., Dillon, T., Eder, J., Bellahsene, Z., Ritter, N., De Leenheer, P., Dou, D. (eds.) ODBASE 2013. LNCS, vol. 8185, pp. 572–587. Springer, Heidelberg (2013)
13. Volz, J., Bizer, C., Gaedke, M., Kobilarov, G.: Discovering and Maintaining Links on the Web of Data. In: Bernstein, A., Karger, D.R., Heath, T., Feigenbaum, L., Maynard, D., Motta, E., Thirunarayan, K. (eds.) ISWC 2009. LNCS, vol. 5823, pp. 650–665. Springer, Heidelberg (2009)

The Role of Design Rationale in the Ontology Matching Step during the Triplification of Relational Databases

Rita Berardi[1], Marcelo Schiessl[2], Matthias Thimm[3], and Marco A. Casanova[1]

[1] Departamento de Informática,
Pontifícia Universidade Católica do Rio de Janeiro,
Rio de Janeiro, RJ – Brazil CEP 22451
{rberardi,casanova}@inf.puc-rio.br
[2] Faculdade de Ciência da Informação,
Universidade de Brasilia,
Brasflia, DF - Brazil CEP 70910-900
schiessl@unb.br
[3] Institute for Web Science and Technologies (WeST),
Universität Koblenz-Landau,
Koblenz, Germany
thimm@uni-koblenz.de

Abstract. A common task when publishing relational databases as RDF datasets is to automatically define a vocabulary using the schema data, called *Database Schema Ontology* (DSO). To automatically reuse terms from existing vocabularies, Ontology Matching (OM) algorithms are used to generate recommendations for the DSO vocabulary. However, when a relational database is partially published due to privacy reasons, the DSO vocabulary loses contextual information and OM results in quite general recommendations, such as *FOAF* or *Dublin Core* vocabularies. To reduce the loss of context and to achieve recommendations better related to the DSO context, this paper proposes to use the design rationale captured during the publication of relational databases to RDF, represented as annotations. The case studies show that this annotation strategy helps find recommendations better related to the DSO context.

Keywords: triplification, private relational databases, design rationale.

1   Introduction

One of the most popular strategies to publish structured data on the Web is to expose relational databases (RDB) in the Linked Data format, that is the RDB-to-RDF process - also called *triplification*. Broadly stated, there are two main approaches for mapping relational databases into RDF: (1) the *direct mapping* approach where the database schema is directly mapped to ontology elements and (2) the *customized mapping* approach where the schema of the RDF may differ significantly from the original database schema. Here, we consider the first approach where a vocabulary is

H. Decker et al. (Eds.): DEXA 2014, Part I, LNCS 8644, pp. 385–393, 2014.

directly defined from the relational database schema, resulting in a Database Schema Ontology (DSO). In order to promote interoperability of Linked Data it is recommended that this vocabulary should reuse terms from existing vocabularies [4]. For that, Ontology Matching (OM) algorithms are useful to find recommendations to reuse terms from existing vocabularies. In this scenario, occasionally, only part of a relational database can be published as RDF, sometimes for privacy reasons or just because the rest of the data is not considered as interesting for other users. In the following, the part of the database that is not to be published is called *private*. In these cases, the DSO may lose important contextual information, leading to an OM recommendation of terms from more general vocabularies, such as FOAF or Dublin Core. For example, let us consider two OWL classes of a DSO with the terms "dso:publication" and "dso:author". It is not possible to identify which kind of publication the term "dso:publication" is referring to: it could refer to a research publication, a book, an article of a newspaper, or even to a song. The same happens with the term "dso:author". Without any additional information it is very hard to recommend terms from vocabularies of some specific context. This additional information could be their related classes or properties in the complete DSO, or their related entities or relationships in the complete RDB, or even typical instances in the RDB. In this case, OM algorithms are able to only recommend general terms like "dc:creator" (from Dublin Core vocabulary) or "foaf:person" (from FOAF vocabulary) for "dso:author". These recommendations are not wrong, but they do not provide the complete advantage of Linked Data since they represent a semantically weak description of these concepts. We argue that design rationale (DR) captured during the RDB-to-RDF mapping can improve the triplification process in many aspects, such as to analyze changes in the RDB context during the triplification, to verify loss of information and to improve the quality of vocabulary recommendations. In general, DR consists of decisions taken during a design process, the accepted and rejected options, and the criteria used. Concretely, we present in this paper a process, called StdTrip 2.0, that captures DR and uses it to reduce the loss of context when relational databases are partially published as RDF. We focus on the DR represented as a traceability of the original RDB form and how to use it to improve the quality of vocabulary recommendations. We assume that the private data must remain private, but the access to the schema information about the private part would not harm privacy policies. The DR is represented by systematically annotating the published data with the private schema information of the RDB. StdTrip 2.0 is an evolution of StdTrip+K [2, 1 12]. It improves StdTrip+K by adding the following features: (i) it enriches the DSO by annotating it with the private part of the relational database using *rdfs:comment*; (ii) it keeps a trace of all RDB-to-RDF transformations by storing the design rationale (DR) in a simpler directed graph and (iii) it includes a specific annotation strategy. Quite a few tools have been developed to the mechanical process of transforming relational data to RDF triples, such as Triplify [1], D2RQ [3], Virtuoso RDF view [9], RDBtoOnto [5] and RBA with support to R2RML [8][14], for both direct and customized mapping. However, they do not provide any design rationale capturing or any discussion for partial triplification of relational databases. The rest of this paper is organized as follows. Section 2 introduces the StdTrip 2.0

process. Section 3 presents case studies. Finally, Section 4 contains the conclusions and suggestions for future work.

2   The StdTrip 2.0 Process

StdTrip 2.0 includes a new step, for the cases where the relational database is partially published. The process receives as input a relational database (RDB), a set of mapping rules (MR), and known vocabularies of domain ontologies in the LOD cloud. It outputs an RDF vocabulary to represent the partial data from the RDB, alignments between DSO terms and terms of known vocabularies according to their context, and the final design rationale captured.

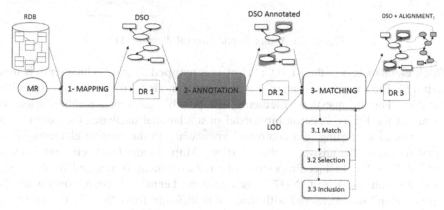

Fig. 1. The StdTrip 2.0 process

After each main step, a DR is recorded. We use the step number to refer to each DR; for instance, in step 1, we use DR1. Figure 2 depicts the publication database that we use as an example. Fig. 2 (b) shows the corresponding Entity Relationship model of the relational database. To exemplify a partially published relational database, we consider as private the grey part of Fig.2. (a) and (b)).

2.1   Step 1 - Mapping

The Mapping step receives an Entity-Relationship (ER) extracted from an RDB schema (Fig.2) and a set of mapping rules (MR), defined by a domain expert. It outputs a DSO and the corresponding design rationale (DR1). The DR1 records the original elements of the ER and how they were mapped to ontology elements. DR1 is created according to the definition of the trace proposed in [2]. The original ER is stored as a graph node and the trace regarding the mapping of each node is stored as node attributes. For example, in Fig. 3, the RE "Publication" is a node with attributes "**Element**", "**Map**" and "**Term**" that represent the trace of its mapping. These node attributes represent the "questions" of the DR1 model defined in [2], respectively:

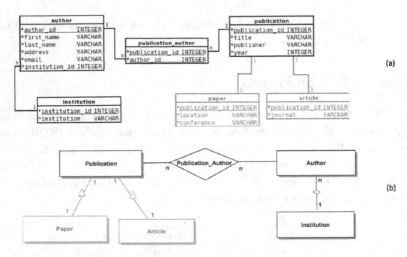

Fig. 2. The Author-Publication ER diagram [11]

"Which **element** is it in the RDB?"; "How is it **mapped**?"; and "Which **term** is used to map it?". The answers to each of these questions (node attributes) obey a controlled vocabulary. For the question "**Element**", the possible answers are abbreviations of elements of the Entity-Relationship model of a relational database. The edges in all DR are named according to a controlled vocabulary to the relation elements in the original database. An answer to the question "**Map**" is any OWL element, such as "owl:Class" or "owl:ObjectProperty", when the element is mapped to the DSO, otherwise the answer is just "NOT". The question "**Term**" is answered only when the question "**Map**" was answered with something different from "NOT". The answer is naturally the term used for the mapping.

2.2   Step 2 – Annotation

The annotation process is a new step in StdTrip 2.0 and aims at using the *rdfs:comment* property to add information about the private database schema in the DSO. The input of this step is: (i) a DSO transformed from a corresponding ER model of an RDB and the DR1 containing the trace of this transformation; and (ii) the private schema data of the original relational database. The output of this step is the annotated DSO and the trace of the annotation in DR2. The DR2 is incrementally built from the preceding DR1. The benefits of using the DR graph instead of directly using the relational database are: (1) Since the DR graph is created for provenance purposes, it can be accessed without having to create a new graph based in the RDB to know what has to be annotated, (2) Since the DR graph is always created in the StdTrip 2.0, it can be consumed when needed, without having to re-execute the mapping step.

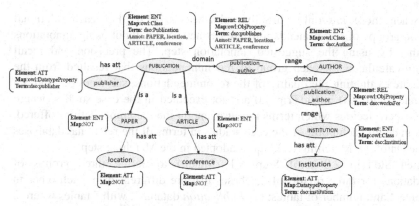

Fig. 3. Design Rationale 2 with annotations marked in red

The annotation is executed according to the *neighboring mapping*, that is: for each mapped element, look for its neighbors in the DR graph that were unmapped (Map:NOT) and, if they exist, the label of the unmapped node is annotated as a literal of the *rdfs:comment* property. The search for unmapped neighbors is executed according to the annotation strategy and can be done at any depth in the DR graph. We adopted, as an initial strategy, the consideration of at most two levels based on empirical observation. More specifically, the empirical evidence is that we noticed that for automatic annotations, including more than two levels becomes superfluous, as the additional levels are more likely out of context. As an example of output, the annotations in the DR2 are ready to be added in the DSO ontology resulting in DSO$_A$ in Fig. 3.

2.3   Step 3 – Matching

The general goal of this step is to find correspondences between the annotated DSO and standard well-known RDF vocabularies that really represent the context of the original database. This step comprises three sub-steps: (3.1) *Matcher execution*, where an ontology matching process is used to execute the matching; (3.2) *Selection of match candidates*, which creates, for each term of the DSO, a list of recommendations of terms from existing ontologies - here user interaction plays an essential role; and (3.3) *Inclusion*, where the domain expert can include other terms if he or she does not agree with the recommendations. Ideally, the user should know the database domain because he or she has to select the vocabulary elements that best represent each concept in the database. Similarly, to the previous DR models, the DR3 of this stage is incrementally stored.

3   Case Studies

The case studies aim at validating whether the annotations generated by StdTrip 2.0 can help in finding recommendations better related to the context of relational databases when those are only partially published as RDF. For each relational database, two groups of recommendations are defined: one without using annotations and the other by using the suggested annotation step. The precision and recall measures are calculated for each group of recommendations that resulted from the OM. As we are evaluating the quality of the recommendations, sub-steps 3.2 and 3.3 (Selection and Inclusion) in StdTrip 2.0 are not executed in the case studies, where the domain expert decides which terms to reuse among the recommendations offered. We discuss in detail each point of the case studies in terms of the relational databases used and the Ontology Matching techniques adopted in the Matching step.

We executed StdTrip 2.0 until Step 3.1 (Match) to define three groups of recommendations for three relational databases that are different from each other in terms of context and number of tables: (1) *Publication* database, with 7 tables where 2 of them were considered as private [7], (2) *osCommerce*[1] database, with 48 tables where 36 of them were considered as private and (3) *phpBB*[2] database. The *OsCommerce* and *phpBB* databases were also used to evaluate Triplify that is one of the most popular tools to transform RDB in RDF [1]. In the Triplify evaluation [1], they affirm that only parts of these databases were interesting for publishing as RDF. Besides that, the vocabulary to publish each one of them was completely defined by a domain expert reusing existing vocabularies, such as FOAF, SIOC and SKOS.

Regarding the Ontology Matching techniques used in the case studies, we propose a similarity measure of the terminological layer of OM. A terminological layer uses similarity measures to discover alignments by comparing labels, comments and definitions (annotations). Using OM algorithms from OAEI challenge [13], the results presented instability for matching a DSO that originally had no comments or definitions and comparing it with ontologies rich with definitions and comments. These measures, such as the Jaccard distance [6], "penalize" groups of strings with very different sizes and, consequently, result in low similarity values for them. We propose a *subset string measure-SbS* (1) that gives the similarity ratio minimizing the influence of the difference of sizes between the groups.

$$SbS(A, B) = \frac{|A \cup B||A \cap B|}{|A||B|} \in [0,1] \tag{1}$$

The Subset String (SbS) compares two lexical non-empty entries A, B and satisfies the following properties (i) SbS(A,B) = 1 iff $A \subseteq B \vee B \subseteq A$; (ii) SbS(A,B) = 0 iff $A \cap B = 0$ and (iii) SbS(A,B) $\uparrow \sim A \cap B \uparrow$, i.e., the SbS value increases as the intersection size also increases.

3.1   Result Analysis

We used the precision and recall measures to evaluate the quality of the results regarding our terminological SbS measure. For each relational database, these measures are calculated according to a Reference Alignment that has more contextual

---

[1] http://www.oscommerce.com/
[2] http://www.phpbbdoctor.com/doc_tables.php

recommendations. The analysis is performed by comparing precision and recall measures for all databases, but when other interesting points appear besides these measures, they are also explored. Fig. 4 shows the results for the three databases.

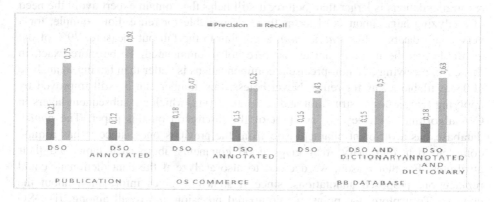

Fig. 4. Database results

The Reference Alignment for the *Publication* database was built according to our empirical knowledge about the publication context. The ontologies candidates for reusing were BIBO and FOAF. As we have previously discussed in Section 1, FOAF can be considered as a "general" context, then the ontology BIBO is more contextually related with the *Publication* database. Analyzing the results of *Publication* database, the recommendations for DSO ANNOTATED obtained lower values of precision value, if compared with DSO without annotation. However, the recall of DSO ANNOTATED got higher values, when compared to DSO without annotations. This characteristic is common for approaches that use extra information [5]. The StdTrip 2.0 still returns, as a recommendation, general terms (like FOAF), but it additionally recommends more contextual terms (like BIBO). Although the recommendations are more related with the context, they do not point exactly the right term, thereby reducing precision. If we consider the complete StdTrip 2.0 process, recall values have more impact. Thus, if they are low, the domain expert has fewer options to find the most representative term for the context, or even no option if the exact term is not recommended. The precision would have more impact, if a domain expert participation is not taken into account in the following steps of the process. The Reference Alignment for the *osCommerce* database was built according to the Triplify indication about which vocabularies for each part were used during its evaluation [1]. Analyzing the precision and recall values, we had the same results as the *Publication* database. Note that, even with a much larger relational database (50 tables), the results are the same. Despite our positive interpretation for the recall values in both the *Publication* and the *osCommerce* databases, there is an important issue to explore. Even though the recommendations include a large group that are not precise (according to the precision values), the number of non-relevant terms not recommended is even greater. To analyze that issue, we used the *fall-out* measure [10] that gives the proportion of non-relevant terms recommended to reuse, out of all non-relevant terms available for reuse. The numbers of fall-out for *Publication* and *osCommerce* are respectively 0.21 and 0.01. It shows that, although the group of recommendations is larger than before, it still helps the domain expert avoid the need of analyzing huge amounts of terms that are available for reuse. For example, for a very small database like *Publication*, according to the fall-out measure, 79% of the available and the non-relevant terms were not recommended. To be more exact, in this case, receiving 217 non-precise recommendations is better than having to analyze 1003 available terms for reuse. Nevertheless, this number can be still improved by applying techniques of structural and semantic layers, which are subsequent layers in OM algorithms but they are not part of the discussion on this paper. The *phpBB* database has a different characteristic from the previous ones, since it has a quite complete data dictionary[3] that gives good comments about each table and their attributes. For that reason, we decided to also analyze if the data dictionary could replace our proposed annotations, since it somehow gives information about the context. To explore that point, we compared precision and recall among: (i) DSO without any annotations or dictionary data; (ii) DSO and DICTIONARY with dictionary data; and (iii) DSO ANNOTATED and DICTIONARY with dictionary data and annotations generated by StdTrip 2.0. If precision and recall would not differentiate between (ii) and (iii), it would mean that the annotations do not make a difference when the database has a dictionary. Analyzing the results, we may conclude that the data dictionary plays an important role, if compared to the DSO, but the annotated DSO, together the dictionary data, still has the best numbers for both precision and recall.

4   Summary and Future Work

In this paper, we introduced StdTrip 2.0 as a process that automatically defines a vocabulary from a relational database (Database Schema Ontology) and supports the generation of design rationale represented by annotations, recorded when they are partially published as RDF. The annotations are the private schema data information that provides more contextual information about the DSO. This contextualization helps in the terminological layer of Ontology Matching techniques for recommending existing vocabularies better related to the database context. The analysis of the case studies showed that these annotations, in different relational databases, provide important contextual information and help finding vocabularies to reuse that are better related to the database. As future work we are interested in achieving even better quality vocabulary recommendations by taking advantage of techniques in the further layers of OM such as syntactic and semantic layers. Thus, we can also achieve a lower number of recommendations besides more contextualized terms.

---

[3] Data dictionary here consists of textual definitions of tables and attributes.

References

1. Auer, S., Dietzold, S., Lehmann, J., Hellmann, S., Aumueller, D.: Triplify: light-weight linked data publication from relational databases. In: WWW 2009, pp. 621–630. ACM, New York (2009)
2. Berardi, R., Breitman, K., Casanova, M.A., Lopes, G.R., de Medeiros, A.P.: StdTrip+K: Design Rationale in the RDB-to-RDF Process. In: Database and Expert Systems Applications, Prague, Czech Republic (2013)
3. Bizer, C. and Seaborne, A.: D2RQ-treating non-RDF databases as virtual RDF graphs. In: Proceedings of the 3rd International Semantic Web Conference ISWC 2004 (2004)
4. Breslin, J.G., Passant, A., Decker, S.: The Social Semantic Web, Inc. (2009)
5. Cerbah, F.: Learning highly structured semantic repositories from relational databases: In: Bechhofer, S., Hauswirth, M., Hoffmann, J., Koubarakis, M. (eds.) ESWC 2008. LNCS, vol. 5021, pp. 777–781. Springer, Heidelberg (2008)
6. Cheatham, M., Hitzler, P.: String Similarity Metrics for Ontology Alignment. In: Alani, H., Kagal, L., Fokoue, A., Groth, P., Biemann, C., Parreira, J.X., Aroyo, L., Noy, N., Welty, C., Janowicz, K. (eds.) ISWC 2013, Part II. LNCS, vol. 8219, pp. 294–309. Springer, Heidelberg (2013)
7. Cullot, N., Ghawi, R., Yétongnon, K.: DB2OWL: A Tool for Automatic Database-to-Ontology Mapping. In: SEBD 2007, pp. 491–494 (2007)
8. Das, S., Sundara, S., Cyganiak, R.: R2RML: RDB to RDF Mapping Language, W3C Working Draft (2012), http://www.w3.org/TR/r2rml/
9. Erling, O., Mikhailov, I.: RDF support in the virtuoso DBMS. In: Networked Knowledge-Networked Media, pp. 7–24 (2009)
10. Euzenat, J., Shvaiko, P.: Ontology Matching. Springer, Heidelberg (2007)
11. Salas, P., Viterbo, J., Breitman, K., Casanova, M.A.: StdTrip: Promoting the Reuse of Standard Vocabularies in Open Government Data. In: Wood, D. (ed.) Linking Government Data, Springer Verlag, pp. 113–134. Springer, Heidelberg (2011)
12. Salas, P.,, B.K., Viterbo, J., Casanova, M.A.: Interoperability by design using the StdTrip tool: an a priori approach. In: I-SEMANTICS (2010)
13. Shvaiko, P., Euzenat, J.: Ontology Matching: State of the Art and Future Challenges. In: IEEE TKDE, Los Alamitos, CA, USA (2013)
14. Vidal, V.M., Casanova, M.A., Neto, L.E., Monteiro, J.M.: Semi-Automatic Approach for Generating Customized R2RML Mappings. In: Accepted to the 29th ACM Symposium on Applied Computing, Gyeongju, Korea, March 24-28 (2014)

SK-Languages as a Comprehensive Formal Environment for Developing a Multilingual Semantic Web

Vladimir A. Fomichov

Department of Innovations and Business in the Sphere of Informational Technologies,
Faculty of Business Informatics, National Research University Higher School of Economics,
Kirpichnaya str. 33, 105187 Moscow, Russia
vfomichov@hse.ru, vfomichov@gmail.com

Abstract. The paper proposes a workable solution to the problem of constructing a comprehensive semantic formal environment (CSFE) for developing a Multilingual Semantic Web. It is to be a collection of the rules enabling us to construct step by step a semantic representation (or text meaning representation) of practically arbitrary sentence or discourse pertaining to mass spheres of human's professional activity. It is grounded that the class of SK-languages (standard knowledge languages) determined by the theory of K-representations (knowledge representations) can be interpreted as the first version of a CSFE for developing a Multilingual Semantic Web. The current version of the latter theory is stated in the V.A. Fomichov's monograph published by Springer in 2010.

Keywords: natural language processing, structured meaning, semantic representation, theory of K-representations, algorithm of semantic-syntactic analysis, multilingual semantic web, bioinformatics.

1   Introduction

During last fifteen years, one has been able to observe a quick expansion of the stock of Web-based informational sources in other natural languages (NLs) than English. That is why many research centres and groups throughout the world have been designing the applied computer systems being able to understand (i.e., to extract partial or complete meanings) the texts in various languages. The activity of all these centres and groups complements very large scale studies aimed at creating a Semantic Web and contributes to developing a Multilingual Semantic Web. The global vector of all these research activities indicates the direction initially outlined in [12].

In this connection, one of the most acute problems is the realization of effective cross-language informational search. Many scholars believe that the key to solving this problem is the creation and use of intermediary formal language for representing in the same way the semantic content of written text and speech utterances in various natural languages (see, e.g., [10]).

The process of endowing the existing Web with the ability of understanding many NLs is an objective ongoing process. The analysis has shown that there is a way to

H. Decker et al. (Eds.): DEXA 2014, Part I, LNCS 8644, pp. 394–401, 2014.

increase the total successfulness, effectiveness of this global decentralized process. It would be especially important with respect to the need of cross-language conceptual information retrieval and question - answering. The proposed way is a possible new paradigm for the mainly decentralized process of endowing the existing Web with the ability of processing many NLs.

The principal idea of a new paradigm is as follows. There is *a common thing* for the various texts in different natural languages. This common thing is the fact that *the natural language texts (the NL-texts) have the meanings.* The meanings are associated not only with NL-texts but also with the visual images (stored in multimedia databases) and with the pieces of knowledge from the ontologies.

That is why the great advantages are promised by the realization of the situation when a unified formal environment is being used in different projects throughout the world for formalizing semantics of lexical items, for reflecting structured meanings of the texts in various NLs, for representing knowledge about application domains, for constructing semantic annotations of informational sources, and for building high-level conceptual descriptions of visual images.

This paper continues the line of the works [2, 3, 4, 5, 6]. *The first subject of the paper* is to set forth the reasonable requirements to a broadly applicable formalism for representing semantic content of sentences and complex discourses in many NLs (Sections 2, 3). *The second, principal subject of the paper* is to state that the expressive mechanisms of the class of SK-languages (standard knowledge languages) determined in [2], [4] satisfy the collection of listed requirements (Section 4). Section 5 contains a discussion of related approaches, Section 6 describes some applications of the theory of K-representations.

2   What Is the Main Obstacle on the Way of Developing a Comprehensive Semantic Formal Environment

The analysis of many relatively recent publications on semantic processing of NL-texts by computer intelligent systems evokes the astonishment concerning a huge gap between the scale of the problems to be solved and the used formal means for reflecting semantics of NL-texts. For instance, Harrington and Clark [7] describe the ASK-Net system designed in the Oxford University Computing Laboratory. This system is able to automatically extract semantic information from the texts in English and, as a result of integrating this information, constructs a large-scale semantic network (SN). For this, the ASKNet system uses a number of existing NL processing tools and an enriched spreading activation theory. The system is able to create SN consisting of over 1.5 million nodes and 3.5 million edges in less than three days.

According to [7], the underlying semantic theory is Discourse Representation Theory (DRT). But DRT is a syntactic version of First Order Logic (FOL), and numerous restrictions of FOL are well known. Due to these restrictions, the authors of the discussed paper are forced to draw a picture (a kind of a semantic net) for representing semantic structure of the discourse Disc1 = "Yesterday John heard that ABC Inc. hired Susan. Bob decided that ABC Inc. will move to London. Susan met Bob twice".

The reason for drawing a picture for this simple text is that FOL allows for constructing the simplest formulas only of the form $P(t_1, ..., t_n)$, where n is not less than 1, P is the name of an n-ary predicate, $t_1, ..., t_n$ are terms (so no one element from $t_1, ..., t_n$ can be a formula). That is why FOL, and, as consequence, DRT, don't provide adequate formal means of describing semantic structure of the sentences with complex direct or indirect speech, etc.

The analysis of the scientific literature on the design of semantics-oriented NL processing systems and a Multilingual Semantic Web provides serious arguments in favour of putting forward the following conjecture: *it is high time for creating a new paradigm* for considering numerous theoretical problems encountered while constructing and processing various conceptual structures associated with Web-based informational sources: semantic representations of written and spoken texts' fragments (in other terms, text meaning representations); high-level conceptual descriptions of visual images; knowledge pieces stored in ontologies; the content of messages sent by computer intelligent agents, etc.

What can be a key to solving this problem? We do know that, using NL, we are able to describe various pieces of knowledge, the content of visual images, the content of a film, etc. That is why it can be conjectured that a key to elaborating a new paradigm of the described kind could be the construction of a Comprehensive (a broadly applicable and flexible) Semantic Formal Environment (CSFE). It is to be a collection of the rules enabling us to construct step by step a semantic representation (or text meaning representation) of practically arbitrary sentence or discourse pertaining to mass spheres of professional activity of people. CSFE is called in [6] a broadly applicable and flexible Conceptual Metagrammar.

The analysis shows that the main obstacle on the way of developing a CSFE is a manner of combining in a new (to be constructed) formalism the expressive possibilities of FOL with the possibilities of: (a) describing semantic structure of sentences with complex direct and indirect speech; (b) building compound designations of objects, notions, and sets; (c) using the logical connectives AND, OR for joining the designations of entities but not only the designations of statements; (d) describing semantic structure of discourses with references to the mentioned entities and with references to the meanings of sentences and larger parts of a discourse; (e) reflecting semantic structure of sentences with infinitives or gerunds; (f) describing semantic structure of sentences with the words "term", "notion"; (g) constructing formal expressions reflecting both the content of a natural language text and its metadata: the list of authors, the language, the year of publishing, the edition, etc.; (h) considering non-traditional relationships with the attributes being sets of things, situations, sets of notions, sets of semantic representations of sentences and discourses in NL.

No one of the principal known formal approaches to representing the content of natural language sentences and discourses possesses all properties listed above. In particular, it applies to FOL, DRT, Theory of Generalized Quantifiers, Theory of Semantic Nets, Theory of Conceptual Graphs, Episodic Logic [11], Natural Logic [1], [9]. That is why it is necessary to invent an original solution to the problem of developing a CSFE.

3   Desirable Properties of a Broadly Applicable Semantic Formalism

Let's try to formulate the reasonable requirements to a broadly applicable formalism for representing semantic content of NL-texts and knowledge items. The considered examples pertain to biology. Semantic representations (SRs) of the texts will be the values of the variable Semrepr.

Property 1: The possibility of describing semantic structure of sentences with indirect speech.

Example: Let T1 = "In 1903, Walter Sutton, an American medical student, and Theodor Boveri, a German biologist, independently conjectured that chromosomes carry the hereditary factors, or genes". Then a possible desirable structure of SemRepr: *Event(e1, conjecturing * (Agent, #Group-of-scholars-description#)(Time, 1903)(Content-inform, #hypothesis-description#)).*

Property 2: The possibility of building compound designations of objects.

Example: Let T2 = "Each cell contains a nucleous within which there are several thread-like structures known as *chromosomes"*. Then a possible desirable structure of SemRepr:

*Contain1(every cell1, certain nucleous1 * (Part-set, several structure1 *(Similar, thread)(Called, chromosome))).*

Property 3: The possibility of building compound designations of sets.

Example: Let T3 = "Type A blood group are persons who possess type A isoantigen on red blood cells and anti-B agglutinin in plasma". Then a possible desirable structure of SemRepr:

*(type-A-blood-group ≡ all person * (Possess1, #components-description#)).*

Property 4: The possibility of using the logical connectives AND, OR for joining the designations of entities but not only the designations of statements.

Example: Let T4 = "Five distinct stages of mitosis (the process of somatic cell division, during which the nucleus also divides) are prophase, prometaphase, metaphase, nanaphase, telophase". Then a possible desirable structure of SemRepr:

Stages(mitosis, (prophase∧ prometaphase ∧ metaphase ∧ nanaphase ∧ telophase)).

Property 5: The possibility of describing semantic structure of discourses with references to the mentioned entities.

Example: Let T5 = "On the basis of Mendel's plant experiments, three main principles were established. These are known as the laws of uniformity, segregation and independent assortment". Then it could be desirable that (1) an SR of Sentence1 includes the fragment

*certain set * (Qualitative-composition, principle *(Attribute, main))(Individs-composition, (x1 ∧ x2 ∧ x3));*

(2) an SR of Sentence2 is as follows: *(Called(x1, law-of-uniformity) ∧ Called(x2, law-of-segregation) ∧ Called(x2, law-of-independent-assortment)).*

Property 6: The possibility of describing semantic structure of discourses with references to the meanings of sentences and larger parts of a discourse.

Example: Let T6 = "All granulocytes are polymorphonuclear; that is, they have multilobed nuclei". Then a possible desirable structure of SemRepr is

$$(SemRepr1 : P1 ∧ Explanation(P1, SemRepr2)),$$

where SemRepr1 is an SR of the first sentence, and SemRepr2 is an SR of the phrase "they have multilobed nuclei" in the context of SemRepr1.

Property 7: The possibility to reflect semantic structure of sentences with infinitives or gerunds.

Example: Let T7 = "With DNA technology rapidly progressing, a group of visionary scientists in the USA persuaded Congress in 1988 to fund a coordinated international program to sequence the entire human genome". Then a possible desirable structure of SemRepr:

*Event(e1, persuasion * (External-conditions, #external-situation-description#)*
*(Agent, #group-description#)(Object-to-influence, certain congress * (Country, USA))*
*(Time, 1988)(Goal, funding * (Activity-role, #program-description#))).*

Property 8: The possibility to describe semantic structure of sentences with the words "term", "notion".

Example: Let T8 = "The term gene was first coined in 1909 by a Danish botanist, Johannsen, and was derived from the term pangen introduced by De Vries". Then it could be desirable to have the possibility of building compound semantic items of the form *certain term * (Called, "gene").*

Property 9: The possibility to construct formal expressions reflecting both the content of a natural language text and its metadata: the list of authors, the language, the year of publishing, the edition, etc.

Example: Let T9 = "Control gene is a gene which can turn other genes on or off, i.e. regulate". Then a desirable structure of the informational object corresponding to T9 could be as follows:

*certain inform-object * (Content1, #content-of-Def1#)(Authors,*
(D.Turnpenny ∧ S.Ellard))(Publishing-house, Elsevier)(Year, 2005)
(Title, "Emery's Elements of Medical Genetics")(Edition-number, 12).

4   The Theory of K-representations as a Discovery in Mathematical Informatics and Mathematical Linguistics

The analysis of the literature shows that now only the theory of K-representations (knowledge representations) possesses all nine properties listed above and is compatible with FOL. It is an original theory of designing semantic-syntactic analysers of NL-texts with the broad use of formal means for representing input, intermediary, and output data [2], [4]. This theory also contributes to the development of logic-informational foundations of (a) Semantic Web of a new generation, (b) E-commerce, and (c) multi-agent systems theory (agent communication languages) [3, 4, 5, 6].

In order to understand the principal distinction of the theory of K-representations from other mentioned approaches to formalizing semantics of NL, let's consider an analogy. Bionics studies the peculiarities of the structure and functioning of the living beings in order to discover the new ways of solving certain technical problems. Such theories as Theory of Generalized Quantifiers, Discourse Representation Theory, Theory of Semantic Nets, Theory of Conceptual Graphs, Episodic Logic (EL) and several other theories were elaborated on the way of expanding the expressive mechanisms of FOL. To the contrary, the theory of K-representations was developed as a consequence of studying the basic expressive mechanisms of NL and putting forward a conjecture about a system of partial operations on conceptual structures underpinning these expressive mechanisms. Of course, the idea was to develop a formal model of this system being compatible with FOL.

The *first basic constituent* of the theory of K-representations is the theory of SK-languages (standard knowledge languages). The kernel of this theory is a mathematical model describing a system of such 10 partial operations on structured meanings (SMs) of natural language texts (NL-texts) that, using primitive conceptual items as "blocks", we are able to build SMs of arbitrary NL-texts (including articles, textbooks, etc.) and arbitrary pieces of knowledge about the world. This model is set forth in Chapters 2 - 4 of [2] and Chapters 3 – 5 of [4].

The expressions of SK-languages will be called the K-strings.   If *Expr* is an expression in natural language (NL) and a K-string *Semrepr* can be interpreted as a semantic representation of *Expr*, then *Semrepr* will be called a K-representation (KR) of the expression *Expr*.

The analysis shows that the class of SK-languages possesses the properties 1 - 9 listed above. It is shown in Section 6.3 of [4] that SK-languages are convenient for building high-level conceptual descriptions of visual images.

The *second basic constituent* of the theory of K-representations is a broadly applicable mathematical model of a linguistic database [2], [4]. The *third basic constituent* of the theory of K-representations is formed by several complex, strongly structured algorithms carrying out semantic-syntactic analysis of texts from some practically interesting sublanguages of NL [2], [4].

5   Discussion

Taking into account the advantages listed above and the content of the considered examples, it is possible to conjecture that the theory of K-representations can be used as an adequate methodological basis for developing a new version of the system ASKNet [7] with an enhanced intelligent power.

The objective of the SemML project [8] is the creation of a standardized semantic markup language. It seems that high expressive possibilities of SK-languages are to urge the authors of SemML project to update the goals of the project in the sense of considering some sublanguages of SK-languages and replacing some designations by the designations being more habitual for the programmers and Web designers.

The advantages of the theory of K-representations in comparison with FirstOrder Logic, Discourse Representation Theory, and Episodic Logic are, in particular, the possibilities: (1) to distinguish in a formal way objects (physical things, events, etc.) and notions qualifying them; (2) to build compound representations of notions; (3) to distinguish in a formal manner objects and sets of objects, concepts and sets of concepts; (4) to build complex representations of sets, sets of sets, etc.; (5) to describe set-theoretical relationships; (6) to effectively describe structured meanings (SMs) of discourses with references to the meanings of phrases and larger parts of discourses; (7) to describe SMs of sentences with the words "concept", "notion"; (8) to describe SMs of sentences where the logical connective "and" or "or" joins not the expressions-assertions but designations of things, sets, or concepts; (9) to build complex designations of objects and sets; (10) to consider non-traditional functions with arguments or/and values being sets of objects, of concepts, of texts' semantic representations, etc.; (11) to construct formal analogues of the meanings of infinitives with dependent words and, as a consequence, to represent proposals, goals, commitments; (12) to build object-oriented representations of information pieces.

The items (3) - (8), (10) – (12) in the list above indicate the principal advantages of the theory of K-representations in comparison with the Theory of Conceptual Graphs. The global advantage of the theory of K-representations is that it puts forward a hypothesis about a system of partial operations on conceptual structures being sufficient and convenient for constructing semantic representations of sentences and discourses in NL pertaining to arbitrary fields of human's professional activity.

6   Applications and Conclusions

The arguments stated above and numerous additional arguments set forth in the monograph [4] give serious grounds to conclude that the class of SK-languages, provided by the theory of K-representations, can be interpreted as the first comprehensive semantic formal environment for studying various semantics-associated problems of developing a Multilingual Semantic Web.

It seems to be reasonable to say about two levels of applying the theory of K-representations to solving practical tasks. The *first level* is the direct use in the design of NL processing systems of a mathematical model of a linguistic database introduced in Chapter 7 of the monograph [4] and of the algorithm of semantic-syntactic analysis *SemSynt1* described in Chapters 9 and 10 of the same monograph. This algorithm is multilingual: its input texts may be the questions of many kinds, statements, and commands from the sublanguages of English, German, and Russian languages. The mentioned model and algorithm were applied by the author and his Ph.D. students to the design of a NL-interface of a recommender system, to the design of an advanced semantic search system (see LNCS 7557, pp. 296-304), and to the design of a NL-interface of an applied intelligent system making easier the interaction of a user with the file system of a computer (see LNCS 8455, pp. 81-84; Procedia Computer Science, 2014, Vol. 31, pp. 1005-1011).

The great advantages of the proposed comprehensive semantic formal environment are promised by the *second level* of applications: it is the case of using SK-languages for describing lexical semantics, representing semantic content of sentences and discourses in NL, and building models of ontologies in numerous scientific centres and research groups throughout the world.

Acknowledgements. I am grateful to the anonymous referees of this paper for precious remarks.

References

1. Andreasen, T., Bulskov, H., Fischer Nilsson, J., Anker Jensen, P., Lassen, T.: Conceptual Pathway Querying of Natural Logic Knowledge Bases from Text Bases. In: Larsen, H.L., Martin-Bautista, M.J., Vila, M.A., Andreasen, T., Christiansen, H. (eds.) FQAS 2013. LNCS, vol. 8132, pp. 1–12. Springer, Heidelberg (2013)
2. Fomichov, V.A.: The Formalization of Designing Natural Language Processing Systems. MAX Press, Moscow (2005) (in Russian)
3. Fomichov, V.A.: A Comprehensive Mathematical Framework for Bridging a Gap between Two Approaches to Creating a Meaning-Understanding Web. International Journal of Intelligent Computing and Cybernetics 1(1), 143–163 (2008)
4. Fomichov, V.A.: Semantics-Oriented Natural Language Processing: Mathematical Models and Algorithms. Springer, Heidelberg (2010a)
5. Fomichov, V.A.: Theory of K-representations as a Comprehensive Formal Framework for Developing a Multilingual Semantic Web. Informatica. An International Journal of Computing and Informatics 34(3), 387–396 (2010b)
6. Fomichov, V.A.: A Broadly Applicable and Flexible Conceptual Metagrammar as a Basic Tool for Developing a Multilingual Semantic Web. In: Métais, E., Meziane, F., Saraee, M., Sugumaran, V., Vadera, S. (eds.) NLDB 2013. LNCS, vol. 7934, pp. 249–259. Springer, Heidelberg (2013)
7. Harrington, B., Clark, S.: ASKNet: Creating and Evaluating Large Scale Integrated Semantic Networks. In: Proceedings of the 2008 IEEE International Conference on Semantic Computing, pp. 166–173. IEEE Computer Society, Washington DC (2008)
8. Harrington, B., Wojtinnik, P.-R.: Creating a Standardized Markup Language for Semantic Networks. In: Proceedings of the 2011 IEEE Fifth International Conference on Semantic Computing, pp. 279–282. IEEE Computer Society, Washington DC (2011)
9. Muskens, R.: An Analytic Tableau System for Natural Logic. In: Aloni, M., Bastiaanse, H., de Jager, T., Schulz, K. (eds.) Logic, Language and Meaning. LNCS, vol. 6042, pp. 104–113. Springer, Heidelberg (2010)
10. Rindflesh, T.C., Kilicoglu, H., Fiszman, M., Roszemblat, G., Shin, D.: Semantic MEDLINE: An Advanced Information Management Application for Biomedicine. Information Services and Use, vol. 1, pp. 15–21. IOS Press (2011)
11. Schubert, L.K., Hwang, C.H.: Episodic Logic meets little red riding hood: A comprehensive, natural representation for language understanding. In: Iwanska, L., Shapiro, S.C. (eds.) Natural Language Processing and Knowledge Representation: Language for Knowledge and Knowledge for Language, pp. 111–174. MIT/AAAI Press, Menlo Park, Cambridge (2000)
12. Wilks, Y., Brewster, C.: Natural Language Processing as a Foundation of the Semantic Web. In: Foundations and Trends in Web Science, vol. 1(3), Now Publ. Inc., Hanover (2006)

intelliGOV – Compliance Verification
of Service-Oriented Architectures with Ontologies
and Semantic Rules and Queries

Haroldo Maria Teixeira Filho[1], Leonardo Guerreiro Azevedo[1,2], and Sean Siqueira[1]

[1] UNIRIO – Federal University of State of Rio de Janeiro, Rio de Janeiro, Brazil
[2] IBM Research, Rio de Janeiro, Brazil
{haroldo.filho,azevedo,sean}@uniriotec.br, LGA@br.ibm.com

Abstract. Organizations are adopting Service-Oriented Architecture (SOA) to simplify system landscape, reduce costs and achieve deadlines. To accomplish these goals, it is necessary to ensure that the architecture and its evolution are compliant with business goals, best practices, legal and regulatory requirements. However, compliance verification of SOA is difficult due to the wide set of domains and the heterogeneity of the elements used to compose a service oriented solution. Although ontologies and rules could provide a solution for this problem, this approach cannot represent and verify a significant set of governance policies. Therefore, we propose intelliGOV, an architecture that gathers data from SOA environment, loads it in an ontology and uses semantic rules and queries to verify compliance. A case study conducted in a global energy company provides evidence of solution expressiveness, low coding demand and independence of methods and tools.

Keywords: Service-Oriented Architecture (SOA), Governance, Ontology, Semantic Rules, Semantic Queries.

1   Introduction

Service-Oriented Architecture (SOA) is an approach for building applications by discovering, invoking and composing distributed services [1]. It aims to reduce costs and schedules by developing applications through service composition [2]. Several authors [3-6] indicate the necessity of SOA Governance to obtain these benefits.

According to Janiesch et al. [6], SOA Governance is the establishment of structures, processes, policies and metrics to ensure the adoption, implementation, operation and evolution of SOA aligned with business objectives and compliant with laws, regulations and best practices.

Academia [3, 5, 7] and industry [8, 9] proposed models defining the necessary processes and components for SOA Governance. One common element in all these models is compliance. According to Tran [10], compliance defines the means to ensure that applications and services are built in an organization aligned with laws, regulations and best practices. The compliance verification activity measures this

H. Decker et al. (Eds.): DEXA 2014, Part I, LNCS 8644, pp. 402–409, 2014.

alignment, providing data to decide actions that increases compliance and reduces risks. Therefore, compliance in SOA context is an important issue.

Approaches for compliance verification [10–12] propose the use of models (e.g., ontologies) to represent SOA domain and rules to represent policies which describe the compliant behavior. However, existing drawbacks in the approaches are: (i) Demand for development of rule validation modules [12]; (ii) Code that is added into services to trigger non-conformity events (i.e., they are intrusive) [10]; or, (iii) Lack of operators in the rule representation language for policy representation [11].

We propose intelliGOV, a new architecture for compliance verification in SOA environments based on ontologies, semantic rules and semantic queries. Our proposal gathers data from SOA environment, load it into an ontology, and verify compliance through semantic rules and queries. IntelliGov architecture was implemented in this work employing a set of open-source software components for ontology handling and rule validation without the necessity of additional coding or intrusion in service implementations to trigger compliance events. The use of semantic queries allowed the representation of a large set of rules that the use of ontology and rules solely was not able to represent. We conducted a case study to evaluate the approach. The results indicated that the approach allows policies specification, and rule validation without errors, providing an assertive and precise description of the policies

The remainder of this work is organized as follows. Section 2 presents the related work. Section 3 presents intelliGov. Section 4 describes the case study used to show the solution in practice. Finally, Section 5 presents the conclusions and future work.

2   Related Work

Several authors address the compliance verification problem in SOA using semantic models and rules.

Zhou et al. [12] propose the use of the Service Modelling Language (SML) [13] and ISO Schematron [14]. SML is used to describe SOA elements (e.g., services and systems) while Schematron is used to describe policies as schema constraints. A set of tools verifies if the elements described in SML are valid considering the Schematron restrictions. This solution has the drawback of demanding the implementation of specific components and the extension of standards to deal with the non-capability of SML and Schematron to work with inference and to handle performance issues.

Tran et al. [10] propose the use of Model Driven Architecture (MDA) [15] for compliance verification. Services' code is generated from models that describe them. Policies are included in these models to represent governance constraints. The design is evaluated prior to code generation upon service creation and service adaption. It avoids the deployment of an invalid service. However, this solution can be applied only to services generated using this method. Besides, it is not able to verify compliances of services consumed from external providers, which service code is encapsulated from the consumers.

Some researches propose the use of ontologies and rules to deal with compliance verification, making easier the process of heterogeneous knowledge representations and interoperability between environments. Spies [11] proposes the use of ontologies to describe an IT governance domain and the use of axioms and rules in the ontology to describe policies. The author proposes the use of SWRL [16] as the rule language for policy description. It allows capturing the knowledge of the domains involved in SOA in a formal language and allows the use of inferences to validate compliance. Hence, it provides a precise specification and the possibility of policy interpretation by automated agents using existing software to handle ontologies and rule evaluation.

We contribute to this line of work, enhancing expressivity of policies by the use of semantic queries. This allows the representation of a larger set of policies, using more operators for data aggregation and grouping, which are provided by a query language. This representation also allows a formal specification of governance policies that can support an accurate implementation.

3   intelliGov – An Approach for Compliance Verification of Service-Oriented Architectures

The intelliGOV approach is based on a modular architecture that combines ontologies, semantic rules and semantic queries with an inference engine and data extraction tools to provide an accurate perspective of the compliance state of a SOA.

According to Gruber [17], an ontology is an explicit representation of a conceptualization composed by objects, concepts and other domain entities and their relationships. In intelliGOV, the use of an ontology leads to a unique vocabulary definition capable to represent formally and unambiguously the SOA domain concepts; thus, providing correct interpretation of data by human or computational agents. The ontology aims at reducing the necessity of experts to define or execute compliance verification. Another important point is the possibility to reuse and extend existing ontologies to reduce the efforts and errors in domain representation.

For policy definition, it is important to use a mechanism capable of dealing with concepts expressed in the ontology. The ontology axioms should be used to simplify the policy description aiming to reuse concepts. Finally, both the ontology axioms and the policy specification have to be interpretable by a machine to automate compliance verification; thus, reducing time and errors when executing this task.

In order to deal with these characteristics, we propose the use of rules and queries to describe governance policies. In this work, rules are assertions that can describe enterprise behavior and are able to express organizational business rules, company policies, external regulations and standards [18].

IntelliGOV contributes to other ontology-based approaches by aggregating queries for policy specification. The use of inference rules combined with queries enhances the expressiveness of the solution. It allows the representation of complex operations like aggregations and cardinality verification, which rules solely are not able to represent.

To evaluate compliance, data representing the SOA environment must be loaded or manually registered as instances of the ontology. It provides information for the compliance verification. In this context, intelliGOV provides data extraction tools for obtaining data from the SOA environment. Besides, there is a user interface module, which provides a user-friendly interface to input data, for policy specification, and to present compliance reports. Finally, we use open standards in the intelliGov implementation, through ontology manipulation, reasoning, and rule evaluation libraries, while queries are also reused. It reduces the necessity of specific code implementation.

IntelliGov has three components (Figure 1): (i) Data Extractor; (ii) User Interface (for data input and report generation); and (iii) Inference Engine. Domain knowledge and governance policies are stored as ontology, semantic rules and queries.

The *Data Extractor* module collects data describing the elements that compose the architecture from the tools that support the *SOA Environment* of the organization, such as *Service Bus*, *Service Registry*, *Artifact Stores*, and *Spreadsheets*. The *Data Extractor* converts the collected data into instances of the ontology that describes the SOA elements. The *SOA Ontology* contains classes and relationships describing the SOA domain, e.g. services, service contracts, systems and other architecture elements.

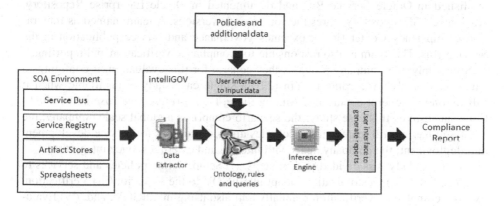

Fig. 1. intelliGOV architecture conceptual view

The *User Interface* provides mechanisms for *data input* and *report generation*. The user can create SOA governance policies as *semantic rules and queries*, and to input data that cannot be gathered from the *SOA Environment* as *Ontology* instances. One example of *Compliance Report* is a list of services and policies they are compliant with.

The *Inference Engine* is responsible to validate the policies considering the ontology instances in order to verify compliant behavior.

IntelliGov uses the Ontology Web Language (OWL) [20] for ontology description, SWRL [16] to describe policy rules and SQWRL [21] to express queries. All of these components are stored in OWL files.

The *User Interface* was implemented in Java and deployed to a JBOSS application server. The *Data Extractor* was implemented in Java, importing APIs for accessing *Service Bus* and repositories. These two modules use Protégé API to manipulate the ontology. Finally, Drools rule engine is accessed by means of Protégé API to execute the queries and validate rules in order to infer results.

4   Case Study

We conducted a case study to evaluate intelliGOV. According to Yin [22], a case study is an empirical approach that evaluates a phenomena in the environment where it occurs, especially when the limits between the phenomena and its context are not evident. In this work, the activity of compliance verification depends on human knowledge and the organizational context, making it hard to perform the evaluation in a controlled environment; thus, a case study is more suitable.

The case study was executed in a Brazilian energy company with global operations, engaged in a SOA initiative. The company has 118 services implemented by distributed teams in different technologies, such as, ABAP, Java, Lotus Notes, and .NET. Organization's applications and services also consume services from external providers, e.g., government agencies and travel companies. Organization services are published in Oracle Service Bus and documented in Oracle Enterprise Repository. A Clear CASE repository stores the source code versions. A team, named as Integration Competency Center (ICC), executes governance and service publication in the Service Bus. This team is also responsible for compliance verification and reporting.

Since only one company provides the scenario for this evaluation, it configures a single case study. According to Yin [22], a single case study is justifiable when it fulfills one of several conditions. Among these, *the representative case* is the condition that justifies this case study: the scenario comprises a global scale company; the company has several teams, methodologies and technologies for service development and deployment; the company deals with internal and external service integration.

The case study was divided in four steps: (i) Setup: define policies and services to be considered; (ii) Customization: adapt intelliGOV to the scenario; (iii) Verification: execute compliance verification manually and also using intelliGOV; and, (iv) Evaluation: analysis and discussion with the ICC team. To avoid biasing results during the execution of the case study, ICC members performed steps (i), (iii) and (iv).

In the first step, ICC members selected policies based on problems and challenges of the organization. To help the selection, a list of common processes for SOA governance [23] was presented and the ICC members selected the processes with high volume of problems, which were: service design and service version control. From this analysis, seven policies were defined. After that, ICC members selected a sample of 25 services from the organization and created a validation template with the expected compliance state for each service according to each policy. Finally, ICC members performed a manual evaluation, which resulted an average error rate of 21%, indicating the necessity of an automated solution.

In the second step, an ontology was modeled according to the elicited policies. To create the ontology, we used the 101 Methodology [24] and extended the Open Group SOA Ontology [19] to fully represent the domain of the organization. Afterwards, the policies were written as SWRL and semantic queries in SQWRL. We wrote rules aiming reuse and simplicity. To finish the customization of intelliGOV, specific *Data Extractors* were implemented to gather data from the *SOA Environment* (*Service Bus* and *Artifact Store*) and to create instances in the ontology.

We executed the compliance verification using intelliGOV. As a result, this verification was performed without errors.

Finally, an analysis of the results was executed, including a discussion with ICC members to identify the reasons for divergences. Some points were raised. The first point is the necessity to use a query language like SQWRL to complement SWRL rules to express governance policies. From the seven policies used in the case study, six demanded SQWRL operators that are not available in SWRL.

The second identified point regards the implementation. No services implementation details were required, except contract and WSDL information. In other words, intelliGOV can handle internal and external services, implemented in any platform.

The third point was the representation of policies using a formal description tool. Using the ontology as a vocabulary for the policies, the use of queries and rules provided an objective description of the rules. Doubts about the policies were answered during the policy specification in SQWRL resulting on policy description refinement. If an organization uses other tools for compliance verification, intelliGOV can act as a tool for specifying and testing the policies, using a formal description.

Finally, available open APIs were used to implement intelliGOV. These APIs are compatible with ontology management functions and inference. Custom code was required only for presenting details and determining parameters for rule execution.

5   Conclusion and Future Work

Compliance verification of SOA is an expensive and complex activity. One challenge is to correctly represent and check policies that describe compliant behavior. In this work, we proposed intelliGOV, an approach based on ontologies, semantic rules and queries that reduces complexity of compliance verification. OWL, SWRL and SQWRL were used to implement a software toolset according to intelliGov architecture aiming to increase interoperability and flexibility.

The contributions are an effective approach to describe and validate governance policies in SOA environments and a toolset that automates the verification process. This approach allows the representation of a wider set of policies without the necessity of code implementation for executing the compliance verification, based on open standards. It also helps to refine the policy descriptions, by means of a unified vocabulary and formalism provided by the rule and query language.

We executed a case study in a global energy company to evaluate the results of intelliGov. Company employees defined policies describing the expected behavior for processes, which were suggested by organization specialists. Afterward, these policies were applied to a set of services of the organization, considering wide technology diversity.

IntelliGOV correctly identified the compliance state. It allows the representation of complex policies which cannot be represented with solely a rule language, like SWRL. The solution can deal with any kind of service deployed in a service bus, independent of the provider.

Several authors [5, 7, 9] propose that SOA governance needs a cyclic implementation based on Plan-Do-Check-Act strategy. IntelliGOV deals with the *Check* part of this cycle. Future work could develop methods for policy identification (the plan stage), methods for translation of policies to rules and queries, and generation of Data Extractors for loading ontology data based on policy and Bus semantics (Do phase), and implementation of agents to react to identified non-conformities (the Act Phase).

References

1. Papazoglou, M.P., Traverso, P., Dustdar, S., Leymann, F.: Service-oriented computing: State of the art and research challenges. Computer 40, 38–45 (2007)
2. Erl, T.: Service-Oriented Architecture (SOA): Concepts, Technology, and Design. Prentice-Hall, Upper Saddle River (2005)
3. Niemann, M., Miede, A., Johannsen, W., Repp, N., Steinmetz, R.: Structuring SOA Governance. International Journal of IT/Business Alignment and Governance 1, 58–75 (2010)
4. Hsiung, A., Rivelli, G., Huttenegger, G.: How to design a global SOA infrastructure:Coping with challenges in a global context. In: Proceedings - 2012 IEEE 19th International Conference on Web Services, ICWS 2012, pp. 536–543 (2012)
5. Schepers, T.G.J., Iacob, M.E., Van Eck, P.A.T.: A lifecycle approach to SOA governance. In: Proceedings of the 2008 ACM Symposium on Applied computing, Fortaleza, CE, pp. 1055–1061 (2008)
6. Janiesch, C., Korthaus, A., Rosemann, M.: Conceptualisation and facilitation of SOA governance. In: Proceedings of: ACIS 2009: 20th Australasian Conference on Information Systems, Melbourne, pp. 154–163 (2009)
7. Hojaji, F., Shirazi, M.R.: A Comprehensive SOA Governance Framework Based on COBIT. In: 2010 6th World Congress on Services (SERVICES-1), pp. 407–414. Miami, FL (2010)
8. Bennett, S.G.: Oracle Practitioner Guide - A Framework for SOA Governance,
   http://www.oracle.com/technetwork/topics/entarch/
   oracle-pg-soa-governance-fmwrk-r3-2-1561703.pdf
9. The Open Group: SOA Governance Framework,
   https://www2.opengroup.org/ogsys/jsp/publications/
   PublicationDetails.jsp?catalogno=c093
10. Tran, H., Zdun, U., Holmes, T., Oberortner, E., Mulo, E., Dustdar, S.: Compliance in service-oriented architectures: A model-driven and view-based ap-proach. Information and Software Technology 54, 531–552 (2012)
11. Spies, M.: Continous Monitoring for IT Governance with Domain Ontologies. In: 2012 23rd International Workshop on Database and Expert Systems Applications (DEXA), pp. 43–47 (2012)
12. Zhou, Y.C., Liu, X.P., Wang, X.N., Xue, L., Tian, C., Liang, X.X.: Context model based SOA policy framework. In: ICWS 2010 - 2010 IEEE 8th International Conference on Web Services, pp. 608–615 (2010)
13. Pandit, B., Popescu, V., Smith, V.: Service Modeling Language, Version 1.1,
   http://www.w3.org/TR/sml/
14. Jelliffe, R.: The Schematron Assertion Language 1.6,
   http://xml.ascc.net/resource/schematron/Schematron2000.html

15. Mellor, S.J., Scott, K., Uhl, A., Weise, D.: Model-driven architecture. In: Bruel, J.-M., Bellahsène, Z. (eds.) OOIS 2002. LNCS, vol. 2426, pp. 290–297. Springer, Heidelberg (2002)
16. Horrocks, I., Patel-Schneider, P.F., Boley, H., Tabet, S., Grosof, B., Dean, M., et al: SWRL: A semantic web rule language combining OWL and RuleML. W3C Member submission 21, 79 (2004)
17. Gruber, T.R.: Toward Principles for the Design of Ontologies Used for Knowledge Sharing. Knowledge Acquisition 5, 199–220 (1993)
18. Bajec, M., Krisper, M.: A methodology and tool support for managing business rules in organisations. Information Systems 30, 423–443 (2005)
19. Service-Oriented Architecture Ontology, https://www2.opengroup.org/ogsys/protected/publications/viewDocument.html?publicationid=12245&documentid=11637
20. Hitzler, P., Krotzsch, M., Parsia, B., Patel-Schneider, P., Rudolf, S.: OWL 2 Web Ontology Language Primer, 2nd edn. (2012), http://www.w3.org/TR/2012/REC-owl2-primer-20121211/
21. O'Connor, M.J., Das, A.K.: SQWRL: A Query Language for OWL. In: OWLED (2009)
22. Yin, R.K.: Case study research: Design and methods. Sage (2009)
23. Teixeira Filho, H.M., Azevedo, L.G.: Governance of Service-Oriented Architecture through the CommonGov Approach. International Journal of Computer Information Systems and Industrial Management Applications 6, 505–514 (2014)
24. Noy, N.F., McGuinness, D.L.: Ontology development 101: A guide to creating your first ontology. Stanford knowledge systems laboratory technical report KSL-01-05 and Stanford medical informatics technical report SMI-2001-0880 (2001)

Early Classification on Multivariate Time Series with Core Features

Guoliang He[1,2], Yong Duan[1,2], Guofu Zhou[1,2], and Lingling Wang[1,2,3,*]

[1] State Key Laboratory of Software Engineering, Wuhan University, China
[2] College of Computer Science, Wuhan University, China
[3] College of Information, Huazhong Agricultural University, China
{Glhe,gfzhou,llwang}@whu.edu.cn, dydm_13128@163.com

Abstract. Multivariate time series (MTS) classification is an important topic in time series data mining, and has attracted great interest in recent years. However, early classification on MTS data largely remains a challenging problem. To address this problem, we focus on discovering hidden knowledge from the data for early classification in an explainable way. At first, we introduce a method MCFEC (Mining Core Feature for Early Classification) to obtain distinctive and early shapelets as core features of each variable independently. Then, two methods are introduced for early classification on MTS based on core features. Experimental results on both synthetic and real-world datasets clearly show that our proposed methods can achieve effective early classification on MTS.

Keywords: Multivariate time series, Early classification, Feature selection.

1   Introduction

Multivariate time series (MTS for short) are widely used in many areas such as speech recognition, anomaly detection of EEG/ECG data, science and engineering. For example, in an intensive care unit (ICU), Patient Monitoring can detect dynamically several physiological parameters, including respiration, ECG, blood pressure, body temperature and the saturation level of blood oxygen over a time interval.

MTS classification is an important problem in time series data mining. Because of its multiple variables and the possibility of different lengths for different components, MTS is difficult for traditional machine learning algorithms to address. Recently, many efficient models and techniques have been represented for the classification of multivariate time series, such as Recent Temporal pattern [1-2], SVMs [3-4], CLeVer [5-6], and metafeatures [7].

At the same time, early classification of time series, which means to classify time series data as early as possible provided that the classification quality meets the demand, is an interesting and challenging topic and has attracted a substantial amount of attention [8-9].

---

* Corresponding author.

H. Decker et al. (Eds.): DEXA 2014, Part I, LNCS 8644, pp. 410–422, 2014.

To the best of our knowledge, the problem of early classification on MTS data largely remains untouched except for [10]. In [10] Ghalwash *et al.* defined a multivariate shapelet, which is composed of multiple segments, and each segment is extracted from exactly one component. For this type of shapelet, different starting and ending positions are not allowed in a segment of each component. However, it has limited capacity to discover distinctive patterns since shapelets of interest often have different intervals for different variables.

In this paper we focus on discovering the internal relationships among variables and enhance the interpretability of early classification. We make several contributions. First, a method MCFEC (Mining Core Feature for Early Classification) is proposed to obtain core features for each variable independently. Second, we present a novel feature evaluation strategy that is more suitable for early classification. Last, two methods are introduced for early classification on multivariate time series based on the core features of all variables.

The remainder of this paper is organized as follows. In Section 2, we review the problem definition and related work. Section 3 introduces concrete algorithms to efficiently mine core features. Section 4 discusses two methods of early classification using these core features as a tool. In Section 5, we perform a comprehensive set of experiments on various problems of different domains. Finally, we conclude our work and suggest directions for future work in Section 6.

2   Background and Related Work

2.1   Background

In this section, we define shapelets and the notations used in this paper.

Definition 1. Univariate time series: a univariate time series $s=t_1,t_2,....,t_L$ is an ordered set of L real-valued readings. For instance, a univariate time series $s_0=(1.2, 2.2, 3,6, 1.3, 5.3, 7.1)$.

Definition 2. Multivariate time series: a multivariate time series is a vector of sequences $X=(x_1, x_2,..., x_T)$, where each component x_j $(1\leq i\leq T)$ is a univariate time series, and the lengths of different components might not be equal.

The MTS object X has T variables, and the corresponding component of the ith variable is x_i $(1\leq i\leq T)$.

For the sake of simplicity, we will use the word "time series" as univariate time series, which is a component of a multivariate time series.

Definition 3. Subsequence: Given a time series s of length L, s_sub= $s[m, m+n-1]$, is a subsequence of length $n<L$ that has a contiguous position from s starting at the mth position and ending at $(m+n-1)^{th}$ position, in other words, s_sub=$t_m,...,t_{m+n-1}$ for $1 \leq m \leq L-n+1$.

Definition 4. Similarity degree: For two time series b and s (assuming that $|b| \leq |s|$), the similarity degree between b and s is calculated by Sim$(b,s)=$ min$\{$dist(b,s_i)$\}$, where s_i is any subsequence of a time series s with $| s_i |=|b|$, and dist(b,s_i) is Euclidean distance between b and s_i.

From this definition, we can see that the smaller the value of Sim (*b,s*) is, the higher the similarity degree between *b* and *s*.

Definition 5. Shapelet/Feature: A shapelet (feature) is a time series subsequence that is representative of a class. Informally, a shapelet (feature) p = (b,δ, C), where b is a subsequence, δ is a threshold, and C is a class label. An unlabeled time series object s is considered to be matching a shapelet p and is labeled as the class C if Sim(b,s) ≤ δ.

For simplicity, later Sim(s, p) is used to represent the similarity degree between the shapelet p and the time series s. Class(*s*) means the class of a time series *s*.

Definition 6. Precision: Given the time series data D and a feature p = (b,δ, C), the precision of p is the ratio of the number of samples that have the class label C and match the feature p and the number of samples that could match the feature p in data D.

$$\text{Precision(p)} = \frac{\|\{ s|\ \text{Sim(s,p)}<\delta\ \ \text{class(s)}=C)\}\|}{\|\{ s|\ \text{Sim(s,p)}<\delta\}\|}, s \in D.$$

Definition 7. Recall: Given time series data D and a feature p = (b,δ, C), the recall of p is the ratio of the number of samples that have the class label C and match the feature p and the number of samples that have the class label C in data D.

$$\text{Recall(p)} = \frac{\|\{ s|\ \text{Sim(s,p)}<\delta\ \ \text{class(s)}=C)\}\|}{\|\{ s|\ \text{class(s)}=C\}\|}, s \in D.$$

Definition 8. Earliness: Given time series data D and a shapelet p = (b,δ, C), the minimal identifiable length (MIL for short) of a time series *s*∈ D MIL(s,p) means that p could classify this sample by scanning its subsequence from the beginning to the position MIL(s, p).The formal definition of MIL(s, p) is the following.

MIL(s,p) = $\arg\min_{|p|\leq i\leq|s|} dist(s[i - |p| + 1, i], p) \leq \delta$, where $s \in D' = \{s|\text{Dist}(s, p) < \delta, s \in D\}$ and s[*i*–|p|+1,*i*] is a subsequence of s starting at the (*i*–|p|+1)th position and ending at *i*th position.

Then, the earliness of p is

$$\text{Earliness(p)} = \sum_{s\in D'}\left(1-\frac{MIL(s)}{|s|}\right)/\|D'\|.$$

MIL(s) is equal to the minimum of all MIL(s, p) for all shapelets. The earliness degree of times series sample s is (1-MIL(s)/|s|) and that of the data D is the average earliness degree of all samples in D.

Definition 9. Accuracy: Given a classifier M to early classify a multivariate time series D, the accuracy of this classifier M is

Accuracy(*M*) = *NM*/‖D‖, where *NM* means the number of objects that are classified correctly by the classifier *M*.

2.2   Related Works

Time series classification is an important problem in time series data mining and has attracted great interest in recent years [11-15]. To mine the hidden knowledge in the time series data, Ye *et al.* [16] introduced a novel primitive for time series and shape mining, time series shapelets, which could gain some insights into the data and make the classification result more explainable.

For early classification, a method called Early Classification on Time Series (ECTS) [9] was proposed to explore the stability of the nearest neighbor relationship in the full space and in the subspaces formed by prefixes of the training examples. To overcome the drawback of ECTS that could not extract and summarize useful information for users, local shapelets [8] were introduced and extracted as features that distinctly manifest the target class locally and early. However, these methods for univariate time series are not suitable for early classification of MTS.

To resolve the issue of MTS, Ghalwash et al. [10] defined multivariate shapelets and proposed a method called Multivariate Shapelets Detection (MSD) for early classification. For a multivariate shapelet, a segment of different variables should be extracted in the same sliding time window. For instance, a multivariate shapelet is shown in Figure 1(a). Therefore, it is incapable of discovering distinctive shapelets of each variable because they are generally in different intervals. A multivariate shapelet could not include distinctive shapelets of all variable unless its length is sufficiently long. When a shapelet is too long, it would not be able to classify the data as early as possible, and the classification costs a substantial amount of storage space and computation time.

Now, we give an example to better illustrate this issue.

Example 1. Four 4-variable time series objects that belong to a class are illustrated in Figure 1(b). To highlight shapelets in time series, symbols "*" are marked in shapelets. We can see that shapelets (starlike curves) of the 1st, 3rd and 4th component are in the time period 1-7, 1-5 and 7-12, respectively, while two shapelets of the 2nd component are in the time period 4-8 and 16-20. Because interesting features of each component are in different intervals, MSD method could not obtain distinctive shapelets. Since multivariate shapelets cannot present non-redundant distinctions of MTS data whether their lengths are long or short, its interpretability is not very good.

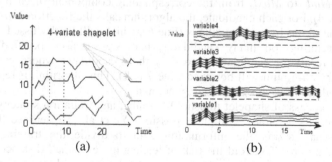

(a)   (b)

Fig. 1. Shapelets of MTS objects. A multivariate shapelet extracted by the MSD method is shown in (a). Shapelets (star-like curves) of each variable of MTS data are shown in (b).

3   Mining Core Features from MTS

In this section, we propose an approach called MCFEC, which mines core features in two distinct stages. First, an algorithm conducts a single scan of the MTS training data to extract all shapelet candidates for each variable. Second, a subset of distinctive shapelets is selected from these shapelet candidates as core features; this subset will be used to build a classifier for early classification in the next section.

3.1   Feature Extraction

The first step in MCFEC extracts shapelet candidates from the training data set. This step takes two parameters: *minL* and *maxL*. Then, it extracts all subsequences of length between *minL* and *maxL* from every training sample as candidates. Here, we adopt a general method for extracting candidates from each component of the MTS data, as shown in Algorithm 1.

```
Algorithm 1: FeatureExtraction
Input: the training data D, a variable t
Output: a set of shapelet candidates for the variable t
1:   output=φ
2:   for each MTS object s in D
3:       for i=0 to length(sₜ)-MinL
//sₜ is the corresponding component of the variable t in
s
4:         for L = MinL to MaxL
5:           candidate = Generatecandidate (sₜ, i, L)
6:           threshold = Learnthreshold (candidate, D)
7:           if (candidate satisfies conditions)
8:               output.add (candidate, threshold)
9:           else eliminate (candidate)
10:      end for
11:    end for
12:  end for
13:  return output
```

Given a variable t, algorithm 1 first generates all possible shapelet candidates with lengths from *MinL* to *MaxL* from the corresponding component of each object s in dataset D (line 4). For each candidate, the algorithm calls the function *Learnthreshold* to obtain its threshold by some methods (line 6), which will be discussed in the next. To avoid generating too many candidates, here we set user-specified minimum precision and recall for candidates, and only those candidates whose precision and recall above the minimum will be saved (line 7 to 9). This algorithm is repeated until shapelet candidates of each variable are generated.

During the process of shapelet threshold learning, an evaluation strategy is usually used to improve its accuracy as high as possible. Ye et al. [16] learned the distance threshold by maximizing the information gain to divide the training data into relatively pure groups. To avoid the risk of leading to very small distance thresholds for features and thus overfitting the training data, Xing et al. [8] adopted Kernel density estimation and Chebyshev's inequality to determine the threshold of a feature, which belongs to the specified class with a high probability.

However, the sensitivity of a shapelet, a very important characteristic during the process of feature threshold learning, is often overlooked. To enhance the precision and sensitivity of a shapelet f simultaneously, we adopt the F-measure method as the following to evaluate to f during the process of learning a threshold.

F-measure(f) $= 2/(1/\text{Precision}(f) + 1/\text{Recall}(f))$

A distance threshold is obtained when the shapelet reaches the highest quality.

Example 2. For the dataset D $= \{s_1(+), s_2(+), s_3(+), s_4(-), s_5(-), s_6(+)\}$, our aim is to learn a threshold of a subsequence b_1, which is assumed to be extracted from a positive sample. First, we calculate the similarity degree between the subsequence b_1 and each sample in D as follows: $Sim(b_1, s_1)=d_1$, $Sim(b_1, s_2)=d_2$, $Sim(b_1, s_3)=d_3$, $Sim(b_1, s_4)=d_4$, $Sim(b_1, s_5)=d_5$ and $Sim(b_1, s_6)=d_6$. Suppose that $d_{min}= Min\{d_1,d_2,d_3,d_4,d_5,d_6\}$ and $d_{max}=Max\{d_1,d_2,d_3,d_4,d_5,d_6\}$; the threshold of b_1 is between d_{min} and d_{max}. Using any possible value within the scope as the temporary threshold, we could further calculate the precision and recall of b_1. Finally, we select a temporary threshold with which b_1 reaches the highest quality as a final threshold of b_1. Because there are an unlimited number of values in the scope, it is impossible to evaluate the quality of a feature using all of these possible thresholds. To simplify this process and not affect the performance in terms of the quality of a feature, we select here only the means of the two nearest values to be the alternative thresholds. Suppose that $d_1<d_2<d_3<d_4<d_5<d_6$; then, the alternative thresholds are $(d_1+d_2)/2$, $(d_2+d_3)/2$, $(d_3+d_4)/2$, $(d_4+d_5)/2$ and $(d_5+d_6)/2$.

3.2   Feature Selection

A good shapelet is critical for early classification. Because of the specific nature of time series, recently some novel approaches have been proposed to select distinctive shapelets for classification, such as information gain [16] and Greedy strategy [8], among others. However, a single class is often composed of various sub-clusters or sub-concepts [17], which is popular in many classification problems. Because the number of samples in each sub-cluster is often different, shapelets of a non-dominate sub-cluster could be easily overlooked. This within-class imbalance is implicit in most cases, and the above methods cannot effectively handle this issue. Therefore, selecting distinctive and stable features is critical in the process of building an effective classifier. However, few existing feature selection methods of time series touched this challenging problem.

To deal with this issue and obtain stable and distinctive shapelets, we first divide shapelet candidates of each variable with a class label into several clusters using the Silhouette Index (SI) method [18], which is based on a compactness-separation measurement. Each shapelet candidate is dynamically adjusted to its nearest cluster according to its Silhouette Index, which is defined as the following:

$SI(i) = (b(i)-a(i))/max\{b(i), a(i)\}$

where *a(i)* is the average distance between the ith sample and each of the same-cluster samples, while *b(i)* is the minimum average distance between the ith sample and each of the different-cluster samples. To obtain a better clustered dataset, this regulation process is repeated until the sum of the Silhouette Indexes of all of the data has converged.

Next, a core feature is selected to present each cluster by ranking their qualities with the GEFM method, which will be represented in Section 3.3. The process of selecting core features is shown in algorithm 2.

Algorithm 2: FeatureSelection
Input: all shapelet candidates (G) of a variable t with a class label
Output: core features
1: output=ϕ

```
2: Shapelet candidates G are divided into m groups G₁,
G₂, ... , Gₘ using SI clustering method
3: for each group Gᵢ
4:  A distinctive shapelet with the highest quality in
Gᵢ is selected as one of the core features
5: end for
6: return
```

This algorithm is repeated for each variable until core features of all variables are selected. In the process of feature selection, shapelet candidates are first clustered in terms of their similarity instead of selecting top-k shapelets directly according to their qualities, and then, core features are collected from each cluster, including some rare clusters. Therefore, this approach considers some distinctive and rare shapelets and can effectively solve the problem of sub-concepts.

3.3   Quality of Feature

During the process of feature selection, a criterion is required to rank the shapelets in each cluster. Since our purpose is to predict the target class of an unlabeled object as early as possible, the evaluation measure should consider the earliness of shapelets. Here, we propose a generalized extended F-measure (GEFM for short) to evaluate the quality of shapelets in terms of the recall, precision and earliness. Moreover, to efficiently handle imbalanced training data and to improve the sensitivity of the classifier in the end, a weight parameter is added to each property. The quality of a shapelet f is

$$GEFM(f) = 1/(w_0/Earliness(f)+w_1/Precision(f)+w_2/Recall(f)).$$

Here, w_0, w_1, and w_2 are weights to control the importance of the earliness, precision and recall. In this paper, we focus on influencing factors that affect the selection strategy of core features. Therefore, for simplicity, each weight parameter is equal to 1. Further analysis of these weights was discussed in [19].

4   Early Classification on MTS

From Figure 1(b), we see that the same-class MTS data have single or even several core shapelets for a variable. However, not all combinations of core shapelets are useful rules for classification. On the one hand, because there is usually some redundant information in the data, we could predict the target class of a MTS object when only some of components match core shapelets with a class label. On the other hand, some deviating values, for instance, which are caused by measurement errors or environmental changes, are unavoidable in the real datasets. If these deviating values of the data occur in the interval where the internal characteristics appear simultaneously, the distribution of this period could not match a core shapelet very well.

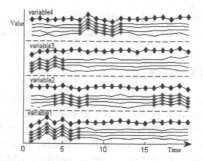

Fig. 2. Core shapelets (star-like) and MTS data (diamond-like curve)

For example, the same-class core shapelets (star-like curves) are illustrated in Figure 2, and our aim is to predict an unlabeled object (a diamond-like curve). From this figure, we can see that the first, second and the fourth component match the corresponding core shapelets very well, whereas the third component does not, which is caused by some errors.

To predict the class of unlabeled MTS objects early and effectively, in this section we introduce two methods.

4.1   MCFEC-Rule Classifier

To discover the hidden knowledge among all variables, in this subsection, we use core shapelets to find coherent rules for early classification from the training data. These rules are explainable and could reflect the relevant connection among the components of the MTS data. Each rule is composed of one or more core shapelets that belong to a class but includes at most one core shapelet of each variable. Because there are too many combinations of core features, we expect to obtain an optimized subset for classification. Here, we continuously select the best rule from the remaining combinations in terms of their qualities until these chosen rules could cover the training data.

To ensure its effectiveness, we only mine those rules whose precision and recall are no less than a user-specified minimum value.

For example, we want to find coherent rules from positive samples in the dataset D (having four variables). Suppose that the first variable has two core shapelets while others has only one; and the core shapelets of the $1^{st}, 2^{nd}, 3^{rd}$ and 4th variable are f_{11} and f_{12}, f_2, f_3 and f_4. Therefore, there are 21 possible rules as following: rule1: f_{11}, rule2: f_{12}, rule3: f_2, rule4: f_3, rule5: f_4, rule6: $f_{11} \wedge f_2$, rule7: $f_{12} \wedge f_2$, rule8: $f_{11} \wedge f_3$, rule9: $f_{12} \wedge f_3$, rule10: $f_{11} \wedge f_4$, rule11: $f_{12} \wedge f_4$, rule12: $f_2 \wedge f_3$, rule13: $f_2 \wedge f_4$, rule14: $f_3 \wedge f_4$, rule15: $f_{11} \wedge f_2 \wedge f_3$, rule16: $f_{12} \wedge f_2 \wedge f_3$, rule17: $f_{11} \wedge f_2 \wedge f_4$, rule18: $f_{12} \wedge f_2 \wedge f_4$, rule19: $f_2 \wedge f_3 \wedge f_4$, rule20: $f_{11} \wedge f_2 \wedge f_3 \wedge f_4$, rule21: $f_{12} \wedge f_2 \wedge f_3 \wedge f_4$. In order to improve the efficiency of classification, we first rank these rules in terms of the precision and recall. Finally, the best rule is selected from the remaining rules repeatedly until the choose rules cover the positive samples in the dataset D. For instance, suppose that rule 12 is selected in the process of rule mining; for an unlabeled object B, the second component matches f_2 at time point 21, and the third component matches f_3 at the time point 27. So the object B matches rule12 ($f_2 \wedge f_3$) and its target class is labeled positive at time point 27. When there are other rules matches this object earlier than rule12, the object B could be identified earlier. Our aim is to classify an unlabeled object as early as possible.

4.2   MCFEC-QBC Classifier

Instead of rule discovery, we present here another classification method, which is based on query by committee (QBC) [20] to classify MTS data as early as possible. Based on the core features of each variable, we classify each component of a sample and count which class is in the majority. Finally, this sample is labeled as the predominant class. More concretely, given an unlabeled MTS object, if a component of this object matches any core feature of its corresponding variable, then this component is labeled as the target class of this matched feature. This process runs concurrently for each component until the number of components that are labeled as being in the same class is in the majority. Then, this MTS object is labeled the same as the target class of the preponderant components. If all of the components are scanned from the beginning to the end and no target class is in the majority, the object is labeled as being in any class, randomly.

5   Experimental Evaluation

In this section, we empirically study the proposed methods for the early classification of MTS. All experiments follow a 10-fold cross-validation scheme, and each runs 5 times. Finally, we take an average result for each experiment. To eliminate classification bias, the features are always selected from the training sets; in other words, features extracted in different folds could be different. All of the experimental results are obtained by using a PC computer with Intel Xeon 2.5GHz CPU and 4GB of main memory. The parameter settings are the same, $MinL = 5$, $MaxL = L/3$, where L is the full length of the time series. For user-specified minimum value of precision and recall, 0.51 and 0.10 respectively is good enough to prune ineffective shapelets or rules while keeping the accuracy of the classification.

The algorithms are implemented in C++ using Microsoft Visual Studio 2010.

5.1   Datasets

To evaluate the ability of our method, here we first generate two synthetic MTS datasets using the probability distribution and joint probability distribution for early classification. The first dataset contains 2 classes and 3 variables, and the length of each sample is 120. The second synthetic dataset includes 3 classes and 4 variables, and its length is 150.

For the real-world MTS datasets, we use here the Wafer and ECG dataset [21] to verify the performance of our proposed methods.

There are 1194 samples in Wafer dataset. The length of a Wafer sample is between 104 and 198. The ECG dataset contains 200 MTS samples, and the length of each sample is between 39 and152.

Table 1 shows the summary of all of the datasets used in the experiments.

Table 1. Summary of datasets used in the experiments

	Syn 1	Syn 2	Wafer	ECG
Num of labels	2	3	2	2
Num of variables	3	4	6	2
Max length	120	150	198	152
Min length	120	150	104	39
Num of samples	1000	900	1194	200

5.2   Analysis of the Feature Selection Method

In Section 3, we defined a method called MCFEC to mine core features for early classification. In this process, feature selection method is the most crucial step. To demonstrate the utility of this technique, our second experiment is designed to verify the validity of our proposed SI clustering-based feature selection method by comparing with a greedy method, which is often used to select the best feature from the remaining candidates iteratively in terms of their qualities until these selected distinctive features cover the training data. Here, we perform experiments on four datasets to analyze the function of classification using both feature selection methods. To be fair, we use a classification method that is similar to MCFEC-Rule except that different feature selection methods are used. The classification results of the accuracy(Acc), and earliness(Eln) using different feature selection methods are shown in Table 2. From this Table we could see that the accuracy and the earliness of the SI-clustering method are higher than that of the greedy method.

Table 2. Summary results of classification using two feature selection methods on data sets

	Syn 1		Syn 2		Wafer		ECG	
	Acc	Eln	Acc	Eln	Acc	Eln	Acc	Eln
SI-Clust	0.98	0.78	0.74	0.47	0.97	0.73	0.78	0.74
Greedy method	0.98	0.77	0.68	0.43	0.93	0.68	0.76	0.61

5.3   Performance of the Feature Evaluation Strategy

To analyze the effectiveness of our proposed feature evaluation strategy, here we compare GEFM with F-score method (FSM) without considering the earliness of a feature f as following.

$$\text{FSM } (f) = 1/(1/\text{Precision}(f) + 1/\text{Recall}(f)).$$

We do experiments on four datasets to analyze the function of classification by both feature evaluation methods. To be fair, here we use a classification method similar to MCFEC-Rule except different feature evaluation methods.

The classification accuracy and earliness with GEFM feature evaluation method on the first synthetic dataset are 0.98 and 0.78, which are equal to that of FSM evaluation method (0.98 and 0.78). For the second synthetic dataset, though the accuracy of GEFM (0.74) is lower than that of FSM (0.83), the differences of the earliness between two evaluation methods are also obvious (the earliness of GEFM is 0.47, and that of FSM is 0.44).

For two real datasets, the classification accuracy of GEFM (0.97 and 0.78, respectively) also approximates to that of FSM method (0.99 and 0.81, respectively), while the classification earliness of GEFM (0.73 and 0.74, respectively) is much higher than that of FSM method (0.60 and 0.49, respectively).

A comparison of the classification accuracy (Acc) and earliness (Eln) on four datasets are shown in Figure 3 and Figure 4 respectively. As a consequence, the GEFM evaluation method could improve the earliness of classification without sacrificing the classification accuracy in a certain degree.

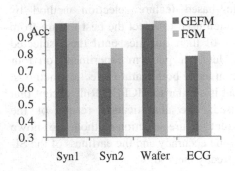 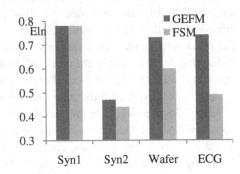

Fig. 3. The comparison of the classification accuracy with two feature evaluation methods

Fig. 4. The comparison of the classification earliness with two feature evaluation methods

5.4   Comparison with the MSD Method

The last experiment analyzes the function of our two proposed classification methods (MCFEC-Rule and MCFEC-QBC, which we call MCFECs) against the MSD method [10] on four datasets, and the results are listed in Table 3.

For the Syn1 and Syn2 data, the MSD has a little higher earliness than two MCFECs. However, its accuracy is much lower than that of two MCFECs. For the Wafer data, the earliness of both MCFECs is a much higher than that of MSD as well as the accuracy. For the ECG data, both MCFECs have a higher earliness than MSD, although their accuracy is slightly lower than that of MSD. Because our aim is to optimize earliness under the constraint of minimum accuracy, both MCFEC-QBC and MCFEC-Rule are competitive in comparison with MSD in terms of their accuracy and earliness.

Table 3. Summary results of MCFECs and MSD on the data sets

	Syn 1		Syn 2		Wafer		ECG	
	Acc	Eln	Acc	Eln	Acc	Eln	Acc	Eln
MCFEC-QBC	0.99	0.43	0.77	0.53	0.90	0.77	0.77	0.76
MCFEC-Rule	0.98	0.78	0.74	0.47	0.97	0.73	0.78	0.74
MSD	0.74	0.68	0.34	0.75	0.74	0.35	0.74	0.59

6   Conclusions and Future Work

To address the problem of early classification on MTS, we introduce and extract core features from multivariate time series, and we develop two simple classification methods for early classification. The experimental results show clearly that core features extracted by our methods are interpretable and can achieve effective early classification.

In reality, collecting lots of labeled time series is highly time consuming and cost lots of human effort and money. Meanwhile, unlabeled time series samples may be relative easy to collect, but how to use them for early classification is a challenge. We plan to research early classification on partially labeled training data in future.

References

1. Batal, I., Sacchi, L., Bellazzi, R., Hauskrecht, M.: Multivariate time series classification with temporal abstractions. In: Proceedings of the Twenty-Second International FLAIRs Conference (2009)
2. Batal, I., Fradkin, D., Harrison, J., et al.: Mining recent temporal patterns for event detection in multivatiate time series data. In: KDD 2012 (2012)
3. Orsenigo, C., Vercellis, C.: Combining discrete SVM and fixed cardinality warping distances for multivariate time series classification. Pattern Recognition 43, 3787–3794 (2010)
4. Chandrakala, S., Chandra Sekhar, C.: Classification of varying length multivariate time series using Gaussian mixture models and support vector machines. Int. J. of Data Mining, Modeling and Management 2(3), 268–287 (2010)
5. Yoon, H., Yang, K., Shahabi, C.: Feature subset selection and feature ranking for multivariate time series. IEEE Transactions on Knowledge and Data Engineering 17(9), 1186–1198 (2005)
6. Yang, K., Yoon, H., Shahabi, C.: CLeVer: A Feature Subset Selection Technique for Multivariate Time Series. In: Ho, T.-B., Cheung, D., Liu, H. (eds.) PAKDD 2005. LNCS (LNAI), vol. 3518, pp. 516–522. Springer, Heidelberg (2005)
7. Kadous, M.W., Sammut, C.: Classification of multivariate time series and structured data using constructive induction. Machine Learning 58, 179–216 (2005)
8. Xing, Z., Pei, J., Yu, P.S., Wang, K.: Extracting interpretable features for early classification on time series. In: SDM 2011 (2011)
9. Xing, Z., Pei, J., Yu, P.S.: Early prediction on time series: a nearest neighbor approach. In: IJCAI (2009)

10. Ghalwash, M.F., Obradovic, Z.: Early classification of multivariate temporal observations by extraction of interpretable shapelets. BMC Bioinformatics (August 2012)
11. Lines, J., Davis, L.M., Hills, J., Bagnall, A.: A shapelet transform for time series classification. In: KDD 2012 (2012)
12. Wei, L., Keogh, E.: Semi-supervised time series classification. In: KDD 2006 (2006)
13. Nguyen, M.N., Li, X.-L., Ng, S.-K.: Positive unlabeled learning for time series classification. IJCAI 2011 (2011)
14. Xi, X., Keogh, E., Shelton, C., Wei, L.: Fast time series classification using numerosity reduction. In: ICML 2006 (2006)
15. Jeong, Y.-S., Jeong, M.K., Omitaomu, O.A.: Weighted dynamic time warping for time series classification. Pattern Recognition 44, 2231–2240 (2011)
16. Ye, L., Keogh, E.: Time series shapelets: a new primitive for data mining. In: KDD 2009 (2009)
17. Sun, Y., Kamel, M.S., Wong, A.K.C., Wang, Y.: Cost-sensitive boosting for classification of imbalanced data. Pattern Recognition 40, 3358–3378 (2007)
18. Rousseeuw, P.J.: Silhouettes: a Graphical Aid to the Interpretation and Validation of Cluster Analysis. Computational and Applied Mathematics 20, 53–65 (1987)
19. He, G., Duan, Y., et al.: Early Prediction on Imbalanced Multivariate Time Series. In: CIKM 2013 (2013)
20. Seung, H.S., Opper, M., Sompolinsky, H.: Query by committee. In: COLT 1992 Proceedings of the Fifth Annual Workshop on Computational Learning Theory,, pp. 287–294 (1992)
21. http://www.cs.cmu.edu/~bobski/

Semantically Consistent
Human Motion Segmentation

Michal Balazia, Jan Sedmidubsky, and Pavel Zezula

Masaryk University, Botanicka 68a, 602 00 Brno, Czech Republic
{xbalazia,xsedmid,zezula}@fi.muni.cz

Abstract. The development of motion capturing devices like Microsoft
Kinect poses new challenges in the exploitation of human-motion data
for various application fields, such as computer animation, visual surveillance, sports or physical medicine. In such applications, motion segmentation is recognized as one of the most fundamental steps. Existing methods usually segment motions at the level of logical actions, like
walking or jumping, to annotate the motion segments by textual descriptions. Although the action-level segmentation is convenient for motion
summarization and action retrieval, it does not suit for general action-
independent motion retrieval. In this paper, we introduce a novel semantically consistent algorithm for partitioning motions into short and
further non-divisible segments. The property of semantic consistency ensures that the start and end of each segment are detected at semantically equivalent phases of movement to support general motion retrieval.
The proposed segmentation algorithm first extracts relative distances between particular body parts as motion features. Based on these features,
segments are consequently identified by constructing and analyzing a
one-dimensional energy curve representing local motion changes. Experiments conducted on real-life motions demonstrate that the algorithm
outperforms other relevant approaches in terms of recall and precision
with respect to a user-defined ground truth. Moreover, it identifies segments at semantically equivalent phases with the highest accuracy.

1 Introduction

Human movement activities such as walking, jumping or exercising, can be
recorded by various motion capturing devices based on a system of synchronized
video cameras, like Vicon, ASUS Xtion or Microsoft Kinect. These devices can
online digitize human movement into so-called *motion capture data*. The motion
capture data model human movements using simplified stick figures of human
skeletons. The stick figure consists of bones that are connected by *joints*. The
positions of joints are estimated for each video frame in form of 3D coordinates.
An example of spatio-temporal motion capture data is illustrated in Figure 1.

The motion capture data have a great potential to be utilized in sports to
compare performance aspects of athletes, in law-enforcement to identify special-
interest persons [9,24], in health care to determine the success of rehabilita-
tive treatments [6], or in computer animation to synthesize credible motions

H. Decker et al. (Eds.): DEXA 2014, Part I, LNCS 8644, pp. 423–437, 2014.

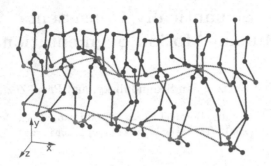

Fig. 1. Motion capture data in seven consecutive video frames. Each frame illustrates a stick figure of a walking human by 3D coordinates of 17 joints. The red and blue lines only emphasize the trajectories of movements of hands and feet.

for production of high-quality games or movies [10]. These applications require to process motion data effectively and efficiently. To do that, they usually exploit specialized techniques for annotating segments of motions by textual descriptions [13,21,22], retrieving (sub-)motions that are similar to a query motion [4,19,23,25,26,27], generating new instances of visually appealing motions by concatenating segments of existing ones [1,13], or for summarizing and compressing motions [15].

Focusing on the implementation of above-mentioned techniques, *motion segmentation* is recognized as one of the most fundamental steps. The segmentation process partitions a recorded motion into a sequence of consecutive segments (sub-motions) that can be separately analysed, indexed, clustered, and compared on the basis of similarity. The way the recorded motions are segmented depends on the purpose of a target application. This brings various levels of segmentation – from the low level of *gestures* (e.g., "raising left arm"), through the middle level of *actions* (e.g., "cartwheel"), to the high level of *activities* (e.g., "gymnastic performance"). Considering segments as non-divisible parts, action-level segmentation methods [13,21,22] are convenient for motion annotation, summarization or action retrieval. However, these methods do not suit for general action-independent motion retrieval since they cannot retrieve such (sub-)motions that only consist of parts of segmented actions. To deal with this problem, we focus on segmentation at the level of gestures which are already perceived as fairly short movements to comply with any query motion.

Motion retrieval applications based on the idea of gesture-level segmentation can primarily benefit from much more efficient segment-level indexing, rather than indexing at the level of individual frames. Nevertheless, segment-level indexing requires a query motion and all similar database (sub-)motions to be segmented in the same way. In other words, segmentation cuts of two similar motions (e.g., a query motion and a similar database sub-motion) have to be identified at *semantically equivalent* phases of movement. For example, considering instances of walking motions, then segmentation cuts detected at moments of passing feet in all walk cycles correspond to semantically equivalent phases of movement. Such *semantically consistent* segmentation guarantees with a high probability that segments of similar sub-motions are identified at equivalent phases of movement disregarding their positions within parent motions.

In this paper, we introduce a semantically consistent segmentation algorithm for partitioning motions into gesture-level segments in which body part movements register a monotonous trend. Taking motion capture data as input, characteristic features in form of relative distances between specific body parts are extracted to represent a motion as a sequence of poses. Local neighbourhoods of individual poses are explored to construct a one-dimensional energy curve that represents significance of local changes in motion. By analyzing this curve, poses of locally maximal energy are selected as key poses, partitioning the processed motion into gesture-level segments. The selected key poses are detected at semantically equivalent phases of movement to support a more efficient development of techniques for motion retrieval, generation or summarization.

The main contributions of this paper are (1) the definition of the property of semantic consistency and (2) the proposal of a semantically consistent algorithm for segmenting motions at the level of short gestures. Moreover, we demonstrate that the proposed segmentation algorithm outperforms other three relevant approaches [2,8,28] in terms of recall and precision of identified key poses with respect to a user-defined ground truth. We also experimentally verify that our algorithm generates key poses at semantically equivalent phases of movement with the highest accuracy.

The rest of the paper is organized as follows. Section 2 describes a broad area of different segmentation methods for motion capture data and introduces relevant gesture-level approaches in more detail. Section 3 mathematically defines the motion data representation and segmentation at semantically equivalent phases of movement. Section 4 presents our segmentation algorithm which is experimentally evaluated and compared with other approaches in Section 5.

2   Related Work

Existing segmentation methods can be generally divided into learning and non-learning approaches. Learning-based algorithms [14,17,18,22] require training data in form of positive and negative examples to be able to segment new motions. In particular, the technique of genetic algorithms [18,22] is commonly utilized to segment motions based on prior knowledge obtained from analysis of training data. However, such supervised segmentation is not applicable to motions not included in the training data. Such limited extensibility is the main reason we do not concentrate on learning-based approaches in the following.

The non-learning approaches [2,3,8,13,16,28] usually extract characteristic motion features, such as distances between joints, bone angles and joint velocity or acceleration, to represent a motion as a sequence of poses. To segment the motion, its extracted poses are processed to identify a collection of *key poses* that play the role of separators between consecutive segments.

Liu et al. [16] introduced one of the first approaches that employs the idea of key poses to estimate similarity of motions. The motion similarity is calculated by a modified version of the Dynamic Time Warping [7] applied to collections of key poses of given motions. The key poses are determined by clustering all motion poses by a weighted Euclidean distance and choosing the centroids of clusters as key poses. However, this approach is not applicable for motion retrieval based on segment-level indexing since periodic activities, such as simple walk, need key poses to be detected periodically in all walk cycles and not only one per cluster.

Recently, Zhou et al. [29] proposed a hierarchical clustering algorithm for segmenting motions. The proposed algorithm internally utilizes K-means clustering, which requires to set the number of clusters separately for each motion and thus limits the practical applicability of this approach. Barbic et al. [3] presented a segmentation method based on the Probabilistic Principal Component Analysis where segmentation cuts are assigned according to the changes of pose distributions. Similarly, Lan et al. [12] identified segmentation cuts by analyzing notable changes of distribution within a local pose neighborhood. However, both these algorithms segment motions at the level of actions, which is not convenient for general segment-based indexing. Sedmidubsky et al. [25] utilized fixed-size query segments for motion retrieval based on frame-level indexing. Despite this approach is efficient, fixed-size segments do not suit for segment-based retrieval.

Assa et al. [2] and Xiao et al. [28] applied modifications of standard curve simplification techniques to detect key poses within a high-dimensional curve of motion features. The motion-feature curve is being gradually approximated by a simple curve until the resulting approximation satisfies an error requirement. Key poses are then identified at positions where the simple curve touches the original feature curve. However, setting of the error requirement threshold strongly depends on a given dataset and thus can be hardly estimated without any information about motion data.

Gong et al. [8] proposed a key-pose detection algorithm which is the most relevant to our approach. They also construct and analyze an energy curve, representing significance of local motion changes. In particular, they calculate the energy of a pose as the Euclidean distance to the previous pose. Key poses are then determined at positions with locally extremal energies. However, taking all local minima and maxima into account contributes to a huge number of identified key poses – a single gesture movement can be separated by several segments. To deal with this problem, we propose constructing the energy curve based on knowledge of not only the previous pose but from a reasonable large neighborhood. Moreover, we determine key poses only at positions with locally maximal energies that, in addition, exceed a predefined energy threshold. The value of this threshold constitutes minimum significance of movement change, which is independent of any prior knowledge of motion data.

To clarify the suitability of our solution for gesture-level segmentation, we experimentally compare our algorithm against relevant algorithms of Assa [2], Xiao [28] and Gong [8]. The experimental evaluation focuses on the quality of segmentation and the accuracy of semantic consistency of identified key poses.

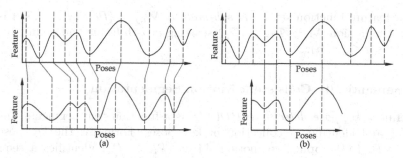

Fig. 2. Dashed-line key poses identified at semantically equivalent phases of movement for (a) two similar motions and (b) motion and its sub-motion. For better clarity, human motion is represented by a single feature only (i.e., $\mathcal{P}_j = (f_j^1)$).

3   Motion Segmentation

The objective of motion segmentation is to identify a collection of *key poses* that partition a motion into a sequence of consecutive segments (sub-motions). The identification of appropriate key poses primarily depends on the purpose of a target application. We especially focus on motion retrieval applications that can benefit from segment-based indexing. The problem is to find a segmentation that partitions motions at the level of gestures (e.g., "raising right leg") and, at the same time, identifies key poses at semantically equivalent phases of movement. Figure 2 illustrates two examples satisfying these requirements.

In this section, we formally define motion data, key-pose detectors and semantically consistent motion segmentation. We also explain how to find out whether two similar motions have their key poses generated at semantically equivalent phases of movement.

3.1   Motion Data Representation

Human motion is modeled using a simplified stick figure represented by bones that are connected by joints, such as left/right hands, knees or feet. The positions of joints are estimated for each video frame in form of 3D coordinates, i.e., a single frame is an element of $\mathbb{R}^{3 \times |joints|}$. To describe characteristic aspects of motions, various kinds of motion features can be extracted from the 3D joint coordinates, such as relative joint distances, bone angles or joint velocities and accelerations [2,8,28].

Specific features are usually extracted on the level of individual frames. Mathematically, we denote F as a set $F = \{\mathcal{F}_1, \ldots, \mathcal{F}_m\}$ of feature functions $\mathcal{F}_i$ ($i \in [1, m]$), where $\mathcal{F}_i : \mathbb{R}^{3 \times |joints|} \to \mathbb{R}$ processes the 3D joint coordinates in j-th frame and extracts the specific *feature value* $f_j^i \in \mathbb{R}$, e.g., an elbow angle or distance between right and left foot. Individual feature values $f_j^1, \ldots, f_j^m$ extracted in j-th frame forms an m-dimensional vector $\mathcal{P}_j = (f_j^1, \ldots, f_j^m)$, called the *pose feature vector*, or simply *pose*. A motion $\mathcal{M}$ of n frames is then represented as a sequence $\mathcal{M} = (\mathcal{P}_1, \ldots, \mathcal{P}_n)$ of poses extracted according to the specific feature function set F. A *sub-motion* $\mathcal{M}_{i,j} = (\mathcal{P}_i, \mathcal{P}_{i+1}, \ldots, \mathcal{P}_j)$ is the part of the motion $\mathcal{M} = (\mathcal{P}_1, \ldots, \mathcal{P}_n)$ starting at i-th frame and ending at j-th frame $(1 \leq i \leq j \leq n)$.

3.2 Semantically Consistent Motion Segmentation

We define a *key-pose detector* $KPD(\mathcal{M})$ as a function that segments the motion $\mathcal{M}$ and identifies a collection of key poses. Formally, the key-pose detector $KPD(\mathcal{M})$ applied to motion $\mathcal{M} = (\mathcal{P}_1, \ldots, \mathcal{P}_n)$ identifies a sequence $(k_1, \ldots, k_s), s \leq n$, of increasing indices $1 \leq k_1 < k_2 < \ldots < k_s \leq n$ that determine the desired set $\{\mathcal{P}_{k_1}, \ldots, \mathcal{P}_{k_s}\}$ of *key poses* $\mathcal{P}_{k_1}, \ldots, \mathcal{P}_{k_s}$. We say that a key-pose detector is *semantically consistent* if for any sub-motion the same key poses are identified disregarding the placement of the sub-motion within the parent motion. Formally, key-pose detector KPD is semantically consistent if the identified set of key poses for any motion $\mathcal{M}$ is a superset of the key-pose set identified for any sub-motion $\mathcal{M}_{i,j}$ of $\mathcal{M}$:

$$KPD(\mathcal{M}) \supseteq KPD(\mathcal{M}_{i,j}), 1 \leq i \leq j \leq n. \tag{1}$$

Semantically consistent key-pose detectors inherently identify key poses at semantically equivalent phases of movements. Note that a trivial approach identifying each 10-th motion pose as key pose is not semantically consistent since it violates Equation 1: $KPD(\mathcal{M}) = \{\mathcal{P}_{10}, \mathcal{P}_{20}, \mathcal{P}_{30}, \ldots\} \not\supseteq \{\mathcal{P}_{15}, \mathcal{P}_{25}\} = KPD(\mathcal{M}_{6,32})$. Equation 1 implies that a semantically consistent key-pose detector has to be even able to decide whether a trivial sub-motion $\mathcal{M}_{i,i} = \{\mathcal{P}_i\}$ – formed by a single pose $\mathcal{P}_i$ – represents the key pose or not. This equation strongly limits detectors to identify key poses on the basis of information about a single pose only and not to exploit information about neighboring poses or the character of motion data. Another example of a semantically consistent detector is one that identifies a pose $\mathcal{P}_i = (f_i^1, \ldots, f_i^5)$ as key pose if the second feature value f_i^2 is greater than a fixed threshold τ. Although such approach is obviously semantically consistent, i.e., it repeatedly marks a series of consecutive poses as key poses $(f_i^2 > \tau)$ and the following series of consecutive poses as non-key poses $(f_i^2 \leq \tau)$, the result segmentation is probably unusable for any application.

3.3 Semantically ϵ-Consistent Motion Segmentation

Semantically consistent detectors are restricted to identify key poses according to a single pose only, which results to semantically hardly usable motion segmentation. To overcome this issue, we allow key-pose detectors to take a local character of motion data into account. Given any motion $\mathcal{M} = (\mathcal{P}_1, \ldots, \mathcal{P}_n)$ and its sub-motion $\mathcal{M}_{i,j}$, we say that a key-pose detector KPD is *semantically ϵ-consistent* if the identified key-pose set for $\mathcal{M}$ is a superset of the key-pose set for $\mathcal{M}_{i,j}$ without its first and last ϵ poses:

$$KPD(\mathcal{M}) \supseteq KPD(\mathcal{M}_{i+\epsilon,j-\epsilon}), 1 \leq i \leq i + 2 \cdot \epsilon \leq j \leq n, \tag{2}$$

where $\epsilon \in \mathbb{N}$ is a user-defined constant, independent of the processed motion $\mathcal{M}$. Equation 2 allows semantically ϵ-consistent detectors to ignore key poses generated at first and last ϵ poses of sub-motion $\mathcal{M}_{i,j}$. Such benevolence enables utilizing the local character of motion data in order to decide whether a given pose is key pose or not. In more detail, detectors can mark the pose $\mathcal{P}_l$ ($i+\epsilon \le l \le j - \epsilon$) as key pose on the basis of information about neighboring poses that are represented by sub-motions $(\mathcal{P}_{l-\epsilon}, \ldots, \mathcal{P}_{l-1})$ and $(\mathcal{P}_{l+1}, \ldots, \mathcal{P}_{l+\epsilon})$ of preceding and succeeding poses, respectively.

Remark that semantically consistent detectors equal to 0-consistent detectors. An example of a semantically 1-consistent detector ($\epsilon = 1$) is the approach of Gong et al. [8]. They analyze just one preceding and succeeding pose to decide whether the current pose is key pose or not.

3.4 Segmentation of Similar Motions

Semantically ϵ-consistent detectors guarantee to identify key poses at the same positions (with an eventual exception of first and last ϵ poses) for motion $\mathcal{M}$ and any its sub-motion $\mathcal{M}_{i,j}$. However, real applications usually require to compare identified motion segments of *similar* motions, rather than motions and their sub-motions. This has led us to propose semantically ϵ-consistent detectors that identify key poses at semantically equivalent phases of movement also for similar motions – see Figure 2a. The crucial issues are (1) how to interpret a similarity of motions and (2) how to determine that two key poses are identified at semantically equivalent phases of movement.

The first issue can be solved by using distance-based functions to measure similarity of motions, like the popular Dynamic Time Warping (DTW), Uniform Time Warping (UTW) [22], Uniform Scaling (US) [11], and Scaled and Warped Matching (SWM) [7]. Note that these functions internally utilize a local cost measure to determine similarity of poses. The examples of cost measures are the Manhattan distance (L_1 metric), Euclidean distance (L_2 metric), Earth mover's distance, Hamming distance, and many others [5].

To deal with the second issue, we adopt an alignment method defined inside the DTW function. The basic DTW variant determines the similarity of two motions $\mathcal{M} = (\mathcal{P}_1, \ldots, \mathcal{P}_n)$ and $\mathcal{M}' = (\mathcal{P}'_1, \ldots, \mathcal{P}'_{n'})$ by constructing a warping matrix $W \in \mathbb{R}^{n \times n'}$ and looking for an optimal path from $W[1, 1]$ to $W[n, n']$. The cost of this path corresponds to the similarity of motions $\mathcal{M}$ and $\mathcal{M}'$. In particular, the optimal path determines the mapping among poses of motions $\mathcal{M}$ and $\mathcal{M}'$. Figure 3 illustrates the example of such mapping between two similar motions. We say that poses $\mathcal{P}_i$ ($1 \le i \le n$) and $\mathcal{P}'_j$ ($1 \le j \le n'$) are situated at *semantically equivalent* phases of movement if they are mapped on each other, i.e., the optimal path traverses through the matrix element $W[i, j]$.

It is important to realize that there is always a mapping between poses when comparing arbitrary motions. If two motions have a completely different length or data character, it is not meaningful to focus on semantically equivalent phases. It is the reason we compare the generated sets of key poses of only such motions which are visually similar, i.e., their DTW distance is smaller than a predefined

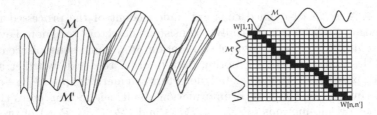

Fig. 3. Visualization of DTW mapping between poses of two similar motions $\mathcal{M}$ and $\mathcal{M}'$ (left) and the warping matrix along with the optimal path corresponding to such mapping (right).

similarity threshold. In the following, we introduce our novel key-pose detector and experimentally demonstrate that the sets of key poses of similar motions are generated at semantically equivalent phases of movement with the highest accuracy in comparison to other existing approaches.

4   Key-Pose Detection Algorithm

We introduce a novel semantically ϵ-consistent algorithm for partitioning motions into gesture-level segments. This algorithm first extracts motion features in form of relative joint distances. Based on these features, an energy curve is then constructed and finally analyzed to identify desired key poses.

4.1   Feature Extraction

We extract characteristic motion aspects by the set $F = \{\mathcal{F}_1, \ldots, \mathcal{F}_5\}$ of 5 feature functions that measure relative distances between specific pairs of joints for each frame separately. In particular, these functions $\mathcal{F}_1, \ldots, \mathcal{F}_5$ calculate the Euclidean distance between 3D coordinates of (1) left hand and right hand, (2) left hand and left foot, (3) right hand and right foot, (4) left knee and right knee, and (5) left foot and right foot. These relative distances are independent of camera viewpoint. Temporal variation in these distances captures the changes in human movement and is primarily exploited as information for key-pose detection. After the feature extraction, a motion $\mathcal{M}$ of n frames is represented as a sequence $\mathcal{M} = (\mathcal{P}_1, \ldots, \mathcal{P}_n)$ of 5-dimensional poses $\mathcal{P}_j = (f_j^1, \ldots, f_j^5)$, where f_j^i ($i \in [1,5]$) represents i-th relative distance extracted in j-th frame.

4.2   Pose Energy

The introduced key-pose detector identifies key poses at semantically equivalent phases where a significant change of motion occurs. To locate poses of significant motion change, we measure the value of *pose energy*, similarly as in [2,8,28]. Nevertheless, our approach fundamentally differs from existing approaches in the way of estimating such energy value for each pose.

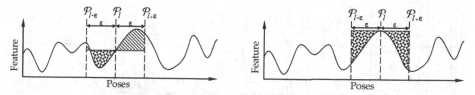

Fig. 4. Visualization of computation of a feature-energy value in l-th frame. Such value is computed as the difference between dashed and dotted areas for (left) pose without significance in motion change and (right) pose where a motion change occurs.

The pose energy indicates how significant the motion change is registered at a given pose. A motion change can be generally detected by examining close neighborhoods of the processed pose, as neither the pose itself nor any far-away pose reveals any information about local changes in motion. We denote the *neighborhood* of a pose $\mathcal{P}_l$ as the sub-motion $\mathcal{M}_{l-\epsilon,l+\epsilon}$. Such neighborhood encloses ϵ preceding and ϵ succeeding poses along with the pose $\mathcal{P}_l$ itself, where $\epsilon \in \mathbb{N}$ represents the neighborhood radius (in the number of poses). The ϵ constant should be set to such value so that a significant motion change could be recognized from $2 \cdot \epsilon + 1$ poses. We have experimentally verified that about 800 ms is sufficient enough to recognize a significant motion change. Having the sampling frequency of motion data 120 Hz, we set $\epsilon = 50$.

The value of pose energy is calculated as the sum of *feature-energy* values estimated for each of 5 features of the processed pose. The feature energy represents how much the value of the given feature differs at the neighboring poses, respecting its increasing or decreasing trend. The energy $E_{feature}(f_l^i)$ of i-th feature in pose P_l is computed as the sum of differences between the processed feature value f_l^i and i-th feature values in neighboring poses – see Figure 4. Mathematically, the feature energy $E_{feature}(f_l^i)$ is defined as:

$$E_{feature}(f_l^i) = \sum_{l'=l-\epsilon}^{l+\epsilon} \left(f_l^i - f_{l'}^i\right).$$

The zero value of feature energy implies that the feature does not change in neighboring poses or has a monotonous trend (i.e., it either increases, or decreases in the same way). Locally extremal feature-energy values can be both positive and negative and constitute significant changes in this feature. Figure 5c illustrates the example of a motion feature along with its corresponding energy.

The local minima and maxima of feature-energy values reflect a motion change from the perspective of a single feature. To be independent of a given feature, we compute the energy $E_{pose}(\mathcal{P}_l)$ of pose $\mathcal{P}_l$ by summing the energy values of its 5 features:

$$E_{pose}(\mathcal{P}_l) = \sum_{i=1}^{5} \left|E_{feature}(f_l^i)\right|.$$

The absolute value of individual feature-energy values give the same weight for local maxima and minima. In this way, the greater the pose-energy value is,

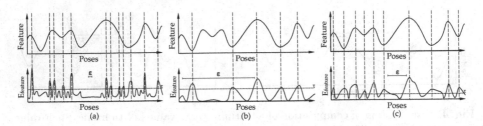

Fig. 5. Visualization of feature-energy values computed on the basis of different sizes of neighborhood: (a) $\epsilon = 10$, (b) $\epsilon = 200$ and (c) $\epsilon = 50$. We can simply deduce that setting (a) $\epsilon = 10$ is too small and generates a lot of irrelevant key poses, (b) $\epsilon = 200$ is too large and omits some poses with a significant change in motion and (c) $\epsilon = 50$ is reasonable and correctly identifies all poses with a significant change in motion.

the more significant motion change is registered. The zero pose-energy value principally constitutes that trajectories of individual joints change in the same way as in the neighboring poses or, in particular, do not change at all.

4.3   Identification of Key Poses

The pose-energy values indicate significance of local changes in motion. It is the reason our detector identifies poses as key poses if they have an energy value greater than a predefined threshold τ. The setting of this threshold should reflect user's viewpoint on "significance" of motion change. Moreover, the key-pose detector only selects such key pose having its energy greater than energies of the preceding and succeeding pose in order to generate key poses at semantically equivalent phases of movement. In other words, the semantically equivalent phases correspond to local maxima of the curve generated by the pose-energy values. Such maxima constitute significant changes in motion. More formally, our detector KPD processes a motion $\mathcal{M} = (\mathcal{P}_1, \ldots, \mathcal{P}_n)$ and identifies the following set of key poses:

$$KPD(\mathcal{M}) = \{\mathcal{P}_l \mid \mathcal{P}_l \in \mathcal{M}, 2 \leq l \leq n - 1 : E_{pose}(\mathcal{P}_l) \geq \tau \wedge$$
$$E_{pose}(\mathcal{P}_{l-1}) \leq E_{pose}(\mathcal{P}_l) \geq E_{pose}(\mathcal{P}_{l+1})\}.$$

The pseudo-code of the whole detector is described in Algorithm 1. We show that this algorithm is semantically ϵ-consistent, i.e., it satisfies Equation 2. Given a motion $\mathcal{M} = (\mathcal{P}_1, \ldots, \mathcal{P}_n)$ and any its sub-motion $\mathcal{M}_{i,j}, 1 \leq i \leq i+2 \cdot \epsilon \leq j \leq n$, pose-energy values computed for sub-motion $\mathcal{M}_{i+\epsilon,j-\epsilon}$ are the same as pose-energy values computed for the whole motion $\mathcal{M}$ without the first $i + \epsilon$ and last $n - j - \epsilon$ values. This observation comes from the fact that the algorithm computes a pose energy on the basis of only a local neighborhood of ϵ poses. Since the algorithm always processes the same pose-energy values in the same way, it also detects key poses at the same positions.

Algorithm 1. Key-pose detection algorithm.

> **function** KPD$(\mathcal{M}, \epsilon, \tau)$ ▷ $\mathcal{M} = (\mathcal{P}_1, \ldots, \mathcal{P}_n)$
> $energy \leftarrow array[1..n]$ ▷ Initialization of pose-energy curve
>   **for** $l \leftarrow 1$ **to** n **do**
> $energy[l] \leftarrow 0$ ▷ Initialization of energy of l-th pose
>   **for** $i \leftarrow 1$ **to** 5 **do**
> $featureEnergy \leftarrow 0$
>   **for** $l' \leftarrow max(1, l - \epsilon)$ **to** $min(l + \epsilon, n)$ **do**
> $featureEnergy \leftarrow featureEnergy + f_l^i - f_{l'}^i$ ▷ $\mathcal{P}_l = (f_l^1, \ldots, f_l^5)$
>   **end for**
> $energy[l] \leftarrow energy[l] + |featureEnergy|$
>   **end for**
>   **end for**
> $keyPoses \leftarrow \emptyset$ ▷ Collection of identified key poses
>   **for** $l \leftarrow 2$ **to** $n - 1$ **do**
>   **if** $energy[l-1] < energy[l] > energy[l+1] \wedge energy[l] \geq \tau$ **then**
> $keyPoses \leftarrow keyPoses \cup \mathcal{P}_l$
>   **end if**
>   **end for**
>   **return** $keyPoses$
> **end function**

5 Experimental Evaluation

We experimentally verify that the proposed algorithm outperforms relevant approaches of Assa [2], Xiao [28] and Gong [8] in terms of recall and precision with respect to a user-defined ground truth (Section 5.2). We also show that key poses are generated at semantically equivalent phases of movement with the highest accuracy (Section 5.3).

5.1 Dataset and Ground Truth

Our experiments have been conducted on one of the most widely used publicly-available motion capture dataset: HDM05 [20]. The HDM05 dataset provides its own ground truth which categorizes 2,089 manually selected sub-motions (i.e., parts of motions) into 118 classes corresponding to the specific motion actions, like *turn left*, *sit down on a chair*, or *clap hands five times*. We have utilized this classification to determine how accurately tested approaches generate key poses at semantically equivalent phases of movement.

From the HDM05 dataset, we have also selected 7 representative motions of different kinds of movements, containing 35,938 frames in total. By manually analyzing videos of skeleton movement frame-by-frame, we have precisely identified 565 key poses as another user-defined ground truth[1]. A pose in a given frame is marked as key pose if there is a change registered in the trend of trajectory direction of any joint, with respect to previous and following frames. Remark that rotating all around one's axis without additional changes in movement does not contribute to new key poses. It is important to realize that it is quite easy for

---

[1] Ground truth is available at: http://fi.muni.cz/~xsedmid/keyFrameGT.csv.

Table 1. Results of quality of generated key poses of tested approaches

	Recall (%)	Precision (%)	F-measure (%)	Ratio (%)
Our approach	52.2	46.5	49.2	112.1
Assa	35.2	39.2	37.1	89.7
Gong	92.1	4.6	8.7	2040.8
Xiao	30.7	45.3	36.6	67.7

a user to register a change in the trend of trajectory direction from video data, however, hard for a computer having only information about specific motion features available.

5.2   Evaluation of Quality of Key-Pose Detection

We evaluate the quality of generated key poses by calculating *recall* and *precision* metrics with respect to the ground truth created for the 7 preselected motions. Recall is measured as the ratio between the number of *relevantly* generated key poses and 565 ground-truth key poses. A key pose is considered as relevant if there exists a ground-truth key pose within 10 frames – tolerance of 10 frames corresponds to 0.08 seconds, which is under a user differentiation ability. Note that we allow only a single key pose to be mapped to a single ground-truth key pose. Precision is then calculated as the ratio between the number of relevantly generated key poses and the number of all generated key poses.

Results of recall and precision are illustrated in Table 1. This table also depicts two additional indicators: (1) *F-measure* as a harmonic mean of recall and precision and (2) *Ratio* as the ratio between the number of generated key poses and the number of ground-truth key poses. The evaluated results clearly demonstrate that our approach outperforms algorithms of Assa, Gong and Xiao. Moreover, our algorithm together with Assa's one generate a similar number of key poses as the number of ground-truth key poses – their *Ratio* metric mostly approaches 100 %.

5.3   Evaluation of Accuracy of Semantic Consistency

We evaluate how accurately tested approaches generate key poses at semantically equivalent phases of movement. As we have explained in Section 3.4, the accuracy of semantic consistency can be verified on similar motions only experimentally. To calculate visual similarity of motions, we implement a different kind of motion features based on joint-angle rotations [25]. For this experiment, we utilize the HDM05 ground truth classifying 2,089 motions into 118 categories. To obtain pairs of visually similar motions, we construct all possible pairs of motions within each category independently. In addition, we further consider only such pairs of motions whose DTW distance is under a predefined threshold.

The accuracy of semantic consistency is calculated using the DTW alignment applied to visually similar motions. First, given a pair of motions, each tested

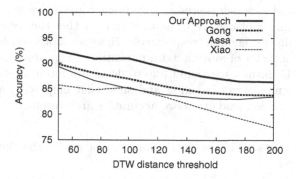

Fig. 6. Accuracy of semantic consistency of tested approaches with different length-normalized DTW distance thresholds.

approach is utilized to generate sets of key poses corresponding to both motions. Second, we employ the warping matrix of these motions to determine the number of mapped poses between two key-pose sets – see Section 3.4. Third, we compute the consistency accuracy as the ratio between the determined number of mapped key poses and the number of all key poses of both motions. To be independent of lengths of motions, we compute this ratio across all motion pairs.

The results of accuracy of semantic consistency are depicted in Figure 6 for different distance thresholds. In particular, the thresholds vary from 50 (corresponding to 241 highly similar motion pairs) up to 200 (corresponding to 21,676 motion pairs). Note that a resulting DTW distance is normalized by dividing the distance by the sum of lengths of motions so that the threshold values do not depend on motion lengths. We can see that with an increasing distance threshold the accuracy of semantic consistency decreases. It is caused by the fact that the more visually similar motions are, the higher number of semantically equivalent phases of movement they have. In general, our approach is superior to all the others, disregarding any distance threshold. The main reason of lower accuracy of Assa's and Xiao's approaches is that their detection of key poses depends on the character of entire motions, which contributes to more inconsistencies even at highly similar motions. Gong's approach improves semantic consistency by taking only a local neighborhood into account, however, many negligible local dissimilarities within the constructed energy signal cause that quite a lot of generated key poses are not mapped on each other.

6   Conclusions

We introduce an unsupervised key-pose detection algorithm for segmentation of motion capture data. The proposed algorithm partitions motions at the level of gestures. Gesture-level segmentation cuts are detected at semantically equivalent phases of movement with the highest accuracy, in comparison to existing approaches of Assa [2], Xiao [28] and Gong [8]. Our algorithm requires to define two constants ϵ and τ, which represent how many consecutive frames are needed to recognize a significant change in motion and whether the recognized change in motion is significant enough, respectively. However, both constants are independent of the character of motion data, in contrast to parameters used in other approaches. In the future, we would like to develop a motion retrieval application based on segment-level indexing and show that employing a semantically consistent segmentation leads to more accurate search results.

Acknowledgements. This research was supported by the national project GAP103/12/G084.

References

1. Arikan, O., Forsyth, D.A.: Interactive Motion Generation from Examples. In: 29th Annual Conference on Computer Graphics and Interactive Techniques, pp. 483–490. ACM (2002)
2. Assa, J., Caspi, Y., Cohen-Or, D.: Action Synopsis: Pose Selection and Illustration. In: SIGGRAPH 2005, pp. 667–676. ACM (2005)
3. Barbič, J., Safonova, A., Pan, J.Y., Faloutsos, C., Hodgins, J.K., Pollard, N.S.: Segmenting Motion Capture Data into Distinct Behaviors. In: Graphics Interface Conference (GI 2004), pp. 185–194. Canadian Human-Computer Communications Society (2004)
4. Choensawat, W., Choi, W., Hachimura, K.: Similarity Retrieval of Motion Capture Data Based on Derivative Features. Journal of Advanced Computational Intelligence and Intelligent Informatics 16(1), 13–23 (2012)
5. Deza, M.M., Deza, E.: Encyclopedia of Distances, 2nd edn. Springer (2013)
6. Fernandez-Baena, A., Susin, A., Lligadas, X.: Biomechanical Validation of Upper-Body and Lower-Body Joint Movements of Kinect Motion Capture Data for Rehabilitation Treatments. In: 4th International Conference on Intelligent Networking and Collaborative Systems (INCoS 2012), pp. 656–661 (2012)
7. Fu, A.W.C., Keogh, E., Lau, L.Y., Ratanamahatana, C.A., Wong, R.C.W.: Scaling and Time Warping in Time Series Querying. The VLDB Journal 17(4), 899–921 (2008)
8. Gong, W., Bagdanov, A.D., Roca, F.X., Gonzàlez, J.: Automatic Key Pose Selection for 3D Human Action Recognition. In: Perales, F.J., Fisher, R.B. (eds.) AMDO 2010. LNCS, vol. 6169, pp. 290–299. Springer, Heidelberg (2010)
9. Josiński, H., Switoński, A., Michalczuk, A., Wojciechowski, K.: Motion Capture as Data Source for Gait-Based Human Identification. Przeglad Elektrotechniczny 88, 201–204 (2012)
10. Kang, J., Badi, B., Zhao, Y., Wright, D.K.: Human Motion Modeling and Simulation. In: 6th International Conference on Robotics, Control and Manufacturing Technology (ROCOM 2006), pp. 62–67. WSEAS (2006)
11. Keogh, E., Palpanas, T., Zordan, V.B., Gunopulos, D., Cardle, M.: Indexing large human-motion databases. In: 30th International Conference on Very Large Data Bases (VLDB 2004), pp. 780–791. VLDB Endowment (2004)
12. Lan, R., Sun, H.: Automated Human Motion Segmentation via Motion Regularities. The Visual Computer, 1–19 (2013)

13. Lan, R., Sun, H., Zhu, M.: Text-Like Motion Representation for Human Motion Retrieval. In: Yang, J., Fang, F., Sun, C. (eds.) IScIDE 2012. LNCS, vol. 7751, pp. 72–81. Springer, Heidelberg (2013)
14. Li, C., Kulkarni, P.R., Prabhakaran, B.: Segmentation and recognition of motion capture data stream by classification. Multimedia Tools and Applications 35(1), 55–70 (2007)
15. Lin, Y., McCool, M.D.: Nonuniform Segment-Based Compression of Motion Capture Data. In: Bebis, G., et al. (eds.) ISVC 2007, Part I. LNCS, vol. 4841, pp. 56–65. Springer, Heidelberg (2007)
16. Liu, F., Zhuang, Y., Wu, F., Pan, Y.: 3D Motion Retrieval with Motion Index Tree. Computer Vision and Image Understanding 92(2-3), 265–284 (2003)
17. Lv, F., Nevatia, R.: Recognition and Segmentation of 3-D Human Action Using HMM and Multi-class AdaBoost. In: Leonardis, A., Bischof, H., Pinz, A. (eds.) ECCV 2006. LNCS, vol. 3954, pp. 359–372. Springer, Heidelberg (2006)
18. Müller, M., Baak, A., Seidel, H.: Efficient and Robust Annotation of Motion Capture Data. In: ACM SIGGRAPH/Eurographics Symposium on Computer Animation (SCA 2009), pp. 17–26 (2009)
19. Müller, M., Demuth, B., Rosenhahn, B.: An Evolutionary Approach for Learning Motion Class Patterns. In: Rigoll, G. (ed.) DAGM 2008. LNCS, vol. 5096, pp. 365–374. Springer, Heidelberg (2008)
20. Müller, M., Röder, T., Clausen, M., Eberhardt, B., Krüger, B., Weber, A.: Documentation Mocap Database HDM05. Tech. Rep. CG-2007-2, Universität Bonn (2007)
21. Müller, M., Baak, A., Seidel, H.P.: Efficient and Robust Annotation of Motion Capture Data. In: ACM SIGGRAPH/Eurographics Symposium on Computer Animation (SCA 2009), p. 10. ACM Press (2009)
22. Park, J.P., Lee, K.H., Lee, J.: Finding Syntactic Structures from Human Motion Data. Computer Graphics Forum 30(8), 2183–2193 (2011)
23. Ren, C., Lei, X., Zhang, G.: Motion Data Retrieval from Very Large Motion Databases. In: International Conference on Virtual Reality and Visualization (ICVRV 2011), pp. 70–77 (2011)
24. Sedmidubsky, J., Valcik, J., Balazia, M., Zezula, P.: Gait Recognition Based on Normalized Walk Cycles. In: Bebis, G., et al. (eds.) ISVC 2012, Part II. LNCS, vol. 7432, pp. 11–20. Springer, Heidelberg (2012)
25. Sedmidubsky, J., Valcik, J., Zezula, P.: A Key-Pose Similarity Algorithm for Motion Data Retrieval. In: Blanc-Talon, J., Kasinski, A., Philips, W., Popescu, D., Scheunders, P. (eds.) ACIVS 2013. LNCS, vol. 8192, pp. 669–681. Springer, Heidelberg (2013)
26. Wang, P., Lau, R.W., Zhang, M., Wang, J., Song, H., Pan, Z.: A Real-time Database Architecture for Motion Capture Data. In: 19th International Conference on Multimedia (MM 2011), pp. 1337–1340. ACM (2011)
27. Wu, S., Wang, Z., Xia, S.: Indexing and Retrieval of Human Motion Data by a Hierarchical Tree. In: 16th Symposium on Virtual Reality Software and Technology (VRST 2009), pp. 207–214. ACM Press (2009)
28. Xiao, J., Zhuang, Y., Yang, T., Wu, F.: An Efficient Keyframe Extraction from Motion Capture Data. In: Nishita, T., Peng, Q., Seidel, H.-P. (eds.) CGI 2006. LNCS, vol. 4035, pp. 494–501. Springer, Heidelberg (2006)
29. Zhou, F., De la Torre Frade, F., Hodgins, J.K.: Hierarchical Aligned Cluster Analysis for Temporal Clustering of Human Motion. IEEE Transactions on Pattern Analysis and Machine Intelligence 35(3), 582–596 (2013)

Active Query Selection for Constraint-Based Clustering Algorithms

Walid Atwa and Kan Li

School of Computer Science and Technology, Beijing Institute of Technology, China
walid_mufic@yahoo.com, likan@bit.edu.cn

Abstract. Semi-supervised clustering uses a small amount of supervised data in the form of pairwise constraints to improve the clustering performance. However, most current methods are passive in the sense that the pairwise constraints are provided beforehand and selected randomly. This may lead to the use of constraints that are redundant, unnecessary, or even harmful to the clustering results. In this paper, we address the problem of constraint selection to improve the performance of constraint-based clustering algorithms. Based on the concepts of Maximum Mean Discrepancy, we select the set of most informative instances that minimizes the difference in distribution between the labeled and unlabeled data. Then, we query these instances with the existing neighborhoods to determine which neighborhood they belong. The experimental results with state-of-the-art methods on different real world dataset demonstrate the effectiveness and efficiency of the proposed method.

Keywords: Semi-supervised clustering, pairwise constrain, active learning.

1   Introduction

Semi-supervised clustering algorithms attempt to partition the unlabeled data into a set of clusters with the help of a small amount of pairwise constraints (must-link and cannot-link). Existing constraint-based clustering algorithms can improve the clustering performance with a suitable passively chosen set of constraints [1, 2]. However, if the constraints are selected improperly, they may also degrade the clustering performance [3, 4]. Moreover, selecting constraints typically requires a user to manually inspect the data instances that can be time consuming and costly. For those reasons, we would like to optimize the selection of the constraints to improve the performance of semi-supervised clustering algorithms.

Active learning is well motivated in many supervised learning scenarios that enable to select the most informative unlabeled instances from enormous amount of unlabeled data for querying. Specifically, the goal of active learning is to query data as little as possible, to achieve a certain performance, thus saving considerable cost for generating good queries.

In this paper, we consider active learning in an iterative manner. In each iteration, set of queries are selected and queried with the existing neighborhoods to improve the clustering results. Specifically, we select a batch of informative query instances such

H. Decker et al. (Eds.): DEXA 2014, Part I, LNCS 8644, pp. 438–445, 2014.
© Springer International Publishing Switzerland 2014

that the distribution represented by the selected query set and the available labeled data is closest to the distribution represented by the unlabeled data. In other words, we select a set of samples S from the unlabeled data, denoted by D_U, such that the probability distributions represented by $D_L \cup S$ and $D_U \setminus S$, where D_L is the set of available labeled data, are similar to each other. We measure the difference in the probability distribution between the two sets of data using the Maximum Mean Discrepancy (MMD) [5, 6]. Once the batch of informative query instances are selected, we query them with the existing neighborhoods to determine which neighborhood they belong. Well-formed neighborhoods can provide valuable information regarding what the underlying clusters look like.

We empirically evaluate the proposed method with baseline and state of the art methods on UCI real datasets. The evaluation results demonstrate that our method achieves consistent improvements over the baseline methods.

The remainder of the paper is organized as follows. Section 2 presents a brief review of the related work. Section 3 introduces our proposed method. Experimental results are presented in Section 4. Finally, we conclude the paper and discuss future directions in Section 5.

2   Related Work

Active learning has a long history in supervised learning algorithms [7, 8, 9, 10]. Recently, active learning approaches are used in constrained-based clustering problem. Few studies reported the result of using active learning for constrained-based clustering. The first study on this topic was conducted by Basu et al. [1] that proposed an active k-means clustering using the farthest-first strategy that has two-phases (*Explore* and *Consolidate*). Mallapragada et al. [11] proposed an improvement to *Explore* and *Consolidate* named Min-Max, which modifies the *Consolidate* phase by choosing the most uncertain point to query (as opposed to randomly).

Xu et al. [12] proposed an active constrained spectral clustering algorithm that examines the eigenvectors to identify the boundary points (of two classes) and sparse points; then, it queries the oracle for constraints based on the these points. It has shown limited applicability because it requires many queries to the oracle and assumes that errors in the clustering result only occur on the boundary points.

Vu et al. [13, 14] proposed an active query selection method (denoted as ASC). This method relies on two aspects: (1) a k-nearest neighbors graph is used to determine the best candidate queries in the sparse regions of the dataset between the clusters where traditional clustering algorithms performs poorly, and (2) a propagation procedure allows each user query to generate several constraints which limits the user intervention. The propagation procedure discovers new constraints from the information stored in already chosen constraints using the notion of strong paths.

Recently, Xiong et al. [15] proposed an active learning method based on the classic uncertainty-based principle. They studied the selection of most informative constraints by selecting the most informative instance to form queries accordingly. The responses to the queries are then used to improve the clustering results. However this method selects only a single instance that can become very slow for retraining with each single instance being queried. Furthermore, if a parallel querying system is available, e.g., multiple annotators working in parallel, these methods would not be able to make the effective use of the resources.

3   Active Constraint Selection Method

Traditional data mining and machine learning algorithms are based on the assumption that the training data (X, Y) represents the true underlying distributions of X and Y where $X=\{x_1, x_2,..., x_n\}$ is the training data and $Y=\{y_1, y_2,..., y_n\}$ is the corresponding labels of X. Hence a model learned on this data works well for the test data (X_{test}, Y_{test}) which is also drawn independently and identically distributed from the same distribution. Thus, a batch of query instances can be selected from unlabeled data such that the distribution represented by the queried and labeled data is similar to the probability distribution of the unlabeled data set. In other words, selecting a batch of instances S from the unlabeled data (denoted by D_U) such that the joint probability distribution represented by $D_L \cup S$ and $D_U \setminus S$ are similar to each other, where D_L is set of available labeled data.

To measure the difference between two distributions, Maximum Mean Discrepancy (MMD) has been shown to be an effective measure of the difference in their marginal probability distributions [5, 6, 7]. Now, let us assume that we have u instances of unlabeled data D_U and l instances of labeled data D_L and we would like to select a batch S of b instances such that the distribution of $D_L \cup S$ is similar to the distribution of $D_U \setminus S$. Thus, the MMD between the sets $D_L \cup S$ and $D_U \setminus S$ defined by $f(S)$, can be computed as follows:

$$f(S) = \left\| \frac{1}{l+b} \Sigma_{j \in D_L \cup S} \Phi(x_j) - \frac{1}{u-b} \Sigma_{i \in D_U \setminus S} \Phi(x_i) \right\|_{\mathcal{H}}^2 \tag{1}$$

Since, we want to select a set S from unlabeled data set D_U to minimize the mismatch between $D_L \cup S$ and $D_U \setminus S$. We define a binary vector α of size u where each entry α_i indicates whether the data $x_i \in D_U$ is selected or not. If a point is selected, the corresponding entry α_i is 1 else 0. Thus the minimization problem reduces to finding α that minimizes the cost function $f(S)$:

$$\min_{\alpha:\alpha_i \in \{0,1\}, \alpha^T 1 = b} \left\| \frac{1}{l+b} \left(\Sigma_{j \in D_L} \Phi(x_j) + \Sigma_{i \in D_U} \alpha_i \Phi(x_i) \right) - \frac{1}{u-b} \Sigma_{i \in D_U} (1 - \alpha_i) \Phi(x_i) \right\|_{\mathcal{H}}^2 \tag{2}$$

where $\mathbf{1}$ is a vector of the same dimension as α with all entries 1 and symbol T is used to represent the matrix or vector transpose operation. Evidently, the cost function in Equation (2) is an alternative (equivalent) representation of the cost function $f(S)$ in Equation (1). The first term denotes the mean of the mapped features of the labeled and selected points. Note that if a point x_i is not selected in the current set then α_i will be 0 and this term would not get added in the summation. The second term is mean of the mapped features of the unlabeled data set minus the selected query set. The first constraint ensures that each entry in α is either 0 or 1 and the second constraint ensures that exactly b entries of α are 1, meaning exactly b instances are selected from the unlabeled data set, where b is specified a priori by the user.

Once the batch of most informative query instances are selected, we query them against the existing neighborhoods to determine to which neighborhood they belong. A neighborhood contains a set of data instances that are known to belong to the same cluster (i.e., connected by must-link constraints). Different neighborhoods are connected by cannot-link constraints and thus are known to belong to different clusters. This naturally motivates us to consider an active learning strategy that incrementally expands the neighborhoods by selecting the most informative instances and querying them against the known neighborhoods. We summarize our active learning method in Algorithm 1.

We should always start by querying s against the neighborhood that has the highest probability of containing s to minimize the total number of required queries. If a must-link is returned, we can stop with only one query. Otherwise, one should ask the next query against the neighborhood that has the next highest probability of containing s. This process is repeated until a must-link constraint is returned or we have a cannot-link constraint against all neighborhoods. If no must-link is achieved, a new neighborhood will be created using the instance s (lines 14-16). Finally, we apply the semi-supervised clustering algorithm using the selected active pairwise constraints to generate the final clusters (line 18). In this paper, we consider the semi-supervised clustering algorithm as a black-box and any existing algorithm can be used here.

Algorithm 1. Active Learning Method

Input: A set of instances D; total number of queries Q; batch size b.
Output: A set of clusters.
1.   Initialization: set $N_1 = \{x\}$, where x is a random instance; $C = \emptyset$; $q = 0$;
2.   **While** $q < Q$
3.   Select the batch query instances S
4.   **For** each instance $s \in S$
5.   **For** each neighborhood $N_i \in N$ in decreasing order of $p(s \in N_i)$
6.   Query instance s against any instance $x_i \in N_i$;
7. q++;
8.   Update the constraint set C based on the results;
9.   **If** a must-link achieved between s and x_i **then**
10.   Add instance s to neighborhood N_i;
11.   Break;
12.   **End if**
13.   **End for**
14.   **If** no must-link is achieved **then**
15.   Create new neighborhood with the instance s;
16.   **End if**
17.   **End for**
18.   Apply semi-supervised clustering(D, C);
19.   **End while**

---

4   Experiments

In this section, we empirically evaluate the performance of our proposed method on real-world datasets from UCI Machine Learning Repository[1] (each with the following number of instances, attributes and clusters): Heart (270/13/2), Breast (683/9/2), Yeast (1484/8/10), Image Segmentation (2310/19/7), Digit-389 (3165/16/3) and Magic (19020/10/2).

In our experiments we assume the availability of constraint-based clustering algorithm. For this purpose, we report the results obtained with MPCKMeans algorithm [2]. When evaluating the performance of particular methods on a given dataset D, we apply it to select up to 150 pairwise queries, starting from no constraint at all. The queries are answered based on the ground-truth class label for the dataset. MPCKMeans algorithm is then applied to the data with the resulting constraints. We repeat this process for 50 independent runs and report the average performance using evaluation criteria described below. For our method we set the parameter batch size $b=10$.

To evaluate the performance of the methods, we used Normalized Mutual Information (*NMI*) as the clustering validation metrics. *NMI* is an external validation metric, which is used to estimate the quality of clustering with respect to the given true labels of the datasets. *NMI* measures how closely the clustering algorithm could reconstruct the underlying label distribution in the data.

4.1   Experimental Results

In this section we present the results of our active learning method compared to the baseline methods. Figure 1 shows the clustering performance (*NMI* values) on the UCI real datasets. It can be observed from Figure 1, that the proposed method generally outperforms other constraint selection methods in conjunction with the clustering algorithm MPCKMeans. This implies that the usefulness of constraints depends on how they are utilized by a clustering algorithm. Furthermore, the proposed method keeps a smooth increase in the clustering performance while the other constraint selection heuristics drop in performance when the number of queries increases.

In comparison, it can be seen from Figures 1, that the ASC method obtains better results than our active learning method with small number of queries (e.g. Yeast, Segment and Digit-389 datasets). However, when increasing the number of queries make the proposed method superior to ASC. The superiority of ASC in a small number of queries comes from its propagation procedure that discovers new constraints from the information stored in previously chosen constraints and gives ASC the capability to select a well propagated set of constraints in a small number of queries.

In comparison with NPU method that generally outperforms Random, Min-Max, and ASC method. NPU is generally able to improve the clustering performance consistently with increasing the number of queries. However, its performance is dominated by our method in most cases.

---

[1] http://www.ics.uci.edu/~mlearn/MLRepository.html

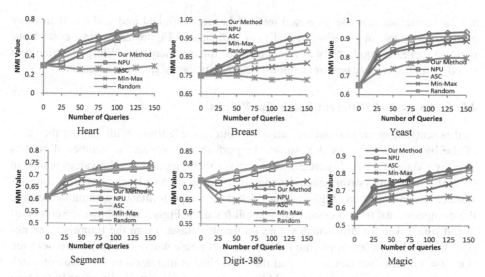

Fig. 1. Comparisons of normalized mutual information over the different number of queries

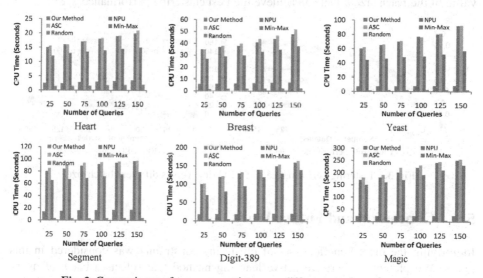

Fig. 2. Comparisons of average run time over different number of queries

Also, the primary motivation of our active learning method is to reduce the amount of computation in iterative active learning. Figure 2 shows the average CPU time of our selection method (using batch size value $b = 10$) with the baseline methods on a 2.4 GHz Intel Core 2 PC and 2 GB main memory. From the results, we clearly see that the proposed method is significantly more efficient and scalable than other methods. As we observe, when the dataset size increases, the time cost of NPU and ASC increases dramatically, while the proposed method increases linearly. Specifically, on the magic dataset with 150 queries, NPU takes about 240 seconds and ASC takes about 260 seconds, while the proposed method needs only about 24 second. Hence, we can conclude that the proposed method is more efficient and scalable than other methods for large applications. Finally, as indicated in Figure 2, the time cost of our method increases moderately as the number of queries increases. While, the time cost of other methods increases tremendously as the number of queries increases.

4.2   Analysis of Different Batch Size Values

In this section, we carried out an analysis of our active learning with varying the value of the batch size b. Figure 3(a) shows the performance versus the number of queries on magic dataset. From Figure 3(a), we can see that selecting small b values results in similar (or better) performance compared to those obtained selecting only one instance. On the contrary, high b values decrease the performance without decreasing the computational time if compared to small b values. Figure 3(b) shows the computational time taken for different b values on magic dataset. From the figure we can notice that the largest learning time is obtained in the case where one instance is selected (i.e. $b = 1$). Also, we can notice that the CPU time is increased as the value of batch size b is increased. Therefore it could be interesting to automatically identify the best value of the batch size b that can achieve the best clustering performance.

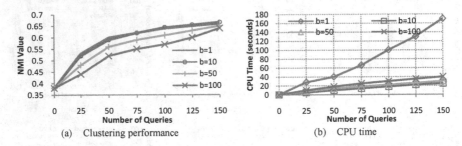

(a)   Clustering performance   (b)   CPU time

Fig. 3. Clustering performance and CPU time versus different batch size

5   Conclusion and Future Work

Identifying the most beneficial set of clustering constraints was considered in this paper. We present an iterative active learning method that selects a batch of query instances from the unlabeled data so that the marginal probability distribution represented by the labeled data after annotation, is similar to the marginal probability distribution represented by the unlabeled data. And, incrementally expands the neighborhoods by using the selected queries. Experiments carried out on different real datasets show that the constraints selected by our active learning process are generally more beneficial for constraint-based clustering algorithms than those provided by the NPU and ASC methods and that our method achieving high clustering performance with minimizing the amount of computation for selecting the active constraints. However, the efficiency of our method strongly depends on the value of the batch size b. In future work, we are interesting to automatically identify the best value of the batch size b that achieves the best clustering performance.

References

1. Basu, S., Banerjee, A., Mooney, R.: Active semi-supervision for pairwise constrained clustering. In: SIAM International Conference on Data Mining, pp. 333–344 (2004)
2. Bilenko, M., Basu, S., Mooney, R.: Integrating constraints and metric learning in semi-supervised clustering. In: International Conference on Machine Learning, pp. 11–18 (2004)
3. Davidson, I., Wagstaff, K.L., Basu, S.: Measuring constraint-set utility for partitional clustering algorithms. In: Fürnkranz, J., Scheffer, T., Spiliopoulou, M. (eds.) PKDD 2006. LNCS (LNAI), vol. 4213, pp. 115–126. Springer, Heidelberg (2006)
4. Greene, D., Cunningham, P.: Constraint selection by committee: An ensemble approach to identifying informative constraints for semi-supervised clustering. In: Kok, J.N., Koronacki, J., Lopez de Mantaras, R., Matwin, S., Mladenič, D., Skowron, A. (eds.) ECML 2007. LNCS (LNAI), vol. 4701, pp. 140–151. Springer, Heidelberg (2007)
5. Borgwardt, K., Gretton, A., Rasch, M., Kriegel, H., Schölkopf, B., Smola, A.: Integrating structured biological data by kernel maximum mean discrepancy. Bioinformatics 22(14), 49–57 (2006)
6. Sriperumbudur, B., Gretton, A., Fukumizu, K., Schölkopf, B., Lanckriet, G.: Hilbert space embeddings and metrics on probability measures. In: JMLR, pp. 1517–1561 (2010)
7. Chattopadhyay, R., Fan, W., Davidson, I., Panchanathan, S., Ye, J.: Joint Transfer and Batch-mode Active learning. In: Proceedings of the 30th International Conference on Machine Learning, USA, pp. 253–261 (2013)
8. Guo, Y., Schuurmans, D.: Discriminative batch mode active learning. In: Advances in Neural Information Processing Systems, pp. 593–600 (2007)
9. Hoi, S., Jin, R., Zhu, J., Lyu, M.: Batch mode active learning and its application to medical image classification. In: International Conference on Machine Learning, pp. 417–424 (2006)
10. Huang, S., Jin, R., Zhou, Z.: Active learning by querying informative and representative examples. In: Advances in Neural Information Processing Systems, pp. 892–900 (2010)
11. Mallapragada, P., Jin, R., Jain, A.: Active query selection for semi-supervised clustering. In: International Conference on Pattern Recognition, pp. 1–4 (2008)
12. Xu, Q., Desjardins, M., Wagstaff, K.: Active constrained clustering by examining spectral eigenvectors. In: Discovery Science, pp. 294–307 (2005)
13. Vu, V.V., Labroche, N., Bouchon-Meunier, B.: Improving constrained clustering with active query selection. Pattern Recognition 45(4), 1749–1758 (2012)
14. Vu, V.V., Labroche, N., Bouchon-Meunier, B.: An efficient active constraint selection algorithm for clustering. In: Proceedings of the 20th International Conference on Pattern Recognition, pp. 2969–2972 (2010)
15. Xiong, S., Azimi, J., Fern, X.Z.: Active Learning of Constraints for Semi-Supervised Clustering. IEEE Transactions on Knowledge and Data Engineering (2013)

Clustering Evolving Data Stream with Affinity Propagation Algorithm

Walid Atwa and Kan Li

School of Computer Science and Technology, Beijing Institute of Technology, China
walid_mufic@yahoo.com, likan@bit.edu.cn

Abstract. Clustering data stream is an active research area that has recently emerged to discover knowledge from large amounts of continuously generated data. Several data stream clustering algorithms have been proposed to perform unsupervised learning. Nevertheless, data stream clustering imposes several challenges to be addressed, such as dealing with dynamic data that arrive in an online fashion, capable of performing fast and incremental processing of data objects, and suitably addressing time and memory limitations. In this paper, we propose a semi-supervised clustering algorithm that extends Affinity Propagation (AP) to handle evolving data steam. We incorporate a set of labeled data items with set of exemplars to detect a change in the generative process underlying the data stream, which requires the stream model to be updated as soon as possible. Experimental results with state-of-the-art data stream clustering methods demonstrate the effectiveness and efficiency of the proposed method.

Keywords: Affinity propagation, data streams, semi-supervised clustering.

1   Introduction

Streaming data is the discipline specifically concerned with handling large-scale datasets in an online fashion [1, 2]. Clustering is one of the most important unsupervised learning methods which partition a given set of objects into subsets called clusters. Clustering data stream continuously produces and maintains the clustering structure from the data stream in which the data items continuously arrive in the ordered sequence [3, 4]. There are many data streaming algorithms have been adapted from clustering algorithms, e.g., the partitioning method k-means [5, 6, 7, 8], the density based method DBSCAN [9, 10], and grid based methods [11, 12].

In this paper, we propose a semi-supervised clustering algorithm for streaming data called SSAPStream. The proposed algorithm proceeds by extending the Affinity Propagation (AP) method, a message passing-based clustering method [13]. The proposed algorithm SSAPStream involves three main steps described as follow: (1) using the AP method on the first bunch of data arrives at time t_0 to identify the exemplars and initialize the stream model; (2) as the stream flows in, each point p is compared to the exemplars; if too far from the nearest exemplar, p is put in the buffer, otherwise the stream model is updated; (3) if the buffer full, the stream model is rebuilt based on the current model and buffer using the set of labeled data.

H. Decker et al. (Eds.): DEXA 2014, Part I, LNCS 8644, pp. 446–453, 2014.

The experimental results on synthetic and real data sets validate the effectiveness of our method in handling dynamically evolving data streams. Also, we study the execution time and memory usage of SSAPStream, which are important efficiency factors for streaming algorithms.

The rest of the paper is organized as follows. Section 2 briefly reviews related work. Section 3 presents the proposed Semi-Supervised Affinity Propagation (SSAP). Section 4 describes SSAPStream algorithm, extending AP to streaming data. Section 5 shows the experimental results. Section 6 presents the conclusion and future work.

2   Related Work

The CluStream framework proposed in [6] is effective in handling evolving data streams. It divides the clustering process into online and offline components. The online component uses microclusters to continuously captures synopsis information from the data stream while the offline component uses this information and other user inputs to yield on-demand clustering results. The main drawback of this algorithm is that the number of micro-clusters needs to be predefined. For high-dimensional data stream clustering, Aggarwal et al. [14] proposed HPStream, which reduces the dimensionality of the data stream via data projection before clustering.

Based on the online maintenance and offline clustering strategy, Cao et al. [9] proposed a DenStream algorithm, which extends DBSCAN [15] by introducing micro-clusters to the density-based connectivity search. It is an algorithm that forms local clusters progressively by detecting and connecting dense data item neighborhoods. Independently, Chen and Tu [11, 12] also proposed a density-based method termed D-Stream. Rather than using microclusters, D-Stream partitions the data space into grids and maps new data points into the corresponding grid to store density information, which are further clustered based on the density.

Zhang et al. [16] presented a version of the AP algorithm called StrAP that is closer to handling data streaming. The StrAP algorithm proceeds by incrementally updating the current model if the current data item fits the model. Otherwise, detecting the data item as an outlier and putting it in a reservoir. Recently, Ackermann et al. proposed StreamKM++, a two-step algorithm that is merge-and reduce [7]. The reduce step is performed by the coreset tree, considering that it reduces $2m$ objects to m objects. The merge step is performed by another data structure, namely the bucket set, which is a set of L buckets (also named buffers), where L is an input parameter. Each bucket can store m objects.

3   Semi-Supervised Affinity Propagation

Affinity Propagation (AP) [13] is a clustering method that takes as input similarities between data points. It aims to identify exemplars among data points and forms clusters of data points around these exemplars. Each exemplar is characterized by a data point representative of the sample itself and some other similar points. The objective of AP is to maximize the sum of similarities between the data points and their exemplars.

Assume that $X = \{x_1, x_2, \ldots, x_n\}$ is a set of distinct data items and let $S(x_i, x_j)$ denote the similarity between the data items x_i and x_j, with $i \neq j$. AP begins by simultaneously considering all data items as potential exemplars, and iteratively exchanges messages between data items until a good set of exemplars and clusters emerges. The messages can be combined at any stage to decide which data items are exemplars and, for every other item, which exemplar it belongs to. There are two kinds of messages exchanged between data items [13]:

1. The responsibility massage $r(i, k)$, sent from data item x_i to candidate exemplar item x_k, reflects the accumulated evidence for how well-suited item x_k is to serve as the exemplar for item x_i.

$$r(i,k) = S(x_i, x_k) - \max_{j \neq k}\{S(x_i, x_j) + a(i,j)\} \tag{1}$$

2. The availability massage $a(i, k)$, sent from candidate exemplar item x_k to item x_i, reflects the accumulated evidence for how appropriate it would be for item x_i to choose item x_k as its exemplar. It is initialized as zero and updated as follows:

$$a(i,k) = \begin{cases} \sum_{i' \neq k} \max[0, r(i', k)] & i = k \\ \min[0, r(k,k) + \sum_{i' \notin \{i,k\}} \max[0, r(i', k)]] & i \neq k \end{cases} \tag{2}$$

After convergence, the exemplars are obtained by calculating the set of positive $a(i, i) + r(i, i)$ messages for each x_i, and the items are assigned to their respective exemplars (clusters) according to the following rule, $c(x_i) = \arg\max_k (a(i,k) + r(i,k))$. The message passing procedure stops after a specific number of iterations or after cluster structure does not significantly change for a given number of iterations.

The original affinity propagation is an unsupervised clustering method. To utilize the partially labeled data, semi-supervised clustering aims to improve the clustering performance by learning from a combination of both labeled samples and unlabeled data.

Assume we have L labeled data set $\{(x_1, y_1), (x_2, y_2), \ldots, (x_l, y_l)\}$, where y_l is the cluster label of data item x_l, and U unlabeled data items $x_{l+1}, x_{l+2}, \ldots, x_{l+u}$. Let C be the set of exemplars in the data set. For a certain labeled sample x_i $(1 \leq i \leq l)$ and unlabeled data item x_j $(l + 1 \leq j \leq u)$ we can have two possible situations where the labeled sample may be associated with the unlabeled data item after a run of the AP algorithm.

1. The unlabeled data item x_j takes the labeled sample x_i as the cluster exemplar. The message $a(x_i, x_i) + r(x_i, x_i)$ is positive, i.e., $x_i \in C$, and if $x_i = \max\{a(x_j, x_k) + r(x_j, x_k)\}$ for each $k = \{1, 2, \ldots, N\}$.
2. The labeled sample x_i takes the unlabeled data item x_j as the cluster exemplar. The message $a(x_i, x_i) + r(x_i, x_i)$ is negative, i.e., $x_i \neq C$, and if $x_j = \max\{a(x_i, x_k) + r(x_i, x_k)\}$ for each $k = \{1, 2, \ldots, N\}$.

If one of the two conditions is satisfied, the unlabeled data item x_j is the most similar to the labeled sample x_i. Then, the unlabeled data item x_j is selected and set to the label of x_i. Therefore, we can select the most similar unlabeled data item x_j as follows:

$$V = \begin{cases} x_i & \text{if } x_i = max_{1 \le k \le N}\{a(x_j, x_k) + r(x_j, x_k)\} \text{ and } x_i \in C \\ x_j & \text{if } x_j = max_{1 \le k \le N}\{a(x_i, x_k) + r(x_i, x_k)\} \text{ and } x_i \notin C \end{cases}$$

where V is the new labeled sample set picked from U. Selecting the unlabeled data items is defined according to the operational mechanism of the AP algorithm. This process repeats until no unlabeled data items left in U. At each iteration, the selection process takes advantage of the updated results of the AP algorithm.

4   Clustering Streaming Data

In clustering data stream, data evolve over time and thus new clusters may appear, clusters may merge or delete. The goal is to identify clusters of data and study the evolution of clusters over time. We consider the problem of clustering a data stream in the damped window model, in which the weight of data items decreases with time t according to the decay function $f(t) = 2^{-\lambda \cdot t}$, where, $\lambda > 0$ is a decay factor. In other words, the weight at time t_k of a data item x_i with time stamp t_i (i.e. an item that arrived at time t_i) will be given by:

$$w(x_i, t_k) = 2^{-\lambda(t_k - t_i)} \tag{3}$$

Data items that have arrived at previous time cannot be assumed to be available at future time. Therefore, we would like to maintain a succinct synopsis of data generated during previous time, taking also into account the weight of these historical data. The exemplars can be considered as representatives of the data in their clusters, and thus we define an exemplar vector to provide a synopsis of underlying data. The exemplar vector is consists of a set of 4-tuple (e_i, n_i, w_i, t_i), where e_i ranges over the exemplars, n_i is the number of items associated to exemplar e_i, w_i is the weight of e_i (see Definition 1), and t_i is the last time stamp when an item was associated to e_i.

Definition 1: *Weight of exemplar (or cluster).* For an exemplar e, at a given time t, let n be the set of data items that are associated to e at or before time t. The weight of e is defined as the sum of the weights of all data items that associated to e. Namely, the weight of e at t is:

$$w(e, t) = \sum_{j=1}^{n} w(x_j, t) \tag{4}$$

We develop SSAPStream a variation of the initially AP clustering algorithm, that is capable of handling sequences of data in an online fashion under the limited memory constraints imposed by streaming applications. We apply the AP clustering algorithm on the first bunch of data items that arrived at time t_0 to identify the exemplars and initialize the stream model. For each new data item p, we try to merge p into its nearest exemplar e_i in the current model (w.r.t. distance d). If $d(p, e_i)$ is less than some threshold ε (heuristically, set to the average distance between data items and exemplars in the initial model), p is affected to the i-th cluster and the model is updated accordingly; otherwise, p is put in the buffer.

In order to prevent the number of exemplars from growing beyond control, one must be able to forget the exemplars that have not been visited for a long time. In other words, for each existing exemplar e_i, if no new data item is merged into it, the weight of e_i will decay gradually and it should be deleted and its memory space released for new exemplar. Accordingly, we compare the weight of each exemplar with its lower limit of weight (denoted as η). If the weight of an exemplar is below its lower limit of weight, we can safely delete it. The lower limit of weight is defined as:

$$\eta(t_c, t_i) = \frac{1 - 2^{-\lambda(t_c - t_i + 1)}}{1 - 2^{-\lambda}} \tag{5}$$

where t_c is the current time and t_i is the last update time of the exemplar e_i. This function $\eta(t_c, t_i)$ is an increasing function for fixed t_i value. Thus, the longer an exemplar exists, the larger its weight is expected to be.

Finally, when the buffer is full or detecting a change in the generative process underlying the data stream, a new model is rebuilt by launching SSAP on the dataset and data items in the buffer. As mentioned previously, we have two possible situations where the labeled samples represented in the exemplars can be associated with the unlabeled data items in the buffer. The cost of selecting current exemplar e_i to be the exemplar of data item x_j is ordinary similarity $-d(x_j, e_i)^2$, while the cost of selecting x_j to be the exemplar of e_i is increased by a factor of n_i. Therefore, the current exemplar e_i will have more chance to be an exemplar again. SSAP accordingly selects the set of data items from buffer that are most similar to the exemplar e_i and thus, merges them to the exemplar e_i. This process repeats until no data items left in the buffer.

5   Experiments

In this section, we evaluate the effectiveness and efficiency of SSAPStream and compare it with several well-known data stream clustering algorithms, including CluStream [6], DenStream [9], StrAP [16] and StreamKM++ [7]. All the experiments are conducted on a 2.4 GHz Intel Core 2 PC with 2 GB main memory.

5.1   Experimental Setup

To evaluate the effectiveness and efficiency of SSAPStream algorithm, both synthetic and real data sets are used. Two synthetic data sets, DS1 and DS2, are generated as shown in Figures 1(a) and (b), respectively. Each of them contains 10000 points. The KDDCUP99 data set is a real data set that evolves significantly over time and has been widely used to evaluate data stream clustering algorithms [6, 7, 9, 16]. It consists of a series of TCP connection records of LAN network traffic managed by MIT Lincoln Labs. The complete data set contains approximately 4.9 million records, and as in the previous work [6, 9, 16], a sub-sampled subset of length 494020 is used.

For the performance measure we used the Rand index [17], which measures how accurately a cluster can classify data items by comparing cluster labels with the underlying class labels. The value of Rand index lies between 0 and 1. A higher Rand index indicates better clustering results [17].

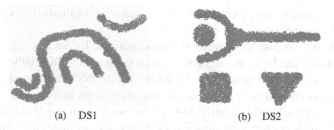

<table>
<tr><td>(a)   DS1</td><td>(b)   DS2</td></tr>
</table>

Fig. 1. Synthetic data sets

The initialization of the SSAPStream algorithm considers the first 1000 points of DS1 and DS2 and the first 1000 connections of the KDDCUP99 data set which totally includes 494,021 network connection records. The distance threshold ε heuristically, set to the average distance between data items and exemplars in the initial model, and the maximum number of data items stored in the buffer is 100 items. The parameter λ is the decay factor that controls the importance of historical data to current clusters, we set $\lambda = 0.98$.

5.2   Experimental Results

At first, we test the clustering quality of SSAPStream algorithm and compare with state-of-the-art data stream clustering methods. Figure 2 shows the average Rand indices by the five algorithms. In general, SSAPStream obtains the highest Rand indices on the three testing streams. On the two synthetic data streams, SSAPStream has significantly outperformed the clustering methods such as CluStream, StrAP and StreamKM++. For instance, on DS1 data set, SSAPStream has obtained the average Rand index as high as 0.95, which is 0.12 higher than the second winner DenStream and is 0.22 higher than StrAP algorithm. Similarly, on DS2 data set, SSAPStream has obtained the average Rand index of 0.97, which is 0.09 higher than the second winner DenStream and is 0.18 higher than StrAP.

On the KDDCUP99 real data streams, SSAPStream also outperforms the other four data stream clustering methods. SSAPStream has obtained the average Rand index of 0.77, which is slightly better than the second winner StrAP by 0.09 and makes a

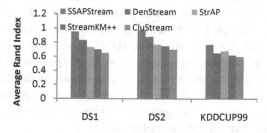

Fig. 2. Clustering quality comparison

significant improvement compared with the three clustering methods (i.e., CluStream, DenStream and StreamKM++).

Also, we test and compare the execution time consumed by the five algorithms on the KDDCUP99 data set. First, the algorithms are tested on the KDDCUP99 data with different sizes as shown in Figure 3(a). From the figure, SSAPStream algorithm is comparable to its counterparts in terms of the execution time, that is, it is faster than some of them, i.e., StrAP, and StreamKM++. It can also be seen that SSAPStream has better scalability since its clustering time grows slower with an increasing data size.

Next, the algorithms are tested on the KDDCUP99 data with different dimensionality. We set the size of data set as 100K and vary the dimensionality in the range of 2 to 40. We list the time costs under different dimensionality by the five algorithms as shown in Figure 3(b). SSAPStream is faster than other algorithms and scales better with an increasing dimensionality. For instance, when the dimensionality is increased from 2 to 40, the time of SSAPStream only increases by 20 seconds which is better than the second faster DenStream that increases by 50 seconds.

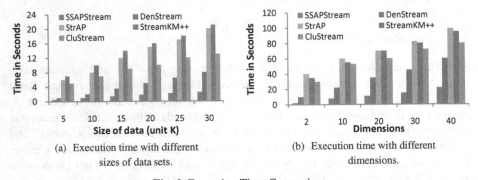

(a) Execution time with different
sizes of data sets.

(b) Execution time with different
dimensions.

Fig. 3. Execution Time Comparison

6   Conclusion and Future Work

In this paper, we propose SSAPStream, a semi-supervised clustering algorithm that extends Affinity Propagation for clustering evolving data streams. The algorithm uses the exemplars to summary information of the historical data items and thus, it has limited memory consumption. Our goal is to make full use of both the existing labeled exemplars and the unlabeled data items to improve the clustering performance in streaming data. Experimental results on KDD'99 data demonstrate the effectiveness and efficiency of the proposed algorithm. This work opens several perspectives for further research. A main limitation of AP is that it cannot model the category consisting of multiple subclasses since it represents each cluster by a single exemplar, such as scene analysis and character recognition. To remedy this deficiency, we extend the single-exemplar model to a multi-exemplar model.

References

1. Aggarwal, C.: Data Streams: Models and Algorithms. Springer (2007)
2. Cormode, G., Muthukrishnan, S., Zhuang, W.: Conquering the divide: Continuous clustering of distributed data streams. In: Proceedings of the International Conference on Data Engineering (ICDE), pp. 1036–1045 (2007)
3. Jonathan, A.S., Elaine, R.F., Rodrigo, C.B., Eduardo, R.H., André, C.P., Gama, J.: Data stream clustering: A survey. ACM Computing Surveys (CSUR) 46(1), 1–31 (2013)
4. Babcock, B., Babu, S., Datar, M., Motwani, R., Widom, J.: Models and issues in data stream systems. In: Proceedings of the 21st ACM SIGMOD-SIGACT-SIGART Symposium on Principles of Database Systems, pp. 1–16 (2002)
5. Guha, S., Meyerson, A., Mishra, N., Motwani, R., O'Callaghan, L.: Clustering data streams: Theory and practice. IEEE Transactions on Knowledge and Data Engineering (TKDE) 15, 515–528 (2003)
6. Aggarwal, C.C., Han, J., Wang, J., Yu, P.S.: A framework for clustering evolving data streams. In: Proceedings of the International Conference on Very Large Data Bases (VLDB), pp. 81–92 (2003)
7. Ackermann, M.R., Maartens, M., Raupach, C., Swierkot, K., Lammersen, C., Sohler, C.: StreamKM++: A clustering algorithm for data streams. Journal on Experimental Algorithmics 17(1) (May 2012)
8. Shindler, M., Wong, A., Meyerson, A.: Fast and accurate k-means for large datasets. In: Advances in Neural Information Processing Systems (NIPS), pp. 2375–2383 (2011)
9. Cao, F., Ester, M., Qian, W., Zhou, A.: Density-based clustering over an evolving data stream with noise. In: SIAM Conference on Data Mining (SDM), pp. 326–337 (2006)
10. Ruiz, C., Menasalvas, E., Spiliopoulou, M.: C-DenStream: Using Domain Knowledge on a Data Stream. In: Gama, J., Costa, V.S., Jorge, A.M., Brazdil, P.B. (eds.) DS 2009. LNCS, vol. 5808, pp. 287–301. Springer, Heidelberg (2009)
11. Chen, Y., Tu, L.: Density-based clustering for real-time stream data. In: Proceedings of the 13th ACM International Conference on Knowledge Discovery and Data Mining (SIGKDD), pp. 133–142 (2007)
12. Chen, Y., Tu, L.: Stream data clustering based on grid density and attraction. ACM Transactions on Knowledge Discovery from Data (TKDD) 3(3), 1–27 (2009)
13. Frey, B., Dueck, D.: Clustering by passing messages between data points. Science, 972–976 (2007)
14. Aggarwal, C.C., Han, J., Wang, J., Yu, P.S.: A Framework for Projected Clustering of High Dimensional Data Streams. In: Proceedings of the 30th Int'l Conf. Very Large Data Bases, VLDB (2004)
15. Ester, M., Kriegel, H.-P., Sander, J., Xu, X.: A Density-Based Algorithm for Discovering Clusters in Large Spatial Databases with Noise. In: Proceedings of the Second Int'l Conf. Knowledge Discovery and Data Mining (1996)
16. Zhang, X., Furtlehner, C., Perez, J., Germain-Renaud, C., Sebag, M.: Toward autonomic grids: Analyzing the job flow with affinity streaming. In: Proceedings of the 15th ACM International Conference on Knowledge Discovery and Data Mining, SIGKDD (2009)
17. Rand, W.M.: Objective Criteria for the Evaluation of Clustering Methods. J. Am. Statistical Assoc. 66(336), 846–850 (1971)

Incremental Induction of Belief Decision Trees in Averaging Approach

Salsabil Trabelsi, Zied Elouedi, and Mouna El Aroui

Larodec, Institut Supérieur de Gestion de Tunis, University of Tunis, Tunisia

Abstract. This paper extends the belief decision tree learning method to an incremental mode where the tree structure could change when new data come. This so-called incremental belief decision tree is a new classification method able to learn new instances incrementally, by updating and restructuring an existing belief decision tree. The induced decision tree is originally built in a batch mode under an uncertain framework, by means of the belief function theory, then updated by incorporating instances in a one-by-one basis once a new training set is available.

Keywords: classification, decision tree, uncertainty, belief function theory, incremental induction.

1 Introduction

A decision tree is a classification model presented in the form of a tree graph built from already classified instances (learning task) to classify new unseen instances (classification task). To deal with the uncertain environment, one idea was to adapt the belief function theory [4], as explained in the Transferable Belief Model (TBM) [5], with the traditional decision tree to reach the so-called belief decision tree in its averaging or conjunctive approaches [1], assuming that the uncertainty is lying on classes of training instances.

Most decision tree induction algorithms have focused on environments that are static, deterministic, and fully observable. What is to be done when, as in the real world, the environment is interactive, incremental, dynamical, and partially observable, where new data could appear at anytime from anywhere and of any type? In this case, the standard decision tree induction algorithms discard the existing tree and regenerate another one. Such reproduction loses the previous knowledge and effort while its cost may be too expensive. Therefore, incremental learning concept is raised to compensate for the limitations of batch mode algorithms, which is able to build and restructure a decision tree once a new training case is added. Such learning strategy should be more economical than rebuilding a new one from scratch. Some works have been done to introduce the incremental mode for learning decision trees [3] [6] [7].

Unfortunately, there are not many algorithms dealing with uncertainty in the incremental learning version, namely, the incremental probabilistic decision trees [9] and the incremental fuzzy decision trees [8]. So, the aim of this research

H. Decker et al. (Eds.): DEXA 2014, Part I, LNCS 8644, pp. 454–461, 2014.

is to develop what we call an incremental belief decision tree in the averaging approach by trying to adapt the incremental mode of learning decision trees under the belief function theory as a framework for dealing with uncertainty.

The objective is to develop an approach which is able to update and restructure a belief decision tree, once a new training set is added. Thus, this method learns new objects incrementally, which are characterized by certain attribute values and an uncertain class, and performs an update of the tree or a reconstruction instead of a rebuilding from scratch. This paper is organized as follows: section 1 provides a brief description of the basics of belief function theory as interpreted by the TBM and belief decision trees. Then, in section 2, we describe our averaging approach developed for building an incremental belief decision tree. Finally, in section 3, we carry simulations to illustrate the efficiency of our incremental method from real world databases using two evaluation criteria: classification accuracy and running time.

2   Theoretical Aspects

2.1   Belief Function Theory as Interpreted by the TBM

The TBM [5] is a model to represent quantified beliefs based on belief functions. Let Θ be a finite set of elementary events, called the frame of discernment. The basic belief assignment (bba) is a function m : $2^{\Theta} \rightarrow [0, 1]$ such that:

$$\sum_{A \subseteq \Theta} m(A) = 1 \tag{1}$$

The value m(A), named the basic belief mass (bbm), represents the portion of belief committed exactly to the event A and nothing more specific. The events having positive bbm's are called focal elements. Associated with m is the belief function [5] defined for $A \subseteq \Theta, A \neq \emptyset$ as:

$$bel(A) = \sum_{\emptyset \neq B \subseteq A} m(B) \quad and \quad bel(\emptyset) = 0 \tag{2}$$

The degree of belief $bel(A)$ given to a subset A of the frame Θ is defined as the sum of all the basic belief masses given to subsets that support A without supporting its negation. The probability functions called the pignistic probabilities are used to make decisions from beliefs denoted BetP [5] and is defined as:

$$BetP(A) = \sum_{B \subseteq \Theta} \frac{|A \cap B|}{|B|} \frac{m(B)}{(1 - m(\emptyset))}, \forall A \in \Theta \tag{3}$$

2.2   Belief Decision Trees (BDT)

A BDT [1] is a decision tree technique adapted in order to handle uncertainty about the actual class of the objects in the training set. We assume that the values of the attributes of each training object are known with certainty, whereas its corresponding class is uncertain and represented by a basic belief assignment defined on the set of possible classes Θ. Two attribute selection measures are proposed to build a belief decision tree [1]:

1. The first one is an extension of the classical approach developed by Quinlan and based on the gain ratio criterion [2]. It is called the averaging approach.
2. The second one represents ideas behind the TBM itself and based on distance criterion. It is called the conjunctive approach.

In our work, we propose the incremental mode of learning BDT for the averaging approach. So, only the attribute selection measure relative to the this approach will be detailed below using these following notations:

- T: a given training set composed by p objects I_j, $j = 1, ..., p$
- S: a set of objects belonging to the training set T,
- A: an attribute,
- $\Theta = \{C_1, C_2, ...,C_n\}$: the frame of discernment made of the n possible classes related to the classification problem.
- $m^{\Theta}\{I_j\}(C)$: the bbm given to the hypothesis that the actual class of object I_j belongs to $C \subseteq \Theta$

In order to define the gain ratio measure of an attribute A over a set of objects S within the averaging approach, we follow the next steps:

1. For each object I_j in a set of instances S, compute the pignistic probability by:

$$BetP^{\Theta}\{I_j\}(C_i) = \sum_{C_i \in C \subseteq \Theta} \frac{1}{|C|} \frac{m^{\Theta}\{I_j\}(C)}{1 - m^{\Theta}\{I_j\}(\emptyset)}, \qquad (4)$$

$\forall C_i \in \Theta$

2. Compute the average pignistic probability function:

$$BetP^{\Theta}\{S\}(C_i) = \frac{1}{|S|} \sum_{I_j \in S} BetP^{\Theta}\{I_j\}(C_i) \qquad (5)$$

3. Compute the entropy of the average pignistic probabilities in S. This Info(S) value is equal to:

$$Info(S) = - \sum_{i=1}^{n} BetP^{\Theta}\{S\}(C_i) log_2 BetP^{\Theta}\{S\}(C_i) \qquad (6)$$

4. Select an attribute A_k. Collect the subset $S_v^{A_k}$ made with the cases of S having v as a value for the attribute A_k.
5. Compute the average pignistic probability for those cases in the subset $S_v^{A_k}$. Let the result be denoted by $BetP^{\Theta}\{S_v^{A_k}\}$ for $v \in D(A_k)$, $A_k \in A$.

6. Compute $Info_{A_k}(S)$ using the same definition as suggested by Quinlan, but using the pignistic probabilities instead of the proportions. We get:

$$Info_{A_k}(S) = \sum_{v \in D(A_k)} \frac{|S_v^{A_k}|}{|S|} Info(S_v^{A_k}) \tag{7}$$

where $D(A_k)$ is the domain of the possible values of the attribute A_k and $Info(S_v^{A_k})$ is computed using Equation 6 using $BetP^{\Theta}\{S_v^{A_k}\}$.
The term $Info_{A_k}(S)$ is equal to the weighed sum of the different $Info(S_v^{A_k})$ relative to the considered attribute. These $Info(S_v^{A_k})$ are weighted by the proportion of objects in $S_v^{A_k}$.

7. Compute the information gain provided by the attribute A_k in the set of objects S such that:

$$Gain(S, A_k) = Info(S) - Info_{A_k}(S) \tag{8}$$

8. Using the Split Info, compute the gain ratio relative to the attribute A_k:

$$GainRatio(S, A_k) = \frac{Gain(S, A_k)}{SplitInfo(S, A_k)} \tag{9}$$

Where

$$Split\,Info(S, A) = - \sum_{v \in D(A_k)} \frac{|S_v^{A_k}|}{|S|} \log_2 \frac{|S_v^{A_k}|}{|S|} \tag{10}$$

9. Repeat for every attribute $A_k \in A$ and choose the one that maximizes the gain ratio

Structure of leaves: Each leaf in the induced tree will be characterized by a bba. According to the averaging approach, the leaf's bba is equal to the average of the bba's of the objects belonging to this leaf.

3 Incremental Belief Decision Trees in Averaging Approach (IBDT)

Inducing a belief decision tree in a batch mode of learning without being able to learn dynamically by adding new objects to the tree causes in most cases the loss of effort, information and time. In this paper, we are interested with the ability of learning incrementally the BDT by updating an existing induced tree built under an uncertain context, once a new training set is added. Thus, we propose a Incremental Belief Decision Trees in averaging approach (IBDT), able to learn new objects incrementally, which are characterized by uncertain classes, and perform a tree updating. Each time a new training instance is presented, we propose the following steps in order to update or restructure the BDT:

1. Select the path which will be traversed by the new instance, until reaching its corresponding leaf.

2. For each visited node according to the new instance's path, add the new object to its corresponding set or subset of training instances (with its corresponding bba)
3. Select the best attribute which has the highest Gain Ratio (see Eq. (9)) for each traversed decision node or leaf, by computing the pignistic probability of each object I_j (see Eq. (4)) then, the average pignistic probability $BetP^\Theta\{S\}$ on each class (see Eq. (5)),
4. If each concerned test attribute still the best one according to his position, update the node information (its belief values: the bba's average),
5. Otherwise
   (a) Restructure the tree, so that an attribute with the highest Gain Ratio is at the root,
   (b) Recursively re-establish a best test attribute in the root of each subtree.

The incremental averaging approach has the interesting property to generate the same decision tree as when all the training instances had been given in one batch.

4   Experimentation and Simulation

In this section, some experimentations and simulations are done to evaluate the performance our IBDT relative to the averaging approach from some U.C.I. repository[1] databases which are characterized by symbolic attributes. Different results will be presented and analyzed in order to evaluate our proposed incremental method compared with non-incremental case using two evaluation criteria: Classification accuracy and time running.

4.1   Description of Databases

These databases are modified in order to include uncertainty in classes. Uncertainty is represented by a bba given on the set of possible classes, taking into account the real class C of an object and the degree of uncertainty, which could be low, middle or high:

- Low degree of uncertainty: we take $0 < P \leq 0.3$
- Middle degree of uncertainty: we take $0.3 < P \leq 0.6$
- High degree of uncertainty: we take $0.6 < P \leq 1$

Each bba has almost 2 focal elements:

1. The first is the actual class C of the object with bbm , m(C)= 1-P, such that P is a randomly generated probability.
2. The second is a subset θ of Θ (randomly generated) such that the actual class of the object under consideration belongs to θ and every class of the others belongs to θ with P probability. m(θ)= P

---

[1] http://www.ics.uci.edu/mlearn/MLRepository.html

In Table 1, we give a brief description of these databases, hence we present the composing parameters: #Data, #Ts, #attributes and #classes denote respectively the total number of instances in the database, the number of testing instances, the number of attributes and the number of classes.

Table 1. Description of databases

Database	#Data	#Ts	#attributes	#classes
W. Breast Cancer	690	69	8	2
Balance Scale Weight	624	62	4	3
Congressional Voting	490	49	16	2
Zoo	100	10	17	7
Nursery	1641	164	8	3
Solar Flares	322	32	12	2
Lung Cancer	26	3	56	2
Hayes-Roth	131	13	5	3
Car evaluation	1728	173	6	4
Lymphography	147	15	18	4
Spect Heart	79	8	22	2
Tic-Tac-Toe Endgame	957	96	9	23

4.2   Evaluation Parameters

Different results carried out from these simulations will be presented and analyzed in order to evaluate the efficiency of the "Incremental Belief Decision Tree (IBDT)" against the "Belief Decision Tree (BDT)". For this reason, two relevant criteria are used to judge the performance of our method which are described as follows:

1. Classification accuracy criterion: we use the PCC representing the percent of the correct classification of the objects belonging to the testing set which are classified according to the induced decision tree.
2. Running time: presents an efficient manner to show how speed is our proposed approach against the non-incremental learning mode. It measures the running time in seconds dedicated for the update of the tree.

The construction of an incremental belief decision tree requires a partition of the training set, the first part will concern the building of a belief decision tree in a non-incremental mode. However, the second part will present the new coming data needed to be additionally learn. A partition of our training set is made with N-1 instances (such that N is the number of instances of a given database) for the non-incremental building of the belief decision tree and the one last instance to be incorporated into our tree, then our method for tree updating is performed.

4.3   Results

Table 2 presents the PCC in the non-incremental and incremental modes resulting from the application of the averaging approach. Running time results are

Table 2. PCC results (%)

Databases	Non-incremental mode and Incremental mode		
	$BDT/IBDT_{low}$	$BDT/IBDT_{middle}$	$BDT/IBDT_{high}$
W. Breast Cancer	67.97	67.83	66.09
Balance Scale Weight	58.68	58.35	63.1
Congressional Voting	94.29	94.08	92.27
Car evaluation	71.32	71.66	70.83
Zoo	85.09	84.42	83.26
Nursery	93.69	93.01	92.27
Solar Flares	78.2	77.9	77.87
Tic-Tac-Toe Endgame	73.92	73.77	73.44
Spect Heart	73.48	73.03	72.20

presented in Table 3 with both modes of learning. In the following, we will denote the batch averaging approach with low degree of uncertainty by BDT_{low}, with middle degree of uncertainty by BDT_{middle} and with high degree of uncertainty by BDT_{high}. In addition, we will denote the incremental averaging approach with low degree of uncertainty by $IBDT_{low}$, with middle degree of uncertainty by $IBDT_{middle}$ and with high degree of uncertainty by $IBDT_{high}$.

From Table 2, results prove the similarity of the two resulting trees in both incremental and non-incremental cases, for all degrees of uncertainty of different real-world databases. For example, for the Breast Cancer database, the induced trees in both incremental and non-incremental modes have the same PCC values equals to 67.97% for the low degree of uncertainty.

Our proposed method shows good results (see Table 3) relative to the second evaluation criterion "Running time". The reduction of the time spent with our incremental method comparing to the batch mode method, seems to be considerable for all degrees of uncertainty. For example, with a high degree of uncertainty, the running time value of the Breast Cancer database is reduced from 3.344 seconds to 0.928 seconds for the averaging approach. The same interpretation is valid for the other databases.

Table 3. Running time results (seconds)

Databases	Non-incremental mode			Incremental mode		
	BDT_{low}	BDT_{middle}	BDT_{high}	$IBDT_{low}$	$IBDT_{middle}$	$IBDT_{high}$
W. Breast Cancer	3.478	3.330	3.344	0.997	0.930	0.928
Balance Scale Weight	2.91	2.890	2.859	0.765	0.5	0.588
Congressional Voting	2.914	2.755	3.012	0.802	0.781	0.986
Car evaluation	5.762	5.602	5.533	2.653	2.562	2.747
Zoo	2.941	2.966	2.806	0.35	0.342	0.349
Nursery	11.092	12.121	12.437	3.367	3.891	4.348
Solar Flares	1.678	1.579	1.486	1.245	1.12	0.812
Tic-Tac-Toe Endgame	7.386	7.237	7.041	2.35	2.192	1.442
Spect Heart	0.934	1.105	0.998	0.523	0.631	0.523

5   Conclusion

In this paper, we have presented our IBDT method in averaging approach with the objective to reduce the time and the effort of rebuilding a BDT from scratch once new objects to be learn are presented. The incremental learning is a way to update a belief decision tree with the guarantee the induction of the same tree as in the batch mode. Then, we have presented the different results obtained from simulations and that have been performed on real databases. These experimentations have shown interesting results for the performance of our incremental averaging approach comparing with the traditional averaging approach, specially for the running time criterion. Regarding the interesting results obtained in this work, we could propose further works to extend the BDT to the incremental mode of learning relative to the conjunctive approach or to handle numeric and uncertain attribute values in training data.

References

1. Elouedi, Z., Mellouli, K., Smets, P.: Belief decision trees: Theoretical foundations. International Journal of Approximate Reasoning, IJAR 28, 91–124 (2001)
2. Quinlan, J.R.: C4.5: Programs for Machine learning. Morgan Kaufman, San Mateo (1993)
3. Kalles, D., Morris, T.: Efficient Incremental Induction of Decision Trees. Machine Learning 1, 2–3 (1995)
4. Shafer, G.: A mathematical theory of evidence. Princeton University Press, Princeton (1976)
5. Smets, P., Kennes, R.: The transferable belief model. Artificial Intelligence 66, 191–236 (1994)
6. Utgoff, P.E.: ID5: An incremental ID3. In: Proceedings of the Fifth International Conference on Machine Learning, pp. 107–120. Morgan Kaufmann (1988)
7. Utgoff, P.E.: An improved algorithm for incremental induction of decision trees. In: Proceedings of the 11th International Conference on Machine Learning, pp. 318–325. Morgan Kaufmann (1994)
8. Wang, T., Li, Z., Yan, Y., Chen, H.: An incremental fuzzy decision tree classification method for mining data streams. In: Perner, P. (ed.) MLDM 2007. LNCS (LNAI), vol. 4571, pp. 91–103. Springer, Heidelberg (2007)
9. Wanke, C., Greenbaum, D.: Incremental, probabilistic decision making for en route traffic management. Air Traffic Control Quarterly 15, 299–319 (2007)

Digraph Containment Query
Is Like Peeling Onions

Jianhua Lu[1], Ningyun Lu[2], Yelei Xi[1], and Baili Zhang[1]

[1] School of Computer Science and Technology, Southeast University
lujianhua@seu.edu.cn
[2] College of Automation Engineering,
Nanjing University of Aeronautics and Astronautics
luningyun@nuaa.edu.cn

Abstract. Graph data is ubiquitous in various data applications, such as chemical compounds, proteins, and social network. Graph containment query processing in large-scale graph databases is one of the key challenges to the database community. The graph feature based index structures are widely used to narrow the isomorphism validating space. However, most of the existing index structures have expensive constructing overheads for the mining of frequent graph features. This paper proposes a novel way to query containment digraphs based on the partial order constraints. Firstly, the partial orders in digraphs, which are easier to obtain, are proved to be capable of filtering graph containment queries. Secondly, the partial orders are converted into layered vertex sequences, which can filter the digraphs in an efficient way. Thirdly, two optimized layered vertex sequences are further introduced to improve the filter ability. Finally, experimental results are presented to show the effectiveness and efficiency of the proposed algorithms.

Keywords: Digraph, Containment Query, Partial Order Constraint, Sequence Matching.

1 Introduction

Graphs have become increasingly important in modeling complicated structures and schema-less data such as proteins, circuits, images, Web, and XML documents. With the increasing usage of graph databases, efficient containment query processing in large-scale graph databases has received more and more attention. In many practical applications, graph databases often update frequently. A typical example is SCF Finder database, where about 4000 newly found compounds are joined in everyday. Therefore, it is of great significance to explore update friendly graph containment query processing algorithms.

Containment query could be very expensive. Suppose there are two objects T and T_1, where T has six parts, a, b, c, d, e and f, and T_1 has three parts, x, y and z, the objective is to determine whether or not T_1 is contained by T. Firstly, x is picked out to match with the six parts of T. If x matches with one part of

H. Decker et al. (Eds.): DEXA 2014, Part I, LNCS 8644, pp. 462–476, 2014.

T, then y is picked out to match with the remaining five parts in T, and so on. The whole process need up to $6 * 5 * 4 = 120$ match operations.

However, if the problem can be formulated as peeling an onion, the containment query becomes much easier. Suppose the object T is considered as an onion O that has six layers, $a/b/c/d/e/f$ (from the skin to the core), and the object T_1 is a part of onion O_1 that has three layers, $x/y/z$, the problem is to determine if O_1 is identical to some part of O. It can be solved by peeling O_1 layer by layer referring to O. Firstly, x is peeled off when x matches with any of a, b, c or d (e and f are out of consideration because there remain only two layers starting from e and one layer from f). Suppose x matches with c, then y is peeled off only if it matches with d. The final step is to check if z matches with e. In the above process,the number of matches is up to $4 * 3 = 12$, which is much less than 120.

Inspired by onions, this paper proposes a simple but efficient way to index digraphs and process containment queries following the **filtering-verification** mechanism[2]. By utilizing the partial order constraints in digraphs, this method can easily transform digraphs into onion-like layered vertex sequences. When the database is updated, only the newly updated graphs are processed without affecting other data or index terms. A sequence matching procedure acts like peeling onions to filter out unqualified digraphs. By introducing two optimized vertex sequences, the filtering capability can be significantly improved.

The main contributions of this paper are summarized as follows. (1)We have proven that, partial order constraints in digraphs are much easier to obtain than the commonly used frequent graph features, and they can be easily used to filter containment queries. (2) The graph vertices are partitioned into different layers and the partial order sets are converted into layered vertex sequences, based on which, the proposed sequence matching based filtering is effective and efficient. (3) Furthermore,two optimized layered vertex sequences are introduced to improve the performance of the proposed sequence matching based filtering algorithm.

The remainder of the paper is organized as follows. The preliminary concepts and some related works are introduced in Section 2. The basic idea of partial order constraint based graph containment query processing is presented in Section 3. The concept of layered vertex sequence is given in Section 4, as well as the demonstration of the effectiveness and efficiency of the sequence matching based filtering. Section 5 presents two optimized vertex sequences to improve the filtering capability. The experiments are conducted in Section 6, and Section 7 summaries the entire work we have done.

2   Preliminaries and Related Works

Graph, especially labeled graph, is widely used to model complicated structures and schemaless data. In a labeled graph, the vertices and edges represent entities and relationships respectively, and the properties associated with the entities and relationships are called labels. For example, XML is a kind of directed labeled graph, while chemical compounds are undirected labeled graphs. Many graph-based applications require efficient processing of containment graph queries. In the following, we first introduce some preliminary concepts and related works.

Definition 1 *(Directed Labeled Graph, Digraph).* *A directed labeled graph is defined as $G = \{V, E, \Sigma, l\}$, where V is a set of vertices, E is a set of edges (vertex pair $<v_i, v_j>$ expresses a directed edge from v_i to v_j), Σ is the set of vertex labels, and the labeling function l defines the mapping: $V \to \Sigma$.*

As a notational convention, the vertex set of a graph G is denoted by V_G, the edge set is denoted by E_G, and the size of G is denoted by $size(G) = \|E_G\|$.

Definition 2 *(Graph Isomorphism).* *Given two graphs, $G = \{V, E, \Sigma, l\}$ and $G' = \{V', E', \Sigma', l'\}$, a graph isomorphism from G to G' is an injective function $f : V \to V'$, which satisfies the following two conditions:*

 1. *$\forall u \in V, l(u) = l'(f(u))$,*
 2. *$\forall u, v \in V, <u, v> \in E$ iff $<f(u), f(v)> \in E'$,*

Definition 3 *(Subgraph Isomorphism, Graph Containment).* *Given two graphs, $G = \{V, E, \Sigma, l\}$ and $G' = \{V', E', \Sigma', l'\}$, a subgraph isomorphism from G to G' (G' contains G) is an injective function $f : V \to V'$, which satisfies the following two conditions:*

 1. *$\forall u \in V, l(u) = l'(f(u))$,*
 2. *$\forall u, v \in V,$ if $<u, v> \in E$, then $<f(u), f(v)> \in E'$,*

Definition 4 *(Graph Containment Query).* *Given a graph database $D = \{G_1, G_2, ..., G_n\}$, and a query graph q, a graph containment query is to find the set $D_q \subseteq D$ that every graph in D_q contains q.*

As shown in Fig. 1, graph database $D = \{G_1, G_2, G_3\}$ and query graph q, the results of subgraph containment query are $\{G_2, G_3\}$.

Graph feature based **filtering-verification** mechanism is widely used to process graph containment queries[2][3][4][5][6]. **Filtering** phase is to filter out a candidate graph set, which contains all the possible query answers, based on the index structures created offline; **verification** phase is to test the subgraph isomorphism for each candidate. Both in **filtering** phase and **verification** phase, subgraph isomorphisms have to be performed. The cost can be modeled as $\|N_f\| * C_f + \|N_c\| * C_c$[9]. $\|N_f\|$ is the number of subgraphs tested in **filtering** phase. $\|N_c\|$ is the size of candidate set. C_f and C_c are average costs of subgraph isomorphism in **filtering** and **verification** stage respectively. To improve graph containment query performance, both $\|N_f\|$ and $\|N_c\|$ should be small since the time complexity of subgraph isomorphism is very high[11]. Furthermore, when a graph is modified or newly inserted into the database, the subgraph isomorphism tests should be conducted, which results in high index structure updating cost and poor online query processing performance.

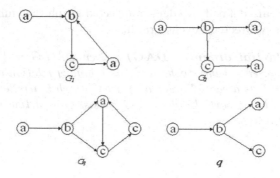

Fig. 1. Sample graph database D

At the offline stage of GraphGrep[1], all paths with length less than lp are listed to form an index. If graph g contains query graph q, the path in q must be in g. The gIndex[2][3] selects a part of frequent subgraphs with high filter ability as index items since paths are not sufficient to describe the graph structures. FG-Index[4] uses all frequent subgraphs as index items. In order to avoid the index size explosion, FG-Index introduces a two-layer indexing strategy. TreePi[5] uses frequent subtrees as index items. Besides, Tree+Δ[6] combines frequent subtrees with some subgraphs as the index items. In addition to the feature-based methods, gString[7] proposes a string coding scheme to show the graph structure. However, this approach is only applicable for molecular structure databases, and it is not easy to be extended for other applications. Closure-Tree[8] produces closure-graphs for graph database and organizes them into a Closure-Tree index. If the sub-Closure-Tree does not contain the query graph q, then no graph in the node branch would contain q. GCoding[9] maps the graph structures into values, and designs a two steps filtering strategy to improve the query performance.

This paper proposes a method that can filter digraphs without subgraph isomorphism checking. It is as simple as peeling onions, and at the same time, it can achive good filtering ability.

3   Filter Digraphs with Partial Order Set

3.1   Filter Directed Acyclic Graphs

Partial order appears in many applications. We use directed acyclic graph (DAG), a typical data model with partial order constraint, to express the basic idea.

Definition 5 (Partial Order). *A partial order ' $<$ ' over a set S is a binary relation that is irreflexive, antisymmetric, and transitive. , i.e., for all a, b, and c in S, we have that:*

1. *Irreflexivity: $a \not< a$,*
2. *Asymmetry: if $a < b$ then $b \not< a$,*
3. *Transitivity: if $a < b$ and $b < c$, then $a < c$.*

If partial orders are defined by edges, a directed graph that fulfills the partial order constraint is a directed acyclic graph.

Definition 6 (Partial order in DAG). *Given DAG $G = \{V, E, \Sigma, l\}$, for vertex set V, define each edge $<a, b> \in E$ as a binary relation $a < b$, then the binary relation " $<$ " is a partial order over set V which is irreflexive, transitive, and asymmetric. All the partial orders in G consist of a partial order set, which is denoted as $POSet(G)$.*

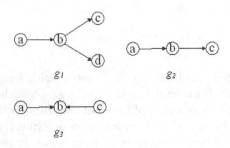

Fig. 2. Sample directed acyclic graphs

As shown in Fig. 2, g_1, g_2 and g_3 are all DAGs. There is a subgraph isomorphism from g_2 to g_1, while no isomorphism relation exists from g_3 and g_1. Tab. 1 lists all the partial orders of g_1, g_2, and g_3.

Table 1. The partial orders of g_1, g_2, and g_3

	g_1	g_2	g_3
	$a < b$	$a < b$	$a < b$
	$a < c$	$a < c$	$c < b$
Partial Order	$a < d$	$b < c$	
	$b < c$		
	$b < d$		

As Tab. 1 shows, partial order set of g_1 contains partial order set of g_2, but does not contain partial order set of g_3. According to Definition 3, if graph g contains g', the vertex set and edge set of g contain those of g'. In Definition 6, partial orders of DAG are defined with edges. Therefore, we have the following theorem:

Theorem 1 *(partial order set based filtering).* *Given two DAGs, $G = \{V, E, \Sigma, l\}$ and $G' = \{V', E', \Sigma', l'\}$, if there exists a subgraph isomorphism from G to G' under function f, then $POSet(G) \subseteq POSet(G')$, i.e., $\forall v_1, v_2 \in V$ and $v_1 < v_2$, we have $f(v_1), f(v_2) \in V'$ and $f(v_1) < f(v_2)$.*

Proof (Proof by contrapositive). For DAGs, G and G', G' contains G under function $f : V \to V'$. Assume $v_1, v_2 \in V$, $v_1 < v_2$, but $f(v_1) \not< f(v_2)$, we have two possible cases. Case I: if edge $<v_1, v_2> \in E$, then according to definition 3 and 6, we have edge $<f(v_1), f(v_2)> \in E'$ and $f(v_1) < f(v_2)$, a contradict occurs. Cast II: if edge $<v_1, v_2> \notin E$, then there exists a path from v_1 to v_2 in G, say $v_1, v_{i+1}, ..., v_{i+k}, v_2$. According to definition 3, there must exist a path from $f(v_1)$ to $f(v_2)$, which is $f(v_1), f(v_{i+1}), ..., f(v_{i+k}), f(v_2)$. Then by definition 6, we have $f(v_1) < f(v_{i+1})$, $f(v_{i+j}) < f(v_{i+j+1})$, $f(v_{i+k-1}) < f(v_{i+k})$, and then the transitivity of partial order leads to $f(v_1) < f(v_2)$, which contradicts our assumption, $f(v_1) \not< f(v_2)$.

With Theorem 1, we can filter DAG containment queries by checking whether the partial order set containment relationship fulfills or not.

3.2 Virtual Vertex

Cycles are ubiquitous in digraphs. Within a cycle, all vertices are reachable mutually. Therefore, reflexivity holds for all cycle vertices, which means Theorem 1 does not work any longer. The strongly connected graphs/subgraphs are general scenarios, where vertices are connected to each other. We introduce a notion of virtual vertex to solve this problem.

Definition 7 *(Virtual vertex).* *A virtual vertex of graph $G = \{V, E, \Sigma, l\}$ is defined as $V_v = <M_v, l_v>$, where $M_v \subseteq V$ and l_v is the label of V_v. For any two vertices $v_1, v_2 \in M_v$, the partial relationship $v_1 < v_2$ holds.*

Clearly, if all the strongly connected components of a digraph are converted and replaced by virtual vertices, the digraph will be converted into a DAG. As a result, Theorem 1 will be still valid.

As Fig. 3(a) shows, there is a strongly connected component containing vertices b, c, d and e. A virtual vertex labeled as $'bcde'$ is constructed to replace it. This virtual vertex includes four original graph vertices, b, c, d and e. An edge

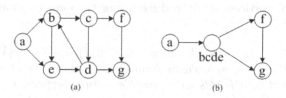

(a)   (b)

Fig. 3. Cycles and virtual vertex

from vertex a to vertex $bcde$ is added since there are edges from vertex a to b and a to e in the original digraph, and the edges $<bcde, f>$ and $<bcde, g>$ are added due to similar reasons.

```
 1: procedure Digraph2DAG(Digraph G = {V, E, Σ, l})
 2:    for each strongly connected component C = {Vc, Ec, Σ, l} of G do
 3:        new a virtual vertex Vv
 4:        V←V ∪ {Vv}; l(Vv) ←label set of Vc
 5:        for each vertex v in Vc do
 6:            for each vertex u in V − Vc do
 7:                if <v, u> ∈ E then
 8:                    E ← E ∪ {(Vv, u)}; E ← E − {<v, u>}
 9:                end if
10:                if <u, v> ∈ E then
11:                    E ← E ∪ {(u, Vv)}; E ← E − {<u, v>}
12:                end if
13:            end for
14:        end for
15:        V←V-Vc; E←E-Ec
16:    end for
17: end procedure
```

Fig. 4. Converting Digraph into DAG with virtual vertices

Fig. 4 gives the algorithm of converting a digraph into a DAG with virtual vertices. Tarjan algorithm[12] can be used to get the strongly connected components (line 2). For every strongly connected component, a virtual vertex is constructed and labeled with all the vertex labels of the component (line 3 and 4). If there exists a fan-in/fan-out edge incident with any vertex of the component, then the edge is removed and a fan-in/fan-out edge incident with the virtual vertex is accordingly added into the digraph (line 5 to line 14). Finally, the component is removed from the original digraph (line 15).

3.3   Filter Digraphs with Virtual Vertices

By converting digraphs with strongly connected components into DAGs with virtual vertices, the partial order sets consist of partial orders both from directed edges and virtual vertices. We introduce a new theorem to guarantee the filter ability of partial order set.

Theorem 2. *Given two digraphs, $G = \{V, E, \Sigma, l\}$ and $G' = \{V', E', \Sigma', l'\}$, if there exists a subgraph isomorphism from G to G' under function f, then after they are converted into DAGs with possible virtual vertices, G_v and G'_v, have $POSet(G_v) \subseteq POSet(G'_v)$, i.e., $\forall v_1, v_2 \in V$ and $v_1 < v_2$, we have $f(v_1), f(v_2) \in V'$ and $f(v_1) < f(v_2)$.*

Proof (Direct proof). There are three cases. Case I: G_v and G'_v do not contain any virtual vertices, for which, Theorem 1 has proved its correctness. Case II: $f(v_1)$ and $f(v_2)$ are both in a virtual vertex of G'_v. In this case, no matter v_1 and v_2 are in a virtual vertex of G_v or not, $f(v_1) < f(v_2)$ holds according to Definition 7. Case III: $f(v_1)$ and $f(v_2)$ are not in a virtual vertex of G'_v, and v_1 and v_2 are in a virtual vertex of G_v. It means v_1 and v_2 are mutually reachable, while $f(v_1)$ and $f(v_2)$ are not, which is in contradiction with the assumption G' contains G. Thus, this case never happens.

As a matter of fact, if digraphs are strongly connected, Theorem 2 could not provide more filter ability than node set test since they have all the possible partial orders. In this paper, we do not consider this scenario.

4   Filter Digraphs with Layered Vertex Sequence

4.1   Layered Vertex Sequence

The partial order sets of digraphs are proved effective in filtering digraph containment queries by Theorem 1 and Theorem 2. However, The size of partial order set is exponential to the graph size, which means the partial order set based filtering is inefficient in practice. In this section, we demonstrate how to convert a digraph into an onion-like layered sequence by introducing the notion of **vertex layer** and **layered vertex sequence**.

Definition 8 *(Vertex Layer).* For a digraph $G = \{V, E, \Sigma, l\}$, its vertex set V is partitioned into several subsets according to the partial order constraints, say $V = V_1 \cup V_2 \cup ... \cup V_n$, if

- $V_i \cap V_j = \emptyset (1 \leq i, j \leq n, i \neq j)$,
- $\forall u, v \in V_i$, then $u \not< v$ and $v \not< u$,
- if $u < v$, $u \in V_i$, and $v \in V_j$, then $i < j$,
- $\forall u \in V_i, \forall v \in V_j$ $(i < j)$, then $u < v$ or $u \not< v$, $v \not< u$,
- $\forall u \in V_i (i > 1)$, $\exists v \in V_{i-1}$ so that $v < u$.

then V_i is called the ith layer of G, noted as $VL(G, i)$.

Definition 9 *(Layered Vertex Sequence, LVS).* Given a digraph $G = \{V, E, \Sigma, l\}$, its layered vertex sequence is defined as follows,
   $LVS(G) = \{v | v \in VL(G, 1)\}/\{v | v \in VL(G, 2)\}/.../\{v | v \in VL(G, n)\}$.
   The layered vertex sequence from layer l to n is defined as follows,
   $LVS(G, l) = \{v | v \in VL(G, l)\}/\{v | v \in VL(G, l+1)\}/.../\{v | v \in VL(G, n)\}$.

After replacing the strongly connected components with virtual vertices, it is easy to get the layered vertex sequence of a digraph by slightly modifing the topology sorting algorithm[13]. The time complexity is $O(\|V_g\| + \|E_g\|)$. Take the graph g_1 in Fig. 2 as an example, there are three layers, $VL(g_1, 1) = \{a\}$, $VL(g_1, 2) = \{b\}$, and $VL(g_1, 3) = \{c, d\}$, respectively. The layered vertex sequence of g_1 is $'a/b/\{c, d\}'$.

4.2   Filter Digraphs with LVS

To filter digraph containment queries with layered vertex sequences, we have the following theorem.

Theorem 3. *Given two digraphs, $G = \{V, E, \Sigma, l\}$ and $G' = \{V', E', \Sigma', l'\}$, together with $LVS(G)$ and $LVS(G')$, if there exists a subgraph isomorphism from G to G' under function f, then*

- $\forall v \in V$ and its mapping node $f(v) \in V'$, if $v \in VL(G, l_G)$ and $f(v) \in VL(G', l_{G'})$, then $l_G \not> l_{G'}$;
- Let $VL(G, l_G) = \{v_1, v_2, ..., v_m\}$, $f(v_i) \in VL(G', l_{G'}^{v_i})$ $(1 \leq i \leq m)$ and $l_{min} = min(l_{G'}^{v_1}, l_{G'}^{v_2}, ..., l_{G'}^{v_m})$, then $\forall u_j \in VL(G, l_G + 1)(1 \leq j \leq k)$ and $f(u_j) \in VL(G', l_{G'}^{u_j})$, have $l_{G'}^{u_j} > l_{min}$ and $min(l_{G'}^{u_1}, l_{G'}^{u_2}, ..., l_{G'}^{u_k}) = l_{min} + 1$.

Proof (Direct proof). For the first statement, vertex $f(v)$ is in layer $l_{G'}$ means that G' has a partial order sequence of length $l_{G'}$ with a tail $f(v)$. Accordingly, G has a partial order sequence of length l_G with a tail v. If $l_G > l_{G'}$, then from Theorem 2, we know that G is not a subgraph of G', which contradicts the assumption that there exists a subgraph isomorphism from G to G'. To the second statement, it is easy to know that $l_{G'}^{v_i} \geq l_G (1 \leq i \leq m)$ and $l_{G'}^{u_j} \geq l_G + 1(1 \leq j \leq k)$, which results to $l_{G'}^{u_j} > min(l_{G'}^{v_1}, l_{G'}^{v_2}, ..., l_{G'}^{v_m}) = l_{min}$ and $min(l_{G'}^{u_1}, l_{G'}^{u_2}, ..., l_{G'}^{u_k}) = l_{min} + 1$.

```
 1: procedure PORDER(Query q, LVSes of Database D_G : D_LVS)
 2:     CanSet ← ∅; lvs_q ← LVS(q)
 3:     for each lvs_g ∈ D_LVS do
 4:         qCurrLyr ← 1
 5:         while true do
 6:             if  ∀ v ∈ LV(q, qCurrLyr),
 7:                 ∃f(v) ∈ LVS(g, gCurrLyr) then
 8:                 qCurrLyr ← qCurrLyr + 1; gCurrLyr ← min(layer_of(f(v))) + 1
 9:             else
10:                 break
11:             end if
12:         end while
13:         CanSet ← Canset ∪ {g}
14:     end for
15:     return CanSet
16: end procedure
```

Fig. 5. LVS matching based digraph containment filtering algorithm

According to Theorem 3, we can filter digraphs by matching layered vertex sequences, like peeling an onion layer by layer. The filtering algorithm **POrder** is shown in Fig. 5. The worst time complexity of POrder is $\|D_G\| * \|q\| * gSize$

where $gSize$ is the average graph size in database D_G. The worst case happens when there are multipule matching vertices for vertices of q. While the best time complexity is $\|D_G\| * (\|q\| + gSize)$ when there is at most only one matching vertex for each vertex of q.

5   Optimized Layered Vertex Sequences

During matching, POrder checks simply whether the vertex labels are the same or not. Two more complicated and powerful layered vertex sequences are introduced as follows.

Definition 10 *(LVS with Vertex Information, LVS_{vi})*. *Given a digraph $G = \{V, E, \Sigma, l\}$, its layered vertex sequence with vertex informaiton is defined as follows,*
$$LVS_{vi}(G)=\{v[u|edge<v,u> \in E]|v \in VL(G,1)\}/.../$$
$$\{v[u|edge<v,u> \in E]|v \in VL(G,n)\}$$

Each vertex in LVS_{vi} defined in Definition 10 contains its fan-out neighbours, which filters out more unwanted graphs for sure. For digraph g_1 in Fig. 2, $LVS_{vi}(g_1)$ is $'\{a[b]\}/\{b[c,d]\}/\{c[],d[]\}'$.

Layered vertex sequence can filter digraphs since it keeps the partial order constraints. As a matter of fact, unlike real onions, partial order does not necessarily exist between vertices in different layers. For q in Fig. 6, b_1 and b_2 are both labeled $'b'$, its LVS is $'\{g,b_1\}/\{f,b_2\}/\{h\}/\{k\}'$. Vertex f and h are in different layers, but there is no partial order between them. However, POrder checks all the vertices to see whether the partial orders hold or not even they are non-direct partial orders. So the worst time complexity is $\|D_G\| * \|q\| * gSize$. In order to improve the filtering ablility, we propose to divide layered vertex sequence into several shorter sequences named as LVS_{dpo}, where non-direct partial orders are eliminated.

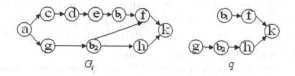

Fig. 6. Example of LVS_{dpo} matching

Definition 11 *(LVS with Direct Partial Order, LVS_{dpo})*. *Given a digraph $G = \{V, E, \Sigma, l\}$, its layered vertex sequence with direct partial order is a set of LVSes. Within each sequence, given any two vertices u and v, there must have a path from u to v or from v to u.*

In Fig. 6, $LVS_{dpo}(q)$ consists of two sequences, $'b_1/f/k'$ and $'g/b_2/h/k'$.

After introducing the two optimized layered vertex sequences, the digraph filtering algorithm should be revised accordingly. Then we have algorithm **POrder+**, where the optimized layered vertex sequences are used, noted as LVS_{opt}.

```
1: procedure PORDER+(Query q, LVSes of Database D_G : D_{LVSopt})
2:     CanSet ← ∅; lvs_q ← LVS_{opt}(q)
3:     for each lvs ∈ lvs_q do
4:         for each lvs_g ∈ D_{LVSopt} do
5:             match lvs with each sequence of lvs_g
6:             if match fails then
7:                 continue
8:             end if
9:         end for
10:        CanSet ← Canset ∪ {g}
11:    end for
12:    return CanSet
13: end procedure
```

Fig. 7. LVS matching based digraph containment filtering algorithm

Theoretically, the time complexity of algorithm **POrder+** is almost the same as algorithm **POrder**. However, due to the characteristics of the digraphs, the optimized vertex sequences can filter out more unqualified graphs.

6   Performance Evaluation

6.1   Data Sets and Setups

The data set is produced by GraphGen[10]. Due to the space limitation, we will only demonstrate a few representative results. The data set consists of 10,000 graphs. The average size is 30, average densities are 0.3, 0.5 and 0.7. There are five kinds of query graphs with different sizes, 4, 8, 12, 16 and 24, respectively. The experiment setup is with CPU of Intel(R) Xeon(R) E5506 @2.13GHz, with memory of 12GB, with OS of Windows 7-x64.

In order to evaluate the performance of the proposed algorithms, we compare them with gIndex, FG-Index, Tree+Δ and GCoding implemented in iGraph[10].

6.2   Performance Analysis

Experiment 1. As Fig. 8 shows, compared with other algorithms, POrder and POrder+ have much less index construction costs. Layered vertex sequences are used as index entries by POrder and POrder+, the computing complexity of layered vertex sequence is linear to the number of graph vertices and edges. The cost of POrder+ is a little bit higher due to the computation of optimized layered vertex sequences. Nevertheless, POrder+ is still hundreds of, even thousands of times faster than the others. Algorithm gIndex needs to mine frequent subgraphs, which is very time consuming. Graph density affects frequent subgraph mining greatly, which leads to the highest cost of gIndex construction at density 0.7.

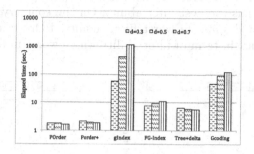

Fig. 8. Index construction time

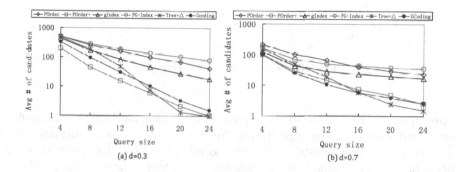

Fig. 9. Comparison of filter capabilities

Experiment 2. Fig. 9 shows the filter capabilities of POrder and POrder+ against the existed four algorithms by measuring the average number of candidates. For all algorithms, more complex query graph leads to more index entries, further leads to stricter filter conditions, and results in smaller candidate sets. The naive sequence (POrder) is not good at filtering because digraphs are far more complex than onions. In a digraph, there exits no explicit layer order relationship between its vertices; while for an onion, it has inherently a linear layer order from the skin to the core. When the query graph size is small (less than 16), the selectivity of POrder is the worst. The filter ability of POrder improves faster than that of gIndex when the query size increases, since the size of partial order constraints is nearly an exponential function of graph size. More partial order constraints means more filter parameters. The filter ability of optimized layered vertex sequences (POrder+) is tens of times better than POrder, and it is similar to the filter ability of the existing best algorithm Tree+Δ, while Tree+Δ has much higher index construction cost due to Fig. 8.

Experiment 3. Fig. 10 shows the average elapsed time of filtering phases of the six algorithms. The query graph size does not affect the filtering time of any algorithm except Tree+Δ. Tree+Δ has the worst performance (exponential to graph size) even it has the best filter ability. On the contrary, FG-Index filters digraphs the fastest despite its worst filter ablility. POrder and POrder+ are better than FG-Index in term of filter ability where worse in filter time.

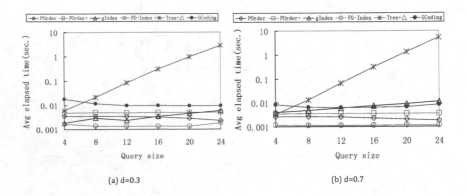

(a) d=0.3

(b) d=0.7

Fig. 10. Performance of filtering phase

Experiment 4. Fig. 11 shows the average elapsed time for digraph containment query processing in the six algorithms. Tree+Δ has the worst performance even it has the best filter ability. The query graph size affects its query time significantly. Due to its simplicity for sequence matching (onion peeling), POrder has the similar time overhead as the best existing ones (FG-Index and GCoding) despite its poor filter ability. POrder+ outperforms all the others with no doubt because it has high filter ability of the optimized layered vertex sequences and high efficiency of sequence matching based filtering. Why both POrder and POrder+ perform consistently when query size increases? The reason is most of the sequence matchings are unsuccessful and terminate in middle way no matter how long the sequences are.

Experiment 5. Fig. 12 shows the average elapsed time for online digraph containment query processing in the six algorithms. Online query means processing queries when the database is updating, where both the index maintaining cost and the querying cost are take into consideration. We can clearly see that POrder and POrder+ are more efficient than the others since their index maintaining costs are much lower according to Fig. 8. For the same reason, POrder outperforms POrder+. From Fig. 11 and Fig. 12, we can also see that the index maintaining cost is the dominent factor of the online query performance of

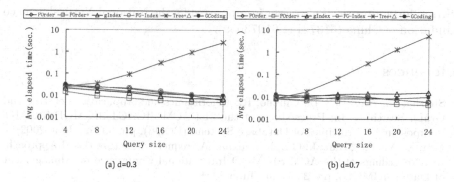

Fig. 11. Performance of digraph containment querying

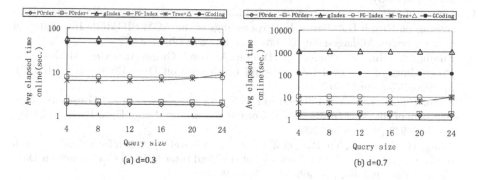

Fig. 12. Performance of online containment querying

gIndex, FG-Index, Tree+Δ and GCoding, while POrder and POrder+ are both update friendly.

7   Conclusions

Graph feature based index structures are widely used to filter out unwanted graphs. However, most of the existing index structures have expensive constructing overheads for the mining of frequent features. We demonstrate that partial order constraints in digraphs are much easier to obtain and can be used to filter containment queries. Based on partial order constraints, we propose to partition graph vertices into different layers and convert partial order sets into layered vertex sequences. The sequence matching based graph containment filtering is proved to be effective and efficient. Two optimized layered vertex sequences, LVS with vertex information and LVS with direct partial orders, are further introduced to improve the filter ability.

Acknowledgements. This paper is supported by the National Natural Science Foundation of China (61374141,61073059).

References

1. Shasha, D., Wang, J.T.L., Giugno, R.: Algorithmics and Applications of Tree and Graph Searching. In: Proceedings of the 21st ACM SIGACT-SIGMOD-SIGART Symposium on Principles of Database Systems (PODS), pp. 39–52 (June 2002)
2. Yan, X., Yu, P.S., Han, J.: Graph Indexing: A Frequent Structure Based Approach. In: Proceedings of the ACM SIGMOD International Conference on Management of Data (SIGMOD), pp. 335–346 (June 2004)
3. Yan, X., Yu, P.S., Han, J.: Graph Indexing Based on Discriminative Frequent Structure Analysis. ACM Transactions on Database Systems (TODS) 30(4), 960–993 (2005)
4. Cheng, J., Ke, Y., Ng, W., et al.: FG-Index: Towards Verification Free Query Processing on Graph Databases. In: Proceedings of the ACM SIGMOD International Conference on Management of Data, pp. 857–872 (June 2007)
5. Zhang, S., Hu, M., Yang, J.: Treepi: A Novel Graph Indexing Method. In: Proceedings of the 23rd International Conference on Data Engineering (ICDE), pp. 966–975 (April 2007)
6. Zhao, P., Yu, J.X., Yu, P.S.: Graph Indexing: Tree + Delta >= Graph. In: Proceedings of the 33rd International Conference on Very Large Data Bases (VLDB), pp. 938–949 (September 2007)
7. Jiang, H., Wang, H., Yu, P.S., et al.: Gstring: A Novel Approach for Efficient Search in Graph Databases. In: Proceedings of the 23nd International Conference on Data Engineering (ICDE), pp. 566–575 (April 2007)
8. He, H., Singh, A.K.: Closure-tree: An Index Structure for Graph Queries. In: Proceedings of the 22nd International Conference on Data Engineering (ICDE), pp. 38–47 (April 2006)
9. Zou, L., Chen, L., Yu, J.X., Lu, Y.: A novel spectral coding in a large graph database. In: Proceedings of the 11th International Conference on Extending Database Technology (EDBT), pp. 181–192 (March 2008)
10. Han, W.S., Lee, J., Pham, M.D., et al.: iGraph: A Framework for Comparisons of Disk-Based Graph Indexing Techniques. In: Proceedings of the 36rd International Conference on Very Large Data Bases (VLDB), pp. 449–559 (September 2010)
11. Huan, J., Wang, W., Prins, J.: Efficient Mining of Frequent Subgraphs in the Presence of Isomorphism. In: Proceedings of the 3rd IEEE International Conference on Data Mining (ICDM), pp. 549–552 (2003)
12. Hacupler, B., Kavitha, T., Mathew, R., Sen, S., Tarjan, R.E.: Incremental Cycle Detection, Topological Ordering, and Strong Component Maintenance. ACM Transactions on Algorithms 8(1), 3 (2012)
13. Corman, T.H., Leiserson, C.E., Rivest, R.L., et al.: Introduction to Algorithms. The MIT Press (2001)

Efficient Multidimensional AkNN Query Processing in the Cloud

Nikolaos Nodarakis[1], Evaggelia Pitoura[2], Spyros Sioutas[3],
Athanasios Tsakalidis[1], Dimitrios Tsoumakos[3], and Giannis Tzimas[4]

[1] Computer Engineering and Informatics Department, University of Patras,
26500 Patras, Greece
{nodarakis,tsak}@ceid.upatras.gr
[2] Computer Science Department, University of Ioannina, Greece
pitoura@cs.uoi.gr
[3] Department of Informatics, Ionian University,
49100 Corfu, Greece
{sioutas,dtsouma}@ionio.gr
[4] Computer & Informatics Engineering Department,
Technological Educational Institute of Western Greece, 26334 Patras, Greece
tzimas@cti.gr

Abstract. A k-nearest neighbor (kNN) query determines the k nearest points, using distance metrics, from a given location. An all k-nearest neighbor (AkNN) query constitutes a variation of a kNN query and retrieves the k nearest points for each point inside a database. Their main usage resonates in spatial databases and they consist the backbone of many location-based applications and not only. In this work, we propose a novel method for classifying multidimensional data using an AkNN algorithm in the MapReduce framework. Our approach exploits space decomposition techniques for processing the classification procedure in a parallel and distributed manner. To our knowledge, we are the first to study the kNN classification of multidimensional objects under this perspective. Through an extensive experimental evaluation we prove that our solution is efficient, robust and scalable in processing the given queries.

Keywords: classification, nearest neighbor, MapReduce, Hadoop, multidimensional data, query processing.

1 Introduction

Classification is the problem of identifying to which of a set of categories a new observation belongs, on the basis of a training set of data containing observations (or instances) whose category membership is known. One of the algorithms for data classification uses the kNN approach [6] as it computes the k nearest neighbors (belonging to the training dataset) of a new object and classifies it to the category that belongs the majority of its neighbors.

H. Decker et al. (Eds.): DEXA 2014, Part I, LNCS 8644, pp. 477–491, 2014.
© Springer International Publishing Switzerland 2014

A k-nearest neighbor query [11] computes the k nearest points, using distance metrics, from a specific location and is an operation that is widely used in spatial databases. An all k-nearest neighbor query constitutes a variation of a kNN query and retrieves the k nearest points for each point inside a dataset in a single query process. Although AkNN is a fundamental query type, it is computationally very expensive. As a result, quite a few centralized algorithms and structures (M-trees, R-trees, space-filling curves, etc.) have been developed towards this direction [4], [7], [22]. However, as the volume of datasets grows rapidly even these algorithms cannot cope with the computational burden produced by an AkNN query process. Consequently, high scalable implementations are required. Cloud computing technologies provide tools and infrastructure to create such solutions and manage the input data in a distributed way among multiple servers. The most popular and notably efficient tool is the *MapReduce* [5] programming model, developed by Google, for processing large-scale data.

In this paper, we propose a method for efficient multidimensional data classification using AkNN queries in a single batch-based process in *Hadoop* [14], [16], the open source MapReduce implementation. More specifically, we sum up the technical contributions of our paper as follows:

- We present an implementation of a classification algorithm based on AkNN queries using MapReduce. We apply space decomposition techniques (based on data distribution) in order to bound the amount of distance calculations needed to reckon the k-NN objects before the classification step. The implementation defines the MapReduce jobs with no modifications to the original Hadoop framework.
- We provide an extension for $d > 3$ in Section 5 (d stands for dimensionality).
- We evaluate our solution through an experimental evaluation against large scale data up to 3 dimensions, that studies various parameters that can affect the total computational cost of our method using real and synthetic datasets. The results prove that our solution is efficient, robust and scalable.

The rest of the paper is organized as follows: Section 2 discusses related work. Section 3 presents the initial idea of the algorithm, our technical contributions and some examples of how the algorithm works. Section 4 presents a detailed analysis of the classification process developed in Hadoop, Section 5 provides an extension for $d > 3$ and Section 6 presents the experiments that where conducted in the context of this work. Finally, Section 7 concludes the paper and presents future steps.

2   Related Work

AkNN queries have been extensively studied in literature. A structure that is popular for answering efficiently to AkNN queries is R-tree [11]. Pruning techniques can be combined with such structures to deliver better results [4], [7]. Moreover, efforts have been made to design low computational cost methods that execute such queries in spatial databases [18]. The works in [17], [20] propose algorithms to answer kNN join.

The methods proposed above can handle data of small size in one or more dimensions, thus their use is limited in centralized environments only. During the recent years, the researchers have focused on developing approaches that are applicable in distributed environments, like our method, and can manipulate big data in an efficient manner. The MapReduce framework seems to be suitable for processing such queries. For example, in [19] the discussed approach splits the target space in smaller cells and looks into appropriate cells where k-NN objects are located, but applies only in 2-dimensional data. Our method speeds up the naive solution of [19] by eliminating the merging step, as it is a major drawback. We have to denote here that in [19] it is claimed that the computation of the merging step can be performed in one node since we just consider statistic values. But this is not entirely true as we are going to see in the experimental evaluation. In addition, the merging step can produce sizeable groups of points, especially as k increments, that can overload the AkNN process. Moreover, our method applies for more dimensions. Especially, for $d >= 3$ the multidimensional extension is not straightforward at all.

In [13], locality sensitive hashing (LSH) is used together with a MapReduce implementation for processing kNN queries over large multidimensional datasets. This solution suggests an approximate algorithm like the work in [21] (H-zkNNJ) but we focus on exact processing AkNN queries. Furthermore, AkNN queries are utilized along with MapReduce to speed up and optimize the join process over different datasets [1], [10] or support non-equi joins [15]. Moreover, [2] makes use of a R-tree based method to process kNN joins efficiently.

In [3] a minimum spanning tree based classification model is introduced and it can be viewed as an intermediate model between the traditional k-nearest neighbor method and cluster based classification method. Another approach presented in [9] recommends parallel implementation methods of several classification algorithms but does not contemplate the perspective of dimensionality.

In brief, our proposed method implemented in the Hadoop MapReduce framework, extends the traditional kNN classification algorithm and processes exact AkNN queries over massive multidimensional data to classify a huge amount of objects in a single batch-based process. The experimental evaluation considers a wide diversity of factors that can affect the execution time such as the value of k, the granularity of space decomposition, dimensionality and data distribution.

3   Overview of Classification Algorithm

In this section, we first define some notation and provide some definitions used throughout this paper. Table 1 lists the symbols and their meanings. Next, we give a brief review of the method our solution relies on and then we extend it for more dimensions and tackle some performance issues.

Table 1. Symbols and their meanings

n	granularity of space decomposition
k	number of nearest neighbors
d	dimensionality
D	a d-dimensional metric space
$dist(r,s)$	the distance from r to s
$kNN(r,S)$	the k nearest neighbors of r from S
$AkNNC(R,S)$	$\forall r \in R$ classify r based on $kNN(r,S)$
I	input dataset
T	training dataset
c_r	the class of point r
C_T	the set of classes of dataset T
S_I	size of input dataset
S_T	size of training dataset
M	total number of Map tasks
R	total number of Reduce tasks

3.1 Definitions

We consider points in a d-dimensional metric space D. Given two points r and s we define as $dist(r,s)$ the distance between r and s in D. In this paper, we used the distance measure of Euclidean distance

$$(r,s) = \sqrt{\sum_{i=1}^{d} (r[i] - s[i])^2}$$

where $r[i]$ (respectively $s[i]$) denote the value of r (respectively s) along the i-th dimension in D. Without loss of generality, alternative distance measures (i.e. Manhattan distance) can be applied to our solution.

Definition 1. kNN: *Given a point r, a dataset S and an integer k, the k nearest neighbors of r from S, denoted as $kNN(r,S)$, is a set of k points from S such that $\forall p \in kNN(r,S)$, $\forall q \in \{S - kNN(r,S)\}, dist(p,r) < dist(q,r)$.*

Definition 2. AkNN: *Given two datasets R,S and an integer k, the all k nearest neighbors of R from S, named $AkNN(R,S)$, is a set of pairs (r,s) such that $AkNN(R,S) = \{(r,s) : r \in R, s \in kNN(r,S)\}$.*

Definition 3. AkNN Classification: *Given two datasets R,S and a set of classes C_S where points of S belong, the classification process produces a set of pairs (r,c_r), denoted as $AkNNC(R,S)$, such that $AkNNC(R,S) = \{(r,c_r) : r \in R, c_r \in C_S\}$ where c_r is the class where the majority of $kNN(r,S)$ belong $\forall r \in R$.*

3.2 Classification Using Space Decomposition

Consider a training dataset T, an input dataset I and a set of classes C_T where points of T belong. First of all, we define as *target space* the space enclosing the points of I and T. The parts that occur when we decompose the target space for 1-dimensional objects are called *intervals*. Respectively, we call *cells* and *cubes* the parts in case of 2 and 3-dimensional objects and hypercubes for $d > 3$. For a new 1D point p, we define as *boundary interval* an interval centred at p that covers k-NN elements. Respectively, we define the *boundary circle* and *boundary sphere* for 2D and 3D points and the *boundary hypersphere* for $d > 3$. The notion of hypercube and hypersphere are analyzed further in Section 5. When the boundary ICSH (interval, circle, sphere or hypersphere) centred in an ICCH (interval, cell, cube or hypercube) $icch_1$, intersects the bounds of an other $icch_2$ we say an *overlap* occurs on $icch_2$. Finally, for a point $i \in I$, we define as *updates* of $kNN(i, T)$ the existence of many different instances of $kNN(i, T)$ that need to be unified to a final set.

We place the objects of T on the target space according to their coordinates. The main idea of equal-sized space decomposition is to partition the target space into n^d equal sized ICCHs where n and the size of each ICCH are user defined. Each ICCH contains a number of points of T. Moreover, we define a new layer over the target space according to C_T and $\forall t \in T, c_t \in C_T$. In order to estimate $AkNNC(I, T)$, we investigate $\forall i \in I$ for k-nearest neighbors only in a few ICCHs, thus bounding the number of computations required.

3.3 Previous Work

A very preliminary study of naive AkNN solutions is presented in [19] and uses a simple cell decomposition technique to process AkNN queries on two different datasets, i.e. I and T. The elements of both datasets are placed on the target space, which comprises of $2^n \times 2^n$ equal-sized cells, according to their coordinate vector and a cell decomposition is applied. $\forall i \in I$ it is expected that its $kNN(i, T)$ will be located in a close range area defined by nearby cells. At first, we look for candidate k-NN points inside the cell (cl) that i belongs in the first place. If we find at least k elements we draw the boundary circle. In case any neighboring cells are overlapped we need to investigate for possible k-NN objects inside them. If no overlap occurs, the k-NN list of i is complete. The algorithm outputs an instance of the k-NN list for every overlapped cell. These instances need to be unified into a final k-NN list.

This approach, as described above, fails to draw the boundary circle if cl contains less than k points. To overcome this issue, before starting calculating $kNN(i, T)$, we need to estimate the number of points that fall into every cell and merge neighboring cells (according to the principles of hierarchical space decomposition used in quad-trees [12]) to assure that all will contain at least k objects. This preprocessing phase induces additional cost to the total computation and the merging step can lead to a bad algorithmic behavior.

3.4 Technical Contributions

In this subsection, we extend the previous method for more dimensions and adapt it to the needs of the classification problem. Moreover, we analyze some drawbacks of the method studied in [19] and propose a mechanism to make the algorithm more efficient. Firstly, we have a training dataset T, an input dataset I and a set of classes C_T where points of T belong. The points in the training dataset have two attributes, the coordinate vector and the class they belong. In order to compute $AkNNC(I, T)$, a classification step is executed after the construction of the k-NN lists. Furthermore, we extend the solution presented in [19] for more dimensions, and now the space is decomposed in 2^{dn} ICCHs.

Figure 1(a) depicts a situation where the merging step of the original method in [19] can significantly increase the total cost of the algorithm. Consider two points x and y entering cells 3 and 2 respectively and $k = 3$. We can draw point's x boundary circle since cell 3 includes at least k elements. On the contrary, we cannot draw the boundary circle of point y, so we need to unify cells 1 through 4 into one bigger cell. Now point y can draw its boundary circle but we overload point's x k-NN list construction with redundant computations and this would happen for all points that would join cells 1,3 and 4 in the first place.

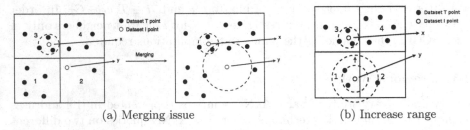

(a) Merging issue   (b) Increase range

Fig. 1. Issue of the merging step before the kNN process and way to avoid it ($k = 3$)

In order to avoid a scenario like above, we introduce a mechanism where only points that cannot find at least k-nearest neighbors in the ICCH in the first place proceed to further actions. Let a point p joining an ICCH $icch$ that encloses $l < k$ neighbors. We draw the boundary ICSH based on these l neighbors and then check if the boundary ICSH overlaps any neighboring ICCHs. In case it does, if the boundary ICSH covers at least k elements in total, then we are able to build the final k-NN list of the point. In case the boundary ICSH does not cover at least k objects in total or does not overlap any ICCHs then we gradually increase its search range (by a fraction of the size of the ICCH each time) until the prerequisites are fulfilled.

Figure 1(b) explains this issue. Consider two points x and y entering cells 3 and 1 respectively and $k = 3$. We observe that cell 3 contains 4 neighbors and point x can draw its boundary circle that covers k-NN elements. However, the boundary circle centred at y does not cover k-NN elements. Consequently, we gradually increase its search range until the boundary circle encloses at least k-NN points. By eliminating the merging step, we also relax the condition of decomposing the target space into 2^{dn} equal-sized splits and generalize it to n^d.

Summing up, our solution can be implemented as a series of MapReduce jobs as shown below. Note, that the first MapReduce job acts as a preprocessing step and its results are provided as additional input in MapReduce Job 3 and that the preprocessing step is executed only once for T.

1. **Distribution Information.** Count the number of points of T that fall into each ICCH. The output of this job is utilized by the third MapReduce job to help determine how much we need to increase the boundary ICSH.
2. **Primitive Computation Phase.** Calculate possible k-NN points $\forall i \in I$ from T in the same ICCH.
3. **Update Lists.** Draw the boundary ICSH $\forall i \in I$ and increase it, if needed, until it covers at least k-NN points of T. Check for overlaps of neighboring ICCHs and derive updates of k-NN lists.
4. **Unify Lists.** Unify the updates of every k-NN list into one final k-NN list $\forall i \in I$.
5. **Classification.** Classify all points of I.

4   Detailed Analysis of Classification Procedure

In this section, we present a detailed description of the classification process as implemented in the Hadoop framework. The records in T have the format <point_id, coordinate_vector, class> and in I have the format <point_id, coordinate_vector>. Furthermore, parameters n and k are defined by the user. In the following subsections, we describe each MapReduce job separately and analyze the Map and Reduce functions that take place in each one of them[1]. Also, we proceed in time and space complexity analysis.

4.1   Getting Distribution Information of Training Dataset

This MapReduce job is a preprocessing step required by subsequent MapReduce jobs that receive its output as additional data. In this step, we decompose the entire target space and count the number of points of T that fall in each ICCH.

The *Map* function takes as input records with the training dataset format, estimates the ICCH id for each point based on its coordinates and outputs a key-value pair where the key is ICCH id and the value is number 1. The *Reduce* function receives the key-value pairs from the Map function and for each ICCH id it outputs the number of points of T that belong to it.

Each Map task needs $O(S_T/M)$ time to run. Each Reduce task needs $O(n^d/R)$ time to run as the total number of ICCHs is n^d. So, the size of the output will be $O(n^d \cdot c_{si})$, where c_{si} is the size of sum and icch_id for an output record.

---

[1] Due to space limitations we do not quote pseudo-code for Map and Reduce functions. Pseudo-code and more details of the current work are available in a technical report in http://arxiv.org/abs/1402.7063

4.2 Estimating Primitive Phase Neighbors of AkNN Query

In this stage, we concentrate all training (L_T) and input (L_I) records for each ICCH and compute possible k-NN points for each item in L_I from L_T inside the ICCH. Below, we condense the Map and Reduce functions. We use two Map functions in this job, one for each dataset.

For each point $t \in T$, *Map1* outputs a new key-value pair in which the ICCH id, where t belongs, is the key and the value consists of the id, coordinate vector and class of t. Similarly, for each point $i \in I$, *Map2* outputs a new key-value pair in which the ICCH id where i belongs is the key and the value consists of the id and coordinate vector of i. The *Reduce* function receives a set of records from both Map functions with the same ICCH ids and separates points of T from points of I into two lists, L_T and L_I respectively. Then, the Reduce function calculates the distance for each point in L_I from L_T, estimates the k-NN points and forms a list L with the format $< p_1, d_1, c_1: \ldots : p_k, d_k, c_k >$, where p_i is the i-th NN point, d_i is its distance and c_i is its class. Finally, for each $p \in L_I$, *Reduce* outputs a new key-value pair in which the key is the id of p and the values comprises of the coordinate vector, ICCH id and list L of p.

Each Map1 task needs $O\left(S_T/M\right)$ time and each Map2 task needs $O\left(S_I/M\right)$ time to run. Suppose u_i and t_i the number of input and training points that are enclosed in an ICCH in the i-th execution of a Reduce function and $1 \leq i \leq n^d/R$. Each Reduce task needs $O\left(\sum_i u_i \cdot t_i\right)$. Let L_s to be the size of k-NN list and ICCH id $\forall i \in I$. The output size is $O\left(S_I \cdot L_s\right)$, which is $O\left(S_I\right)$.

4.3 Checking for Overlaps and Updating k-NN Lists

In this step, at first we gradually increase the boundary ICSH (how much depends on information from the first MapReduce job), where necessary, until it includes at least k points. Then, we check for overlaps between neighboring IC-CHs and derive updates of the k-NN lists. The Map and Reduce functions of this job are outlined next (again, we have two Map functions).

The *Map1* function is exactly the same as *Map1* function in the previous job. For each point $i \in I$, function *Map2* computes the overlaps with neighboring ICCHs. If no overlap occurs, it does not need to perform any additional steps and outputs a key-value pair in which ICCH id is the key and the value consists of id, coordinate vector and list L of i and a flag *true* which implies that no further process is required. Otherwise, for every overlapped ICCH it outputs a new record where ICCH id' (id of an overlapped ICCH) is the key and the value consists of id, coordinate vector and list L of i and a flag *false* that indicates we need to search for possible k-NN objects inside the overlapped ICCHs. The *Reduce* function receives a set of points with the same ICCH ids and separates the points of T from points of I into two lists, L_T and L_I respectively. After that, the Reduce function performs extra distance calculations using the points in L_T and updates k-NN lists for the records in L_I. Finally, for each $p \in L_I$ it generates a record in which the key is the id of p and the values comprises of the coordinate vector, ICCH id and list L of p.

Each Map1 task needs $O\left(S_T/M\right)$ time to run. Consider an unclassified point p initially belonging to an ICCH $icch$. Let r be the number of times we increase the search range for p and $icchov$ the number of ICCHs that may be overlapped for p. For each Map2 task the i-th execution of the Map function performs $icchov_i + r_i$ steps, where $1 \leq i \leq S_I/M$. So, each Map2 task runs in $O\left(\sum_i (icchov_i + r_i)\right)$ time. Suppose u_i and t_i the number of points of I and T respectively that are enclosed in an ICCH in the i-th execution of a Reduce function and $1 \leq i \leq n^d/R$. Each Reduce task needs $O\left(\sum_i u_i \cdot t_i\right)$. The size of updated records is a fraction of S_I. So, the size of the output is also $O\left(S_I\right)$.

4.4 Unifying Multiple k-NN Lists

During the previous step it is possible that multiple updates of a point's k-NN list might occur. This MapReduce job tackles this problem and unifies possible multiple lists into one final k-NN list for each point $i \in I$.

The *Map* function receives the records of the previous step and extracts the k-NN list for each point. For each point $i \in I$, it outputs a key-value pair in which the key is the id of i and the value is the list L. The *Reduce* function receives as input key-value pairs with the same key and computes $kNN(i, T), \forall i \in I$. The key of an output record is again the id of i and the value consists of $kNN(i, T)$.

Each Map task runs in $O\left(S_I/M\right)$. For each Reduce task, assume $updates_i$ the number of updates for the k-NN list of an unclassified point in the i-th execution of a Reduce function, where $1 \leq i < |N_I|/R$ and $|N_I|$ the number of points in input dataset. Then, each Reduce task needs $O\left(\sum_i updates_i\right)$ to run. Let, I_{id} the size of ids of all points in I and L_{final} is the size of the final k-NN list $\forall i \in I$. The size of L_{final} is constant and I_{id} is $O\left(S_I\right)$. Consequently, the size of the output is $O\left(S_I\right)$.

4.5 Classifying Points

This is the final job of the whole classification process. It is a Map-only job that classifies the input points based on the class membership of their k-NN points. The Map function receives as input records from the previous job and outputs $AkNNC(I, T)$. Each Map task runs in $O\left(S_I/M\right)$ time and output size is $O\left(S_I\right)$.

5 Extension for $d > 3$

Here we provide the extension of our method for $d > 3$. In geometry, a hypercube is a n-dimensional analogue of a square ($n = 2$) and a cube ($n = 3$) and is also called a n-cube (i.e. 0-cube is a hypercube of dimension zero and represents a point). It is a closed, compact and convex figure that consists of groups of opposite parallel line segments aligned in each of the space's dimensions, perpendicular to each other and of the same length. Figure 2 displays how to create a hypercube for $d = 4$ (4-cube) from a cube for $d = 3$. Respectively, an n-sphere is a generalization of the surface of an ordinary sphere to a n-dimensional space.

Spheres of dimension $n > 2$ are called hyperspheres. For any natural number n, an n-sphere of radius r is defined as a set of points in $(n+1)$-dimensional Euclidean space which are at distance r from a central point and r may be any positive real number. So, the n-sphere centred at the origin is defined by:

$$S^n = \{x \in \Re^{n+1} : \| x \| = r\}$$

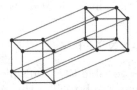

Fig. 2. Creating a 4-cube from a 3-cube

6   Experimental Evaluation

In this section, we conduct a series of experiments to evaluate the performance of our method under many different perspectives such as the value of k, the granularity of space decomposition, dimensionality and data distribution.

Our cluster includes 32 computing nodes (VMs), each one of which has four 2.1 GHz CPU processors, 4 GB of memory, 40 GB hard disk and the nodes are connected by 1 gigabit Ethernet. On each node, we install Ubuntu 12.04 operating system, Java 1.7.0_40 with a 64-bit Server VM, and Hadoop 1.0.4. To adapt the Hadoop environment to our application, we apply the following changes to the default Hadoop configurations: the replication factor is set to 1; the maximum number of Map and Reduce tasks in each node is set to 3, the DFS chunk size is 256 MB and the size of virtual memory for each Map and Reduce task is set to 512 MB. We evaluate the following approaches in the experiments: a) *kdANN*, which is the solution proposed in [19] along with the extension (which invented and implemented by us) for more dimensions, in order to be able to compare it with our solution and b) *kdANN+*, which is our solution for d-dimensional points without the merging step as described in Section 3.

We evaluate our solution using both real[2] and synthetic datasets. We create 1D and 2D datasets from the real dataset keeping the x and the (x, y) coordinates respectively. We process the dataset to fit into our solution (i.e. normalization) and we end up with 1D, 2D and 3D datasets that consist of approximately 19,000,000 points and follow a power law like distribution. Respectively, we create 1, 2 and 3-dimensional datasets containing 19,000,000 uniformly distributed points. From each dataset, we extract a fraction of points (10%) that are used as

---

[2] The real dataset is part of the Canadian Planetary Emulation Terrain 3D Mapping Dataset and is available in http://asrl.utias.utoronto.ca/datasets/3dmap/

a training dataset. For each point in a training dataset we assign a class based on its coordinate vector. The file sizes (in MB) of real datasets are a) Input: {(1D, 309.5), (2D, 403.5), (3D, 523.7)} and b) Training: {(1D, 35), (2D, 44.2), (3D, 56.2)}. The file sizes (in MB) of synthetic datasets are a) Input: {(1D, 300.7), (2D, 359.2), (3D, 478.5)} and b) Training: {(1D, 33.9), (2D, 39.8), (3D, 51.7)}.

One major aspect in the performance of the algorithm is the tuning of granularity parameter n. Each time the target space is decomposed into 2^{dn} equal parts in order for kdANN to be able to perform the merging step, as described in Section 3. The values of n that were chosen for the rest of the experiments are: a) Real dataset: (1D, 18), (2D, 9), (3D, 7) and b) Synthetic dataset: (1D, 16), (2D, 7), (3D, 5). The procedure that was carried out in order to end up with these values is described in the aforementioned technical report.

6.1 Effect of k and Effect of Dimensionality

In this experiment, we evaluate both methods using real and synthetic datasets and record the execution time as k increases for each dimension. Then, we study the effect of dimensionality on the performance of kdANN and kdANN+.

Figure 3(a) presents the results for kdANN and kdANN+ by varying k from 5 to 20 on 1D real and synthetic datasets. In terms of running time, kdANN+ always perform best, followed by kdANN and each method behave in the same way for both datasets, real and synthetic.

In Fig. 3(b), we demonstrate the outcome of the experimental procedure for 2D points when we alter k value from 5 to 20. No results of kdANN are included for the real dataset since the method only produced results for $k = 5$ and needed more than 4 hours. Beyond this, the merging step of kdANN derived extremely sizeable cells leading to a bottleneck to some nodes that strangled their resources, thus preventing them to derive any results. Overall, in the case of power law distribution, kdANN+ behaves much better than kdANN since the last one fails to process an AkNN query as k increases. Also, kdANN+ is faster and in case of synthetic dataset, especially as k grows.

Figure 3(c) displays the results generated from kdANN and kdANN+ for 3D points when we increase k value from 5 to 20. Once again, kdANN could not produce any results for any value of k in the case of real dataset. Table 2 is pretty illustrative in the way the merging step affects the AkNN process. The computational cost is far from negligible if performed in a node (in contrary with the claim of the authors in [19]). Apart from this, the largest merged cube consists of 32,768 and 262,144 initial cubes for $k = 5$ and $k > 5$ respectively. In the case of kdANN+ for the real dataset, it is obvious that the total computational cost is much larger compared to the one shown in Figs. 3(a) and 3(b). Finally, kdANN+ outperforms kdANN, in the case of synthetic dataset, and the gap between the curves of running time tends to be bigger as k increases.

Overall, looking at Figs. 3(a), 3(b) and 3(c), we observe that as k increases the execution time augments. This occurs because we need to perform more distance calculations and the size of intermediate records becomes larger respectively as the value of k rises.

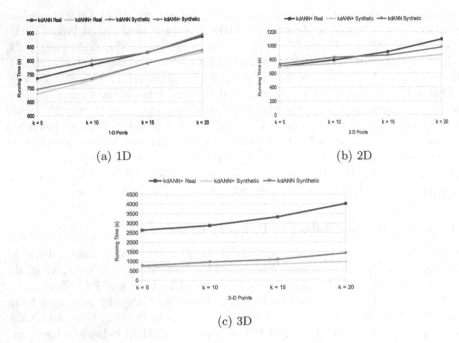

(a) 1D

(b) 2D

(c) 3D

Fig. 3. Effect of k for $d \in \{1, 2, 3\}$

Now, we evaluate the effect of dimensionality for both real and synthetic datasets. Figure 4 presents the running time for $k = 20$ by varying the number of dimensions from 1 to 3. From the outcome, we observe that kdANN is more sensitive to the number of dimensions than kdANN+ when we provide a dataset with uniform distribution as input. In particular, when the number of dimensions varies from 2 to 3 the divergence between the two curves starts growing faster. In the case of power law distribution, we notice that the execution time of kdANN+ increases exponentially when the number of dimensions varies from 2 to 3. This results from the curse of dimensionality. As the number of dimensions increases, the number of distance computations as well as the number of searches in neighboring ICCHs increases exponentially. Nevertheless, kdANN+ can still process the AkNN query in a reasonable amount of time in contrast to kdANN.

Table 2. Statistics of merging step for kdANN

	$k = 5$	$k = 10$	$k = 15$	$k = 20$
Time (s)	271	675	962	1,528
# of merged cubes	798,032	859,944	866,808	870,784
% of total cubes	38%	41%	41.3%	41.5%
Max merged cubes	32,768	262,144	262,144	262,144

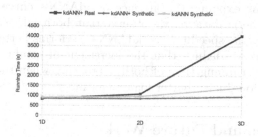

Fig. 4. Effect of dimensionality for $k = 20$

6.2 Scalability and Speedup

In this subsection, we investigate the scalability and speedup of the two approaches. In the scalability experiment we utilize the 3D datasets, since their size is bigger than the others, and create new chunks smaller in size that are a fraction F of the original datasets, where $F \in \{0.2, 0.4, 0.6, 0.8\}$. Moreover, we set the value of k to 5. Figure 5(a) presents the scalability results for real and synthetic datasets. In the case of power law distribution, the results display that kdANN+ scales almost linearly as the data size increases. In contrast, kdANN fails to generate any results even for very small datasets since the merging step continues to be an inhibitor factor in its performance. In addition, we can see that kdANN+ scales better than kdANN in the case of synthetic dataset and the running time increases almost linearly as in the case of power law distribution. Regarding kdANN, the curve of execution time is steeper until $F = 0.6$ and after that it increases more smoothly.

In our last experiment, we measure the effect of the number of computing nodes. We test four different cluster configurations and the cluster consist of $N \in \{11, 18, 25, 32\}$ nodes each time. As before, we use the 3D datasets when $k = 5$. Figure 5(b), displays that total running time of kdANN+, in the case of power law distribution, tends to decrease as we add more nodes to the cluster. Due to the increment of number of computing nodes, the amount of distance calculations and update steps on k-NN lists that undertakes each node decreases respectively. Moreover, it is obvious that kdANN will fail to produce any results

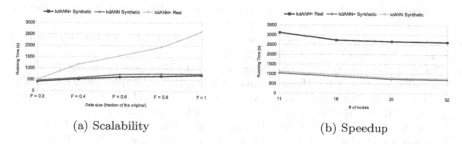

(a) Scalability   (b) Speedup

Fig. 5. Scalability and speedup results

when $N < 32$. This explains the absence of kdANN's curve from Fig. 5(a). In the case of synthetic dataset, we observe that both kdANN and kdANN+ achieve almost the same speedup; still kdANN+ performs betters than kdANN. Observing Fig. 5(b) we deduce that the increment of computing nodes has a greater effect on the running time of both approaches when the datasets follow a uniform distribution due to better load balancing.

7   Conclusions and Future Work

In the context of this work, we presented a novel method for classifying multidimensional data using AkNN queries in a single batch-based process in Hadoop. To our knowledge, it is the first time a MapReduce approach for classifying multidimensional data is discussed. By exploiting equal-sized space decomposition techniques we bound the number of distance calculations needed to compute $kNN(i, S), \forall i \in I$. We conduct a variety of experiments using real and synthetic datasets and prove that our system is efficient, robust and scalable.

In the near future, we plan to extend and improve our system in order to become more efficient and flexible. At first, we have in mind to implement a technique that will allow us to have unequal splits that will contain approximately the same number of points. In this way we will achieve distribution independence and better load balancing between the nodes. In addition, we intend to apply a mechanism in order for the cluster to be used in a more elastic way, by adding/removing nodes as the number of dimensions increase/decrease. Finally, we plan to use indexes in order to prune any points that are redundant and cumber additional cost to the method.

Acknowledgements. This work was partially supported by Thales Project entitled "Cloud9: A multidisciplinary, holistic approach to internet-scale cloud computing". For more details see the following URL:
https://sites.google.com/site/thaliscloud9/home

References

1. Afrati, F.N., Ullman, J.D.: Optimizing Joins in a Map-Reduce Environment. In: Proceedings of the 13th International Conference on Extending Database Technology, pp. 99–110. ACM, New York (2010)
2. Böhm, C., Krebs, F.: The k-Nearest Neighbour Join: Turbo Charging the KDD Process. Knowl. Inf. Syst. 6, 728–749 (2004)
3. Chang, J., Luo, J., Huang, J.Z., Feng, S., Fan, J.: Minimum Spanning Tree Based Classification Model for Massive Data with MapReduce Implementation. In: Proceedings of the 10th IEEE International Conference on Data Mining Workshop, pp. 129–137. IEEE Computer Society, Washington, DC (2010)
4. Chen, Y., Patel, J.M.: Efficient Evaluation of All-Nearest-Neighbor Queries. In: Proceedings of the 23rd IEEE International Conference on Data Engineering, pp. 1056–1065. IEEE Computer Society, Washington, DC (2007)

5. Dean, J., Ghemawat, S.: MapReduce: Simplified Data Processing on Large Clusters. In: Proceedings of the 6th Symposium on Operating Systems Design and Implementation, pp. 137–150. USENIX Association, Berkeley (2004)
6. Dunham, M.H.: Data Mining, Introductory and Advanced Topics. Prentice Hall, Upper Saddle River (2002)
7. Emrich, T., Graf, F., Kriegel, H.-P., Schubert, M., Thoma, M.: Optimizing All-Nearest-Neighbor Queries with Trigonometric Pruning. In: Gertz, M., Ludäscher, B. (eds.) SSDBM 2010. LNCS, vol. 6187, pp. 501–518. Springer, Heidelberg (2010)
8. Gkoulalas-Divanis, A., Verykios, V.S., Bozanis, P.: A Network Aware Privacy Model for Online Requests in Trajectory Data. Data Knowl. Eng. 68, 431–452 (2009)
9. He, Q., Zhuang, F., Li, J., Shi, Z.: Parallel implementation of classification algorithms based on MapReduce. In: Yu, J., Greco, S., Lingras, P., Wang, G., Skowron, A. (eds.) RSKT 2010. LNCS, vol. 6401, pp. 655–662. Springer, Heidelberg (2010)
10. Lu, W., Shen, Y., Chen, S., Ooi, B.C.: Efficient Processing of k Nearest Neighbor Joins using MapReduce. Proc. VLDB Endow. 5, 1016–1027 (2012)
11. Roussopoulos, N., Kelley, S., Vincent, F.: Nearest Neighbor Queries. In: Proceedings of the 1995 ACM SIGMOD International Conference on Management of Data, pp. 71–79. ACM, New York (1995)
12. Samet, H.: The QuadTree and Related Hierarchical Data Structures. ACM Comput. Surv. 16, 187–260 (1984)
13. Stupar, A., Michel, S., Schenkel, R.: RankReduce - Processing K-Nearest Neighbor Queries on Top of MapReduce. In: Proceedings of the 8th Workshop on Large-Scale Distributed Systems for Information Retrieval, pp. 13–18 (2010)
14. The apache software foundation: Hadoop homepage, http://hadoop.apache.org/
15. Vernica, R., Carey, M.J., Li, C.: Efficient Parallel Set-Similarity Joins Using MapReduce. In: Proceedings of the ACM SIGMOD International Conference on Management of Data, pp. 495–506. ACM, New York (2010)
16. White, T.: Hadoop: The Definitive Guide, 3rd edn. O'Reilly Media / Yahoo Press (2012)
17. Xia, C., Lu, H., Chin, B., Hu, O.J.: Gorder: An efficient method for knn join processing. In: VLDB, pp. 756–767. VLDB Endowment (2004)
18. Yao, B., Li, F., Kumar, P.: K Nearest Neighbor Queries and KNN-Joins in Large Relational Databases (Almost) for Free. In: Proceedings of the 26th International Conference on Data Engineering, pp. 4–15. IEEE Computer Society, Washington, DC (2010)
19. Yokoyama, T., Ishikawa, Y., Suzuki, Y.: Processing All k-Nearest Neighbor Queries in Hadoop. In: Gao, H., Lim, L., Wang, W., Li, C., Chen, L. (eds.) WAIM 2012. LNCS, vol. 7418, pp. 346–351. Springer, Heidelberg (2012)
20. Yu, C., Cui, B., Wang, S., Su, J.: Efficient index-based KNN join processing for high-dimensional data. Information & Software Technology 49, 332–344 (2007)
21. Zhang, C., Li, F., Jestes, J.: Efficient Parallel kNN Joins for Large Data in MapReduce. In: Proceedings of the 15th International Conference on Extending Database Technology, pp. 38–49. ACM, New York (2012)
22. Zhang, J., Mamoulis, N., Papadias, D., Tao, Y.: All-Nearest-Neighbors Queries in Spatial Databases. In: Proceedings of the 16th International Conference on Scientific and Statistical Database Management, pp. 297–306. IEEE Computer Society, Washington (2004)

What-if Physical Design for Multiple Query Plan Generation

Ahcène Boukorca, Zoé Faget, and Ladjel Bellatreche

LIAS/ISAE-ENSMA, Futuroscope, Poitiers, France
{boukorca,bellatreche,zoe.faget}@ensma.fr

Abstract. The multiple query optimization problem (MQO) has been largely studied in traditional and advanced databases. In scientific and statistical databases, queries are complex, recurrent and share common sub-expressions. As a consequence, the MQO problem re-emerges in this context. An important characteristic of the MQO problem is that its result (usually represented by a unified plan merging isolated plans of the workload queries) may be used to select optimization structures (OS) such as materialized views. By examining the literature, we discover that the interconnection between these two problems is often neglected. Ignoring what-if questions about selecting the unified plan can result in disastrous consequences when used as a basis for selecting OS. In this paper, we first exhibit the link between global plans and optimization structures. Secondly, we give a formalization of the OS-oriented unified plan generation problem. Thirdly, a generic approach for plan generation is given. Finally, we instantiate our formalization to deal with the problems of selecting materialized views and horizontal data partitioning and we show its effectiveness and efficiency through intensive experiments.

1 Introduction

Finding an optimal global execution plan for multiple queries is one of the cornerstones for MQO. The main idea behind a global plan is to capitalize on shared intermediate results instead of executing a sequence of queries. MQO is especially profitable when queries are highly connected; hence share a lot of intermediate results. Data warehousing applications are examples of situations where queries will most certainly have high interaction, since all queries will have a join with the fact table. Instead of building individual query plans and execute them in an isolated fashion, research introduced the notion of a Multiple-Query Plan (MQP) which focuses on re-using intermediate results [24,27]. The problem of selecting the best MQP for a query workload (in the sense that this MQP will give the best response time for this workload) is known to be NP-hard [23].

In parallel, the selection of Optimization Structures (OS) is at the heart of the physical design problem. DataBase Administrators (DBA) have a wide choice of OS to select from, either in isolate (each type of OS is treated independently) or multiple fashion. MQP have been used as data-structures to select optimization structures such as materialized views [11,27,5], and, although less frequently,

H. Decker et al. (Eds.): DEXA 2014, Part I, LNCS 8644, pp. 492–506, 2014.

horizontal fragmentation schema [18] (to our knowledge, there is no similar work dealing with indexes).

To summarize, a *MQP* finds itself at the intersection between two worlds, the world of Multiple Query Optimization and the world of Physical Design. A natural question then emerges : *how does one choose a solution of the MQO-problem as an input of the Physical Design problem?*

Several strategies are possible. The first (naive) approach is to take an OS-selection algorithm and feed it a random *MQP* chosen amid the solution space, produced either through a cost-model or a rule-based approach. It is clear that this approach will only lead to poor results, since the interconnection of the two worlds is completely neglected. The first work to acknowledge the connection between the two problems is the work of Yang et al [27] which produces a *MQP* (called *Multi-View Processing Plan (MVPP)*), the *MQP* is used to select materialized view. However, Yang et al.'s approach is not generic and does not easily translate to other optimization structures. Moreover, his approach does not scale and the experiments were run on a very small set of six queries. In this paper, we propose an approach inspired by *what-if* questions. *What-if* principles have been introduced in the database community for questions such as modeling, analysis, or index selection [12,10,6] and study impact of hypothetical configurations. Our approach overcomes both problems encountered in Yang's work while dealing with the connected worlds. Precisely, we propose a generic approach to produce OS-oriented *MQP*, *i.e.* *MQP* tailored for a specific OS-selection algorithm, in a reasonable time even for a large set of queries.

Our contributions are: (i) establishment of strong coupling between MQO problem and OS-selection problem, (ii) the use of hypergraph to manage the gigantic solution space of all possible *MQP* and (iii) elaboration of what-if physical design for *MQP* .

The paper is structured as follows. Section 2 overviews the most important existing work related to MQO and the OS-selection using the unified graphs issued from the MQO Problem and summarizes the interaction between these two worlds. In Section 3, basic concepts and definitions are given to facilitate the formalization of our what-if question related to the selection of OS. Section 4 presents our hyper-graph driven approach to selection OS. In Section 5, intensive experiments are conducted to show the interest of What-if physical design. Section 6 concludes the paper.

2   Related Work

In the context of star-join queries executed over a star-schema datawarehouse, queries are likely to have high interaction, *i.e.* share common sub-expressions. Multiple-Query Optimization (MQO) is the tool that has been used to capture interaction between queries. The MQO problem is formalized as follows: given a set of queries to evaluate, find a set of common sub-expressions between queries to share or materialize in order to reduce the total execution time processing of queries. The MQO problem has been widely studied for traditional databases

[21,22,24,25,17,29], and recently for semantic databases [19]. Some works are focused on the common sub-expression identification problem by exploiting the interaction between queries and find the intermediate nodes that have possible benefits of their sharing [7,14]. Other works use the global access plan to identify intermediates results with maximum reuse [24,27,5]. A solution of the MQO problem is a *Multiple-Query Plan (MQP)*, which is the result of merging individual query plans in order to maximize the reuse of common nodes between queries to globally optimize the workload. Getting the optimal *MQP* is a NP-hard problem [23]. Sellis [24], proposed an A* algorithm search algorithm to produce a global execution plan based on all possible local plans of each query, which generation is a hard task [8]. To reduce this complexity, several heuristic algorithms were proposed, of mainly four types: dynamic programming algorithms [27,26], evolutionary approach [29], genetic [3] and randomized algorithms [13]. MQO can be divided into two phases: (1) identify the common nodes among a group of queries and prepare a set of alternative plans for each query; (2) use the results of the previous phase to select exactly one plan for each query. As a result, a global execution plan is obtained.

The physical design phase offers a choice of Optimization Structures (OS) for the DBA to select from, such as materialized views, indexes, horizontal data partition, etc. In DW design, identifying good optimization structures under a set of constraints (disk space, maintenance cost, fragments number, etc.), is an important task to optimize query processing cost. Selecting a candidate for each OS is a combinatory problem, hence the OS selection problem is considered a hard task. Most works focus on one type of OS, although multiple selection can significantly optimize query performance. We mention [1] for combined selection of materialized views and indexes.

Both MQO and OS selection aim at optimizing query performance and use intermediate results. Since the 1990's, *MQP* have been used in the physical design phase as a data structure for OS-selection: Materialized View selection [27,5,11,28,2,29], and more recently Horizontal Data Partitioning [4] or Buffer Management and Query Scheduling [17]. It is clear that the *MQP* selected as input for the OS-selection algorithm will impact the solution's quality. Yang et al [27] is one of the sole works in the context of DW that deals with the two interdependent problems: (a) constructing a global plan of queries and (b) use this plan to select views to be materialized. Construction of the global plan is performed following the bottom up scenario. Initially, the authors select the join local plans of each query (logical plans that have only join operations). These plans are merged in a single plan, called Multi-Views Processing Plan (MVPP). Two algorithms ('a feasible solution' and '0-1 integer programming') are proposed for selecting the best MVPP which has the minimum cost. The view selection algorithm is performed in two steps: (1) generate materialized views candidates which have positive benefit between query processing and view maintenance. This benefit corresponds to the sum of query processing using the view minus the maintenance cost of this view, (2) select only candidate nodes with positive benefit to be materialized. The main drawback of this approach is scalability. Also, this approach is for materialized views only. Zhang et al [29], proposed an hybrid evolutionary algorithm to solve three problems: (1) queries optimization, (2) finding the best global processing plan from multiple global processing plans and (3) selecting materialized views from a given global processing plan. The main drawback of this approach is their excessive computation time which makes it impractical for a large number of queries. To overcome this scalability obstacle, Boukorca et al [5], proposed an approach to generate the *MQP* without using individual query plans, and groups the queries in several disjoint components where each component contains queries that highly share common sub-expressions between them, and where shared sub-expressions between components are minimal.

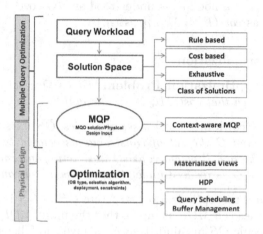

Fig. 1. Interaction between the MQO problem and the Physical Design phase

Figure 1 shows the interaction between the two problems, where a *MQP* is at the same time a solution of the MQO problem and an input of OS-selection algorithm for Physical Design. To our knowledge, the question of how *MQP* and optimization structures are bounded has not yet been investigated.

3 Problem Formalization

3.1 Notations, Definitions and Preliminary Results

In what follows, we call a *class of optimization structures* a family of optimization structures, namely, materialized views, horizontal data partitioning or index, denoted by $\overline{OS} = \{\mathcal{MV}, \mathcal{HDP}, \mathcal{I}\}$. An *optimization structure* is an instance of the chosen class, namely a set of views to be materialized or a fragmentation schema. For example, if a class $OS \in \overline{OS}$ is $\mathcal{MV}$, then the optimization structure will be a set of views to be materialized $MV = \{v_1, \ldots, v_n\}$. Identically, if the chosen OS is $\mathcal{HDP}$, then the optimization structure will be a fragmentation schema $FS = \{F_1, \ldots, F_N\}$.

To each class of $\overline{OS}$ is associated a specific set of constraints denoted by $C(OS)$. Precisely, fragmentation schema usually come with a threshold W for the maximum number of fragments to be generated. Materialized views and indexes have a storage constraint. Further constraints may appear such as a price limit for a paid deployment (such as cloud).

In a MQP, we call *pivot node* the first direct join between the facts table and a dimension table. Several queries sharing a pivot node are called a *connected component* of the MQP. Queries which belong to the same connected component have interactions, since they share at least the first join operation. A MQP is a union of several connected components.

The cost $cost(n_i)$ of a node n_i is understood as I/O cost. The cost of a plan $cost(P)$ is defined as $cost(P) = \Sigma_{v_i \in P} cost(n_i)$.

3.2   Problem Formalization

Multiple Query Optimization Problem. *Given Q a set of queries, find a global plan MQP such that $cost(MQP)$ is minimum.*

Physical Design Problem. *Let Q a set of queries, $\overline{OS}$ the set of optimization structures classes, and C the set of constraints associated to each $OS \in \overline{OS}$. Then the Physical Design Problem consists in selecting instances of $\overline{OS}$ so that the execution cost of Q is reduced under the corresponding constraints.*

In the case where the optimization structure $OS \in \overline{OS}$ is selected with an algorithm g which needs a MQP as input, then the input MQP should be generated with that specific OS in mind. Hence, we give the following formalization of the conditional MQP generation.

OS-oriented MQP Generation Problem. *Given Q a set of queries, OS a class of optimization structure, $C(OS)$ a set of constraints for OS, and g a selection algorithm for OS taking an MQP as input : find a global plan $MQP(OS)$ and an optimization structure respecting the problem constraints such that the cost $cost(MQP(OS))$ is minimum.*

4   Generating OS-Oriented MQP

In order to outline a generic approach for OS-oriented MQP generation, we first explain our motivation to use hypergraphs for representing a workload of queries. We then give definitions and properties related to hypergraphs, and how they can be used for our problem of MQP generation.

4.1   Using an Hypergraph to Represent a Class of MQP

Various graph representation of query workload can be found in classical [22] or semantic [19,9] databases. Graph representation of a workload is intuitive, since

MQP are basically oriented graphs. Here, queries operations are represented by vertices of the graph, and the set of edges joining vertices gives the MQP . Due to algebraic properties of relational operators, there are several possible individual plans for a given query. There is then an exponential number of possible combinations between these local plans, therefore there is a gigantic number of possible execution plans for a given workload. We choose to represent the class of all possible MQP by a hypergraph data structure. The hypergraphs have the advantage to enable representation of overlapping queries (queries sharing operations) without forcing an order between joins. Indeed, contrary to graphs where edges can only connect two vertices (hence defining an order between them), hyper-edges contains as many vertices as needed without any order. Hence, all potential MQP are virtually contained in a single hypergraph, since two MQP for the same workload differ only by the order between operations.

It is well known that join operations are usually the most expensive since they involve the (usually huge) fact table, therefore we focus our effort on optimizing join operations. We can then reduce our MQP to a join plan, *i.e.* a plan featuring only join operations. To this end, we pre-define the order between operations, where selections are executed first, then join operations, followed by projections and finally aggregation operations. We focus on the set of unordered join nodes of the workload that need to be ordered to produce the workload's join plan. The hypergraph has a set of vertices and a set of hyperedges. The vertices represents the set of join nodes such that to each $v_i \in V$ corresponds a join node n_i. Similarly, to each hyperedge $h_j \in \mathcal{H}_A$ corresponds a query q_j of the workload W. Symmetry between hypergraphs and queries is given in table 1.

Table 1. Graph -Query Symmetry

Hypergraph vision	Query vision
$V = \{v_i\}$ set of vertices	$\{n_i\}$ set of join nodes
Hyper-edge h_j	Query q_j
Hypergraph	Workloads of queries
Oriented graph	Multiple-Query Plan

4.2   From Hypergraphs to MQP

Transforming a hypergraph into an oriented graph MQP is a two-part process. On one hand, we select a vertex from the hypergraph to be removed, while on the other hand the corresponding join node is added to the MQP . This process is repeated until there are no more vertices in the hypergraph. Before further detailing in which order nodes are chosen depending on the targeted OS, we must first understand how removing a vertex impacts the hypergraph. In what follows, we focus on *connected sections* since they represent interacting queries (reciprocally, non-interacting queries do not belong to the same section). A hypergraph is the disjoint union of *connected section*. Two hyper-edges belong to the same connected section if there is a connected path of hyper-edges between them. Let S a section of the hypergraph and $v \in S$ a vertex being removed. Then for a vertex $v' \in S$, v' belongs to one and only one of the following three categories:

Type 1: v and v' belongs to the same hyper-edge and there is no hyper-edge to which v' belongs and not v.
Type 2: v and v' belongs to the same hyper-edge and there exists an hyper-edge to which v' belongs and not v.
Type 3: v and v' do not belong to the same hyper-edge.

When a vertex is removed from a section, the remaining nodes of the section are affected in different ways depending on their type.

1. If v' is of type 1, v' is renamed with v added as an index.
2. If v' is of type 2, v' is duplicated, one unchanged and one renamed with v added as an index.
3. If v' is of type 3, v' is unchanged.

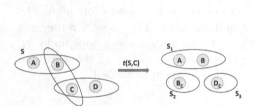

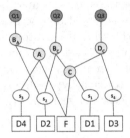

Fig. 2. Example of transformation   **Fig. 3.** A resulting MQP

We denote by $t(S, v)$ the transformation that removes the vertex v from section S. In figure 2, we apply $t(S, C)$, which means we remove C from S. If C is the vertex being removed, then D is type 1 (there is an hyper-edge to which both C and D belong, but no hyper-edge to which D belongs but not C), B is type 2 (there is a hyper-edge to which both C and B belong, and there is a hyper-edge to which B belongs but not C), and A is type 3 (there is no hyper-edge to which A and C belong). Figure 2 shows the effect of $t(S, C)$ on section S, the result being three new sections S_1, S_2, S_3. Figure 3 shows the result of transformations $t(S, C), t(S_1, A), t(S_1, B_A), t(S_2, B_C), t(S_3, D_C)$ on the example of Figure 2.

4.3   Step-by-Step OS-Oriented MQP Generation

In the context of MQP used for OS selection, we want a trade-off between time to produce the MQP and quality of the solution. Following our previous discussion, we want to inject the knowledge of the targeted optimization structure characteristics in the process of MQP generation for better selection. MQP generation has three main steps (for more details see [5]): In the first step, the queries and their join nodes are represented by an hypergraph. An hypergraph is constituted of two sets, a set of vertices that represent the join nodes, and a set of hyperedges that represent the queries. An example of hypergraph is represented by Figure 4-a. In the second step, we groupe queries in several disjoint sections where each section contains queries that highly share common sub-expressions between them, and where shared sub-expressions between sections are minimal. Figure 4-b, show the partition result of the hypergraph presented in Figure 4-a. Using a hypergraph representation allows us to benefit from decades of research efforts carried in a wide panel of domains which have developed and improved all sorts of algorithms for hypergraph manipulation [15]. Partitioning an hypergraph is a well-studied task in domains such as Electronic Design Automation and a choice of algorithms is available. For example, one can use the HMETIS programs [16] which allows parallel or serial partitioning. The third step is ordering of nodes in each sections. Ordering is made as explained previously, by removing vertices from the section and adding them to the *MQP* one by one. Figure 4-c, illustrates the transformation steps of a section (sub-hypergraph) to a *MQP* . The choice of which vertex is removed from the hypergraph at a given point is made through a *benefit* function which takes into account the characteristics of the specific problem. The function computes the benefit relative to the targeted OS for each remaining node of the section and adds the node with maximum benefit.

Sensitivity to the targeted OS is featured in the third step of this *MQP* generation through the choice of a benefit function. The purpose of a benefit function is to select which vertex should be added to the *MQP* at a given moment. The choice is made according to the targeted OS characteristics. Next we instantiate our generic approach by two examples of benefit functions for two types of optimization structures ($\mathcal{MV}$ and $\mathcal{HDP}$).

4.4   Application to Materialized Views and Horizontal Data Partitioning Selection

Here we present two examples of *benefit* functions for two types of optimization structure, namely materialized views and horizontal data partitioning. The benefit function appears in step 3 of the *OS*-oriented *MQP* generation.

Materialized View-Oriented MQP Generation. In both works using an *MQP* for Materialized View selection ([27,5]), candidates for materializations selected by the OS-selection algorithm are nodes with positive benefit, *i.e.* where the processing cost times the number of reuse exceeds the materialization cost. Hence the benefit function for Materialized Views selection is

$$benefit(n_i) = (\text{nbr-1}) * cost(n_i) - cost_{mat}(n_i)$$

where *nbr* is the number of queries using n_i, $cost(n_i)$ is the processing cost and $cost_{mat}(n_i)$ is the materialization cost of n_i . Vertices of the hypergraph

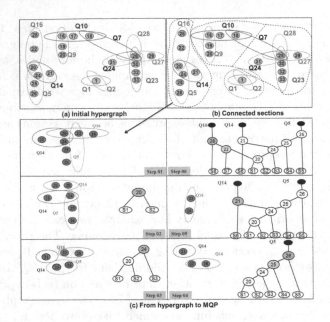

(a) Initial hypergraph   (b) Connected sections

(c) From hypergraph to MQP

Fig. 4. MQP generation steps

which maximizes the benefit function are selected first so that their benefit will propagates towards as much queries as possible.

Horizontal Data Partitioning-Oriented MQP Generation. To our knowledge, the only work using *MQP* as a data structure for $\mathcal{HDP}$ selection is [4]. The selection algorithm, called *Elected Query for Horizontal Data Partitioning (EQHDP)*, takes the *MQP* as input and aims at optimizing each connected component by identifying in each component a query (the elected query) and using its predicates for partitioning. Predicates of extreme selectivity (too high or too low) are ignored. The process is repeated until the constraint (tolerated total number of fragments) is attained. EQHDP has shown better results than state-of-the-art algorithms such as Simulated Annealing or Genetic algorithms, because, contrary to classic algorithms, it does not ignore query interaction. However, tests on EQHDP were run on a *Rule-Based* generated *MQP*. We claim that we can further improve the resulting fragmentation schema by feeding EQHDP with a HDP-oriented *MQP* .

To this end, we push down join operations for which the gain if we partition on its selection predicate is maximum. This will ensure that the fragment that will bring the best gain to the most queries is indeed chosen by the partitioning algorithm before the constraint is exhausted, and that the benefit is propagated throughout the workload.

Hence the benefit function for Horizontal Data Partitioning is

$$benefit(n_i) = (\text{cost without fragment}(n_i) - \text{cost with fragment}(n_i)) * \text{nbr}$$

where nbr is the number of queries using n_i and $cost(n_i)$ is the processing cost.

5   Experimentation

In this section, we present an experimental validation of our approach. We developed a simulator tool in Java Environment. This tool consists in the following modules.
1. Two *MQP* generation modules, one for each optimization structure (materialized views and horizontal data partitioning). Each module contains several functions : (1) a parsing SQL-queries function to get all selection, join, projection and aggregation nodes, (2) hypergraph generation function to represent the queries by an hypergraph of nodes, (3) hypergraph partitioning function that uses *HMETIS* tools [16] to partition the hypergraph into several connected sections, (4) a transformation function, to transform each hypergraph into an *MQP* using the appropriate benefit function according to the targeted OS, (5) a merging function that assembles all selection, join, projection and aggregation nodes to generate the final *MQP*, (6) a display function that uses *Cytoscape*[1] plug-in to display the global processing plan.
2. A Materialized Views selection module, that takes an *MQP* as input and produces candidate views to be materialized.
3. A Horizontal Partitioning module, that takes a *MQP* as input and generates a Horizontal Partition Schema.

We also implement Yang et al.'s two algorithms [27]: feasible solution and 0-1 integer programming, and *Sellis*'s algorithm for *MQP* generation [24].

5.1   MQP Generation Speed

A series of tests was applied to test our *MQP* generation module against *Yang*'s and *Sellis*'s modules. In each test, we change the number of input queries to monitor the behavior of each algorithm. Figure 5 show that it takes Sellis' algorithm up to five days to generate a *MQP* for 30 queries. In Figure 6, we compare generation times for our approach against Yang's method. It takes 10 hours for Yang's algorithm to generate a *MQP* for 2000 queries. On the other hand, our algorithm takes about five minutes to generate a *MQP* for 10 000 queries. This proves that our approach is scalable.

5.2   OS-Sensitivity of MQP

We now exhibit how the choice of *MQP* for OS-selection purposes influences the quality of the resulting OS. We use a cost model developed in [17] to estimate

---

[1] http://www.cytoscape.org

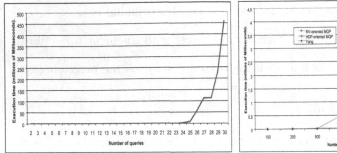 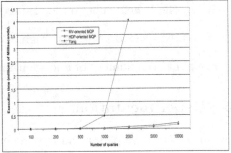

Fig. 5. Execution time to generate an MQP using Sellis's method

Fig. 6. Execution time to generate an MQP using 3 methods

query processing and views maintenance cost. This model estimates the number of Inputs/Outputs pages required for executing a given query. To perform this experiment, we developed a simulator tool using Java Environment. The tool can automatically extract the data warehouse's meta-data characteristics. The data warehouse used in our test is SSB (Start Schema Benchmark) [20]. Its size is 100Go, with a facts table *Lineorder* of 600 millions of tuples and four dimension tables: *Part, Customer, Supplier* and *Dates*. We used a SSB query Generator to generate 10000 queries for *SSB* data warehouse. Our evaluation is conducted using those queries, which cover most types of *OLAP* queries.

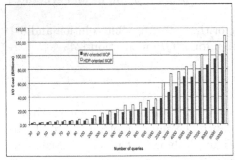 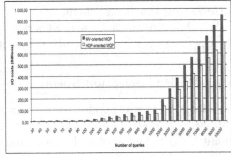

Fig. 7. Total cost of workload using MV selected using MV-MQP and HDP-MQP

Fig. 8. Total cost of workload using HDP selected using MV-MQP and HDP-MQP

We then proceed to generate MV-oriented *MQP* and HDP-oriented *MQP* for different workloads of queries varying from 30 to 10000 queries. The resulting *MQP* are then used by two different OS-selection algorithms : the MV-selection algorithm and the HDP-selection algorithm described in section 4.4. As shown in Figure 7 we see that the MV-oriented MQP always produces a better set of candidate views to materialize than the HDP-oriented MQP. Reciprocally, Figure 8 shows that the HDP-oriented MQP always leads to a better fragmentation schema than the MV-oriented MQP. The difference in performance gain increases with the number of queries.

5.3 Materialized Views Quality

To evaluate the quality of our selected materialized views, we compare it to the set of materialized views selected by Yang's algorithm. We run tests using 30 queries and a data warehouse SSB of 100 Gb saved in Oracle 11g. We use a server of 32 GB of RAM with Intel Xeon Processor E5530 (8M Cache, 2.40 GHz). Figure 9 compare individual query execution times for the selected materialized views of Yang's approach and MV-oriented MQP approach. Figure 10 shows better performance for workload execution using the set of materialized views selected using a MV-oriented MQP.

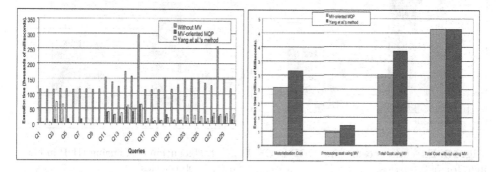

Fig. 9. Individual queries execution time   **Fig. 10.** Workload execution times

5.4 Validation Using Oracle

Finally, in order to validate our theoretical results, we select materialized views candidates and fragmentation schemas using our OS-oriented MQP generation modules for two workloads of 500 and 3000 queries respectively, and deploy the results in Oracle data base (previous configuration). Figures 11 and 12 compare maintenance costs, query processing costs and total execution costs using materialized views selected using either a HDP-oriented or a MV-oriented MQP. MV-oriented MQP always give better results. Similarly, in Figures 13 and 14 we compare execution costs using a fragmentation schema obtained by using either a HDP-oriented or a MV-oriented MQP. The HDP-oriented MQP gives better results. Results are summarized in Figures 15 and 16. Theoretical results are confirmed by Oracle experiments.

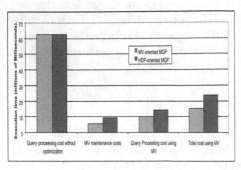
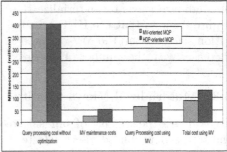

Fig. 11. Different costs using MV in Oracle for 500 queries

Fig. 12. Different costs using MV in Oracle for 3000 queries

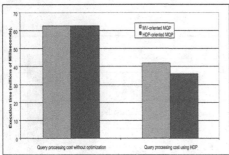
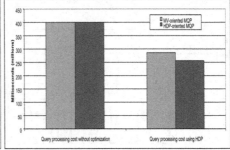

Fig. 13. Execution costs using HDP in Oracle for 500 queries

Fig. 14. Execution costs using HDP in Oracle for 3000 queries

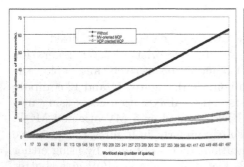
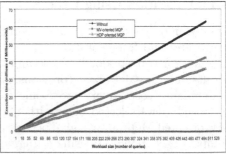

Fig. 15. Validation of MV in Oracle for 3000 queries

Fig. 16. Validation of HP in Oracle for 3000 queries

6 Conclusion

In this paper, we discussed the what-if question in the physical design phase of advanced databases such as data warehouses. We first considered two most important problems related to query performance which are the problem of MQO and the problem of selecting optimization structures. We figured out that solution for the first problem can be used either to optimize queries without using optimization structures or to selection optimization structures such as materialized views. Based on this finding, what-if questions have to be considered. As a consequence, we propose to consider one question which is what-if physical design for multiple query plan generation. A formalization of our problem is given and instantiated with two optimization structures which are materialized views and horizontal data partitioning. The hypergraph data structure used for generating the best multiple query plan facilitate the development of algorithms for selecting our optimization structures, where metrics related to each structures are injected into the process of multiple query plan generation. We conducted intensive experiments using theoretical and a real validation in Oracle 11g. The results comfort our what-if issue.

Currently, we are considering other optimization structures such as indexes and considering the multiple selection of optimization structures.

References

1. Agrawal, S., Chaudhuri, S., Narasayya, V.R.: Automated selection of materialized views and indexes in sql databases. In: Proceedings of the 26th International Conference on Very Large Data Bases, VLDB 2000, pp. 496–505. Morgan Kaufmann Publishers Inc., San Francisco (2000)
2. Baralis, E., Paraboschi, S., Teniente, E.: Materialized view selection in a multidimensional database. In: Proceedings of the International Conference on Very Large Databases (VLDB), pp. 156–165 (August 1997)
3. Bayir, M., Toroslu, I., Cosar, A.: Genetic algorithm for the multiple-query optimization problem. IEEE Transactions on Systems, Man, and Cybernetics, Part C: Applications and Reviews 37, 147–153 (2007)
4. Bellatreche, L., Kerkad, A., Breß, S., Geniet, D.: RouPar: Routinely and mixed query-driven approach for data partitioning. In: Meersman, R., Panetto, H., Dillon, T., Eder, J., Bellahsene, Z., Ritter, N., De Leenheer, P., Dou, D. (eds.) ODBASE 2013. LNCS, vol. 8185, pp. 309–326. Springer, Heidelberg (2013)
5. Boukorca, A., Bellatreche, L., Senouci, S.-A.B., Faget, Z.: SONIC: Scalable multi-query optimizatioN through integrated circuits. In: Decker, H., Lhotská, L., Link, S., Basl, J., Tjoa, A.M. (eds.) DEXA 2013, Part I. LNCS, vol. 8055, pp. 278–292. Springer, Heidelberg (2013)
6. Chaudhuri, S., Narasayya, V.R.: Autoadmin 'what-if' index analysis utility. In: Proc. ACM SIGMOD Int. Conf. on Management of Data, SIGMOD 1998, pp. 367–378. ACM Press (1998)
7. Finkelstein, S.: Common expression analysis in database applications. In: Proceedings of the 1982 ACM SIGMOD International Conference on Management of Data, SIGMOD 1982, pp. 235–245. ACM (1982)
8. Galindo-Legaria, C.A., Grabs, T., Gukal, S., Herbert, S., Surna, A., Wang, S., Yu, W., Zabback, P., Zhang, S.: Optimizing star join queries for data warehousing in microsoft sql server. In: Proceedings of the International Conference on Data Engineering (ICDE), pp. 1190–1199 (2008)

9. Goasdoué, F., Karanasos, K., Leblay, J., Manolescu, I.: View selection in semantic web databases. Proceedings of the VLDB Endowment 5(2), 97–108 (2011)
10. Golfarelli, M., Rizzi, S.: What-if simulation modeling in business intelligence. IJDWM 5(4), 24–43 (2009)
11. Gupta, H.: Selection and maintenance of views in a data warehouse. Ph.d. thesis, Stanford University (September 1999)
12. Herodotou, H., Babu, S.: Profiling, what-if analysis, and cost-based optimization of mapreduce programs. PVLDB 4(11), 1111–1122 (2011)
13. Ioannidis, Y.E., Kang, Y.C.: Randomized algorithms for optimizing large join queries. In: Garcia-Molina, H., Jagadish, H.V. (eds.) ACM SIGMOD, pp. 312–321 (1990)
14. Jarke, M., Koch, J.: Query optimization in database systems. ACM 16(2), 111–152 (1984)
15. Karypis, G., Aggarwal, R., Kumar, V., Shekhar, S.: Multilevel hypergraph partitioning: applications in vlsi domain. IEEE Transactions on Very Large Scale Integration Systems 7(1), 69–79 (1999)
16. Karypis, G., Kumar, V.: Multilevel k-way hypergraph partitioning. In: ACM/IEEE Design Automation Conference (DAC), pp. 343–348. ACM, New York (1999)
17. Kerkad, A., Bellatreche, L., Geniet, D.: Queen-bee: Query interaction-aware for buffer allocation and scheduling problem. In: Cuzzocrea, A., Dayal, U. (eds.) DaWaK 2012. LNCS, vol. 7448, pp. 156–167. Springer, Heidelberg (2012)
18. Kerkad, A., Bellatreche, L., Geniet, D.: La fragmentation horizontale revisitée: Prise en compte de l'interaction des requétes. In: 9èmes Journées Francophones sur les Entrepôts de Données et Analyse en Ligne, EDA 2013 (2013)
19. Le, W., Kementsietsidis, A., Duan, S., Li, F.: Scalable multi-query optimization for sparql. In: Proceedings of the International Conference on Data Engineering (ICDE), pp. 666–677. IEEE (2012)
20. O'Neil, P., O'Neil, B., Chen, X.: Star schema benchmark (2009)
21. Park, J., Segev, A.: Using common subexpressions to optimize multiple queries. In: Proceedings of the Fourth International Conference on Data Engineering, pp. 311–319 (1988)
22. Roy, P., Seshadri, S., Sudarshan, S., Bhobe, S.: Efficient and extensible algorithms for multi query optimization. In: Proceedings of the 2000 ACM SIGMOD International Conference on Management of Data, pp. 249–260. ACM (2000)
23. Sellis, T., Ghosh, S.: On the multiple query optimization problem. IEEE Transaction on Knowledge and Data Engineering, 262–266 (1990)
24. Sellis, T.K.: Multiple-query optimization. ACM Transactions on Database Systems 13(1), 23–52 (1988)
25. Shim, K., Sellis, T., Nau, D.: Improvements on a heuristic algorithm for multiple-query optimization. Data, Knowledge Engineering, 197–222 (1994)
26. Toroslu, I.H., Cosar, A.: Dynamic programming solution for multiple query optimization problem. Information Processing Letters 92(3), 149–155 (2004)
27. Yang, J., Karlapalem, K., Li, Q.: Algorithms for materialized view design in data warehousing environment. In: Proceedings of the International Conference on Very Large Databases (VLDB), pp. 136–145. Morgan Kaufmann Publishers Inc., San Francisco (1997)
28. Yang, J., Karlapalem, K., Li, Q.: A framework for designing materialized views in data warehousing environment. In: ICDCS, p. 458 (1997)
29. Zhang, C., Yao, X., Yang, J.: An evolutionary approach to materialized views selection in a data warehouse environment. IEEE Transactions on, Systems, Man, and Cybernetics, Part C: Applications and Reviews 31(3), 282–294 (2001)

Author Index